Somatic Genome Manipulation

Xiu-Qing Li • Danielle J. Donnelly
Thomas G. Jensen
Editors

Somatic Genome Manipulation

Advances, Methods, and Applications

Springer

Editors
Xiu-Qing Li
Potato Research Centre
Agriculture and Agri-Food Canada
Government of Canada
Fredericton
New Brunswick
Canada

Thomas G. Jensen
Department of Biomedicine
Aarhus University
Herning
Denmark

Danielle J. Donnelly
Department of Plant Science
McGill University
Sainte-Anne-de-Bellevue, Québec
Canada

ISBN 978-1-4939-2388-5 ISBN 978-1-4939-2389-2 (eBook)
DOI 10.1007/978-1-4939-2389-2

Library of Congress Control Number: 2015931020

Springer New York Heidelberg Dordrecht London

Printed on acid-free paper

Springer is part of Springer Science+Business Media (www.springer.com)

Preface

Somatic genome manipulation is revolutionizing medical and biological sciences. This has applications when the conventional sexual crossing approach cannot be used in breeding or genetic treatment of an individual organism. Examples can include gene or cell therapy of a person to correct disease, genetic improvement of vegetatively propagated plants, and genetic replacement of cytoplasm without significantly modifying the nuclear genome. The advantage of somatic genome manipulation is preservation of the genotype while improving select trait(s). Somatic genome manipulation is also an option for genetic improvement of seed propagated plants in overcoming issues of sexual incompatibility or infertility.

Our aim in writing this book was to bring together previously fragmented information on novel technologies in somatic genome manipulation. These technologies are developing quickly across a broad range of disciplines affecting humans and animals, plants, and microorganisms. This book represents the first attempt to assemble updated reviews, detailed protocols, and far-reaching applications in somatic genome manipulation. The chapters are written by 34 experts (physicians, professors, research scientists, research chairs, Doctors of Philosophy, Doctors of Medicine, etc.) on the topic with ready-to-use protocols that were originally developed or adapted from the literature in their laboratories. The book is divided into three major sections: I. Humans and animals; II. Plants; and III. General experimental and bioinformatic technologies.

Section I. Humans and Animals

Drug and gene delivery by electroporation is described for delivery of chemotherapeutic agents in cancer trials. This method (electrochemotherapy) may increase local effects on tumors, locally activate the immune system, or produce transgenic proteins followed by secretion to the systemic circulation. Gene therapy offers a relatively easier and less expensive strategy for therapy than pharmaceutical approaches, since DNA may be produced more easily than formulation of protein drugs; enabling a higher level of access to new potential pharmaceuticals.

Targeted porcine genome engineering with transcription activator-like effector nucleases (TALENs) enables precise editing (e.g., mutations or indels) or insertion of a functional transgenic cassette to user-designed loci without disturbing the general gene background of the individual. The three most promising approaches are reviewed, including TALENs, zinc-finger nucleases (ZFNs), and clustered regularly interspaced short palindromic repeats (CRISPR)/CRISPR-associated (Cas) systems using a model of genetically modified pigs.

Somatic gene therapy can also be effective using viral vectors. For example, lentivirus and adeno-associated virus-based technology has applications in clinical trials for treatment of inherited eye diseases, immunodeficiency syndromes, and hemophilia. The gene therapy community has succeeded in turning infectious agents into vehicles of therapeutics for treatment or amelioration of the disease phenotype, giving significant reassurance that gene therapy will become standard care for a number of individual disorders.

Nonviral gene delivery methods are also in use. The current status of nonviral gene transfer is reviewed, focussing on DNA and its mobilization. The major barriers to nonviral gene delivery are discussed and the potential of minicircle DNA, devoid of bacterial DNA, and the adaptation of DNA transposable elements for genomic gene insertion, are described. A short glimpse into current nonviral gene therapy trials, with particular focus on current attempts to treat cystic fibrosis, is also provided.

Stem cells, because of their nature, are currently considered the most suitable cells for cell therapy. Combining gene therapy with stem cell therapy provides an additional useful dimension to the use of stem cells for treatment. The potential use of gene-modified stem cells, in particular gene-modified mesenchymal stem cells (MSC), in therapy and the challenges facing their use in clinical practice are reviewed.

Transgenic animals, particularly genetically modified mice, have been instrumental in biomedical and genetic research. To facilitate translational research to humans, the development of larger species of transgenic animals is necessary. These have numerous possible applications including the development of higher quality production animals, creation of stem cells for tissue repair (therapeutic cloning), production of protein-based pharmaceuticals (animal pharming), creation of organ donors for xenotransplantation, and the creation of large animal models for biomedical research. Various techniques to produce genetically altered animals are reviewed.

Section II. Plants

Apomixis is the clonal production of a plant through seed; a naturally occurring trait. Studies of both naturally occurring apomicts and mutants of sexual species that mimic the component events of apomixes have revealed potential mechanisms of control, the possible evolutionary origins of apomixes, and the impact it has had on the evolution of species and genomes. Apomixis leads to the formation of genetically uniform populations that can persist over many seedling generations. Important applications for agricultural crops are discussed.

The use of somatic embryogenesis in potato improvement programs is highlighted and discussed with emphasis on variants identified in cultivar Russet Burbank. Potato somatic embryogenesis is reviewed, including explant types, media components, effect of various growth regulators on the initiation and production of somatic embryos, and genes known to control somatic development.

The history of somatic hybridization use for modification of the cytoplasmic male sterility (CMS)-inducing Ogura radish cytoplasm and its application in hybrid seed production for *Brassica* crops is recounted. Highlights include cybrid production from early protoplast fusion experiments and identification of the mitochondrial gene causing CMS. This fascinating story fully explains the Ogu-INRA system, which is now widely used in agriculture.

Protoplast fusion has emerged as an exceptional breeding tool that has successfully produced a large number of intergeneric, intertribal, or interfamily somatic hybrids. The fusion of isolated protoplasts from somatic cells and regeneration of hybrid plants from the fusion products (somatic hybrids) allow combining of complete genomes of two desirable parents, irrespective of their taxonomic relationship. Procedures for protoplast fusion of potato to produce somatic hybrids are included, along with detailed lists of materials, and full descriptions of techniques.

Virus diseases inflict substantial economic losses to major crops by reducing yield and compromising quality. RNA silencing using, e.g., self-complementary hairpin RNA (hpRNA) or artificial microRNA (amiRNA), is an effective method to produce plants that are resistant to specific viruses. By targeting highly conserved viral sequences or several virus genes simultaneously using chimeric constructs, this method can counter multiple viruses and minimize any loss of viral resistance resulting from viral mutation. Due to public concerns about transgenic plant safety, a nontransgenic RNA silencing approach was used to directly deliver hpRNA into plant tissues to induce plant resistance to viruses.

Recent advances in genome engineering provide plant biologists with an important tool for understanding gene function and developing new traits. Three powerful techniques, including TALENs, ZFNs, and RNA-guided endonucleases (RGENs), have been developed for targeted DNA sequence modifications in plants. These sequence-specific nucleases create double-strand breaks (DSBs) in the genomic target sites that are primarily repaired by the nonhomologous end joining (NHEJ) or homologous recombination (HR) pathways, which can be employed to achieve targeted genome modifications such as gene mutations, insertions, replacements, or chromosome rearrangements. Considerable efforts have been made to understand the mechanisms governing gene targeting and to establish efficient DNA delivery systems to achieve precise gene targeting in plants.

Section III. General Experimental and Bioinformatic Technologies

Mitochondria are key players in cellular metabolism and energy production. Mitochondrial DNA mutations or rearrangements cause incurable neurodegenerative diseases in humans or cytoplasmic male sterility in plants, so manipulation of mitochondrial genetics is of particular relevance. The current challenge in the field is to define consensus biotechnological tools. Various procedures for manipulating mitochondrial genomes are reviewed and their promise discussed.

A wide range of laboratory techniques in somatic genome research are described, for research and training purposes, including: (1) in situ hybridization for studying tissue-specific gene expression; (2) mitochondrial visualization using rhodamine staining and confocal microscopy; (3) differential preparation of *Agrobacterium* Ti plasmid and binary plasmid using a noncommercial kit; (4) *Agrobacterium* binary plasmid DNA preparation using a commercial kit; (5) isolation of nuclei for DNA preparation; (6) chloroplast DNA extraction; (7) mitochondrial DNA extraction; (8) total DNA/RNA preparation; (9) enriched mitochondrial RNA preparation; (10) high-resolution DNA melting analysis for studying gene expression; and (11) transcriptome electrophoretic fingerprinting.

Bioinformatic analysis is critical for studies using huge amounts of DNA, RNA, and protein sequences. Various bioinformatics approaches developed or tested in the author's laboratory are described. These approaches include: (1) a statistical method for gene direction analysis, (2) some technical highlights for genome and chromosome base composition analysis, (3) some technical highlights on RNA polyadenylation site analysis; (4) allele comparison for protein domains, and (5) protein network analysis. Unsolved technical issues are highlighted and potential future research directions are discussed.

Acknowledgement

This book represents the cumulative work of 34 authors from 18 research institutes and universities in 6 countries. We wish to express our gratitude to all authors who have contributed manuscripts to this book and to Dr. Yonglun Luo, Department of Biomedicine, Aarhus University, for his assistance to Dr. Thomas G. Jensen for the editing. We thank Mr. Kevin Wright, developmental editor, Kenneth K. Teng, publishing editor, Ms. Aiswarya Narayanan, production editor, Ms. Melissa Higgs, assistant editor, Ms. Hannah Smith (former associate editor), and the entire production team of Springer for their ongoing support.

Contents

Contributors

Prof. Luis B. Agellon School of Dietetics and Human Nutrition, McGill University, Ste-Anne-de-Bellevue, QC, Canada

Dr. Anne Louise Askou Department of Biomedicine, Faculty of Health, Aarhus University, Aarhus C, Denmark

Dr. Ross Bicknell Plant and Food Research, Christchurch, New Zealand

Prof. Lars Bolund Department of Biomedicine, Faculty of Health, Aarhus University, Aarhus C, Denmark

Prof. Vilceu Bordignon Department of Animal Science, McGill University, Ste-Anne-de-Bellevue, QC, Canada

Dr. Francoise Budar Institut Jean-Pierre Bourgin, INRA-AgroParisTech UMR 1318, Versailles, France

Dr. Andrew Catanach Plant and Food Research, Christchurch, New Zealand

Dr. Kunling Chen Institute of Genetics and Developmental Biology, Chinese Academy of Sciences, Beijing, China

Assoc. Prof. Thomas J. Corydon TJC, Department of Biomedicine, Faculty of Health, Aarhus University, Aarhus C, Denmark

Dr. Naomi Dicks Department of Animal Science, McGill University, Ste-Anne-de-Bellevue, QC, Canada

Dr. André Dietrich Institut de Biologie Moléculaire des Plantes, CNRS and Université de Strasbourg, Strasbourg, France

Prof. Danielle J. Donnelly Plant Science Department, McGill University, Sainte Anne de Bellevue, QC, Canada

Prof. Caixia Gao State Key Laboratory of Plant Cell and Chromosome Engineering, Institute of Genetics and Developmental Biology, Chinese Academy of Sciences, Beijing, China

Dr. Julie Gehl Center for Experimental Drug and Gene Electrotransfer, Department of Oncology, Copenhagen University Hospital Herlev, Herlev, Denmark

Assoc. Prof. Monika M. Golas Department of Biomedicine, Faculty of Health, Aarhus University, Aarhus C, Denmark

Dr. José Gualberto Institut de Biologie Moléculaire des Plantes, CNRS and Université de Strasbourg, Strasbourg, France

Prof. Moustapha Kassem Department of Endocrinology and Metabolism, Endocrine Research Laboratory (KMEB), Odense University Hospital, University of Southern Denmark, Odense, Denmark

Prof. Stan Kubow School of Dietetics & Human Nutrition, McGill University, Sainte Anne de Bellevue, QC, Canada

Dr. Xiu-Qing Li Potato Research Centre, Agriculture and Agri-Food Canada, Fredericton, NB, Canada

Dr. Lin Lin Department of Biomedicine, Faculty of Health, Aarhus University, Aarhus C, Denmark

Assist. Prof. Yonglun Luo Department of Biomedicine, Faculty of Health, Aarhus University, Aarhus C, Denmark

Dr. Jin Ma Forestry and Environmental Management, University of New Brunswick, Fredericton, NB, Canada

Assoc. Prof. Jacob Giehm Mikkelsen Department of Biomedicine, Faculty of Health, Aarhus University, Aarhus C, Denmark

Dr. Daria Mileshina Institut de Biologie Moléculaire des Plantes, CNRS and Université de Strasbourg, Strasbourg, France

Dr. Atef M. K. Nassar Plant Science Department, McGill University, Sainte Anne de Bellevue, QC, Canada

Dr. Adnan Khan Niazi Institut de Biologie Moléculaire des Plantes, CNRS and Université de Strasbourg, Strasbourg, France

Dr. Georges Pelletier Institut Jean-Pierre Bourgin, INRA-AgroParisTech UMR 1318, Versailles, France

Dr. Veli-Matti Rokka Green Technology, Natural Resources Institute Finland, Jokioinen, Finland

Assoc. Prof. Charlotte Brandt Sørensen Department of Clinical Medicine, Faculty of Health, Aarhus University, Aarhus C, Denmark

Dr. Frédérique Weber-Lotfi Institut de Biologie Moléculaire des Plantes, CNRS and Université de Strasbourg, Strasbourg, France

Prof. Fujiang Wen State Key Laboratory of Crop Biology, Shandong Key Laboratory of Crop Biology, Shandong Agricultural University, Tai'an, People's Republic of China

Dr. Huimin Xu Canadian Food Inspection Agency, Charlottetown Laboratory, Charlottetown, PE, Canada

Dr. Walid Zaher Department of Endocrinology and Metabolism, Endocrine Research Laboratory (KMEB), Odense University Hospital, University of Southern Denmark, Odense, Denmark

Dr. Changxiang Zhu State Key Laboratory of Crop Biology, Shandong Key Laboratory of Crop Biology, Shandong Agricultural University, Tai'an, People's Republic of China

About the Editors

Dr. Xiu-Qing Li, Docteur d'Etat Dr. Li earned his degree of Docteur d'Etat (State Doctor) of France in natural science from Université de Paris-Sud (1987). Dr. Li has served as an associate professor of biotechnology at Peking University (1987–1993), an invited professor of genetics (Université de Paris-Sud, 1988), a research scientist of molecular genetics (Chargée de recherche, level II) at the Centre National de la Recherche Scientifique, CNRS, Gif-sur-Yvette, France (1987), a visiting professor at Purdue University (1991–1992), and a researcher at McGill University and DNA Landmarks Inc. (1992–1997), and a senior level research scientist of molecular genetics at Agriculture and Agri-Food Canada, Government of Canada (1997 – present). XQL is the organizer and chair for the annual Somatic Genome Workshops at the International Plant and Animal Genome Conferences (San Diego, 2010– present). XQL has extensive research experience and numerous publications on plant somatic genome characterization and manipulation, plant genetic engineering, plant cell technology, somaclonal variation, RNA polyadenylation, potato carbohydrate metabolism and potato processing quality improvement, plant male sterility, bioinformatics, and genome evolution.

Prof. Danielle J. Donnelly, PhD Dr. Donnelly earned a doctorate in plant physiology from Simon Fraser University (1984) and a postdoctoral fellowship (PDF) at AAFC-Vancouver (1985). She has been an invited researcher, for periods of 1 year each, at Agriculture and Agri-Food Canada (AAFC)-Vancouver, BC, The Plant Propagation Centre, NB, Université de Montreal, QC, and AAFC-NB. She currently directs the Plant Tissue Culture Facility on the Macdonald Campus and the Barbados Interdisciplinary Tropical Studies (BITS) field study semester at the Bellairs Research Station in Barbados. DJD conducts research on in vitro technologies for genetic improvement of potato. Field-based selection of somaclonal variants is used for improvement of tuber yield and type while lab-based selections are applied for phytochemical characters for improved human health. DJD has extensive research experience in micropropagation and hydroponic technologies, germplasm storage, and certification programs for clonally propagated species.

Prof. Thomas G. Jensen, MD, DMSc He graduated as medical doctor (MD) from the University of Aarhus (1987) and started his research career in human genetics as a medical student working on molecular genetics of the skin. Following MD graduation, he was a research fellow, moving into functional genetics and experimental gene therapy research. In 1995, TGJ was invited to work in the laboratory of the gene therapy pioneer Michael Blaese at The National Institutes of Health, Bethesda, USA. TGJ returned to Aarhus in 1997 as an associate professor and in 2001 was elected as head of the institute. In 2003, TGJ was appointed as research leader and professor at the Kennedy Institute in Glostrup, Denmark. In 2007, TGJ returned to Aarhus University as full professor, and in 2011 was appointed as head of the newly formed department of biomedicine at Aarhus University.

Part I
Humans and Animals

Chapter 1
Drug and Gene Electrotransfer in Cancer Therapy

Julie Gehl

1.1 Gene Therapy as a Strategy in Personalized Cancer Therapy

The human genome project (Collins et al. 2004; Altshuler et al. 2010) and the many ensuing projects to describe cancer genomes for different individual cancer diseases (Sjoblom et al. 2006; Greenman et al. 2007) have led to an unprecedented understanding of pathways important in cellular function, and how pathways are altered in different cancer forms. Out of this comes the concept of personalized medicine; that we should leave the "one drug fits all" strategy previously used in oncology, and start looking at exactly what pathways are deregulated and find the drug(s) that would be exactly right for the situation. As an example, some breast cancer patients benefit from drugs blocking the human epidermal growth factor receptor 2 (HER-2) receptor (Slamon et al. 2001; Slamon et al. 2011), and actually some salivary gland tumors also overexpress this receptor and may be treated with what we today would call a breast cancer treatment regimen (Limaye et al. 2013). In future, cancers will likely be defined more by their functionally important mutations, and less by the site of origin.

This change of treatment paradigm also brings a drive for new drugs being able to target these newly identified pathways, and indeed, oncology today is seeing a plethora of new agents, with many more on the way.

After identification of suitable targets and development of new drug candidates for cancer therapy, formulation of the drug can be a challenge which is in some cases a prohibitive or limiting factor. If a drug is not possible to deliver orally due to, e.g., first-pass metabolism in the liver, the drug will need to be administered as an intravenous or other route (e.g., intranasal, pulmonary aerosol). But not all drugs can actually be formulated in a way that allows safe and efficient treatment.

J. Gehl (✉)
Center for Experimental Drug and Gene Electrotransfer, Department of Oncology, Copenhagen University Hospital Herlev, Herlev Ringvej 75, 2730 Herlev, Denmark
e-mail: Julie.Gehl@regionh.dk

X.-Q. Li et al. (eds.), *Somatic Genome Manipulation*,
DOI 10.1007/978-1-4939-2389-2_1

Here, gene therapy may offer promise to bring more drugs to patients as drugs will be produced from transgenes by the patients' own cells and then secreted to the bloodstream.

Another issue is the astronomic cost of producing all these different pharmaceutical agents in various formulations, which is a burden for health economies even in wealthy countries (Sullivan et al. 2011)—and may be prohibitive in less affluent communities. Production of DNA may be performed by the same production facility, whichever protein the DNA is coding for. The *patient* actually becomes the producer of medicines after the DNA has been transferred. Furthermore, it will be possible to do co-transfection of two or more plasmids, enabling combination treatment. Thus, it can be proposed that gene therapy may have the perspective of bringing affordable cancer treatment in a world where many cancer patients are in need of accessible treatments, and also that DNA drugs may enter the market much quicker than traditional medicinal products due to similar production requirements for the different DNA drugs.

Vectors and methods of transfer have been an ongoing discussion; if only we could get the gene there, then great things could happen! Debates have been ongoing whether one method supersedes another. But what is more interesting is to look at the possibilities offered by the different methods. For example, when long-term expression of the therapeutic gene is desired, in which case there may be a specific point in not alerting the immune system to the production of the medicinal product taking place in host cells. Here, nonviral methods such as gene electrotransfer may be a very good option. In other cases, viral vectors will be a good option.

The term "gene therapy" covers the use of DNA for therapeutic purposes, and the term "genetic therapy" also encompasses therapies using RNA, microRNA (miRNA), oligonucleotides, etc. It is likely that future developments may come not least from the use of various oligonucleotide constructions.

1.2 Cancer is a Challenging Target for Gene Therapy

Cancer cells basically have sociopathic behavior, not participating in functions of importance for the overall survival of the host organism, and working for their own benefit. Furthermore, these cells may vary considerably in size and shape, be functionally quite different and be in tumors that have varying local environments, e.g., hypoxic, acidic, necrotic, etc. This has a number of consequences for cancer cells as targets for gene delivery: Will tumor cells in a poorly perfused environment be exposed to the transgene? Will tumor cells be able to take up the genetic material? Is expression hampered due to a deranged genome and a cell functioning on the limit of the possible? Is the transgenic protein going to be expressed? Will the expressed transgene lead to a functional protein? And will there be excretion from the cell of this protein? If particular receptors are used for targeting, will heterogeneity in tumors mean that only some cells will be targeted? Will there be development of resistance?

Cancer is a particularly challenging goal for gene therapy, in that cancer cells by themselves are, genetically speaking, highly unstable, with varying amounts of chromosomes, and prone to mutations often with a lowered level of reparation. Targeting "cancer cells" may be difficult both because the target is moving (mutations occur changing the phenotype), and because a tumor cell population is often highly heterogeneous. Finally, as gene therapy is usually not first-line therapy for cancer, but rather an experimental treatment after standardly approved therapies have been tried, the tumors have survived rounds of various genotoxic agents, such as radiotherapy and chemotherapy, and this makes the tumor cell population even more varied.

1.3 Targeting at the Single-Cell Level, as a Paracrine, or Systemic Approach

Initial gene therapy trials focused on "correcting the wrong" in cancer cells, in order to try to revert the malignant phenotype. A typical example was early trials aiming at reinstating normal function of a key cell cycle regulator often mutated in cancer, namely p53 (Swisher et al. 1999). In order for this strategy to work in cancer, practically *every* cell needs to be successfully transfected. In somatic disease, correcting a certain percentage of normally functioning cells might be sufficient to restore a missing function. However, even a few cancer cells left untransfected may allow immediate tumor regrowth, and so this particular approach may actually be a less promising avenue in the treatment of cancer.

A different approach has been aiming at a paracrine effect, where a higher concentration of a transgenic protein in a local environment may have an effect. A nice example of this is the trial on melanoma by Daud et al. from 2008 (Daud et al. 2008a), where DNA coding for the cytokine interleukin-12 was injected and transferred to melanoma tumors. Although interleukin-12 is quickly eliminated in the systemic circulation, concentrations in the local tumor environment, where the transfection took place, may be high. And this can in turn support an immune response against tumor cells in the transfected area—which may translate to a systemic antitumor immune response. Another example of the paracrine approach is secretion of angiogenic factors locally, to alleviate ischemia or assist healing after surgery with flaps (Ferraro et al. 2009).

Finally, a systemic approach may be pursued. In this approach, the host tissue expresses a transgenic protein with systemic effects. One example would be that the erythropoietin gene is transferred to a small amount of muscle tissue, after which the transgenic protein is secreted to the bloodstream and the protein may exert effects in the organism (Hojman et al. 2007; Gothelf et al. 2010). In some cases, only small amounts of protein will be needed to exert an effect, e.g., in the case of hormones such as erythropoietin (Hojman et al. 2007; Gothelf et al. 2010) or growth hormone (Draghia-Akli et al. 2006). In other cases, a limited amount of protein may be sufficient to improve a medical condition, e.g., in the case of hemophilia where even a limited improvement in clotting efficiency may make a medically important difference. In some cases, larger amounts of a protein drug may be warranted, as, for example, when administering the gene as a strategy to treat cancer through production of a protein with properties inhibiting cancer growth. For the systemic approach, it will also be of importance to be able to halt gene expression, in particular when the transfer has taken place in muscle fibers where expression is long-lived. It has been shown that expression of the transgene is quite localized in muscle (Spanggaard et al. 2012), and also that the transgenic expression may be quickly eliminated using electroporation with calcium (Hojman et al. 2011).

Figure 1.1 depicts these different strategies, from Gothelf and Gehl (2011). Figure 1.2 shows the example of gene electrotransfer to skin, from Gothelf and Gehl (2010), with different uses of the technology. Thus, examples for gene transfer for local treatment and the paracrine strategy would include tumor treatment or working with angiogenic factors for wound healing, vaccinations are an example of paracrine/ systemic therapy, and, finally, systemic expression from skin may be used to target anemia, protein deficiency disorders, or cancer.

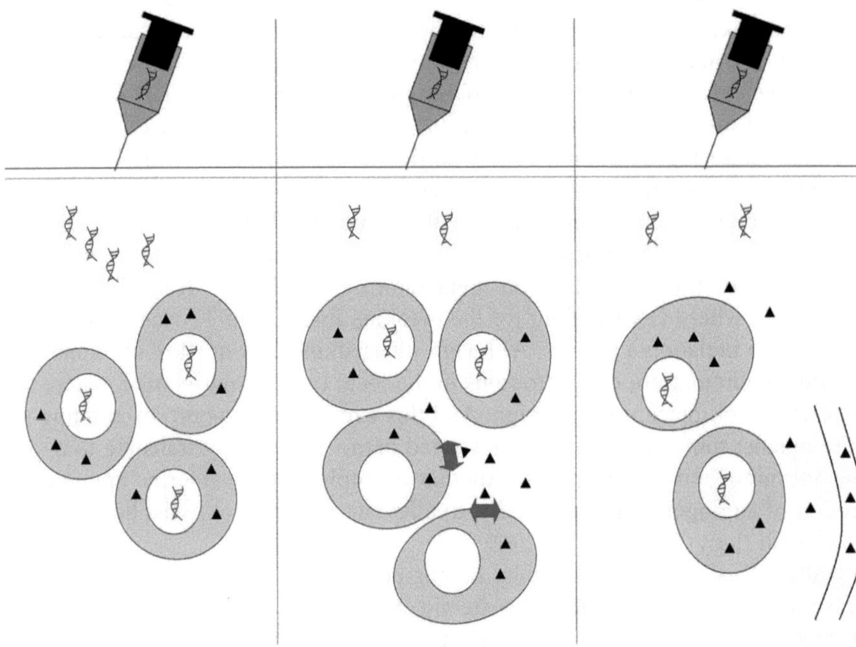

Fig. 1.1 Three approaches of gene therapy. **a** Single-cell approach, in which every single cell in a population has to be transfected with the plasmid in order to benefit from the transgene. An example could be *ex vivo* transfection of keratinocytes. **b** Paracrine approach, in which few cells in a population are transfected with the plasmid and the produced protein then acts locally, e.g., by eliciting an immune response. **c** Systemic approach, in which few cells in a population is transfected with the plasmid. The produced protein is transferred to, e.g., the bloodstream, where it can create a systemic response. The distinction of these three approaches is of course theoretical and the borders between them are arbitrary. From (Gothelf et al. 2010)

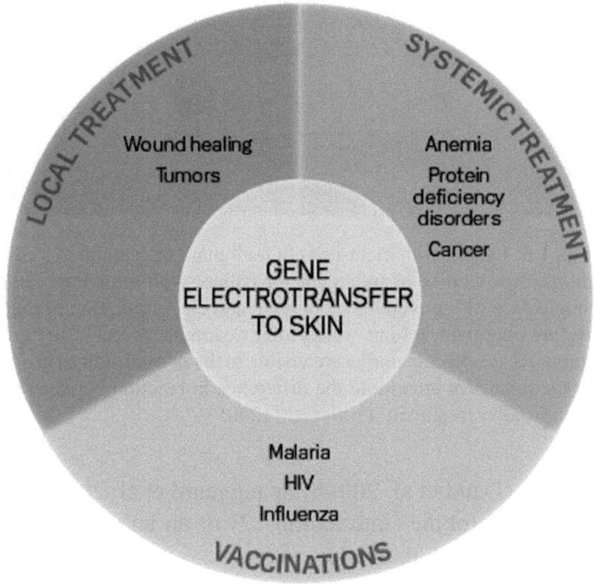

Fig. 1.2 Different objectives for gene electrotransfer to skin. Gene electrotransfer to skin has perspectives in "systemic treatment," "local treatment," and "DNA vaccinations" (Gothelf et al. 2010)

1.4 Electroporation for Nonviral Gene Delivery

As will be familiar to the reader of this book, gene therapy has principally been divided into viral or nonviral vector approaches. The present chapter deals in greater detail with a particular nonviral delivery method, namely electrotransfer.

After the classic 1982 paper by Neumann et al. (Neumann et al. 1982) demonstrating gene transfer to cells, electroporation has become a standardly used technology for gene delivery to bacteria and to cells in culture. A multitude of studies have demonstrated the use of gene electrotransfer to various animal tissues (Mir et al. 2005). In order to perform clinical studies, production of good manufacturing practice (GMP) DNA, as well as approved clinical equipment, was necessary, and the first clinical gene therapy study using electrotransfer was published in 2008 (Daud et al. 2008a).

Introduction of gene therapy into the clinical was helped by the use of electroporation to deliver drugs for treatment of malignant tumors, electrochemotherapy, which is routinely used in the clinic today (Marty et al. 2006; Mir et al. 2006; Matthiessen et al. 2011; Matthiessen et al. 2012). Equipment for delivery of pulses is readily available, and experience on delivery of pulses is at hand. A broader description of equipment for drug and gene electrotransfer may be found in Staal and Gilbert (2011). An example of the efficacy of drug electrotransfer in the treatment of cutaneous metastases of malignant melanoma is shown in Fig. 1.3, from Gehl (2005).

The availability of approved clinical equipment greatly facilitates gene electrotransfer clinical protocols. Clinical gene electrotransfer has been performed in

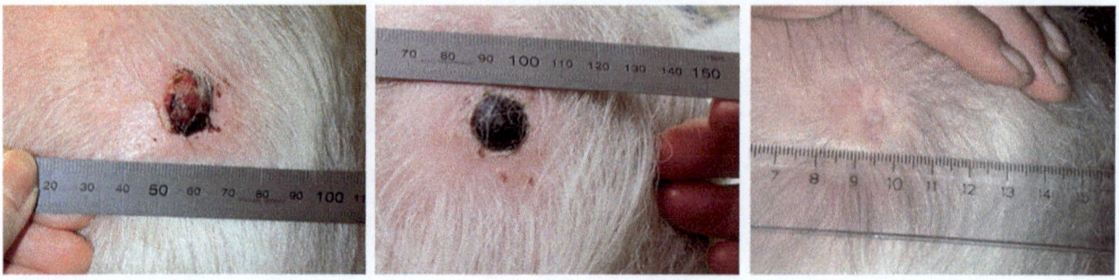

Fig. 1.3 Patient with metastasis of malignant melanoma in a course of treatment using electrochemotherapy. Course of treatment for patient with malignant melanoma metastases, treated with a single dose of intravenous bleomycin and electroporation of metastases using needle electrodes. **a** Before treatment. **b** Four weeks after treatment. A crust can be seen corresponding to the necrotic tumor tissue. Needle marks are visible in the normal skin as the margins of the tumor were included in the treatment area; note the difference in reaction between normal and malignant tissue. **c** Six months after treatment. From Gehl (2005)

tumors (Daud et al. 2008b; Spanggaard et al. 2013), using genes coding for respectively a cytokine (interleukin-12) or an antiangiogenic, antiproliferative molecule (Spanggaard et al. 2013).

1.5 The Technicalities of Electroporation for Drug and Gene Delivery

When brief high-voltage pulses are applied to the cell membrane, rearrangement of the lipid molecules will take place (Ziegler and Vernier 2008), enabling formation of transient pore-like structures (permeabilization), which again may enable passage of otherwise nonpermeant drugs (Orlowski et al. 1988; Gehl et al. 1998; Jaroszeski et al. 2000). For delivery of drugs, typically chemotherapy, a short series of high-voltage pulses are used, as the goal of therapy is to allow passage of molecules through a permeabilized area in the cell membrane. The more permeabilization, the better diffusion of the drug, and if a cancer cell becomes irreversibly permeabilized, this will not be conceived as a problem but as another road to achieve cancer cell death.

For gene delivery, the scenario is quite different: After administration of DNA to tissues, DNA will adsorb to cell membranes (Neumann et al. 1996). Then, electric pulses may be applied, which cause both a destabilization of the cell membrane, and exert an electrophoretic effect on the DNA molecule (Mir et al. 1999a). This may allow movement of the DNA molecule across the cell membrane and subsequent internalization (Golzio et al. 2002).

Once the DNA is inside the cell, various transport mechanisms may assist the journey to the nucleus, e.g., active transport via the microtubule system (Vaughan and Dean 2006). Integration will only rarely take place unless, e.g., nucleases are included in the vector. Expression will depend on a number of factors (of which vector design is of great importance), but posttranscriptional processing will also

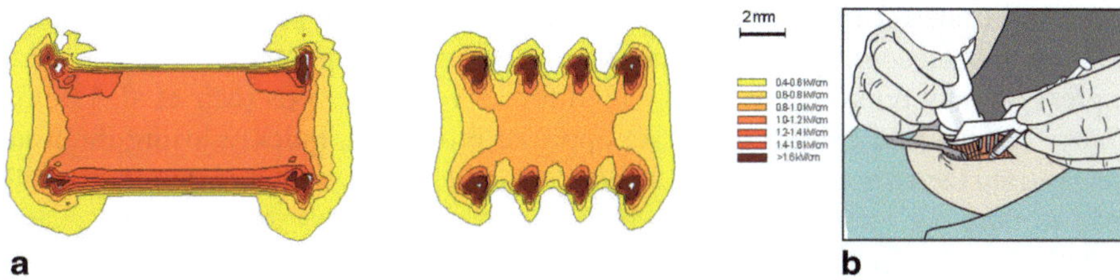

Fig. 1.4 Field distribution calculated for needle array electrode. (*top panel*) Calculated E-field distribution at an applied voltage of 1.2 kV/cm (**a**) for plate electrodes with 1-cm-wide electrodes placed 4 mm apart and (**b**) for needle electrodes at an applied voltage of 1.2 kV/cm with two arrays of each four needles placed 4 mm apart. The spacing between the needles in the array is 2 mm. The small nonsymmetric contour lines are due to limitations in the boundary conditions of the program. (Gehl et al. 1999). (*bottom panel*) Gene electrotransfer to muscle using the depicted electrode. The needle with DNA is placed between the electrodes, and injection is performed. Thereafter, the needle is withdrawn and electric pulses delivered in less than a minute. This procedure has been used in clinical studies (Spanggaard et al. 2012), and (www.clinicaltrials.gov; NCT01664273)

have an effect on measured expression. Finally, the transgenic protein may need to be excreted to exert the desired effect, and after excretion to the extracellular volume also, immunogenicity of the protein may play a role as an immune response towards the protein may decrease presence in circulation. Another situation arises when oligonucleotides are used, and careful optimization may be necessary as the oligonucleotides may differ in, e.g., charge and molecular weight (Joergensen et al. 2011).

When performing gene electrotransfer, the desirable situation is to disturb the cell minimally, and expression of the transgene becomes less likely if cells are stressed by, e.g., high degrees of permeabilization (Bureau et al. 2000; Gehl et al. 2002). Therefore, optimization of pulsing parameters to be just at the threshold for permeabilization, and not beyond, is important for successful gene electrotransfer. Clearly, when using electrodes in tissues, there will be areas subjected to higher field strengths where higher degrees of permeabilization are seen (Gehl et al. 1999), but again most gene electrotransfer protocols will plan on a paracrine or systemic secretion, where a host tissue secretes a transgene. And this transgenic expression may then happen from the area between the electrodes where the field is homogenous and adequate (Gehl et al. 1999; Fig. 1.4).

1.6 Different Tissues: Different Story

Cell type, size, shape, as well as tissue composition matter both for the range of electric parameters to be used and for the distribution of DNA in the tissue.

Cell type: The most important feature of importance for gene delivery by non-viral methods is the longevity of the cell. For example, myofibers have a long life span, and transient transfections may become functionally permanent, in that the fiber neither dies nor divides. It has been shown in several studies that transgene expression may remain consistently high for very long periods (Mir et al. 1999b). Tumor cells would exhibit the exact opposite properties, quickly losing the transiently transfected transgene as cells divide and die. Cells with an intermediate life span to that, e.g., keratinocytes or fibroblasts, may express the transgene over a period of weeks, enabling a short-term expression. This may be very useful for, e.g., the purpose of vaccination (Gothelf and Gehl 2010).

Cell size and shape: The cell size matters for the field at the level of the membrane; thus, a larger cell will be more easily electroporated than smaller cells (Gehl 2003). The field may determine that certain cells in a tissue will have a field more appropriate for transfection than others. Likewise, the field may exert different effects on cells depending on the field direction with respect to cell shape (Kotnik and Miklavcic 2000). A neuron is a good example of a cell which may experience the field differently depending on the field direction respective to the cell, and a myofiber will be much more sensitive to the field if applied along the long axis of the fiber rather than perpendicular to it (Gaylor et al. 1988).

Tissue composition: Electric fields and DNA distribution will also change with tissue composition; the conductance, density, and perfusion, whether there are necrotic areas in the tumor tissue or not, may alter uptake and expression of the transgene.

1.7 Current Clinical Experience

The current clinical experience on electrochemotherapy is now quite extensive. Initially, only small tumors were treated (Belehradek et al. 1993; Glass et al. 1996; Heller et al. 1998; Marty et al. 2006), but now quite extensive lesions are being treated (Whelan et al. 2006; Matthiessen et al. 2011; Matthiessen et al. 2012; see Fig. 1.5). Whereas only a few centers were active in 2006, when standard operating procedures were first published (Mir et al. 2006), today over 100 centers are offering electrochemotherapy in Europe alone. The electrochemotherapy concept is now being taken to internal organs as well, notably clinical trials on the liver (Edhemovic et al. 2011); bone (Fini et al. 2010), clinical trial ongoing; brain (Linnert and Gehl 2009; Agerholm-Larsen et al. 2011; Linnert et al. 2012), and colorectal cancer as described in review by Miklavcic et al. (2012).

The experience on electrochemotherapy for both superficial and internal tumors has had, and will have, great impact on the use of gene therapy. As just one example, the electrode developed for drug delivery in the brain (Mahmood and Gehl 2011), which has been approved for clinical trial, may at a later time be used for gene delivery in the brain, e.g., for the treatment of Parkinson's disease. Current gene therapy trials employing gene delivery to, e.g., tumor or muscle are also making use of approved standard equipment for electrochemotherapy, which has greatly eased the process of approval of the gene therapy trials.

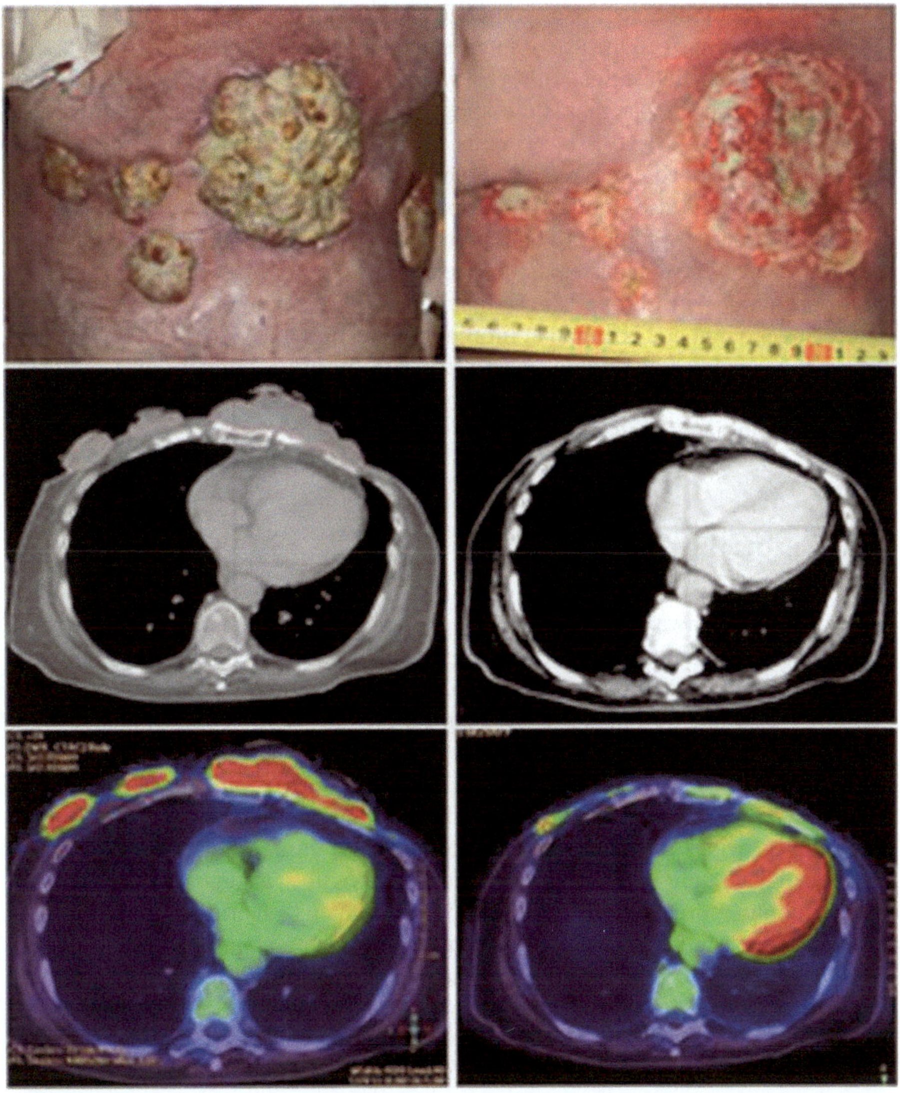

Fig. 1.5 Patient treated with electrochemotherapy against extensive chest wall recurrence. Sixty-four-year-old woman with locoregional recurrence of bilateral receptor negative, HER2-negative breast cancer. Previous treatments over a period of 5 years included radiotherapy 48 Gy in 24 fractions on both sides and reirradiation with 30 Gy in 10 fractions on the left side, systemic therapy (cyclophosphamide, epirubicin, fluorouracil, docetaxel, gemcitabine, vinorelbine, and capecitabine). Despite all these treatments, there was continuous progression of the cutaneous lesions. The column on the left shows image of lesions, CT scan, and PET/CT scan before treatment scan; column on the right shows image of lesion, CT scan, and PET/CT scan after two sessions with electrochemotherapy. (HER2 human epidermal growth factor receptor 2, CT computerized tomography, PET positron emission tomography). From Matthiessen et al. (2012)

The first pivotal clinical paper on gene electrotransfer (Daud et al. 2008a) published in 2008 described the use of intralesional injection of interleukin-12 into malignant melanoma metastases, followed by high-voltage electric pulses. The thought behind this approach was that stimulating a local immune response towards melanoma antigens could lead to a systemic immunologic response. Indeed, 2 (10 %) of 19 patients showed complete regression of all metastases, whereas 8 additional patients showed disease stabilization or partial response (Daud et al. 2008a).

Clinical trials are required to be reported to a recognized public database before they commence; the database most frequently used is clinicaltrials.gov in the USA, but also EudraCT in Europe is very much used. The requirement to register trials before they commence is aimed at limiting publication bias, i.e., that only successful trials get published leading to an artificially positive estimate of a particular treatment. This requirement has the very positive effect that it is easy to get a current overview of initiated trials.

1.8 Future Perspectives

Novel cellular pathways are rapidly being uncovered, and cancer genomes are being sequenced, enabling an unprecedented insight into cell biology, as well as the malignant genotype and phenotype. With these discoveries, novel targets are being identified, enabling new approaches to cancer therapy. A particular challenge—and promise—is that as these new targets are identified, novel drugs may be invented. Gene therapy may offer very interesting new perspectives; thus, a range of DNA drugs may be developed much more easily and at more affordable cost. This again may allow a wider range of treatment possibilities aiming at different targets in cancer cells and their supportive structures, and possibly also at a more accessible price, enabling more patients to benefit. Cancer remains a worldwide leading cause of death and morbidity, and progress in the prevention and treatment of this disease continues to be a very high priority in research and development.

Acknowledgments The author was a Royal Swedish Academy of Sciences Research Fellow supported by a grant from the Acta Oncologica Foundation (2009–2014).

References

Agerholm-Larsen B, Iversen HK, Ibsen P, Moller JM, Mahmood F, Jensen KS, Gehl J (2011) Preclinical validation of electrochemotherapy as an effective treatment for brain tumors. Cancer Res 71:3753–3762

Altshuler D, Durbin R M, Abecasis GR, Bentley DR, Chakravarti A, Clark AG, Collins FS et al (2010) A map of human genome variation from population-scale sequencing. Nature 467:1061–1073

Belehradek M, Domenge C, Luboinski B, Orlowski S, Belehradek J Jr, Mir LM (1993) Electrochemotherapy, a new antitumor treatment. First clinical phase I-II trial. Cancer 72:3694–3700

Bureau MF, Gehl J, Deleuze V, Mir LM, Scherman D (2000) Importance of association between permeabilization and electrophoretic forces for intramuscular DNA electrotransfer. Biochim Biophys Acta 1474:353–359

Collins FS, Lander ES, Rogers J, Waterston RH (2004) Finishing the euchromatic sequence of the human genome. Nature 431:931–945

Daud AI, DeConti RC, Andrews S, Urbas P, Riker AI, Sondak VK, Munster PN, Sullivan DM, Ugen KE, Messina JL, Heller R (2008a) Phase I trial of interleukin-12 plasmid electroporation in patients with metastatic melanoma. J Clin Oncol 26:5896–5903

Daud AI, DeConti RC, Andrews S, Urbas P, Riker AI, Sondak VK, Munster P N, Sullivan DM, Ugen KE, Messina JL, Heller R (2008b) Phase I trial of interleukin-12 plasmid electroporation in patients with metastatic melanoma. J Clin Oncol 26:5896–5903

Draghia-Akli R, Pope MA, Brown PA, Khan AS (2006) Plasmid-based expression technology using growth hormone releasing hormone: a novel method for physiologically stimulating long-term growth hormone secretion. Comb Chem High Throughput Screen 9:181–185

Edhemovic I, Gadzijev EM, Brecelj E, Miklavcic D, Kos B, Zupanic A, Mali B, Jarm T, Pavliha D, Marcan M, Gasljevic G, Gorjup V, Music M, Vavpotic TP, Cemazar M, Snoj M, Sersa G (2011) Electrochemotherapy: a new technological approach in treatment of metastases in the liver. Technol Cancer Res Treatment 10:475–485

Ferraro B, Cruz YL, Coppola D, Heller R (2009) Intradermal delivery of plasmid VEGF(165) by electroporation promotes wound healing. Mol Ther 17:651–657

Fini M, Tschon M, Ronchetti M, Cavani F, Bianchi G, Mercuri M, Alberghini M, Cadossi R (2010) Ablation of bone cells by electroporation. J Bone Joint Surg Br 92:1614–1620

Gaylor DC, Prakah-Asante K, Lee RC (1988) Significance of cell size and tissue structure in electrical trauma. J Theor Biol 133:223–237

Gehl J (2003) Electroporation: theory and methods, perspectives for drug delivery, gene therapy and research. Acta Physiol Scand 177:437–447

Gehl J (2005) Investigational treatment of cancer using electrochemotherapy, electrochemoimmunotherapy and electro-gene transfer. Ugeskr Laeger 167:3156–3159

Gehl J, Skovsgaard T, Mir LM (1998) Enhancement of cytotoxicity by electropermeabilization: an improved method for screening drugs. Anticancer Drugs 9:319–325

Gehl J, Sørensen TH, Nielsen K, Raskmark P, Nielsen SL, Skovsgaard T, Mir LM (1999) In vivo electroporation of skeletal muscle: threshold, efficacy and relation to electric field distribution. Biochim Biophys Acta 1428:233–240

Gehl J, Skovsgaard T, Mir LM (2002) Vascular reactions to in vivo electroporation: characterization and consequences for drug and gene delivery. Biochim Biophys Acta 1569:51–58

Glass LF, Fenske NA, Jaroszeski M, Perrott R, Harvey DT, Reintgen DS, Heller R (1996) Bleomycin-mediated electrochemotherapy of basal cell carcinoma. J Am Acad Dermatol 34:82–86

Golzio M, Teissie J, Rols MP (2002) Direct visualization at the single-cell level of electrically mediated gene delivery. Proc Natl Acad Sci U S A 99:1292–1297

Gothelf A, Gehl J (2010) Gene electrotransfer to skin; review of existing literature and clinical perspectives. Current Gene Ther 10:287–299

Gothelf A, Gehl J (2011) Gene electrotransfer to skin. In Kee S, Gehl J, Lee E (eds) Clinical aspects of electroporation pp 189–202. Springer, New York

Gothelf A, Hojman P, Gehl J (2010) Therapeutic levels of erythropoietin (EPO) achieved after gene electrotransfer to skin in mice. Gene Ther 17:1077–1084

Greenman C, Stephens P, Smith R, Dalgliesh GL, Hunter C, Bignell G et al (2007) Patterns of somatic mutation in human cancer genomes. Nature 446:153–158

Heller R, Jaroszeski M J, Reintgen DS, Puleo CA, DeConti RC, Gilbert RA, Glass LF (1998) Treatment of cutaneous and subcutaneous tumors with electrochemotherapy using intralesional bleomycin. Cancer 83:148–157

Hojman P, Gissel H, Gehl J (2007) Sensitive and precise regulation of haemoglobin after gene transfer of erythropoietin to muscle tissue using electroporation. Gene Ther 14:950–959

Hojman P, Spanggaard I, Olsen CH, Gehl J, Gissel H (2011) Calcium electrotransfer for termination of transgene expression in muscle. Hum Gene Ther 22:753–760

Jaroszeski MJ, Dang V, Pottinger C, Hickey J, Gilbert R, Heller R (2000) Toxicity of anticancer agents mediated by electroporation in vitro. Anticancer Drugs 11:201–208

Joergensen M, Agerholm-Larsen B, Nielsen PE, Gehl J (2011) Efficiency of cellular delivery of antisense peptide nucleic acid by electroporation depends on charge and electroporation geometry. Oligonucleotides 21:29–37

Kotnik T, Miklavcic D (2000) Analytical description of transmembrane voltage induced by electric fields on spheroidal cells [In Process Citation]. Biophys J 79:670–679

Limaye SA, Posner MR, Krane JF, Fonfria M, Lorch JH, Dillon DA, Shreenivas AV, Tishler RB, Haddad RI (2013) Trastuzumab for the treatment of salivary duct carcinoma. Oncologist 18:294–300

Linnert J, Gehl J (2009) Bleomycin treatment of brain tumors. An evaluation. Anticancer Drugs 20:157–164

Linnert M, Iversen HK, Gehl J (2012) Multiple brain metastases—current management and perspectives for treatment with electrochemotherapy. Radiol Oncol 46:271–278

Mahmood F, Gehl J (2011) Optimizing clinical performance and geometrical robustness of a new electrode device for intracranial tumor electroporation. Bioelectrochem 81:10–16

Marty M, Sersa G, Garbay JR, Gehl J, Collins CG, Snoj M, Billard V, Geertsen PF, Larkin JO, Miklavcic D, Pavlovic I, Paulin-Kosir SM, Cemazar M, Morsli N, Rudolf Z, Robert C, O'Sullivan GC, Mir LM (2006) Electrochemotherapy—an easy, highly effective and safe treatment of cutaneous and subcutaneous metastases: results of ESOPE (European Standard Operating Procedures of Electrochemotherapy) study. Ejc Suppl 4:3–13

Matthiessen LW, Chalmers RL, Sainsbury DC, Veeramani S, Kessell G, Humphreys AC, Bond JE, Muir T, Gehl J (2011) Management of cutaneous metastases using electrochemotherapy. Acta Oncol 50:621–629

Matthiessen LW, Johannesen HH, Hendel HW, Moss T, Kamby C, Gehl J (2012) Electrochemotherapy for large cutaneous recurrence of breast cancer: a phase II clinical trial. Acta Oncologica 51:713–721

Miklavcic D, Sersa G, Brecelj E, Gehl J, Soden D, Bianchi G, Ruggieri P, Rossi CR, Campana LG, Jarm T (2012) Electrochemotherapy: technological advancements for efficient electroporation-based treatment of internal tumors. Med Biol Eng Comput 50:1213–1225

Mir LM, Bureau MF, Gehl J, Rangara R, Rouy D, Caillaud JM, Delaere P, Branellec D, Schwartz B, Scherman D (1999a) High-efficiency gene transfer into skeletal muscle mediated by electric pulses. Proc Natl Acad Sci U S A 96:4262–4267

Mir LM, Bureau MF, Gehl J, Rangara R, Rouy D, Caillaud J-M, Delaere P, Branellec D, Schwartz B, Scherman D (1999b) High efficiency gene transfer into skeletal muscle mediated by electric pulses. Proc Natl Acad Sci U S A 96:4262–4267

Mir LM, Moller PH, Andre F, Gehl J. (2005) Electric pulse-mediated gene delivery to various animal tissues. Adv Genet 54:83–114

Mir LM, Gehl J, Sersa G, Collins CG, Garbay JR, Billard V, Geertsen PF, Rudolf Z, O'Sullivan GC, Marty M (2006) Standard operating procedures of the electrochemotherapy: instructions for the use of bleomycin or cisplatin administered either systemically or locally and electric pulses delivered by the CliniporatorTM by means of invasive or non-invasive electrodes. Eur J Cancer Suppl 4:14–25

Neumann E, Schaefer-Ridder M, Wang Y, Hofschneider PH (1982) Gene transfer into mouse lyoma cells by electroporation in high electric fields. EMBO J 1:841–845

Neumann E, Kakorin S, Tsoneva I, Nikolova B, Tomov T (1996) Calcium-mediated DNA adsorption to yeast cells and kinetics of cell transformation by electroporation. Biophys J 71:868–877

Orlowski S, Belehradek J Jr, Paoletti C, Mir LM (1988) Transient electropermeabilization of cells in culture. Increase of the cytotoxicity of anticancer drugs. Biochem Pharmacol 37:4727–4733

Sjoblom T, Jones S, Wood LD, Parsons DW, Lin J, Barber TD, Mandelker D, Leary RJ, Ptak J, Silliman N, Szabo S, Buckhaults P, Farrell C, Meeh P, Markowitz SD, Willis J, Dawson D, Willson JKV, Gazdar AF, Hartigan J, Wu L, Liu CS, Parmigiani G, Park BH, Bachman KE, Papadopoulos N, Vogelstein B, Kinzler KW, Velculescu VE (2006) The consensus coding sequences of human breast and colorectal cancers. Science 314:268–274

Slamon DJ, Leyland-Jones B, Shak S, Fuchs H, Paton V, Bajamonde A, Fleming T, Eiermann W, Wolter J, Pegram M, Baselga J, Norton L (2001) Use of chemotherapy plus a monoclonal antibody against HER2 for metastatic breast cancer that overexpresses HER2. New Engl J Med 344:783–792

Slamon D, Eiermann W, Robert N, Pienkowski T, Martin M, Press M, Mackey J, Glaspy J, Chan A, Pawlicki M, Tamas P, Valero V, Liu MC, Sauter G, von Minckwitz G, Visco F, Bee V, Buyse M, Bendahmane B, Tabah-Fisch I, Lindsay MA, Riva A, Crown J (2011) Adjuvant Trastuzumab in HER2-positive breast cancer. New Engl J Med 365:1273–1283

Spanggaard I, Corydon T, Hojman P, Gissel H, Dagnaes-Hansen F, Jensen TG, Gehl J (2012) Spatial distribution of transgenic protein after gene electrotransfer to porcine muscle. Hum Gene Ther Methods 23:387–392

Spanggaard I, Snoj M, Cavalcanti A, Bouquet C, Sersa G, Robert C, Cemazar M, Dam E, Vasseur B, Attali P, Mir L, Gehl J (2013) Gene electrotransfer of plasmid AMEP in disseminated melanoma: safety and efficacy results of a phase I first-in-man study. Hum Gene Ther Clin Dev 24:99–107. (In Press)

Staal LG, Gilbert R (2011) Generators and applicators: equipment for electroporation. In Kee S, Gehl J, Lee E (eds) Clinical aspects of electroporation pp. 45–65. Springer, New York

Sullivan R, Peppercorn J, Sikora K, Zalcberg J, Meropol NJ, Amir E, Khayat D, Boyle P, Autier P, Tannock IF, Fojo T, Siderov J, Williamson S, Camporesi S, Mcvie JG, Purushotham AD, Naredi P, Eggermont A, Brennan MF, Steinberg ML, De Ridder M, McCloskey SA, Verellen D, Roberts T, Storme G, Hicks RJ, Ell PJ, HirschBR, Carbone DP, Schulman KA, Catchpole P, Taylor D, Geissler J, Brinker NG, Meltzer D, Kerr D, Aapro M (2011) Delivering affordable cancer care in high-income countries. Lancet Oncol 12:933–980

Swisher SG, Roth JA, Nemunaitis J, Lawrence DD, Kemp BL, Carrasco CH, Connors DG, El-Naggar AK, Fossella F, Glisson BS, Hong WK, Khuri FR, Kurie JM, Lee JJ, Lee JS, Mack M, Merritt JA, Nguyen DM, Nesbitt JC, Perez-Soler R, Pisters KMW, Putnam JB, Richli WR, Savin M, Schrump DS, Shin DM, Shulkin A, Walsh GL, Wait J, Weill D, Waugh MKA (1999) Adenovirus-mediated p53 gene transfer in advanced non-small-cell lung cancer. J Nat Cancer Inst 91:763–771

Vaughan EE, Dean DA (2006) Intracellular trafficking of plasmids during transfection is mediated by microtubules. Mol Ther 13:422–428

Whelan MC, Larkin JO, Collins CG, Cashman J, Breathnach O, Soden DM, O'Sullivan GC (2006) Effective treatment of an extensive recurrent breast cancer which was refractory to multimodal therapy by multiple applications of electrochemotherapy. Ejc Supplements 4:32–34

Ziegler MJ, Vernier PT (2008) Interface water dynamics and porating electric fields for phospholipid bilayers. J Phys Chem B 112:13588–13596

Chapter 2
Targeted Porcine Genome Engineering with TALENs

Yonglun Luo, Lin Lin, Monika M. Golas, Charlotte Brandt Sørensen and Lars Bolund

2.1 Introduction

The domestic pig, as an economically and biologically important animal in agricultural, pharmacological, and biomedical applications, is well known for its difficulty in targeted genome engineering. This is mostly due to a lack of established authentic embryonic stem cells and the relatively low efficiency of targeted engineering by homologous recombination (HR). The first targeted gene (alpha-1,3-galactosyltransferase) knockout (KO) pig was generated by conventional gene targeting (vector-mediated HR) and nuclear transfer one decade ago (Lai et al. 2002). However, the conventional targeting method is inefficient and laborious. Another method, based on engineered recombinant adeno-associated virus (rAAV) vector, has greatly improved the targeting efficiency and has been applied to produce several genetically modified (GM) pig models for, e.g., cystic fibrosis (Rogers et al. 2008), breast cancer (Luo et al. 2011b), and hereditary tyrosinemia type 1 (Hickey et al. 2011). Nevertheless, the broad use of rAAV for generating targeted GM pig models for biomedical research is still hampered by its relatively low targeting frequency, although ~1000-fold higher than the conventional targeting method, along with its high rate of random insertion and laboratory requirement such as a biosafety level 2 laboratory.

During the last few years, DNA nuclease technologies, including, e.g., zinc-finger nucleases (ZFNs; Kim et al. 1996) and particularly the transcription activator-like

Y. Luo (✉) · L. Lin · M. M. Golas · C. B. Sørensen · L. Bolund
Department of Biomedicine, Faculty of Health, Aarhus University, Aarhus C, Denmark
e-mail: ALUN@BIOMED.AU.DK

M. M. Golas
Center for Stochastic Geometry and Advanced Bioimaging, Aarhus University, Aarhus C, Denmark

C. B. Sørensen
Department of Clinical Medicine—The Department of Cardiological Medicine B, Aarhus University, Aarhus N, Denmark

X.-Q. Li et al. (eds.), *Somatic Genome Manipulation*,
DOI 10.1007/978-1-4939-2389-2_2

effector nucleases (TALENs; Boch et al., 2009; Bogdanove and Voytas, 2011), have emerged as exciting tools for targeted genome editing in a wide range of cell types and organisms. To date, the efficacy of TALEN-mediated genome engineering has been proven in approximately 20 species (Table 2.1), also see review by Joung and Sander 2013). Moreover, the feasibility of creating biallelic KO F0 founders has greatly facilitated the production procedure and reduced the cost for generating GM large animals such as pigs and cows that have longer pregnancy periods than rodents (Hauschild et al. 2011; Yu et al. 2011; Carlson et al. 2012). In this chapter, we focus on describing TALENs and the use of TALENs for targeted genome editing in primary porcine fibroblasts.

As illustrated in Fig. 2.1a, one TALEN comprises three domains: (i) a modular sequence-specific DNA-binding domain (DBD) core, which is engineered from the bacterial TAL effector protein; (ii) a catalytic domain (CD), which is derived from the *FokI* nuclease and confers double-stranded DNA breaks (DSBs) upon dimerization with another TALEN; and (iii) a nuclear localization signal, which homes the fusion proteins to the nucleus (see also review by Joung and Sander (2013)). The nucleotide-binding specificity of the TAL effector protein is determined by a modularly repeated domain, which employs a simple protein–DNA code (Boch et al. 2009; Fig. 2.1b 2.1a), and thus greatly facilitated the generation of synthetic TALENs (Cermak et al. 2011; Li et al. 2011b; Reyon et al. 2012). Several methods have been developed for engineering TALENs based on modular assembling (Li et al. 2011b), ligation-independent cloning (Schmid-Burgk et al. 2013), "Golden Gate" cloning (Cermak et al. 2011), or high-throughput solid-phase assembling (Reyon et al. 2012). Unlike ZFNs, the engineering of TALENs by these methods can be accomplished with general laboratory reagents, limited budget, and time (i.e., within a few days).

Several kinds of genomic alterations, such as mutations, small and/or large deletions and insertions (indels), translocations, and inversions, have been reported in a wide range of cells and organisms using TALENs (Bogdanove and Voytas 2011; Miller et al. 2011; Carlson et al. 2012; Joung and Sander 2013). The principle that DSBs enhance cellular DNA damage repairing machineries by error-prone nonhomologous end joining (NHEJ) or error-free HR has been known for three decades (Szostak et al. 1983). TALEN-mediated genomic manipulation shares the same principle as DSBs. Upon being introduced into targeted cells, two TALENs bind to their recognition sites (TS1 and TS2 as illustrated in Fig. 2.1a 2.1b). Upon dimerization of the *FokI* CD, an active endonuclease is formed which creates DSBs within the spacer region. An efficient spacer is between 12 and 31 bp depending on the choice of TALE architectures (Cermak et al. 2011; Li et al. 2011a; Miller et al. 2011; Mussolino et al. 2011). Random mutations, indels, and insertions will be introduced at the DSB sites by NHEJ, while site-specific mutagenesis, indels, and/or insertion can be introduced in or near the DSB sites by using a donor oligo or template (Fig. 2.1b). More advanced genomic alterations such as deletions of large fragments, translocations, or inversions can be achieved by using multiple pairs of TALENs (Carlson et al. 2012).

Table 2.1 List of TALEN-mediated genome engineering (until July 2013)

Species	TALEN delivery (cell type)	Genes	Reference
Human (*Homo sapiens*)			
Human (*Homo sapiens*)	Electroporation (myoblasts)	*DMD*	(Ousterout et al. 2013)
Human (*Homo sapiens*)	Transfection (HEK293)	*VKOR*	(Tie et al. 2013)
Human (*Homo sapiens*)	Transfection (HEK 293T)	*miR-155*, miR-155, miR-146a and miR-125b*	(Hu et al. 2013)
Human (*Homo sapiens*)	Transfection (Jurkat and RPE-1 cells)	*NPM1 and ALK*	(Piganeau et al. 2013)
Human (*Homo sapiens*)	Transfection (HEK 293)	*Genome wide*	(Kim et al. 2013)
Human (*Homo sapiens*)	Transfection (HEK293), electroporation (iPSCs)	*HPRT, CFTR, eGFP*	(Sakuma et al. 2013)
Human (*Homo sapiens*)	Transduction (HeLa cells)	*AAVS1*	(Holkers et al. 2013)
Human (*Homo sapiens*)	Transfection (fibroblasts)	*15 genes (AKT2, ANGPTL3, APOB, ATGL, C6orf106, CIITA, CELSR2, CFTR, GLUT4, LINC00116, NLRC5, PLIN1, SORT1, TRIB1, TTN)*	(Ding et al. 2013)
Human (*Homo sapiens*)	Transfection (hESCs)		
Human (*Homo sapiens*)	Transfection (iPSCs)		
Human (*Homo sapiens*)	Ligation-independent cloning (LIC)	*Genome wide*	(Schmid-Burgk et al. 2013)
Human (*Homo sapiens*)	Transfection (HEK 293T)	*NDUFA9*	(Stroud et al. 2013)
Human (*Homo sapiens*)	FLASH assembly	*Genome wide*	(Reyon et al. 2012)
Human (*Homo sapiens*)	Transfection (Hela cells)	*HBB*	(Sun et al. 2012)
Human (*Homo sapiens*)	Transfection (human ESCs and iPSCs)	*PPP1R12C, OCT4, and PITX3*	(Hockemeyer et al. 2011)
Cow (*Bos taurus*) and *pig* (*Sus scrofa*)			
Livestock: Cow (*Bos taurus*) and pig (*Sus scrofa*)	Transfection (fibroblasts) and microinjection (embryos)	*Porcine: Ldlr, DMD, and Ghrhr* Bovine: *Acan* and *GDF8,*	(Carlson et al. 2012)
Rabbit (*Oryctolagus cuniculus*)			
Rabbit (*Oryctolagus cuniculus*)	Microinjection (pronuclear)	*Rag1* and *Rag2*	(Song et al. 2013)

Table 2.1 (continued)

Species	TALEN delivery (cell type)	Genes	Reference
Mouse (*Mus musculus*)			
Mouse (*Mus musculus*)	Microinjection (pronuclear or zygote)	*Mlkl*	(Wu et al. 2013)
Mouse (*Mus musculus*)	Microinjection (zygote)	*Lepr, Pak1ip1, Gpr55, Rprm, Fbxo6, Smurf1, Dcaf13, Fam73a, Wdr20a*, and *Tmem74*	(Qiu et al. 2013)
Mouse (*Mus musculus*)	Microinjection (CD1, C3H, and C57BL/6J oocyte)	*Zic2*	(Davies et al. 2013)
Mouse (*Mus musculus*)	Microinjection (one-cell embryo)	*Rab38*	(Wefers et al. 2013)
Rat (*Rattus norvegicus*)			
Rat (*Rattus norvegicus*)	Electroporation (fibroblasts) and microinjection (zygotes)	*Try*	(Mashimo et al. 2013)
Rat (*Rattus norvegicus*)	Transfection (ES cells)	*Bmpr2*	(Tong et al. 2012)
Rat (*Rattus norvegicus*)	Microinjection (1-cell-stage embryos)	*IgM*	(Tesson et al. 2011)
Xenopus			
Xenopus (*Xenopus laevis*)	Microinjection (one-cell stage)	*Tyr and Pax6*	(Suzuki et al. 2013)
Xenopus (*Xenopus tropicalis*)	Microinjection (one-cell stage)	*Tyr, noggin,* and *MMP-9TH*	(Nakajima et al. 2013)
Xenopus (*Xenopus tropicalis*)	Microinjection (2-cell-stage blastomere)	*Tyr*	(Ishibashi et al. 2012)
Xenopus (*Xenopus laevis*)	Microinjection (1-cell-stage embryos)	*Noggin, Ptf1a/p48*, and *Ets1*	(Lei et al. 2012)
Xenopus (*Xenopus laevis*)	Microinjection (1-cell-stage embryos)	*eGFP*	(Sakuma et al. 2013)
Zebra fish (*Danio rerio*)			
Zebra fish (*Danio rerio*)	Microinjection (one-cell stage)	*Sema3fb, dre-mir-126a, dre-mir-126b, Cluster Chr. 1, from dre-mir-17a-1 to dre-mir-92a-1, Cluster Chr. 9, from dre-mir-17a-2 to dre-mir-92a-2*	(Xiao et al. 2013)
Zebra fish (*Danio rerio*)	Microinjection (one-cell stage)	*idh1, flt3, npm1b, jak2a, npm1a,*	(Ma et al. 2013)

Table 2.1 (continued)

Species	TALEN delivery (cell type)	Genes	Reference
Zebra fish (*Danio rerio*)	Microinjection (one-cell-stage embryos)	*apoea, flt4, lepa, linc-birc6 [megamind] lincRNA gene and the globin locus control region*	(Gupta et al. 2013)
Zebra fish (*Danio rerio*)	Microinjection (1- to 4-cell-stage embryos)	*eGFP*	(Sakuma et al. 2013)
Zebra fish (*Danio rerio*)	Microinjection (1-cell-stage embryos)	*ponzr1, msna, ppp1cab and cdh5*	(Bedell et al. 2012)
Zebra fish (*Danio rerio*)	Microinjection (1-cell-stage embryos)	*golden*	(Dahlem et al. 2012)
Zebra fish (*Danio rerio*)	Microinjection (1-cell-stage embryos)	*elmo1, epas1b, fh, hif1ab, ptpmt1, and scl6a3*	(Cade et al. 2012)
Zebra fish (*Danio rerio*)	Microinjection (1-cell-stage embryos)	*bmi1, ikzf1, phf6, jak3, and myoD*	(Moore et al. 2012)
Zebra fish (*Danio rerio*)	Microinjection (1-cell-stage embryos)	*tnikb*	(Huang et al. 2011)
Zebra fish (*Danio rerio*)	Microinjection (1-cell-stage embryos)	*gria3a* and *hey2*	(Sander et al. 2011)
Fruit fly (*Drosophila melanogaster*)			
Fruit fly (*Drosophila melanogaster*)	Microinjection (embryos)	*egfp*	(Sakuma et al. 2013)
Fruit fly (*Drosophila melanogaster*)	Microinjection (embryos)	*yellow*	(Liu et al. 2012)
Medaka (*Oryzias latipes*)			
Medaka (*Oryzias latipes*)	Microinjection (fertilized eggs)	*DJ-1*	(Ansai et al. 2013)
Silkworm (Bombyx mori)			
Silkworm (*Bombyx mori*)	Microinjection (syncytial-preblastoderm-stage embryo)	*BmBlos2*	(Sajwan et al. 2013)
Silkworm (*Bombyx mori*)	Microinjection (oviposited embryos ≤2 h)	*BmBlos2*	(Ma et al. 2012)
Plants			
Rice (*Oryza sativa*)	Polyethylene glycol (protoplast)	*OsDEP1* (*LOC_Os09g26999*), *OsBADH2* (*LOC_Os08g32870*), *OsCKX2* (*LOC_Os01g10110*), and *OsSD1* (*LOC_Os01g40720*)	(Shan et al. 2013)

Table 2.1 (continued)

Species	TALEN delivery (cell type)	Genes	Reference
Brachypodium (*Brachypodium distachyon*)	Polyethylene glycol (protoplast)	*BdABA1* (*Bradi5g11750*), *BdCKX2* (*Bradi2g06030*), *BdSMC6* (*Bradi4g08527*), *BdSPL* (*Bradi2g03740*), *BdSBP* (*Bradi4g33770*), *BdCOI1* (*Bradi2g23730*), *BdRHT* (*Bradi1g11090*), and *BdHTA1* (*Bradi1g25390*)	(Shan et al. 2013)
Barley (*Hordeum vulgare* L.)	Transformation (12–14-day embryos)	*PAPhy_a*	(Wendt et al. 2013)
Tobacco (*Nicotiana tabacum*)	Transformation (protoplasts)	*ALS*	(Zhang et al. 2013)
Other species			
Mosquitoes (*Aedes aegypti*)	Microinjection (1-h-old preblastoderm embryo)	*kmo*	(Aryan et al. 2013)
Nematode (*Caenorhabditis elegans* and *C. briggsae*)	Microinjection (1–12 h after the L4 adult molt)	*ben-1*	(Wood et al. 2011)
Yeast (*Saccharomyces cerevisiae*)	Transformation	*URA3, LYS2, and ADE2*	(Li et al. 2011b)

Highly efficient TALEN-mediated gene-specific manipulations have been reported in both primary porcine fibroblasts and porcine zygotes (Carlson et al. 2012). In Carlson et al.'s study, the authors achieved modifications up to 75 % of oocytes injected with TALENs as well as a high frequency of monoallelic and biallelic modifications in selected/nonselected fibroblsts clones (Carlson et al. 2012). In addition, the authors further showed that the TALEN-mediated GM cells are suitable for cloning: Biallelic *Ldlr* modified pigs were generated as a model for atherosclerosis (Carlson et al. 2012). Herein, we briefly describe the protocol for engineering TALENs using the "Golden Gate" assembling method adapted from Cermak et al. (2011) and the delivery of TALENs together with a donor plasmid into primary porcine fibroblasts by the 4D-Nucleofector System (Lonza, Basel, Switzerland). As a detailed protocol for assembling TALENs by Golden Gate cloning was introduced by Cermak et al. in their article (Cermak et al. 2011), we mainly focus on describing important *notes* with respect to the design of TALENs, increase of the Golden Gate cloning efficiency, and the efficient delivery of TALENS with or without donor template.

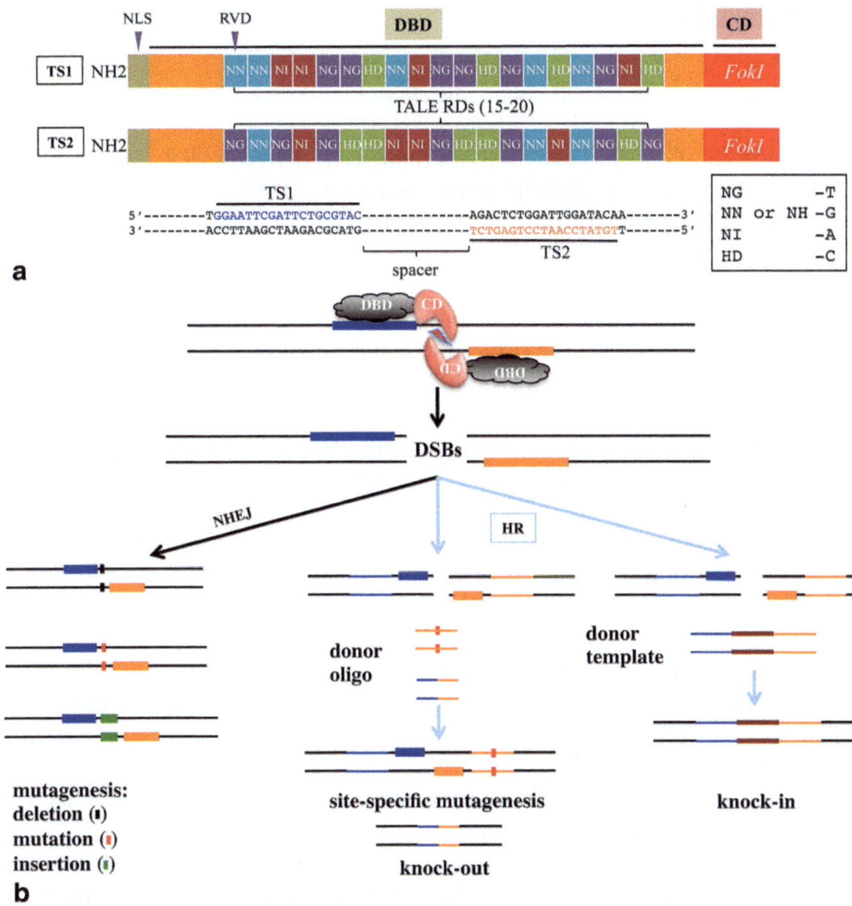

Fig. 2.1 TALEN-mediated genome editing. **a** Schematic diagram of one pair of transcription activator-like effector nuclease (TALEN) that specifically binds to a user-designed genomic locus. One engineered TALEN contains three functional domains: a nuclear localization signal (NLS) at the amino terminus, a truncated transcription activator-like effector (TALE) DNA-binding domain (DBD), and a catalytic domain (CD) consisting of a ubiquitous DNA endonuclease (*FokI*) at the carboxy terminus. The repeat variable di-residues (RVDs), which confer nucleotide specificity, are represented with the two hypervariable residues at position 12–13 (NG, NN or NH, NI, HD). The TALE repeat domain (RD) usually contains 15–20 repeats (however, also longer and shorter versions can be produced). The RD is indispensible for mediating sequence-specific binding of TALENs (TS1 and TS2) to the targeted sites (highlighted in blue and orange). Binding can be predicted using a simple protein-DNA code: NG, NN or NH, NI, and HD recognizing thymine (T), guanine (G), adenine (A), and cytosine (C), respectively. The functionally efficient spacer between the two TALEN recognition sites (TS1 and TS2) ranges from 12 to 31 bp depending on the architecture of TALE used (Cermak et al. 2011; Li et al. 2011a; Miller et al. 2011; Mussolino et al. 2011). **b** Once TALENs bind to the targeted DNA strands, the *FokI* nuclease domains dimerize. This leads to catalytic function of the *FokI* nuclease domains which creates double-stranded DNA breaks (DSBs) within the spacer. DSBs activate the cellular DNA damage repair machinery either preferentially by nonhomologous end joining (NHEJ) or by homologous recombination (HR) in the presence of a homologous donor oligo or template. Many genomic alterations, e.g.,

2.2 Materials

Unless indicated otherwise, all chemicals should be ordered from standard suppliers.

1. Plasmids: a. Golden Gate TALEN and TAL Effector Kit 2.0 (Addgene ID1000000024); b. GoldyTALEN (Addgene ID: 38143).
2. Restriction enzymes: BsaI (Eco31I), AflII (BspT1), XbaI, BsmBI (Esp3I), AatII, and StuI. Note: Enzymatic activity of BsaI is crucial for the first-round Golden Gate assembly reaction.
3. T4 DNA ligase.
4. Plasmid-safe DNA nuclease (Epicentre Biotechnologies, cat# E3110K).
5. T7 Endonuclease I (T7E1; NEB, M0302).
6. Chemically competent Escherichia coli cells.
7. Gel and polymerase chain reaction (PCR) purification kits.
8. LB medium and plates with tetracycline (10 mg/l), spectinomycin (50 mg/l), or ampicillin (50 mg/l).
9. X-gal and dependent on the type of competent cells used also IPTG, if needed.
10. Plasmid mini/midiprep kit. *Note:* Plasmid quality will affect the assembling efficiency; use, e.g., QIAprep® Miniprep from Qiagen.
11. Platinum *pfx* DNA polymerase (Life Technologies, Cat# 11708039).
12. Primary porcine fibroblasts.
13. Cell culture wares: 25 cm^2 and 75 cm^2 flasks, 6-well, 24-well, and 96-well plates.
14. Gelatin (0.1 %).
15. Complete medium for primary porcine fibroblasts (DMEM supplemented with 15 % FBS, 1 % P/S, 1 % glutamine).
16. G418 selection medium (complete medium supplemented with 800 ng/ml G418 and 5 ng/ml bFGF2). *Note:* The use of bFGF2 is to increase the doubling time of primary porcine cells in vitro.
17. Cell lysis buffer: (50 mM KCl, 1.5 mM MgCl2, 10 mM Tris-Cl, pH 8.5, 0.5 % Nonidet P40, 0.5 % Tween, 400 μg/ml Proteinase K PCR grade).
18. P1 Primary Cell 4D-Nucleofector® X Kit S (V4XP-1032, Lonza).
19. PCR thermal cycler.
20. 4D-Nucleofector™ System with X unit (Lonza).

deletions, insertions, and/or mutations, can be created at the DSB site by NHEJ. In the presence of a short donor oligo, user-designed mutations can be introduced such as site-specific mutations or deletion of predetermined sequences (knockout). In the presence of a DNA template comprising two homology arms and a functional transgene cassette, the transgene can specifically be inserted into or near the DSBs with or without disrupting the endogenous gene function (knockin)

2.3 Engineering of TALENs

Herein, we provide notes that we found important for increasing the TALEN engineering efficiency by Golden Gate cloning. Refer to Cermak's study for detailed TALENs engineering protocols (Cermak et al. 2011).

Note 1 Before designing TALENs for your targets sites, make sure to analyze your potential targeted genomic region with RepeatMasker (http://www.repeatmasker.org/cgi-bin/WEBRepeatMasker), and choose a targeted region with no repetitive sequences for designing TALEN target sites using the TALEN targeter 2.0 (https://tale-nt.cac.cornell.edu/). The average success rate of TALENs is approximately 60 %. We select in average four pairs of TALENs for each targeted locus (gene).

Note 2 To lower the handling time for the first Golden Gate reaction, dilute all plasmids from the Golden Gate kit to 150 ng/μl.

Note 3 For the first Golden Gate assembling reaction, alternatively, we predigest pFUS_A plasmid with *BsaI* and gel purify the plasmid backbone (2564 bp) to increase the efficiency.

Note 4 If FastDigest enzyme from Fermentas are used, we noticed that using double amounts (2 μl per reaction) of *BsaI* (*Eco31I*) and increasing the digestion–ligation reaction to 13 cycles can increase the frequency of correct clones (proportion of white colonies in the blue/white selection).

Note 5 To check the success of the first Golden Gate assembling reaction, run 3 μl of the plasmid-safe nuclease (PSN)-treated ligation reaction on 1 % agarose gel as presented in Fig. 2.2a.

Note 6 It is crucial that your final TALEN plasmids are checked by restriction enzyme digestion (Fig. 2.2b) and sequencing before used for transfection.

2.4 Delivery of TALENs into Primary Porcine Fibroblasts

We use the 4D-Nucleofector™ System with X unit for transfection of primary porcine fibroblasts established from two breeds of miniature pigs (Yucatan and Göttingen).

1. Culture primary porcine fibroblasts in complete medium until 80 % confluence.
2. Dissociate cells by trypsinization (0.05 % trypsin-EDTA at 37 °C for 3 min) and stop trypsin with complete medium.
3. Count cells and use 5 × 500,000 cells per reaction.
4. Spin down cells at 200 g for 5 min.
5. Aspirate the medium as complete as possible without disturbing the cell pellet.
6. Resuspend cells in reconstructed P1 solution using 20 μl P1 solution per 5 × 500,000 cells.

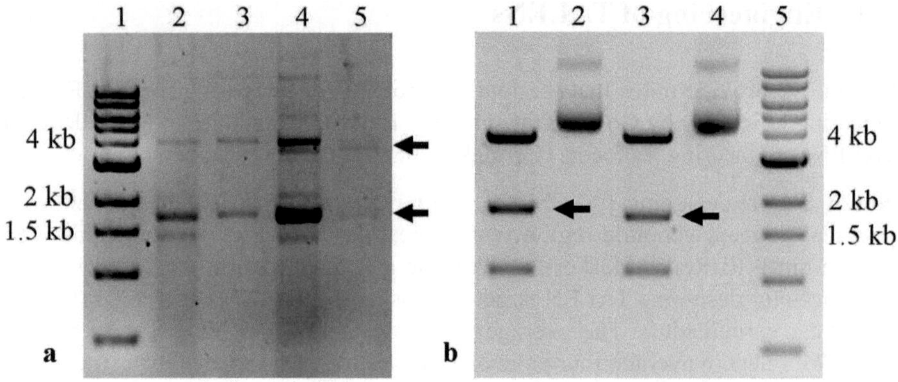

Fig. 2.2 Engineering TALENs by Golden Gate Reaction. **a** Gel electrophoresis of first Golden Gate reaction product. Three-microliter PSN-treated ligation products were separated in a 1 % agarose gel. Lane 1, size marker; Lane 2–5, pFUS_A_10RVDs, pFUS_B5_5RVDs, pFUS_A_10RVDs, pFUS_B4_4RVDs; Expected ligation patterns are indicated by arrows. **b** pcGoldyTALEN plasmids digested with *AtaII* and *StuI*. Lane 1–4, digested pcGoldyTALEN_GOI_16RVDs, un-digested pcGoldyTALEN_GOI_16RVDs, digested pcGoldyTALEN_GOI_15RVDs, un-digested pcGoldyTALEN_GOI_15RVDs; Lane 5, size marker. *Arrows* indicate RVDs containing fragments. *TALEN* transcription activator-like effector nuclease, *PSN* plasmid-safe nuclease, *RVD* repeat variable di-residue

7. Aliquot 20 µl P1 cell solution to sterilized 1.5 ml microcentrifuge tubes which are prefilled with 400 ng TALENs (per pair). Note: If a donor template is needed, we add 400 ng donor plasmid (linear or circular) plus 400 ng TALENs (per pair), though the recommended amount of total substrate DNA by the supplier is 400 ng.
8. Transfer the P1 cell–DNA mixture to a 20 µl Nucleocuvette™ Strip.
9. Run nucleofection using the Nucleofector program CM-137. Note: We tested the nucleofection efficiency in primary fibroblasts with the primary cells optimization kit and found that CM-137 gave the highest transfection efficiency (~60 %) and with a viability of ~20 % (unpublished results). Although this optimization was only tested in Yucatan primary fibroblasts, this protocol should be compatible for fibroblasts from other breeds.
10. Following nucleofection, remove the cuvette from the nucleofector and incubate the transfected cells for 10 min at room temperature.
11. Add 80 µl pre-warmed complete medium to each reaction and transfer to appropriate cell culture wares. *Note:* For analysis of TALEN cleavage, we plate the cells to one well of a 6-well plate coated with 0.1 % gelatin, followed by culturing in 3 ml complete medium. For selection of gene KO and/or knockin (KI) clones, resuspend the nucleofected cells in 10 ml pre-warmed complete medium (supplemented with 5 ng/ml bFGF2) and plate the cells in one 96-well plate coated with 0.1 % gelatin.
12. Culture cells at 37 °C or 30 °C (cold shock) for 72 h. *Note:* Changing medium during the first 72-h incubation period is not necessary.

13. For TALEN-mediated cleavage assay, harvest cells from 6-well plates and extract genomic DNA. Alternatively, cells can be resuspended in 100 μl cell lysis buffer (stated above) and incubated at 65 °C for 30 min and afterwards at 95 °C for 10 min. Use 50 ng genomic DNA or 1 μl cell lysate for PCR.
14. For gene KO or KI, cells are cultured in selection medium (e.g., for neo-resistant cell clones: complete medium supplemented with 0.8 μg/ml G418 and 5 ng/ml bFGF2 if the donor template contains a neo-selection marker) for 48 h after nucleofection. Change medium every 2–3 days.
15. Two weeks after G418 selection, screen-resistant clones for gene KO or KI as described by others and us previously (Rogers et al. 2008; Luo et al. 2011b).
16. Generate gene KO or KI pigs by somatic cells nuclear transfer (Lai and Prather 2003; Du et al. 2007).

2.5 Assessing TALEN-Mediated Cleavage by *T7E1* Assay

1. Design primers flanking the TALEN targeted sites. The amplicon size should be 200–500 bp.
2. Amplify the DNA fragment using your proofreading PCR polymerase of choice and other substitutes with the optimized PCR program. *Note:* In our laboratory, we commonly use the proofreading platinum pfx DNA polymerase from Life Technologies which works very well with almost all tested PCRs (Luo et al. 2011a, 2012a, b). Avoid designing primers that recognize a repetitive region.
3. Purify your PCR product with a quick PCR cleanup kit. *Note:* Your PCR reaction should be specific and only have one band. Do not use unpurified PCR product for the following annealing process.
4. Hybridize about 200–400 ng purified PCR product to itself using a thermocycler or a heating block: For the thermocycler protocol, the purified PCR product is diluted in 1X NEB buffer 2 solution to a final volume of at 10 μl.
   Hybridize the PCR production with the following program:

1 Hold	95 °C for 10 min
70 Cycles	95 °C for 30 s, 94 °C for 30 s (decrease 1 °C for each hold temperature for every subsequent set of cycle; ramp rate: 50 %, or 0.1 °C/s)
1 Hold	4 °C

For the heating block protocol, dilute ~200 ng of purified PCR product in at least 20 μl 1X NEB buffer 2. Denature the PCR product at 95 °C. Turn off the heating block and let cool down to room temperature while leaving the sample in the heating block.

5. T7E1 cleavage. Add 1 µl T7E1 to the annealed PCR product and incubate at 37 °C for 20 min.
6. Analyze digested PCR products using a 2 % agarose gel or a 8–10 % polyacrylamide gel.

2.6 Future Perspectives

Future developments for TALENs are still required to increase efficiency and specificity. A wide range of TALEN-mediated cleavage efficiency has been reported depending on the TALENs used, cells, or organisms. It is known that the choice of TALE protein scaffolds (Miller et al. 2011; Mussolino et al. 2011; Carlson et al. 2012; Kim et al. 2013), spacer lengths (Miller et al. 2011), cell culture status (Doyon et al. 2010), as well as epigenetic modifications can affect TALENs' activity (Kim et al. 2013). It was shown that TALENs with certain N- and C-terminal truncated TALE protein, e.g., GoldyTALEN, had significantly higher TALEN activity compared to the native form (Miller et al. 2011; Mussolino et al. 2011; Carlson et al. 2012). Other strategies such as creating new fusion proteins with highly active DNA nuclease domains have been suggested (Joung and Sander 2013). In addition, methods for efficient delivery of TALENs in forms of plasmid DNA, messenger RNA (mRNA), or protein, to targeted cells, especially when delivering into primary cells or organisms, should be developed. Viral-mediated TALEN delivery is one of such promising methods, as it can efficiently transduce a wide range of non- or poorly transfectable cells. Adenoviral vectors, but not lentiviral vectors, can be used to deliver functional TALENs into cells (Holkers et al. 2013). rAAV vectors are another attractive viral vehicle for delivering TALENs into cells and organisms due to their various advantages such as, e.g., capacity to transduce several kinds of dividing or nondividing cell types, predominantly non-integrating profile, possibility of high titer for production, and low immunogenicity (Hirata and Russell 2000; Baker 2003; Cathomen 2004). A limitation of using rAAV for TALENs delivery is linked to its relatively small genome size (~4.7 kb max; Wu et al. 2010). However, this genome size is still sufficient for generating TALENs with a maximum of 20 RVDs. TALENs with 15 RVDs have been shown to be effective in vivo genome engineering (Ma et al. 2013). Besides, donor oligo or template could be co-delivered with one rAAV-TALEN, similar to the study conducted with rAAV-ZFNs (Ellis et al. 2012), which makes rAAV an attractive method for testing in the future. The reader is referred to Chap. 3 of this book for more information about somatic gene therapy using viral vectors.

However, TALEN-mediated targeting specificity should not be sacrificed in favor of efficiency. Unknown off-target effect by TALEN-mediated genomic engineering is one of the most important difficulties that hamper their broad application in biomedical research, especially for, e.g., gene therapy. Several strategies to lower the off-target effect have been investigated such as using a heterodimeric *FokI* nuclease domain (Doyon et al. 2011), choosing more nucleotide-specific RVDs (Cong et al. 2012), and selecting TALEN targeted sites with minimum off-target sites across the genome (Doyle et al. 2012). It has been revealed by genome-wide analysis that ZFNs could hit several off targets across the genome (Gabriel et al. 2011; Pattanayak et al. 2011). Likewise, next-generation sequencing and aligning the reads to the reference genome could in future be applied to test for specificity of designed TALENs.

Recently, the clustered regularly interspaced short palindromic repeats (CRISPR) and CRISPR-associated (Cas) proteins system, which use small base-pairing RNAs to target and cleave DNA in a sequence-specific manner and provide most bacteria and archaea adaptive immunity to invading virus and plasmids, has evolved as another exciting tool for targeted genome editing (Bhaya et al. 2011; Terns and Terns 2011; Wiedenheft et al. 2012). The CRISPR/Cas system has now been successfully applied to genome engineering in many cells and organisms, such as bacterial (Jiang et al. 2013), zebra fish (*Danio rerio*; Hwang et al. 2013), mouse (Cong et al. 2013), and human cells (Cong et al. 2013; Jinek et al. 2013; Mali et al. 2013) *Arabidopsis and Nicotiana benthamiana* (Li et al. 2013), cynomolgus monkeys (Niu et al. 2014) etc. many more successful applications are expected in the near future. The CRISPR, together with the DNA nuclease systems (ZFNs and TALENs), will greatly enhance the feasibility and efficiency for targeted genome editing in animals and many other potential applications in the future.

Acknowledgments This work has been supported by grants from the Fogh-Nielsen Prize to YL, the AUFF AU IDEAS Programme and Karen Elise Jensen Foundation to CBS, the Sapere Aude Program of the Danish Council for Independent Research to MMG, and the Lundbeck Foundation Fellowship Programme to MMG.

References

Ansai S, Sakuma T, Yamamoto T, Ariga H, Uemura N, Takahashi R, Kinoshita M (2013) Efficient targeted mutagenesis in medaka using custom-designed transcription activator-like effector nucleases. Genetics 193:739–749

Aryan A, Anderson MA, Myles KM, Adelman ZN (2013) TALEN-based gene disruption in the dengue vector *Aedes aegypti*. PLoS One 8:e60082

Baker AH (2003) Targeting AAV vectors. Mol Ther 7:433–434

Bedell VM, Wang Y, Campbell JM, Poshusta TL, Starker CG, Krug RG, 2nd, Tan W, Penheiter SG, Ma AC, Leung AY, Fahrenkrug SC, Carlson DF, Voytas DF, Clark KJ, Essner JJ, Ekker SC (2012) In vivo genome editing using a high-efficiency TALEN system. Nature 491:114–118

Bhaya D, Davison M, Barrangou R (2011) CRISPR-Cas systems in bacteria and archaea: versatile small RNAs for adaptive defense and regulation. Annu Rev Genet 45:273–297

Boch J, Scholze H, Schornack S, Landgraf A, Hahn S, Kay S, Lahaye T, Nickstadt A, Bonas U (2009) Breaking the code of DNA binding specificity of TAL-type III effectors. Science 326:1509–1512

Bogdanove AJ, Voytas DF (2011) TAL effectors: customizable proteins for DNA targeting. Science 333:1843–1846

Cade L, Reyon D, Hwang WY, Tsai SQ, Patel S, Khayter C, Joung JK, Sander JD, Peterson RT, Yeh JR (2012) Highly efficient generation of heritable zebrafish gene mutations using homo- and heterodimeric TALENs. Nucleic Acids Res 40:8001–8010

Carlson DF, Tan W, Lillico SG, Stverakova D, Proudfoot C, Christian M, Voytas DF, Long CR, Whitelaw CB, Fahrenkrug SC (2012) Efficient TALEN-mediated gene knockout in livestock. Proc Natl Acad Sci U S A 109:17382–17387

Cathomen T (2004) AAV vectors for gene correction. Curr Opin Mol Ther 6:360–366

Cermak T, Doyle EL, Christian M, Wang L, Zhang Y, Schmidt C, Baller JA, Somia NV, Bogdanove AJ, Voytas DF (2011) Efficient design and assembly of custom TALEN and other TAL effector-based constructs for DNA targeting. Nucleic Acids Res 39:e82

Cong L, Zhou R, Kuo YC, Cunniff M, Zhang F (2012) Comprehensive interrogation of natural TALE DNA-binding modules and transcriptional repressor domains. Nat Commun 3:968

Cong L, Ran FA, Cox D, Lin S, Barretto R, Habib N, Hsu PD, Wu X, Jiang W, Marraffini LA, Zhang F (2013) Multiplex genome engineering using CRISPR/Cas systems. Science 339:819–823

Dahlem TJ, Hoshijima K, Jurynec MJ, Gunther D, Starker CG, Locke AS, Weis AM, Voytas DF, Grunwald DJ (2012) Simple methods for generating and detecting locus-specific mutations induced with TALENs in the zebrafish genome. PLoS Genet 8:e1002861

Davies B, Davies G, Preece C, Puliyadi R, Szumska D, Bhattacharya S (2013) Site specific mutation of the Zic2 locus by microinjection of TALEN mRNA in mouse CD1, C3H and C57BL/6J oocytes. PLoS ONE 8:e60216

Ding Q, Lee YK, Schaefer EA, Peters DT, Veres A, Kim K, Kuperwasser N, Motola DL, Meissner TB, Hendriks WT, Trevisan M, Gupta RM, Moisan A, Banks E, Friesen M, Schinzel RT, Xia F, Tang A, Xia Y, Figueroa E, Wann A, Ahfeldt T, Daheron L, Zhang F, Rubin LL, Peng LF, Chung RT, Musunuru K, Cowan CA (2013) A TALEN genome-editing system for generating human stem cell-based disease models. Cell Stem Cell 12:238–251

Doyle EL, Booher NJ, Standage DS, Voytas DF, Brendel VP, Vandyk JK, Bogdanove AJ (2012) TAL effector-nucleotide targeter (TALE-NT) 2.0: tools for TAL effector design and target prediction. Nucleic Acids Res W117–22

Doyon Y, Choi VM, Xia DF, Vo TD, Gregory PD, Holmes MC (2010) Transient cold shock enhances zinc-finger nuclease-mediated gene disruption. Nat Methods 7:459–460

Doyon Y, Vo TD, Mendel MC, Greenberg SG, Wang J, Xia DF, Miller JC, Urnov FD, Gregory PD, Holmes MC (2011) Enhancing zinc-finger-nuclease activity with improved obligate heterodimeric architectures. Nat Methods 8:74–79

Du Y, Kragh PM, Zhang Y, Li J, Schmidt M, Bogh IB, Zhang X, Purup S, Jorgensen AL, Pedersen AM, Villemoes K, Yang H, Bolund L, Vajta G (2007) Piglets born from handmade cloning, an innovative cloning method without micromanipulation. Theriogenology 68:1104–1110

Ellis BL, Hirsch ML, Porter SN, Samulski RJ, Porteus MH (2012) Zinc-finger nuclease-mediated gene correction using single AAV vector transduction and enhancement by Food and Drug Administration-approved drugs. Gene Ther 20:35–42

Gabriel R, Lombardo A, Arens A, Miller JC, Genovese P, Kaeppel C, Nowrouzi A, Bartholomae CC, Wang J, Friedman G, Holmes MC, Gregory PD, Glimm H, Schmidt M, Naldini L, von Kalle C (2011) An unbiased genome-wide analysis of zinc-finger nuclease specificity. Nat Biotechnol 29:816–823

Gupta A, Hall VL, Kok FO, Shin M, McNulty JC, Lawson ND, Wolfe SA (2013) Targeted chromosomal deletions and inversions in zebrafish. Genome Res 23:1008–1017

Hauschild J, Petersen B, Santiago Y, Queisser AL, Carnwath JW, Lucas-Hahn A, Zhang L, Meng X, Gregory PD, Schwinzer R, Cost GJ, Niemann H (2011) Efficient generation of a biallelic knockout in pigs using zinc-finger nucleases. Proc Natl Acad Sci U S A 108:12013–12017

Hickey RD, Lillegard JB, Fisher JE, McKenzie TJ, Hofherr SE, Finegold MJ, Nyberg SL, Grompe M (2011) Efficient production of Fah-null heterozygote pigs by chimeric adeno-associated virus-mediated gene knockout and somatic cell nuclear transfer. Hepatology 54:1351–1359

Hirata RK, Russell DW (2000) Design and packaging of adeno-associated virus gene targeting vectors. J Virol 74:4612–4620

Hockemeyer D, Wang H, Kiani S, Lai CS, Gao Q, Cassady JP, Cost GJ, Zhang L, Santiago Y, Miller JC, Zeitler B, Cherone JM, Meng X, Hinkley SJ, Rebar EJ, Gregory PD, Urnov FD,

Jaenisch R (2011) Genetic engineering of human pluripotent cells using TALE nucleases. Nat Biotechnol 29:731–734

Holkers M, Maggio I, Liu J, Janssen JM, Miselli F, Mussolino C, Recchia A, Cathomen T, Goncalves MA (2013) Differential integrity of TALE nuclease genes following adenoviral and lentiviral vector gene transfer into human cells. Nucleic Acids Res 41:e63

Hu R, Wallace J, Dahlem TJ, Grunwald DJ, O'Connell RM (2013) Targeting human microRNA genes using engineered Tal-effector nucleases (TALENs). PLoS ONE 8:e63074

Huang P, Xiao A, Zhou M, Zhu Z, Lin S, Zhang B (2011) Heritable gene targeting in zebrafish using customized TALENs. Nat Biotechnol 29:699–700

Hwang WY, Fu Y, Reyon D, Maeder ML, Tsai SQ, Sander JD, Peterson RT, Yeh JR, Joung JK (2013) Efficient genome editing in zebrafish using a CRISPR-Cas system. Nat Biotechnol

Ishibashi S, Cliffe R, Amaya E (2012) Highly efficient bi-allelic mutation rates using TALENs in *Xenopus tropicalis*. Biology Open 1:1273–1276

Jiang W, Bikard D, Cox D, Zhang F, Marraffini LA (2013) RNA-guided editing of bacterial genomes using CRISPR-Cas systems. Nat Biotechnol

Jinek M, East A, Cheng A, Lin S, Ma E, Doudna J (2013) RNA-programmed genome editing in human cells. eLife 2:e00471

Joung JK, Sander JD (2013) TALENs: a widely applicable technology for targeted genome editing. Nat Rev Mol Cell Biol 14:49–55

Kim YG, Cha J, Chandrasegaran S (1996) Hybrid restriction enzymes: zinc finger fusions to Fok I cleavage domain. Proc Natl Acad Sci U S A 93:1156–1160

Kim Y, Kweon J, Kim A, Chon JK, Yoo JY, Kim HJ, Kim S, Lee C, Jeong E, Chung E, Kim D, Lee MS, Go EM, Song HJ, Kim H, Cho N, Bang D, Kim S, Kim JS (2013) A library of TAL effector nucleases spanning the human genome. Nat Biotechnol 31:251–258

Lai L, Prather RS (2003) Creating genetically modified pigs by using nuclear transfer. Reprod Biol Endocrinol 1:82

Lai L, Kolber-Simonds D, Park KW, Cheong HT, Greenstein JL, Im GS, Samuel M, Bonk A, Rieke A, Day BN, Murphy CN, Carter DB, Hawley RJ, Prather RS (2002) Production of alpha-1,3-galactosyltransferase knockout pigs by nuclear transfer cloning. Science 295:1089–1092

Lei Y, Guo X, Liu Y, Cao Y, Deng Y, Chen X, Cheng CH, Dawid IB, Chen Y, Zhao H (2012) Efficient targeted gene disruption in *Xenopus* embryos using engineered transcription activator-like effector nucleases (TALENs). Proc Natl Acad Sci USA 109:17484–17489

Li JF, Norville JE, Aach J, McCormack M, Zhang D, Bush J, Church GM, Sheen J (2013) Multiplex and homologous recombination-mediated genome editing in arabidopsis and nicotiana benthamiana using guide RNA and Cas9. Nat Biotechnol 31(8):688–691

Li T, Huang S, Jiang WZ, Wright D, Spalding MH, Weeks DP, Yang B (2011a) TAL nucleases (TALNs): hybrid proteins composed of TAL effectors and FokI DNA-cleavage domain. Nucleic Acids Res 39:359–372

Li T, Huang S, Zhao X, Wright DA, Carpenter S, Spalding MH, Weeks DP, Yang B (2011b) Modularly assembled designer TAL effector nucleases for targeted gene knockout and gene replacement in eukaryotes. Nucleic Acids Res 39:6315–6325

Liu J, Li C, Yu Z, Huang P, Wu H, Wei C, Zhu N, Shen Y, Chen Y, Zhang B, Deng WM, Jiao R (2012) Efficient and specific modifications of the *Drosophila* genome by means of an easy TALEN strategy. J Genet Genomics 39:209–215

Luo Y, Bolund L, Sorensen CB (2011a) An improved PCR strategy for fast screening of specific and random integrations in rAAV-mediated gene targeted cell clones. BMC Res Notes 4:246

Luo Y, Li J, Liu Y, Lin L, Du Y, Li S, Yang H, Vajta G, Callesen H, Bolund L, Sorensen CB (2011b) High efficiency of BRCA1 knockout using rAAV-mediated gene targeting: developing a pig model for breast cancer. Transgenic Res 20:975–988

Luo Y, Bolund L, Sorensen CB (2012a) Pig gene knockout by rAAV-mediated homologous recombination: comparison of BRCA1 gene knockout efficiency in Yucatan and Gottingen fibroblasts with slightly different target sequences. Transgenic Res 21:671–676

Luo Y, Kofod-Olsen E, Christensen R, Sorensen CB, Bolund L (2012b) Targeted genome editing by recombinant adeno-associated virus (rAAV) vectors for generating genetically modified pigs. J Genet Genomics 39:269–274

Ma S, Zhang S, Wang F, Liu Y, Liu Y, Xu H, Liu C, Lin Y, Zhao P, Xia Q (2012) Highly efficient and specific genome editing in silkworm using custom TALENs. PLoS ONE 7:e45035

Ma AC, Lee HB, Clark KJ, Ekker SC (2013) High efficiency in vivo genome engineering with a simplified 15-RVD GoldyTALEN design. PLoS ONE 8:e65259

Mali P, Yang L, Esvelt KM, Aach J, Guell M, Dicarlo JE, Norville JE, Church GM (2013) RNA-guided human genome engineering via Cas9. Science 339:823–826

Mashimo T, Kaneko T, Sakuma T, Kobayashi J, Kunihiro Y, Voigt B, Yamamoto T, Serikawa T (2013) Efficient gene targeting by TAL effector nucleases coinjected with exonucleases in zygotes. Sci Rep 3:1253

Miller JC, Tan S, Qiao G, Barlow KA, Wang J, Xia DF, Meng X, Paschon DE, Leung E, Hinkley SJ, Dulay GP, Hua KL, Ankoudinova I, Cost GJ, Urnov FD, Zhang HS, Holmes MC, Zhang L, Gregory PD, Rebar EJ (2011) A TALE nuclease architecture for efficient genome editing. Nat Biotechnol 29:143–148

Moore FE, Reyon D, Sander JD, Martinez SA, Blackburn JS, Khayter C, Ramirez CL, Joung JK, Langenau DM (2012) Improved somatic mutagenesis in zebrafish using transcription activator-like effector nucleases (TALENs). PLoS ONE 7:e37877

Mussolino C, Morbitzer R, Lutge F, Dannemann N, Lahaye T, Cathomen T (2011) A novel TALE nuclease scaffold enables high genome editing activity in combination with low toxicity. Nucleic Acids Res 39:9283–9293

Nakajima K, Nakai Y, Okada M, Yaoita Y (2013) Targeted gene disruption in the *Xenopus tropicalis* genome using designed TALE nucleases. Zoolog Sci 30:455–460

Niu Y, Shen B, Cui Y, Chen Y, Wang J, Wang L, Kang Y, Zhao X, Si W, Li W, Xiang AP, Zhou J, Guo X, Bi Y, Si C, Hu B, Dong G, Wang H, Zhou Z, Li T, Tan T, Pu X, Wang F, Ji S, Zhou Q, Huang X, Ji W, Sha J (2014) Generation of gene-modified cynomolgus monkey via Cas9/RNA-mediated gene targeting in one-cell embryos. Cell 156(4):836–843

Ousterout DG, Perez-Pinera P, Thakore PI, Kabadi AM, Brown MT, Qin X, Fedrigo O, Mouly V, Tremblay JP, Gersbach CA (2013) Reading Frame Correction by Targeted Genome Editing Restores Dystrophin Expression in Cells From Duchenne Muscular Dystrophy Patients. Mol Ther

Pattanayak V, Ramirez CL, Joung JK, Liu DR (2011) Revealing off-target cleavage specificities of zinc-finger nucleases by in vitro selection. Nat Methods 8:765–770

Piganeau M, Ghezraoui H, De Cian A, Guittat L, Tomishima M, Perrouault L, Rene O, Katibah GE, Zhang L, Holmes MC, Doyon Y, Concordet JP, Giovannangeli C, Jasin M, Brunet E (2013) Cancer translocations in human cells induced by zinc finger and TALE nucleases. Genome Res 23:1182–1193

Qiu Z, Liu M, Chen Z, Shao Y, Pan H, Wei G, Yu C, Zhang L, Li X, Wang P, Fan HY, Du B, Liu B, Liu M, Li D (2013) High-efficiency and heritable gene targeting in mouse by transcription activator-like effector nucleases. Nucleic Acids Res 41:e120

Reyon D, Tsai SQ, Khayter C, Foden JA, Sander JD, Joung JK (2012) FLASH assembly of TALENs for high-throughput genome editing. Nat Biotechnol 30:460–465

Rogers CS, Hao Y, Rokhlina T, Samuel M, Stoltz DA, Li Y, Petroff E, Vermeer DW, Kabel AC, Yan Z, Spate L, Wax D, Murphy CN, Rieke A, Whitworth K, Linville ML, Korte SW, Engelhardt JF, Welsh MJ, Prather RS (2008) Production of CFTR-null and CFTR-DeltaF508 heterozygous pigs by adeno-associated virus-mediated gene targeting and somatic cell nuclear transfer. J Clin Invest 118:1571–1577

Sajwan S, Takasu Y, Tamura T, Uchino K, Sezutsu H, Zurovec M (2013) Efficient disruption of endogenous *Bombyx* gene by TAL effector nucleases. Insect Biochem Mol Biol 43:17–23

Sakuma T, Hosoi S, Woltjen K, Suzuki K, Kashiwagi K, Wada H, Ochiai H, Miyamoto T, Kawai N, Sasakura Y, Matsuura S, Okada Y, Kawahara A, Hayashi S, Yamamoto T (2013) Efficient TALEN construction and evaluation methods for human cell and animal applications. Genes Cells 18:315–326

Sander JD, Cade L, Khayter C, Reyon D, Peterson RT, Joung JK, Yeh JR (2011) Targeted gene disruption in somatic zebrafish cells using engineered TALENs. Nat Biotechnol 29:697–698

Schmid-Burgk JL, Schmidt T, Kaiser V, Honing K, Hornung V (2013) A ligation-independent cloning technique for high-throughput assembly of transcription activator-like effector genes. Nat Biotechnol 31:76–81

Shan Q, Wang Y, Chen K, Liang Z, Li J, Zhang Y, Zhang K, Liu J, Voytas DF, Zheng X, Zhang Y, Gao C (2013) Rapid and efficient gene modification in rice and *Brachypodium* using TALENs. Mol Plant 6:1365–1368

Song J, Zhong J, Guo X, Chen Y, Zou Q, Huang J, Li X, Zhang Q, Jiang Z, Tang C, Yang H, Liu T, Li P, Pei D, Lai L (2013) Generation of RAG 1- and 2-deficient rabbits by embryo microinjection of TALENs. Cell Res

Stroud DA, Formosa LE, Wijeyeratne XW, Nguyen TN, Ryan MT (2013) Gene knockout using transcription activator-like effector nucleases (TALENs) reveals that human NDUFA9 protein is essential for stabilizing the junction between membrane and matrix arms of complex I. J Biol Chem 288:1685–1690

Sun N, Liang J, Abil Z, Zhao H (2012) Optimized TAL effector nucleases (TALENs) for use in treatment of sickle cell disease. Mol Biosyst 8:1255–1263

Suzuki KT, Isoyama Y, Kashiwagi K, Sakuma T, Ochiai H, Sakamoto N, Furuno N, Kashiwagi A, Yamamoto T (2013) High efficiency TALENs enable F0 functional analysis by targeted gene disruption in *Xenopus laevis* embryos. Biol open 2:448–452

Szostak JW, Orr-Weaver TL, Rothstein RJ, Stahl FW (1983) The double-strand-break repair model for recombination. Cell 33:25–35

Terns MP, Terns RM (2011) CRISPR-based adaptive immune systems. Curr Opin Microbiol 14:321–327

Tesson L, Usal C, Menoret S, Leung E, Niles BJ, Remy S, Santiago Y, Vincent AI, Meng X, Zhang L, Gregory PD, Anegon I, Cost GJ (2011) Knockout rats generated by embryo microinjection of TALENs. Nat Biotechnol 29:695–696

Tie JK, Jin DY, Tie K, Stafford DW (2013) Evaluation of warfarin resistance using TALENs-mediated vitamin K epoxide reductase knockout HEK293 cells. J Thromb Haemost

Tong C, Huang G, Ashton C, Wu H, Yan H, Ying QL (2012) Rapid and cost-effective gene targeting in rat embryonic stem cells by TALENs. J Genet Genomics 39:275–280

Wefers B, Meyer M, Ortiz O, Hrabe de Angelis M, Hansen J, Wurst W, Kuhn R (2013) Direct production of mouse disease models by embryo microinjection of TALENs and oligodeoxynucleotides. Proc Natl Acad Sci U S A 110:3782–3787

Wendt T, Holm PB, Starker CG, Christian M, Voytas DF, Brinch-Pedersen H and Holme IB (2013) TAL effector nucleases induce mutations at a pre-selected location in the genome of primary barley transformants. Plant Mol Biol

Wiedenheft B, Sternberg SH and Doudna JA (2012) RNA-guided genetic silencing systems in bacteria and archaea. Nature 482:331–338

Wood AJ, Lo TW, Zeitler B, Pickle CS, Ralston EJ, Lee AH, Amora R, Miller JC, Leung E, Meng X, Zhang L, Rebar EJ, Gregory PD, Urnov FD, Meyer BJ (2011) Targeted genome editing across species using ZFNs and TALENs. Science 333:307

Wu Z, Yang H, Colosi P (2010) Effect of genome size on AAV vector packaging. Mol Ther 18:80–86

Wu J, Huang Z, Ren J, Zhang Z, He P, Li Y, Ma J, Chen W, Zhang Y, Zhou X, Yang Z, Wu SQ, Chen L, Han J (2013) Mlkl knockout mice demonstrate the indispensable role of Mlkl in necroptosis. Cell Res

Xiao A, Wang Z, Hu Y, Wu Y, Luo Z, Yang Z, Zu Y, Li W, Huang P, Tong X, Zhu Z, Lin S and Zhang B (2013) Chromosomal deletions and inversions mediated by TALENs and CRISPR/Cas in zebrafish. Nucleic Acids Res

Yu S, Luo J, Song Z, Ding F, Dai Y, Li N (2011) Highly efficient modification of beta-lactoglobulin (BLG) gene via zinc-finger nucleases in cattle. Cell Res 21:1638–1640

Zhang Y, Zhang F, Li X, Baller JA, Qi Y, Starker CG, Bogdanove AJ, Voytas DF (2013) Transcription activator-like effector nucleases enable efficient plant genome engineering. Plant Physiol 161:20–27

Chapter 3
Somatic Gene Therapy Using Viral Vectors: Theoretical and Clinical Implications in Relation to Treatment of Genetic Conditions in Humans

Anne Louise Askou and Thomas J. Corydon

Abbreviations

Ad	Adenovirus
AAV	Adeno-associated virus
AMD	Age-related macular degeneration
AIDS	Acquired immunodeficiency syndrome
BIV	Bovine immunodeficiency virus
BM	Bruch's membrane
CAEV	Caprine arthritis-encephalitis virus
cap	Gene encoding AAV structural proteins
CBA	Chicken ß-actin
CGD	Chronic granulomatous disorder
CNV	Choroidal neovascularization
CTLs	Cytotoxic T lymphocytes
E2A, *E4* and *VA*	Ad genes that mediate AAV vector replication
EIAV	Equine infectious anemia virus
env	Gene encoding the HIV-1 surface glycoprotein gp160
FVIII	Factor VIII
FIX	Factor IX
gag	Gene encoding the LV core proteins
GOI	Gene of interest
HEK293	Human embryonic kidney 293
HIV	Human immunodeficiency virus
HSC	Hematopoietic stem cell
HSV-1	Herpes simplex virus-1

T. J. Corydon (✉)
TJC, Department of Biomedicine, Faculty of Health, Aarhus University,
Wilhelm Meyers Allé 4, 8000 Aarhus C, Denmark
e-mail: corydon@biomed.au.dk

A. L. Askou
Department of Biomedicine, Faculty of Health, Aarhus University,
Wilhelm Meyers Allé 4, 8000 Aarhus C, Denmark

X.-Q. Li et al. (eds.), *Somatic Genome Manipulation*,
DOI 10.1007/978-1-4939-2389-2_3

ICLV	Integration-competent LV
IL-2Rγ	Interleukin-2 receptor common γ–chain
IPM	Inter-photoreceptor matrix
IR	Inherited retinopathies
ITR	Inverted terminal repeats
IDLV	Integration-deficient LV
ILM	Inner limiting membrane
IN	Integrase
LCA	Leber's congenital amaurosis
LP1	Liver promoter 1
LTR	Long terminal repeats
LV	Lentivirus
mFLT1	Membrane-bound VEGF-receptor 1
miRNA	microRNA
MLV	Murine leukemia virus
pol	Gene encoding a set of enzymes required for LV replication
PR	Photoreceptor
Prom	Promoter
rAAV	Recombinant AAV
RCL	Replication-competent lentivirus
RCR	Replication-competent retrovirus
rep	Genes encoding nonstructural AAV proteins
rev	Gene encoding a regulatory protein important for nuclear export of mRNA transcripts
RNAi	RNA interference
RPE	Retinal pigment epithelial
RRE	Rev-responsive element
scAAV	Self-complementary AAV
sFLT1	Soluble VEGF-receptor 1
SCID	Severe combined immunodeficiency
SeV	Sendai virus
shRNA	Short-hairpin RNA
SIN	Self-inactivating
siRNA	Short interfering RNA
SIV	Simian immunodeficiency virus
ssAAV	Single-stranded AAV
TLR	Toll-like receptor
TU	Transducing units
VEGF	Vascular endothelial growth factor
vg	Vector genomes
VMV	Visna/maedi virus
VSV-G	Vesicular stomatitis virus envelope glycoprotein G
WAS	Wiskott–Aldrich syndrome

3.1 Introduction to Vector Technology

The most important challenges in somatic gene therapy are finding safe ways to deliver the transgenes to appropriate cells and the subsequent maintenance and expression (Fig. 3.1). For decades, scientists have asked for gene delivery vehicles for gene replacement and gene silencing that can transduce the cells of interest, avoid the immune response, transduce both dividing and nondividing cells, remain episomal or in other cases insert their cargo in the genome without the risk of insertional mutagenesis. Efficient and long-term expression of transgenes is also required. At present, no single delivery system can fulfill all these criteria, but collectively the viral vectors display all of these sought-after features.

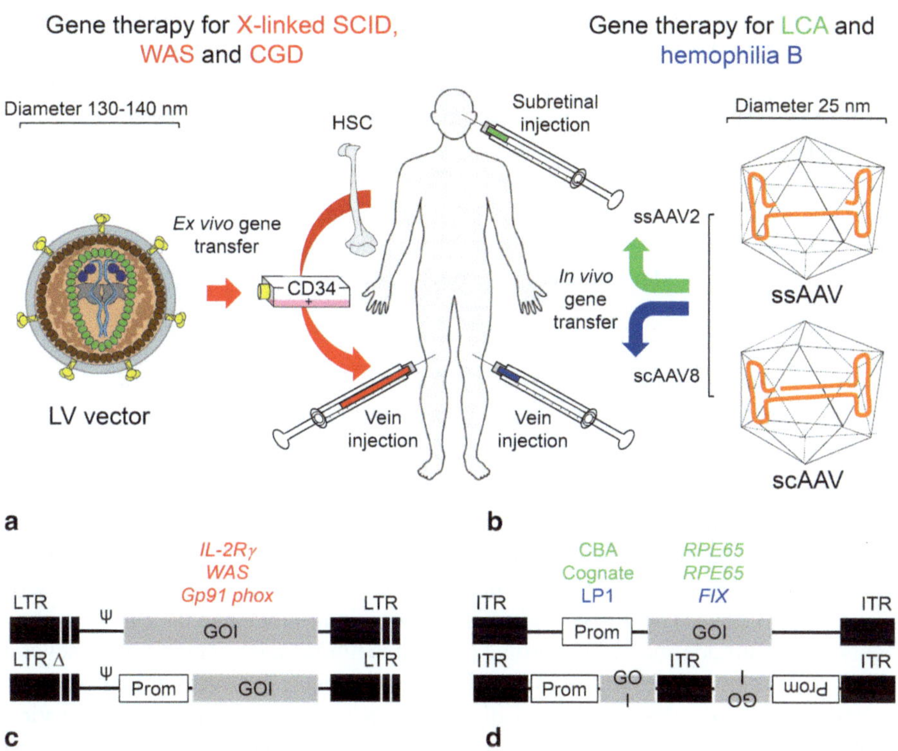

Fig. 3.1 Examples of ex vivo and in vivo gene transfer for the treatment of inherited diseases in humans. **a** Gene therapy for X-linked SCID, WAS, and CGD using retroviral vectors. $CD34^+$ HSCs are transduced ex vivo and following cultivation returned to the patients. **b** rAAV-mediated in vivo gene therapy of the eye disease LCA and bleeding disorder hemophilia B. ssAAV2-RPE65 or scAAV8-FIX are delivered by subretinal injection or peripheral-vein infusion, respectively. **c** Schematic illustration of the utilized γ-retroviral (*upper*) and LV (*lower*) constructs. In the γ-retroviral vectors transgene expression is driven from the LTR. For treatment of X-linked SCID, WAS, and CGD the vector contained the *IL-2Rγ*, *WAS*, and *GP91 phox* gene sequences, respectively. To improve safety and efficacy, third generation LV-based SIN vectors have been employed in recent gene therapy trials. In this vector system, expression is driven from an internal promoter (Cartier et al. 2009; Cavazzana-Calvo et al. 2010). ψ = packing signal. **d** Graphic depiction of the ssAAV2 (*upper*) and scAAV8 (*lower*) vectors used for ocular or hemophilia B gene therapy. See text for details

In the development of gene delivery systems based on natural viruses, several members of different viral families have been investigated, optimized, and utilized, such as adenovirus (Ad), Herpes simplex virus-1 (HSV-1), retroviruses such as Moloney murine leukaemia virus (MLV), and lentivirus (LV), and adeno-associated virus (AAV). Each type of virus has its own niche and has evolved to maximize its chances of replication by transferring its viral genome into host cells and exploiting cellular machinery to assure viral propagation. The main interest of viral vectors for gene therapy lies in gene transfer, a process that can be defined as gene transduction, whereas viral propagation is avoided by deletion of the coding region of the viral genome that leads to replication and toxicity. The viral genome is replaced with the desired transgene, which is packaged and delivered to cells via the natural transduction pathway of the virus.

Given the diversity of disease targets that are potentially amenable to gene therapy, it has become clear that the notion of "best vector" depends on the particular disease or disorder in question. Recently, most strategies for treating genetic diseases have primarily revolved around two viral vectors: The lenti- and the adeno-associated viral vectors (Fig. 3.1). They belong to different viral groups according to whether their genomes integrate into host cellular chromatin or persist as episomes in the cell nucleus, respectively. The most promising results to date have been achieved using AAV, which has been appreciated as the vector of choice for the transduction of post-mitotic cells in vivo, whereas LVs are the primary vector for ex vivo gene transfer into stem or progenitor cells.

3.1.1 From Wild Pathogenic Virus to Sophisticated Gene Delivery Vehicles—The Lentiviral Vectors

Lentiviral vectors have been constructed from several types of LVs such as the human immunodeficiency virus (HIV), simian immunodeficiency virus (SIV), equine infectious anemia virus (EIAV), feline immunodeficiency virus (FIV), Visna/maedi virus (VMV), caprine arthritis-encephalitis virus (CAEV), and bovine immunodeficiency virus (BIV). The viruses induce a variety of pathologies in their host, such as acquired immunodeficiency syndrome (AIDS) in humans, but they all share several common features. Here, we focus on the lentiviral vector derived from HIV-1.

HIV-1 has a single-stranded positive-sense RNA genome encoding nine viral proteins. The *gag* gene encodes the viral core proteins, the *env* gene encodes the viral surface glycoprotein and the *pol* gene encodes a set of enzymes required for viral replication. Besides the major proteins Gag, Env, and Pol, the viral genome also encodes Tat and Rev, regulatory proteins, as well as four accessory proteins, Vif, Vpr, Vpu, and Nef. Long terminal repeats (LTRs), of about 640 nucleotides segmented into U3, R, and U5 regions like other retroviral LTRs, flank the viral genome and are required for viral transcription, reverse transcription, and integration. In lentiviral vectors based on HIV-1, all viral genes have been deleted and even the LTRs have been mutated to improve safety. LVs have a size of approx. 130–140 nm in diameter (Fig. 3.1a). The different generations of LV vectors are further described in the Sect. 3.6. The latest generation of LV vectors requires four different plasmids for packaging and production of infectious virions (see Fig. 3.2a) and in general, the vector yield of third-generation vectors is lower than previous generations. The vector yield of LVs is generally not higher than 10^{10} transducing units (TU)/mL compared to AAVs 10^{12} TU/mL. Enhanced transduction efficiency can be obtained by introduction of a cis-acting central polypurine tract (Zennou et al. 2001). Introduction of a cis-acting Woodchuck hepatitis virus posttranscriptional regulatory element (Zufferey et al. 1999) or incorporation of chromatin-insulator and -opening elements (Aker et al. 2007; Arumugam et al. 2007) have been shown to increase transgene expression.

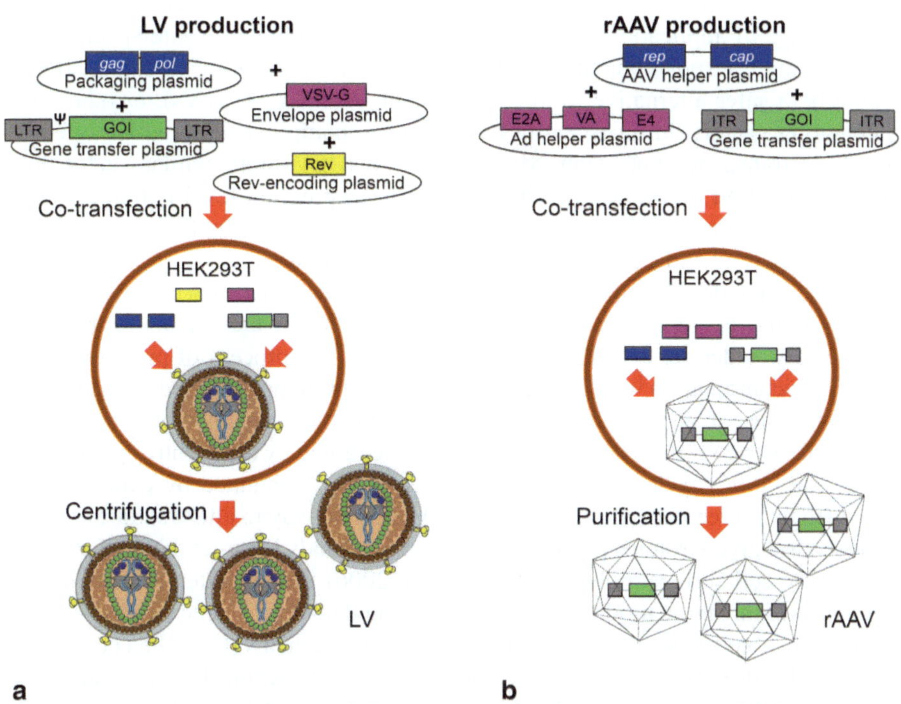

Fig. 3.2 LV and AAV vector production at a glance. **a** LV vectors produced by quadruple transfections (Dull et al. 1998). Pre-seeded HEK-293T cells are transfected by means of the calcium-co-precipitation method. Two days posttransfection, the virus-containing medium is harvested and viral particles are either stored as crude samples or further purified by ultracentrifugation. LV vectors produced by the quadruple transfection procedure, known as third generation LV vectors, are replication-incompetent, and hence, safer to use compared to previous LV systems. For estimation of the titer various methods, including colony-forming assay and ELISA-based assessment of the amount of surface located viral protein (such as p24), can be applied. **b** rAAVs produced by triple transfections. Following triple transfection of pre-seeded HEK-293T cells (Xiao et al. 1998) the cells are maintained for a few days at standard cultivation conditions. Then, the medium is collected and AAV particles purified, e.g., by utilizing a sequential process of nuclei isolation, density gradient centrifugation, and heparin sulfate affinity column chromatography. The latter step is included in order to ensure further purity of the AAV samples. Titration by dot blot analysis is normally performed in order to determine the concentration of the virus particles. See text for details

Upon target cell infection, the viral RNA is reverse transcribed into linear double-stranded DNA. A part of this DNA is integrated into the host cell genome, either by integrase(IN)-dependent or IN-independent mechanisms, and a significant amount is converted into stable episomal circles. The integrated provirus represents the major source of transgene expression, whereas the episomes are a weaker template for effective transgene expression. Insertion mediated by the virus-encoded IN can be greatly reduced by introducing a missense amino acid in the catalytic triad of the IN and thereby favoring the formation of episomal DNA. These IN-mutants are also known as integration-deficient LVs (IDLVs). In most cases, the lentiviral episomes lack a replication signal, and compared to integrating LVs, the gene expression is transient in proliferating cells, where they are progressively diluted out. IDLVs can mediate stable expression in nondividing cells and are proficient vehicles for a broad range of applications such as homologous recombination, site-specific integration, and transposition, and can be converted into replicating episomes. IDLVs could be a safer alternative to integration-competent LVs (ICLVs), or other non-integrating vectors, in quiescent tissue, but gene therapy vectors require effective levels of transgene expression. Most reports on IDLVs note similar vector titers and transduction percentages as ICLVs, but the expression level is reduced from approximately one-half (Apolonia et al. 2007; Coutant et al. 2008) to one-tenth that of ICLVs (Cornu and Cathomen 2007; Bayer et al. 2008).

In the early developmental phase, HIV-1 Env was replaced with vesicular stomatitis virus envelope glycoprotein G (VSV-G) to allow the vectors to transduce a markedly wider set of cells (Burns et al. 1993). The process of producing viral vectors in combination with foreign viral envelope proteins is termed pseudotyping, and in the case of VSV-G, the pseudotyped viral vectors were also found to be more stable, allowing viral particles to be highly concentrated by ultracentrifugation (Akkina et al. 1996). Selective lentiviral delivery can be obtained by exploiting the knowledge of receptor specific viral entry. A different way to alter vector tropism is to fuse viral glycoproteins to a ligand protein or antibody to target the viral particles to specific cell-surface molecules (Yang et al. 2006). Another step in lentiviral infection, which can be modified for targeted gene delivery, is gene expression. Use of a tissue or cell-type specific promoter/regulatory elements (De Palma et al. 2003; Park et al. 2003; Hioki et al. 2007; Lopes et al. 2008; Schmeckpeper et al. 2009) or incorporation of micro-(mi)RNA target sequences can target gene expression to the cells of interest (Brown et al. 2006; Brown et al. 2007). Recently, the research group of Alberto Auricchio showed that miRNA-mediated regulation of transgene expression by adding miRNA target sequences also can be applied in the retina to either restrict expression to a specific cell type or provide an additional layer of gene expression regulation (Karali et al. 2011)

More than two decades have passed since genetically modified HIV was used for gene delivery for the first time. As opposed to the γ-retroviruses, LVs have a desirable, inherent property, namely, the ability to transduce nondividing cells and this discovery was the breakthrough for LVs. Continuous improvements have developed HIV into safer and more effective lentiviral vectors that offer several attractive properties as gene-delivery vehicles. These include persistent gene delivery through stable vector integration into the host cell genome, broad tissue tropism including the possibility to be pseudotyped, the ability to incorporate relatively sophisticated genetic elements such as polycistronic or intron-containing sequences and a relatively easy system for vector manipulation and production. The LV vectors also offer attractive biosafety features such as the split-genome design (see Sect. 3.6), no expression of viral proteins after transduction, and a potentially safer integration site profile than other retroviruses. Production of third-generation LV vectors is briefly described in Fig. 3.2a.

3.1.2 From Dependent Virus to Desirable Vector—The Adeno-Associated Viral Vector

Whereas development of the LV virus has been constrained by the intrinsic pathogenicity of the virus the major hurdle for in vivo gene transfer using AAV has been limited efficacy caused by host immune response due to prior AAV infections. Approximately 80 % of humans are seropositive for AAV2, but despite the high seroprevalence, the virus has not been linked to any human diseases. This provides the derived recombinant AAV (rAAV) vectors with an excellent safety profile compared to other viral systems.

AAV is a small (~25 nm) nonenveloped virus that packages a linear single-stranded DNA genome of ~4.7 kb (Fig. 3.1b). AAV is naturally replication deficient, it belongs to the genus *Dependovirus* of the Parvovirus family because it is dependent on co-infection of a helper-virus such as the adenovirus or herpesvirus for replication. The structure of the wildtype AAV is simple and consists of two viral genes, the *rep* and *cap*, flanked by 145 nucleotide-long inverted terminal repeats (ITR). The segment of 125 nucleotides of the ITR constitutes a palindrome folding upon itself to maximize Watson–Crick base pairing and form a T-shaped hairpin structure with a key role in DNA replication (Fig. 3.1b). The secondary structure of the ITR provides a free 3' hydroxyl group for initiation of viral replication and serves as priming sites for host cell DNA polymerase to begin second-strand synthesis. The ITRs are the only cis-acting elements necessary for genome replication, integration, and packaging.

The viral genes encoding both structural (*cap*) and nonstructural proteins (*rep*) are replaced with the transgene of interest in the rAAV, which subsequently do not express any viral genes and therefore rAAV is unable to replicate and integrate into host cell chromosome. In wildtype AAV, host cell integration is a site-specific reaction with preferential integration of the virus genome into a region designated *AAVS1* on human chromosome 19 (Kotin et al. 1991). Targeted integration is a unique phenomenon among all known eukaryotic viruses and its exploitation is highly attractive for achieving safe and stable transgene expression. The question remains, whether *AAVS1* is suitable for human gene therapy applications, but tissue culture experiments performed on multiple cell types and with multiple transgenes have validated this site for effective and safe transgene insertion in human cells (Lombardo et al. 2011).

The molecular fate of the rAAV vectors lacking the *rep* genes, once in the nucleus, is dependent on host cell activities. Incoming rAAV DNA needs to be converted into double-stranded, transcriptionally functional templates, which happens either by de novo second-strand synthesis or by annealing single-stranded molecules with opposing polarities. The latter is possible when the multiplicity of infection is sufficient to promote base-pairing because AAV genomes with sense and antisense orientation are packaged equally well. Through intra- or intermolecular recombination at the ITRs, rAAV genomes can originate circular forms or linear concatemers, respectively. Both extrachromosomal DNA forms are transcription-competent templates and believed to be responsible for the majority of transgene expression. In addition to the episomal forms, 1 % of the rAAV genomes are integrated into chromosomal DNA by a passive process relying on pre-existent chromosomal breaks and host cell machinery. Concerns about insertional mutagenesis is less for rAAV compared to retroviral vectors because they do not create but instead insert into existing chromosomal breaks and do not display "outward" promoter activity (see Sect. 3.6).

One of the rate-limiting steps for rAAV transduction is the conversion of the single-stranded vector genome into double-stranded DNA prior to gene expression (Ferrari et al. 1996; Fisher et al. 1996). This critical step was circumvented through the use of self-complementary AAV (scAAV) vectors (McCarty et al. 2001). In the scAAV vector, both strands are packaged as a single molecule with an inverted repeat genome that can fold upon itself and mimic the structure of conventional single-stranded AAV (ssAAV) after double-stranded DNA conversion by DNA synthesis (Fig. 3.1b). This enhances formation of transcription-competent molecules, thus improving expression kinetics and renders transduction with scAAV independent of host cell DNA synthesis. scAAV vector genomes have also been found to be more stable and more prone to circularization upon transduction of in vivo tissues than ssAAV vector genomes (Wang et al. 2003). With these attractive features, scAAVs have the potential to attain therapeutic levels of transgene expression at a minimal dose. scAAV vectors were shown to transduce cells in vitro 5–140-fold more effective than the corresponding ssAAV (McCarty et al. 2001) and cells that did not show transgene expression from single-stranded genomes showed expression when transduced with scAAV with the same serotype. The tradeoff for fast expression kinetics is the need to limit the size of the transgene that can be packaged in scAAVs to approximately one half of the ssAAV genome. Heterodimerization or trans-splicing strategies can tackle this drawback and increase vector capacity (Sun et al. 2000). With this strategy, the transgene cassette is split between two rAAV vectors and after coinfection of target cells, the AAVs form head-to-tail concatamers via recombination at the ITRs fusing the 5′ and 3′ ends of the split gene of interest. The fused ITR sequence is removed during mRNA maturation by the use of splicing signals (intron). The split gene approach has the drawback of decreased efficacy.

AAVs gain entry into target cells by the use of different cellular receptors. Different serotypes have unique binding properties and the capsid possesses the specificity for binding cell surface receptors. Over 100 natural serotypes have been isolated, but most clinical trials have been performed with rAAV encapsulated with serotype 2 capsid proteins, the first serotype to be isolated and characterized. Tropism as well as gene expression kinetics differ between the serotypes (Zincarelli et al. 2008).

Transencapsidation or pseudotyping is a concept that refers to the possibility to pack viral genomes of one serotype, mostly AAV2, in the capsid of another serotype (Rabinowitz et al., 2002). At present, the crystal structure of multiple serotypes has been solved and the structures have aided in the re-engineering of the prototype vector AAV2. Capsid engineering is a valuable tool for both tropism modification and alleviation of capsid-specific immunogenicity. Capsid engineering has been performed by epitope insertion, chemical modification, and virus-directed evolution. Rational design has been used to generate chimeric capsids, where important functional domains or amino acids from one serotype are incorporated into another, and the derived mutant will have properties in common with both parental serotypes. Mosaic capsids is another approach where capsids are created made up of protein subunits from more than one serotype by mixing plasmids encoding capsid subunits of each serotype in specific ratios. Capsid engineering has been described in a review by Bartel et al. which is recommended for further reading on the subject (Bartel et al. 2011).

As an alternative to rational design of AAV capsids, directed evolution has emerged. This strategy emulates the process of natural evolution to improve the function of rAAVs and is a powerful tool for rapid selection and isolation of AAV variants with novel valuable properties. Directed evolution is exploited to generate vectors with enhanced tropism for specific cell types while detargeting them from undesired cells (Maheshri et al. 2006). This approach can also be used to isolate vectors that can escape from immune responses. However, one challenge is to translate the properties of vectors generated from selection in cell lines and animal models to the increasingly complex challenges experienced in humans.

In 2008, Zhong and coworkers documented a series of interesting findings that have led to the development of next-generation AAV2 vectors capable of high-efficiency transduction at lower vector doses. The authors showed that site-directed mutagenesis of surface-exposed tyrosine residues results in production of AAV2s that transduce both cells in vitro and murine hepatocytes in vivo more efficiently compared to conventional AAV2 vectors (Zhong et al. 2008). In addition, they showed that therapeutic levels of human Factor IX (FIX) could be produced at a ~10-fold reduced vector dose. It was suggested that the increased transduction efficiency of tyrosine-mutant AAV2s is due to lack of capsid ubiquitination, which will lead to reduced proteasome-mediated degradation and hence, enhanced intracellular trafficking to the nucleus. In another recent study, enhanced gene delivery to neonatal retina was obtained by intravascular administration of AAV9 vectors containing double-mutated tyrosine residues (Dalkara et al. 2012). These results may thus have important implications for the use of tyrosine-mutant AAV vectors in human gene therapy.

At present, more than 80 clinical trials have been initiated with rAAV-based vectors. More than 60 utilize the serotype 2 capsid and all include a genome with rAAV serotype 2 ITRs. The limitations of rAAV learned from these studies have given momentum to efforts to improve this vector. Great efforts have been placed on the challenges of immunogenicity and cell/tissue specificity to increase efficiency of rAAV vector therapy. Still, rAAV vectors are the most attractive vectors for in vivo somatic gene therapy, primarily because they are the safest viral vectors and in addition, gene expression from rAAV vectors has been found to be stable for years in postmitotic tissue (Testa et al. 2013). The success of the rAAV clinical trials has led to increasing effort in developing a scalable manufacturing process that can generate highly pure, high titer, and potent amounts of rAAV vectors. A simplified overview of the production of rAAV vectors is outlined in Fig. 3.2b.

3.2 Ocular Gene Therapy Vectors

3.2.1 Development of Gene Therapy for Inherited Retinal Disorders

Ocular gene therapy has gained tremendous attention despite the facts that most inherited ocular diseases are rare and none of them are fatal, but due to the attractive features of the eye as a target for gene therapy and because most inherited retinopathies (IR) are the perfect candidates for gene therapy (Fig. 3.1b). IRs are mostly monogenic, many of the causative genes have been identified, and rodent and larger animal models that mimic human IRs are available. The eye is small, accessible, anatomically compartmentalized, immune-privileged, and the contralateral eye serves as the perfect control. Retinal cells are postmitotic and transducable by both AAV and LV. Depending on i.a. route of administration and serotype AAVs can transduce all layers in the retina including rod and cone cells, whereas LVs, probably due to the size of the particles, are limited to the more accessible cell layers such as the retinal pigment epithelial (RPE) cells following subretinal administration (Fig. 3.3). On the other hand AAV vectors are limited by their packaging capacity. However, both vectors have great potential as ocular gene therapy vehicles.

Gene therapy of inherited retinal disorders is based on different strategies depending on the effect of the disease-causing mutation. Mutations can result in either a gain-of-function or a loss-of-function of the protein product. Gain-of-function mutations are often seen in dominant diseases. In the case of dominant negative mutations, the protein product is not only nonfunctional but interferes with the wildtype protein, and toxic gain-of-function mutations create protein products that are themselves harmful for the cell. Loss-of-function mutations are mostly seen in recessive diseases. In the latter case, gene therapy requires delivery of wildtype cDNA (also known as gene replacement). Different examples of using the gene replacement strategy in treatment of diseases in humans are reviewed in a later section of this chapter. In the case of gain-of-function mutations, gene therapy is more complicated, since the dominant allele requires silencing. This is mainly done by RNA technology such as antisense-RNA, ribozymes, or RNA interference (RNAi) targeting regions of the mRNA not affected by the mutation to overcome mutational heterogeneity. This leads to knock-down of both mutant and wildtype allele gene product, and co-delivery and expression of cDNA bearing silent mutations, which renders them resistant to degradation/silencing, are required. In addition, studies using RNAi for selective targeting of the mutant allele leaving expression of the wildtype allele unaltered have also been employed (Rodriguez-Lebron et al. 2009). Another approach, described by Mussolino et al. is a repression–replacement

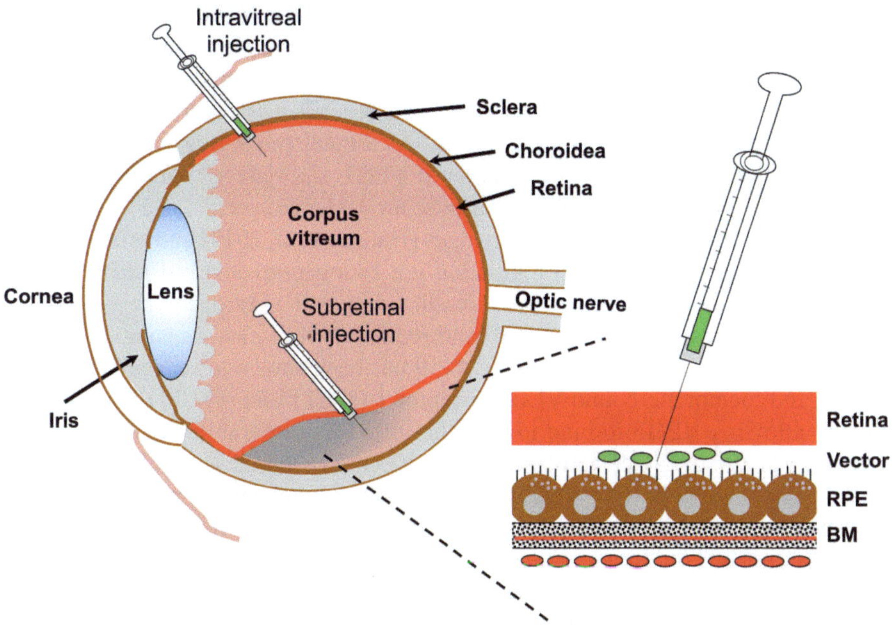

Fig. 3.3 Illustration of intravitreal and subretinal injection strategies used in ocular gene therapy. Depending on the anatomic localization of the target cells intravitreal or subretinal injection routes are chosen. In the case of the latter a small volume of vector particles is injected between the PR and the RPE cells creating a retinal detachment (often referred to as "bleb formation") thereby allowing efficient transduction of retinal cells. Following detachment the retina folds back on the RPE cells. See text for details

strategy, where repression is mediated by an AAV-delivered zinc-finger artificial transcriptional repressor targeted to the mutated *hRHO* transgene in a transgenic mouse model mimicking autosomal dominant retinitis pigmentosa and results in significant reduction of its expression (Mussolino et al. 2011). In the case of dominant negative mutations, delivering the wildtype cDNA can alter the balance between mutant and wildtype protein and effect treatment.

3.2.2 RNA Interference-Based Ocular Therapy

Gene silencing by RNAi has already been employed in ocular gene therapy related to development of new treatment strategies for exudative age-related macular degeneration (AMD), the most frequent cause of blindness among elderly in the Western world (Leung et al. 1989; Aiello et al. 1994; Colquitt et al. 2008; Chadderton et al. 2009; Yuan et al. 2010; Askou et al. 2012). Preliminary results of phase I and II clinical trials of bevasiranib, a small interfering (si)RNA-based antiangiogenic agent targeting vascular epithelial growth factor (VEGF), have shown promising results for the treatment of AMD (Singerman 2009). VEGF stimulates vasculogenesis and angiogenesis, and therefore are involved in restoration of the oxygen supply to tissues when blood circulation is inadequate. However, when VEGF is overexpressed, it can contribute to disease, including AMD and CNV. Therefore, anti-VEGF therapies are among the treatments of these diseases. shRNA-mediated silencing of an endogenous gene in RPE cells has also been described (Paskowitz et al. 2007). Recently, the utilization of shRNAs for scAAV8-mediated anti-VEGF therapy has been described showing potent reduction of VEGF and reduced choroidal neovascularization (CNV) in a laser-induced mouse model (Askou et al. 2012). The identification of cellular micro (mi)RNA sponges (Hansen et al. 2013) and miRNA target sequences (Brown et al. 2007) regulating the endogenous level of miRNAs, and hence, the expression of genes down-regulated by miRNAs opens new strategies for targeted expression (see Sect. 3.6) and reduction of disease-causing gene products. For instance, expression of miRNA-10 has been shown to reduce the level of soluble and membrane-bound VEGF-receptor 1 (sFLT1 and mFLT1) in zebra fish, thereby allowing VEGF to bind to the otherwise less preferable VEGF-receptor 2 (Hassel et al. 2012). Since activation of the VEGF-receptor 2 is involved in angiogenesis, and thereby development of AMD, expression of ectopic miRNA-10 target sequence molecules may thus be used to reduce CNV. For efficient elimination of a specific miRNA, or for simultaneous targeting of several different miRNAs, the target sequences could be expressed either as clustered tough decoys (Bak et al. 2013; Hollensen et al. 2013) or circular miRNA sponges (Hansen et al. 2013). These examples of using shRNA and miRNA target sequences expressed from a viral vector may contribute to the development of effective anti-VEGF therapy (Askou et al. 2012; Pihlmann et al. 2012).

3.2.3 AAV- and LV-Based Vectors Mediate Efficient Transduction of Target Cells

AAV vectors are to date the most widely used vehicles for transgene delivery to the retina. Targeted retinal gene expression can be obtained by pseudotyping, and the variety of natural serotypes provides access to a broad range of novel recombinant AAV vectors. Following subretinal delivery, several rAAV serotypes have been found to transduce RPE cells (AAV1, 2, 4, 5, 7, 8, and 9), photoreceptor (PR) cells (AAV2, 5, 7, 8, and 9), and Müller glia (AAV8 and 9) (Auricchio et al. 2001; Weber et al. 2003; Allocca et al. 2007). AAV5, 7, 8, and 9 were found to exhibit more efficient transduction and faster transgene expression compared with serotype 2 (Auricchio et al. 2001; Allocca et al. 2007; Natkunarajah et al. 2008), where AAV8 has been demonstrated to provide the highest levels of PR transduction in several species.

A recent review by Balaggan and Ali includes a comprehensive overview of the transduction properties of various LV vectors incorporating several different promoters and envelope pseudotypes evaluated after in vivo and ex vivo delivery in a variety of animal species (Balaggan and Ali 2012). LV vectors primarily target RPE cells following subretinal delivery, regardless of surface proteins but cell differentiation and retinal degeneration stage alters this tropism as discussed in the following section.

Route of administration can be exploited to obtain transduction of specific target cells and for retinal diseases mostly include either an intravitreal or a subretinal injection (Fig. 3.3). Intravitreal vector administration is less invasive, but vector diffusion to the outer retinal cells is limited by physical barriers such as the inner retina and the inner limiting membrane. The subretinal injection releases the vector in the subretinal space between the PR cells and the RPE formed by reversible, injection-induced detachment. The subretinal injection is more invasive, but is the preferred route for PR and RPE transduction. In the LCA2 3-year-follow-up study, patients experienced a loss of foveal thickness following subretinal vector delivery suggested to be due to foveal cone loss and that cones may be particularly sensitive to subretinal vector-mediated foveal detachment accompanying the injection (Jacobson et al. 2012). Evidence of efficient transduction of PR or RPE cells following intravitreal injection is nonexisting and efforts in developing vectors that can penetrate the dense inter-photoreceptor matrix (IPM) have been undertaken. AAV2 was the only serotype that had been demonstrated to be able to transduce inner retinal cells from the vitreous (Surace and Auricchio 2008) until the advent of tyrosine-mutant AAV serotype vectors (Zhong et al. 2008). Petrs-Silva et al. were able to demonstrate that point mutations in surface-exposed capsid tyrosine residues in AAV2, 8 and 9 lead to a strong and widespread transgene expression in many retinal cells after intravitreal delivery compared with their wildtype counterparts, where one of the serotype 2 mutants displayed widespread transduction throughout the retinal layers (Petrs-Silva et al. 2009). An alternative approach is co-administration of AAV vectors with a nonspecific protenase. Dalkara et al. investigated the retinal transduction profile of five relevant serotypes (AAV1, 2, 5, 8, and 9) following intravitreal delivery and identified the ILM as a barrier to AAV-mediated retinal transduction (Dalkara et al. 2009). Mild digestion of the ILM combined with AAV intravitreal delivery enabled substantially enhanced transduction of various cell types of the retina, including clinical relevant RPE and PRs. In particular, AAV5 led to robust transgene expression. The authors found a serotype-dependent transduction efficiency and argued that the barrier function of the ILM is unlikely to be purely diffusional or physical.

Recently, Agostina Puppo and coworkers tested vectors derived from five different AAV serotypes isolated from porcine tissues for their ability to mediate gene transfer to both mouse and pig retina. The studies showed that all tested AAV serotypes effectively transduced RPE and PR cells following subretinal injections, showing porcine AAVs to be promising vectors for retinal gene therapy (Puppo et al. 2013).

PR transduction by LV vectors based on HIV-1 has been found to inversely correlate with PR differentiation. Miyoshi et al. demonstrated similar transduction profiles for pups and adult rats following subretinal injection of LV, but in adult rats, expression of the transgene was restricted to the area around the injection site (Miyoshi et al. 1997). They explain this discrepancy by the size of the vector and tightness of the IPM in adult rats. Bainbridge et al. described a similar expression pattern in mice but found an even lower efficiency of PR transduction in pups, and transgene expression was primarily observed in areas where the PR layer was disturbed by the surgical procedure (Bainbridge et al. 2001). Grüter et al. investigated whether a physical barrier between PR and RPE cells and between PRs themselves may prevent transduction of PRs from the subretinal side. They did this by co-injection of neuraminidase X or chondroitinase ABC with the LV particles, enzymes that can change the IPM structure, and found that neuraminidase X significantly improved PR transduction (Gruter et al. 2005). These results clearly show that LV diffusion in intact, adult retina is impaired by components of the IPM. Pang et al. found an age-dependent transduction of PR cells during the first postnatal (P) week and that transgene expression in PR was virtually absent in eyes injected after P14 (Pang et al. 2006). Rod birth in neonatal mice continues until P11 and is completely differentiated by P21, coinciding with the significant shift in tropism. Another explanation for this shift could be a down-regulation of the respective viral receptors present on the photoreceptor plasma membrane (Kumar-Singh 2008). In the human retina, cell division ceases by fetal week 12 and the retina is fully developed in utero. Based on these studies, LVs based on HIV-1 are not predicted to be very useful in human gene therapy requiring efficient transduction of PR cells. Balaggan et al. demonstrate efficient and sustained gene expression in RPE cells and variable expression in PR cells following subretinal delivery of nonprimate LVs based on EIAV in adult mice (Balaggan et al. 2006). PR transduction efficiency was estimated to range between nondetectable to 30%, was not confined to areas of retinal trauma, and was substantially higher than mediated by HIV-1 based vectors (Bainbridge et al. 2001). One clinical trial based on EIAV transgene delivery to PR has been initiated and shows great promise (NCT01367444; Clinicaltrials.gov (Kong et al. 2008)).

Most of the studies investigating LV vector tropism have been carried out on normal retinas or at an early stage of retinal degeneration. Diffusion of viral particles to the outer retina from the vitreous has also been shown to be enhanced if the retinal architecture is altered by a degenerative process, which is often the case for IRs and age-related retinal degenerative disorders such as AMD. Calame et al. showed that LV vectors pseudotyped with the Mokola envelope efficiently transduce Müller cells in a degenerate retina and that the rhodopsin promoter appeared less efficient than the ubiquitous elongation factor-1 promoter to drive transgene expression in degenerate PRs, suggesting that PR-specific promoter activity changes during late stages of PR degeneration (Calame et al. 2011).

Although spatial transduction patterns vary to a great extend with the LV vectors, the expression kinetics is more consistent and expression has been reported to be evident as early as 3 days post-injection following subretinal injection of an EIAV vector in mice (Balaggan et al. 2006). Another study in nonhuman primates reported expression up to at least 4 years without significant decline following subretinal injection of a SIV vector (Ikeda et al. 2009). A recent study by Murakami et al. showed that SIV vectors pseudotyped with Sendai virus (SeV) envelope proteins can achieve rapid, efficient, and long-lasting gene transfer in the mouse retina (Murakami et al. 2010). Subretinal exposure to the vector for only a few minutes resulted in high-level gene transfer to RPE cells continuing over a 1-year-period. The rapid transduction ability of the SeV pseudotyped vectors is in clear contrast to findings obtained with the conventional VSV-G pseudotyped SIV vectors requiring more than 24 h of cell-vector interaction to achieve maximal gene transfer. AAVs have also been reported to show an interaction time-dependent increase in transgene expression, requiring more than 12 h to reach maximum as described for transduction of rat cardiac myocytes (Maeda et al. 1998) and human cystic fibrosis airway epithelial cells (Teramoto et al. 1998). The brief cell-vector interaction required for efficient transduction with SeV pseudotyped SIV vectors enables removal of the vector solution and resolution of retinal detachment during surgery, thereby relieving the stress on vulnerable retinal neurons and providing safer gene transfer.

3.3 Recent Advances in Clinical Gene Therapy

During the past two decades, ocular gene therapy has moved from bench to bedside. One of the most notable clinical trials and successes in gene therapy is restoration of vision in patients suffering from an inherited and incurable childhood blinding disorder, Leber's congenital amaurosis (LCA). Contemporary reviews (Ortiz et al. 2012; Seymour and Thrasher 2012; Naim et al. 2013) have carefully described some of the more notable advances in somatic gene therapy using viral vectors for treatment of diseases in humans, such as cancer, heart illnesses, and degenerative diseases. Here we highlight recent effective therapies of serious genetic conditions, including inherited eye diseases, immunodeficiency syndromes, and hemophilia (Fig. 3.1).

3.3.1 Ocular Diseases

Although conventional medicine has little to offer to patients suffering from retinal diseases, the eye represents one of the most accessible targets for localized delivery and therefore for applications of gene therapies. In addition, the relative small size of the target organ being treated implies the production of a clinical vector for a large number of individuals will be reasonably forthright. One of the most promising and encouraging examples of successful gene therapy for a genetic disease is the treatment of LCA. LCA is a severe eye disease resulting in vision loss, and appears at birth or in the first few months of life. It is an autosomal recessive disorder thought to be caused by abnormal development of PR cells. LCA belongs to a large group of congenital retinal-blinding conditions affecting approximately 1 in 2000 people worldwide. Recent findings from three independent studies show remarkable clinical improvements after treatment of LCA patients for which alternative therapies are unavailable. In this form of LCA (called LCA2) patients are bearing a mutation in the *RPE65* gene, encoding a protein required for the isomerohydrolase activity of the retinal pigment epithelium. This activity produces the 11-*cis*-retinal chromophore from all-*trans*-retinyl esters (Moiseyev et al. 2005). Absence or improper function of RPE65 lead to accumulation of all-trans retinyl esters in the RPE cells promoting PR degeneration. In addition, in the absence of 11-*cis*-retinal the rods and cones of the PRs cannot capture light and transduce it into electrical responses (Perrault et al. 1999). Thus, the binary hallmark of the pathology of LCA2 is a combined dysfunction and degeneration of PR cells, after which retinal dystrophy and vision loss are inevitable (Hartong et al. 2006; Bramall et al. 2010). The hope and expectation were that *RPE65* gene replacement therapy would ameliorate both the dysfunction and the degeneration resulting in improved vision as well as protection of PR cells. In all the three studies the efficient delivery of wildtype *RPE65* cDNA to RPE cells was achieved by accurate, surgical subretinal injection of AAV2 particles (see Fig. 3.1b and 3.1d).

In one study the *RPE65* expression was driven by a 1400 bp fragment of the RPE65 promoter (Le Meur et al. 2007) and administrated to three young adult patients (17 to 23 years old) (NCT00643747, Clinicaltrials.gov (Bainbridge et al. 2008)). Approximately 10^{11} vector genomes (vg) were injected in each patient. There were no serious adverse events. Even though no change in retinal response to visual acuity or retinal response was observed, the authors found that gene therapy led to improved visual function and mobility in one patient, suggesting that the clinical effect of this experimental approach should be further analyzed in other LCA2 patients. In parallel, Maguire and coworkers (NCT00516477, Clinicaltrials.gov (Maguire et al. 2008)) investigated the safety and efficacy of subretinal delivered AAV2 expressing *RPE65* under the transcriptional control of the chicken β-actin (CBA) promoter (Bennicelli et al. 2008). Each of the three patients received 1.5×10^{10} vg. Although normal vision was not achieved, the three treated LCA2 patients (19 to 26 years old) had modest improvements in terms of retinal function in subjective tests of visual acuity. In the third study, a phase-1 dose-escalation trial was performed showing age-dependent effects of *RPE65* gene therapy in 12 LCA2 patients (8 to 44 years old) (Maguire et al. 2009). The patients received either 1.5×10^{10}, 4.8×10^{10}, or 1.5×10^{11} vg of AAV2 expressing *RPE65* under the transcriptional control of the CBA promoter. The retinal and visual function were assessed for up to 2 years. The authors documented that the vector was well tolerated and that all patients showed sustained improvements in both subjective and objective measurements of vision, such as multifocal electroretinography, nystagmus, and ambulatory behavior. Notably, an 8-year-old child obtained a light sensitive level comparable to that of age-matched individuals with normal vision. This study clearly demonstrates that the safety and tenacious improvement in vision in all the treated patients reinforce the use of AAV-mediated gene therapy for treatment of inherited retinal disease. Moreover, and maybe more importantly, early intervention seems to result in the best potential gain.

In 2012, another dose-escalation phase-I study of 15 patients (11 to 30 years old) showed that gene therapy for LCA2 is sufficiently safe and substantially efficacious (Jacobson et al., 2012), thereby supporting the previous findings. In this study the patients were divided into five cohorts based on the applied AAV2-RPE65 vector dose ranging from 5.96×10^{10} to 17.88×10^{10} vg (NCT00481546, Clinicaltrials.gov).

Recently Jean Bennetts group demonstrated safe and efficient readministration of an AAV vector to the contralateral eye of three patients, who had previously received a single, unilateral subretinal injection of AAV2 carrying the *RPE65* gene under the transcriptional control of the CBA promoter (Bennett et al. 2012).

Efficacy of AAV after re-administration in humans has only been described in one study (Vardas et al. 2010) and the major concern was whether the re-administration would elicit a harmful immune response. In the follow-up study by Bennett and coworkers, only modest immunological responses were measured likely due to the immune-privileged nature of the eye and the low dose and high purity of the vector. Careful analysis, including evaluations of the immune response, and retinal and visual function testing suggest that subretinal re-administration is both safe as well as effective despite previous exposure to the AAV2 RPE65 vector.

Aside evaluation of efficacy Cideciyan and coworkers also investigated the effect of *RPE65* gene therapy on the PR cell degeneration in a follow-up study (Cideciyan et al. 2013). Despite the fact that gene therapy leads to long-term improvement of vision, the authors found that the decline of the photoreceptor cell layer continued and gene augmentation does not protect from degeneration, suggesting a need for a combinatory strategy in LCA therapy and a deeper understanding of the underlying disease processes to not only improve vision but also slow retinal degeneration (Cepko and Vandenberghe 2013; Cideciyan et al. 2013). Taken together, AAV delivery of a RPE65 transgene to RPE cells showed proof-of-concept for ocular gene replacement therapy mediated by viral vectors and is presently moving to a phase-III clinical trial (NCT00999609; clinicaltrials.gov).

Another rare disease benefitting from AAV-mediated gene therapy is the MERTK-associated form of autosomal recessive RP. MERTK is a member of the Mer/Axl/Tyro3 receptor tyrosine kinase family, is expressed in the RPE cells and required for phagocytosis of PR outer segments. Absent of MERTK leads to profound retinal degeneration, subsequently leading to PR loss via apoptosis. In the most successful early study, MERTK was expressed from a LV vector (Tschernutter et al. 2005), but coadministration of the LV-MERTK with an AAV vector expressing a neurotrophic factor was found to be superior. At this point in time, a phase-I clinical trial has been initiated (NCT014822195; Clinicaltrials.gov), where three patients have been treated by subretinal injection of AAV2 harboring an RPE specific promoter driving MERTK expression.

As previously mentioned, inhibition of VEGF for the management of pathological ocular neovascularization such as AMD is a proven paradigm. However, monthly intravitreal injections are required for optimal treatment. Besides anti-VEGF gene therapy based on RNAi (see above) it has recently been shown that a soluble anti-VEGF molecule (sFLT01) has antiangiogenic effect in mice and nonhuman primates following intravitreal delivery of AAV2-sFLT01 (Lukason et al. 2011). Based on these results, suggesting that intravitreal delivery of AAV2-sFLT01 may be an effective long-term treatment for AMD, a safety and tolerability study of AAV2-sFLT01 in patients with neovascular AMD has been initiated (NCT01024998; Clinicaltrials.gov).

In conclusion, the potential for successful development of genetically based therapies into conventional medicine for ocular diseases, including LCA and AMD seems particularly good.

3.4 Immunodeficiency Syndromes

As highlighted in the previous section, AAV vectors have successfully been used in ocular gene therapy clinical trials, but during recent years, several clinical trials have been initiated employing retroviral vectors, including LVs, mainly due to their relatively large transgene carrying capacity. In this section, the retroviral vector-based therapies demonstrating the strongest potential for clinical application are described.

Gene therapy of severe combined immunodeficiency (SCID) was initiated more than 10 years ago. SCID is a recessive genetic disorder characterized by the absence of functional T-lymphocytes. Consequently, both B cells and T cells of the adaptive immune system are impaired due to defects in one of nine possible genes, such as the *IL-2Rγ ((insert gamma symbol not y))* gene located on the X-chromosome. The prevalence of SCID is around 1 in 65,000–100,000 births. If untreated, babies typically die within the first year of life due to recurrent infections. The most common treatment of SCID is bone marrow transplantation. However, this treatment may have adverse side effects and in a number of cases it is not possible to find human leukocyte antigen (HLA)-matched hematopoietic stem cell donors, thereby leaving the patients without alternative therapies other than gene therapy. In the first trials developed for the treatment of X-linked SCID, conventional γ-retroviral vectors were used for ex vivo transduction of $CD34^+$ hematopoietic stem and progenitor cells resulting in expression of the interleukin-2 receptor gene under transcriptional control of the retroviral long terminal repeat (LTR)s (see Fig. 3.1a and c). In total, 19 patients were treated in two separate trials and $CD34^+$ cells were infused at doses of 60×10^6 to 207×10^6 and 1×10^6 to 22×10^6 cells per kg, respectively (Hacein-Bey-Abina et al. 2010; Gaspar et al. 2011). For all patients the functional polyclonal T-cell response was restored and transduced T cells were detected for more than 10 years after gene therapy. Despite the fact that five of the patients developed lymphoblastic leukemia due to insertional mutagenesis, these trials have paved the way for utilizing gene therapy applications in the treatment of other hematological diseases such as β thalassemia (Cavazzana-Calvo et al. 2010) and Wiskott-Aldrich syndrome (Boztug et al. 2010; Galy and Thrasher 2011). Interestingly, this ex vivo application regime has also successfully been used to treat metabolic diseases including metachromatic leukodystrophy and X-linked adrenoleukodystrophy (Cartier et al. 2009), illustrating the broad applicability of ex vivo gene therapy. Strikingly, the data from long-term follow-up studies in around 100 patients suffering different forms of inherited immunodeficiencies treated with γ-retroviral mediated gene therapy show an impressive overall survival greater than 90 %. Most of these patients are experiencing substantial clinical benefit regardless of the fact that toxicities were observed in ten individuals (Booth et al. 2011). To improve the treatment, much effort has been invested in solving the toxicity issue. It is now known that the adverse side effects (development of leukemia) share the same molecular mechanism involving activation of proto-oncogene expression caused by insertion of potent promoter sequences contained in the LTRs of the applied γ-retroviral vectors into patient genomes. To overcome insertional mutagenesis the vector strategy was therefore changed by substituting the γ-retroviral vector with the self-inactivating (SIN) vector systems, including LV vectors. The reasons for choosing LV vectors are many, but the two most important benefits are (i) their reduced ability to interfere with host genes and decreased capability of depositing genetic material in promoter regions of genes, thereby reducing the genotoxicity (see Sect. 3.6), and (ii) their capacity to transduce nondividing cells, rendering LV vectors to transduce hematopoietic stem cells (HSCs) more effectively compared to other vector systems, thus reducing the ex vivo transduction period for conservation of engraftment ability (Fig. 3.1c). One important challenge that has to be resolved in terms of using LV vectors as a widespread application for different diseases in many patients is large-scale manufacturing of high quality vectors. Today it is not possible, as in the case of the conventional γ-retroviral vector, to obtain stable packing cell lines for the production of LV vectors, leaving LV vector production laborious and expensive. Despite this challenge LVs have successfully been used in a number of trials, including treatment of X-linked adrenoleukodystrophy (Cartier et al. 2009), and presently trials for Wiskott-Aldrich syndrome (NCT01515462, ClinicalTrials.gov), chronic granulomatous disease (NCT00394316, ClinicalTrials.gov), and X-linked SCID (NCT01306019, ClinicalTrials.gov) are moving into the clinic (Fig. 3.1a).

3.5 Hemophilia

As exemplified by ocular gene therapy localized gene delivery of viral vectors has clear applications for some disorders. On the other hand, preclinical studies have documented that specific AAV serotypes followed by intravascular injection will predominantly target hepatocytes, thereby providing a beneficial strategy using this organ for expression of therapeutic agents needed either locally or systemically. Even though several attempts for the treatment of, e.g., hemophilia and metabolic disorders have resulted in disappointing results due to toxic effects or prevention of long-term expression caused by immunological responses to the viral vector, it has been suggested that stable expression of therapeutic levels of the transgene may as mentioned above be limited by a capsid-specific cytotoxic T-cell response (against the transduced hepatocytes) (Manno et al. 2006). New serotypes have been developed to circumvent the possibility of humoral immunity to AAV. Recently, Nathwani and coworkers have attempted to develop an optimized therapy for hemophilia B, which is an X-linked bleeding disorder resulting from a defect in the gene encoding coagulation factor IX (FIX), a serine protease that is pivotal for blood clotting. Hemophilia B has a prevalence of around 1 in 25,000 to 30,000 males. It has been estimated that individuals with severe hemophilia B only have 1 % of normal values of FIX in the circulation, resulting in frequent bleeding periods, which are associated with early death (Nathwani and Tuddenham 1992; Darby et al. 2007). Even though the present treatment, which involves frequent intravenous injections of recombinant FIX, is well tolerated and results in clinically relevant levels of circulating FIX, it is associated with unwanted effects like inhibition formation and is very expensive.

The Nathwani group replaced the AAV2 vector (used in previous experimental treatments) with AAV8, which has a lower seroprevalence in humans than does AAV2 (Gao et al. 2002; Calcedo et al. 2009). In addition, the authors included two other clever steps: One is the development of a codon-optimized FIX (the protein factor missing in hemophilia patients) expression cassette that is packaged as scAAV, enabling substantially higher-expression levels due to the complementary dimers in each viron (Nathwani et al. 2006). The human FIX transgene was expressed from the synthetic liver promoter 1 (LP1), which consists of liver-specific elements from the human apo-lipoprotein E/C-I gene locus control region and the human α1-antitrypsin promoter. The LP1 promoter has the advantages of being compact and highly liver-specific. The other step was the application of a simple, noninvasive, peripheral-vein-based administration approach that is safe for patients with a bleeding disorder (Nathwani et al. 2007). This step is possible since AAV8 has a strong tropism for the liver (Fig. 3.1b and d).

Six patients (27 to 64 years old), selected on the basis of having low levels of antibodies against AAV8, were enrolled sequentially into one of three cohorts according to dose levels: Low (2×10^{11} vg per kg of body weight), medium (6×10^{11} vg per kg) and high (2×10^{12} vg per kg). In all patients AAV-mediated expression of FIX was detected at 2 to 11 % of normal levels. In addition, decreased frequency (2 of 6 patients) or cessation of spontaneous hemorrhage (4 of six patients) was detected. Both patients receiving the high dose developed transient, asymptomatic side effects related to peripheral AAV8-capsid-specific T cells or an unexplainable slightly increase in liver-enzyme levels. The study demonstrates that (i) peripheral-vein infusion of scAAV8 encoding FIX results in transgene expression at levels sufficient to improve the bleeding phenotype, and (ii) immune-mediated clearance of transduced hepatocytes may be controlled by temporary steroid treatment without loss of FIX expression.

Gene therapy studies have also been initiated to treat hemophilia A, the most common inherited bleeding disorder, caused by deficiency of factor VIII (FVIII). As in the case of hemophilia B, hemophilia A is well suited for gene replacement since a modest rise (> 1 %) in the level of circulation FVIII can ameliorate the bleeding phenotype. However, several obstacles have challenged AAV-mediated gene transfer for hemophilia A gene therapy. One is the distinct biochemical and molecular properties of FVIII resulting in highly inefficient expression of the factor. Optimization of the FVIII coding sequence (termed FVIII-N6), including partial deletion of the B-domain (Miao et al. 2004) (Cerullo et al. 2007), has significantly reduced the tendency to evoke an unfolding protein response and thus improved the levels of expression and secretion, thereby rendering FVIII-N6 a useful variant for additional assessments in the context of gene therapy (Malhotra et al. 2008). Another challenge is the size of the FVIII coding sequence, which at 7.0 kb far surpasses the normal packing capacity of AAV vectors. To solve this issue McIntosh and coworkers recently developed a novel 5.2-kb AAV expression cassette containing a hybrid liver-specific promoter driving the expression of a codon optimized variant, shown to have improved expression when encoded from a lentiviral context (Nathwani et al. 2006; Radcliffe et al. 2008; Ward et al. 2011). In addition, the 226 amino acid long N6 B-domain was replaced with a short 17 amino acid spacer peptide containing the signal required for efficient expression of human FVIII (McIntosh et al. 2013). The authors showed efficient packaging of the cassette into AAV8 particles. Importantly, expression in mice and nonhuman primates was more potent than is possible with previous FVIII variants, thereby providing further support for safe and effective gene therapy for hemophilia A in humans.

3.6 Safety

Safety is of the utmost importance in development of gene therapy vectors. At the infant stage of viral gene therapy, the success of the first cure was tempered by the fact that 5 out of 20 patients who were successfully treated for X-linked severe combined immunodeficiency (SCID) developed leukemia (Hacein-Bey-Abina et al. 2003). As previously described, the treatment was based on reinfusion of haematopoietic stem cells that were transduced ex vivo with a γ-retrovirus MLV vector expressing the common cytokine receptor interleukin-2 receptor gamma (*IL-2Rγ*)(see Fig. 3.1a). A few years earlier, in September 1999, 18-year-old Jesse Gelsinger died of multiorgan failure, 4 days after he received a high dose of adenovirus vectors. Gelsinger took part in a gene therapy clinical trial intended for babies suffering from severe and fatal ornithine transcarbamylase deficiency. The vector was infused directly into the liver but had disseminated into the circulation and triggered a massive inflammatory response (Marshall 1999; Report 2002). These cases serve as constant reminders of the need to develop vectors with better safety profiles in terms of insertional mutagenesis and host immune response. In addition to these risks, the possibility to generate replication-competent viruses will also be assessed in the following paragraph. Clinical trials have revealed different toxicity profiles for the LV and rAAV vectors; the major concern for LV vectors being genotoxicity whereas rAAV vector gene transfer are limited by immunotoxicity. Phenotoxicity is mainly related to the transgene being expressed, and a discussion on the topic is beyond the scope of this chapter.

3.6.1 Insertional Mutagenesis

Genotoxicity caused by insertional mutagenesis is less of a risk for rAAV vectors compared with the retroviral vectors. Integration into the host cell genome is for rAAV a passive process with a frequency $<1\%$ in vivo (Nakai et al. 2001; Inagaki et al. 2008) and since most rAAV applications target nondividing cells, the risk associated with rAAV integration is much less than with retroviral vectors, but not zero. Each vector system has a characteristic genomic integration pattern. rAAV genomes share with the retroviral genome a preference for integrating into genes, but as opposed to the integrase-dependent integration of retroviral genomes, rAAV are believed to be fused by nonhomologous end-joining to pre-existing chromosomal breaks (Miller et al. 2002). γ-Retroviruses preferentially target regulatory sequences, growth-regulatory genes, and are in general more prone to insert near proto-oncogenes than lentiviral vectors (Cattoglio et al. 2007). The lentiviral insertion pattern has been found to be 10 times safer than the γ-retroviral (Montini et al., 2006) and LV vectors have been found to integrate more randomly throughout the entire genome (Schroder et al. 2002).

Insertional mutagenesis can either cause disruption or an abnormal activation of host genes, where the primary concern is oncogenesis caused by the latter. Disruption of a gene is less of a concern, since retroviral vectors often only integrate in one allele leaving the other intact. In rare cases with haploinsufficiency, insertional disruption can become problematic. SIN-vectors have been developed, where the 3'LTR U3 region, which contains all the viral enhancer and promoter activities, is partially deleted (Yu et al. 1986; Miyoshi et al. 1998). Another approach to lower the risk of activation of proto-oncogenes near the site of integration is utilizing chromatin insulator elements. These elements are believed to form expression boundaries, not only preventing interference between promoters and enhancers of adjacent genes, but also overcoming negative chromosome positional effects that can cause silencing of transgene expression (Emery et al. 2000; Labrador and Corces 2002).

Several different approaches can potentially reduce the risk of genotoxicity: Using vectors with random or more neutral integration profiles, targeting cells that lack sustained proliferation potential, using SIN-vectors, using more efficient vectors to avoid multiple integrations in a single cell, and by thoughtful design of expression cassettes that avoid long-distance enhancer interactions or incorporate insulator sequences.

Targeting vector integration is the ultimate solution to the problem of insertional mutagenesis. The integration pattern of retroviruses has been shown to be closely related to the integrase, but not the *gag*, *env*, or *pol* genes (Shibagaki and Chow 1997; Harper et al. 2003; Metais et al. 2010). Fusion of the integrase with sequence-specific DNA binding domains such as a zinc finger protein has also been attempted but with limited success (Bushman and Miller 1997). In nondividing cells, IDLVs represent an interesting alternative approach. These vectors typically have a mutation in the IN gene amplifying episomal formation whereas integration events are reduced up to 10^4 fold (Leavitt et al. 1996).

3.6.2 The Risk of Generation of Replication Competent Viruses

Multiplasmid transient transfection of adherent HEK-293T cells is the most widely used method for production of both rAAV and LV (see Fig. 3.2). Replication competent viruses results from recombination of the helper plasmids that supply the genes for efficient packaging with the vector plasmid. In addition, rAAV depends on Ad helper genes for packaging (Fig. 3.2b). Development of safer packaging systems for rAAV involves cell lines expressing the *rep* and the *cap* gene, plasmids encoding Ad helper genes, and hybrid vectors expressing both AAV and Ad helper genes. All methods yield a minimum of replication competent viruses, but differences in vector yield has been found.

To make the HIV vector prototypes safer in terms of reducing the probability of generating replication competent LVs, the viral genes needed for production and packaging were divided from those encoding viral proteins into two plasmids; one encoding proviral DNA with a deletion in the *env* gene and one expressing Env (Helseth et al. 1990; Page et al. 1990). First-generation HIV-1-based vectors increased safety by splitting vector components into three plasmids; a packing construct expressing HIV Gag, Pol, and regulatory/accessory proteins, an Env plasmid encoding a viral glycoprotein, and a transfer vector containing all of the essential cis-acting elements for packaging, reverse transcription, and integration. In the second-generation LV vectors HIV-1 accessory proteins have been deleted to further increase safety (Zufferey et al. 1997). The third-generation LV vectors have been made Tat-independent by replacing the U3 promoter region of the 5′-LTR in the transfer vector with a strong viral promoter such as the cytomegalovirus promoter (Dull et al. 1998; Kim et al. 1998). The *rev* gene has been removed from the packaging construct and is provided from a separate plasmid. This vector system consists of four plasmids with only three out of the nine original genes of HIV-1 (Fig. 3.2a). To reduce the likelihood of generating replication competent LV vectors even further, the SIN vectors were developed, that also reduce the possibility of activating proto-oncogenes after vector integration (Yu et al. 1986; Miyoshi et al. 1998). In order to increase safety during clinical treatment, analysis of virus preparations using long-term cultures and PCR assays should be used to assess the possibility of replication competent retrovirus (RCR) and replication competent lentivirus (RCL) contamination (Sastry and Cornetta 2009). It is believed that the risk of insertional mutagenesis is greatest if RCR is present, since ongoing viral infection is likely to result in a higher number of insertional events (Sastry and Cornetta 2009). In support to this notion RCR contaminating retroviral vector preparations have been shown to cause malignancy in both mice and nonhuman primates (Donahue et al. 1992; Cornetta et al. 1993). Whether RCL can produce a HIV-1 like syndrome is presently unknown, although HIV-1 accessory proteins normally required for virulence have been removed from vector constructs to improve their safety profile. Although new generations of LV vectors are generated to reduce the likelihood of producing replication competent LVs, the problem has not been completely eliminated.

Vectors derived from nonprimate LVs such as FIV and EIAV cannot replicate in human cells, and could constitute a safer alternative than HIV-based vectors. Due to highly restricted tropism, pseudotyping with the VSV-G glycoprotein has furthered vectors with expanded tropism, but inclusion of VSV-G in nonprimate LV has been proposed to increase the possibility of creating replication-competent viruses.

Another major concern in the use of viral vectors is the possibility of recombination between the therapeutic vector and either endogenous viral elements or exogenous infections in the target cell. Recombination and mobilization can in nondividing cells occur many years after the initial transduction, and homology between vector and WT viruses increase the risk of generating replication competent or recombinant viruses. It has been demonstrated, that LV vectors can be rescued by an exogenous HIV-1 infection in tissue cultures (Evans and Garcia 2000), which can limit the clinical use of LV vectors. Due to low homology, recombination between different viral strains (e.g., retroviral vectors with influenza virus) is unlikely, but the risk of generating recombinant viruses exists, if two viral strains are present in the same target cell.

3.6.3 Host Immune Response

When viral particles are generated and injected in vivo, all components of the virus can induce both innate and adaptive immune responses. Compared to the Gelsinger-case, activation of host immune response by LV or rAAV particles has not been found to pose a fatal risk but constitute a tremendous challenge in development of safe and efficient gene delivery.

The innate pathway is induced through pattern recognition receptors, where the most characterized family consists of the Toll-like receptors (TLRs). After LV transduction of the cell, studies have demonstrated that the critical events are viral entry and reverse transcriptase since RNA and DNA viral genomes are recognized by several TLRs (Breckpot et al. 2010; Agudo et al. 2012). Activation of the TLRs lead to rapid secretion of inflammatory cytokines that can interfere with the efficiency of transduction, as well as prime the adaptive immune response. The VSV-G capsid protein can also initiate an immune response, either by activating the TLR pathway or as a target of the human complement system (DePolo et al. 2000). The adaptive immunity is divided into cell-mediated and humoral immunity. The cell-mediated immunity is comprised of T helper cells and cytotoxic T lymphocytes (CTLs), whereas the humoral response is mediated by antibodies secreted by B cells. Administration of viral particles can result in development of antibodies targeting viral particle components as well as the transgene product. Capsid-specific antibodies have not been found to affect long-term expression, but re-administration of the same vector serotype is blocked (Abordo-Adesida et al. 2005). Development of neutralizing antibodies together with CTLs targeting the transgene are major concerns and rate-limiting factors for the success of in vivo LV gene therapy.

The biggest challenge in rAAV-mediated gene therapy is the presence of pre-existing neutralizing antibodies against the capsid due to the widespread exposure to numerous AAV serotypes in the human population. Serotype-switching, capsid-engineering, and shielding by chemical modifications can potentially overcome this problem, and have been discussed in the AAV section above. Immunosuppression by pharmacological elimination of antibody producing cells (Mingozzi et al. 2012a; Mingozzi et al. 2012b) and plasmapheresis (Monteilhet et al. 2011) have also been suggested to modulate humoral immune response prior to rAAV mediated gene therapy. Cellular immune response against the AAV capsid, especially in the case of ocular gene therapy, has been found to be limited and vary depending on amount, serotype, and route of administration. This problem can be solved by transient immunosuppression (Nathwani et al. 2011). T-cell response against the AAV2 capsid might be avoidable by using surface-exposed tyrosine-mutant AAV2 vectors (Zhong et al. 2008) and/or scAAV genomes, both leading to high-efficiency transduction at lower doses.

3.7 Conclusions

The viral vector-based gene delivery platform is remarkably versatile, and vector advancements for viral gene therapy aim at developing vectors that can safely achieve the highest percentage of therapeutically modified cells with the lowest amount of viral vectors. The viral vector tool box is neither complete nor perfect, but the advent of tyrosine-mutants and scAAVs illustrate what vector manipulation can accomplish in terms of reducing the therapeutic viral load. Inability to predict the immune response in humans is still the biggest impediment in the effort of developing therapeutic viral vectors, since it cannot be completely replicated in animal models. This emphasizes the importance of evaluating novel viral vectors in the final context of use, not only because of uncertainty with immune responses but also because target cells and their environment may be subject to various alterations that can impact on transduction capacities as well as targeted transgene expression. Another important challenge is to reduce the risk of genotoxicity. The review shows that many different approaches have been suggested or even implemented in upcoming gene therapy trials. However, in order to improve safety the field needs to develop even more sophisticated expression cassettes without reducing efficacy.

In conclusion, these highly promising data clearly demonstrate that viral-mediated gene therapy as such is way beyond the proof-of-concept threshold and will, in combination with the growing interest of applying DNA- or RNA-based technologies into the clinic, undoubtedly pave the way for development of efficacious long-term gene therapy treatment of inherited diseases. This resurgence of viral-based gene therapy is supported by substantial investments by the pharmaceutical industry, thereby promising to further accelerate clinical translation. The presented studies also demonstrate (i) that presently available AAV vectors safely and effectively transduce post-mitotic cells for targeted delivery and treatment of inherited diseases, such as LCA and hemophilia B, and (ii) that third generation LV vectors successfully ameliorate the disease phenotype for a number of immunodeficiency syndromes, including X-linked SCID, giving substantial reassurance that gene based technologies will soon enter in mainstream medicine.

References

Abordo-Adesida E, Follenzi A, Barcia C, Sciascia S, Castro MG, Naldini L, Lowenstein PR (2005) Stability of lentiviral vector-mediated transgene expression in the brain in the presence of systemic antivector immune responses. Hum Gene Ther 16:741–751

Agudo J, Ruzo A, Kitur K, Sachidanandam R, Blander JM, Brown BD (2012) A TLR and non-TLR mediated innate response to lentiviruses restricts hepatocyte entry and can be ameliorated by pharmacological blockade. Mol Ther 20:2257–2267

Aiello LP, Avery RL, Arrigg PG, Keyt BA, Jampel HD, Shah ST, Pasquale LR, Thieme H, Iwamoto MA, Park JE et al (1994) Vascular endothelial growth factor in ocular fluid of patients with diabetic retinopathy and other retinal disorders. N Engl J Med 331:1480–1487

Aker M, Tubb J, Groth AC, Bukovsky AA, Bell AC, Felsenfeld G, Kiem HP, Stamatoyannopoulos G, Emery DW (2007) Extended core sequences from the cHS4 insulator are necessary for protecting retroviral vectors from silencing position effects. Hum Gene Ther 18:333–343

Akkina RK, Walton RM, Chen ML, Li QX, Planelles V, Chen IS (1996) High-efficiency gene transfer into CD34 + cells with a human immunodeficiency virus type 1-based retroviral vector pseudotyped with vesicular stomatitis virus envelope glycoprotein G. J Virol 70:2581–2585

Allocca M, Mussolino C, Garcia-Hoyos M, Sanges D, Iodice C, Petrillo M, Vandenberghe LH, Wilson JM, Marigo V, Surace EM, Auricchio A (2007) Novel adeno-associated virus serotypes efficiently transduce murine photoreceptors. J Virol 81:11372–11380

Apolonia L, Waddington SN, Fernandes C, Ward NJ, Bouma G, Blundell MP, Thrasher AJ, Collins MK, Philpott NJ (2007) Stable gene transfer to muscle using non-integrating lentiviral vectors. Mol Ther 15:1947–1954

Arumugam PI, Scholes J, Perelman N, Xia P, Yee JK, Malik P (2007) Improved human beta-globin expression from self-inactivating lentiviral vectors carrying the chicken hypersensitive site-4 (cHS4) insulator element. Mol Ther 15:1863–1871

Askou AL, Pournaras JA, Pihlmann M, Svalgaard JD, Arsenijevic Y, Kostic C, Bek T, Dagnaes-Hansen F, Mikkelsen JG, Jensen TG, Corydon TJ (2012) Reduction of choroidal neovascularization in mice by adeno-associated virus-delivered anti-vascular endothelial growth factor short hairpin RNA. J Gene Med 14:632–641

Auricchio A, Kobinger G, Anand V, Hildinger M, O'Connor E, Maguire AM, Wilson JM, Bennett J (2001) Exchange of surface proteins impacts on viral vector cellular specificity and transduction characteristics: the retina as a model. Hum Mol Genet 10:3075–3081

Bainbridge JW, Stephens C, Parsley K, Demaison C, Halfyard A, Thrasher AJ, Ali RR (2001) In vivo gene transfer to the mouse eye using an HIV-based lentiviral vector; efficient long-term transduction of corneal endothelium and retinal pigment epithelium. Gene Ther 8:1665–1668

Bainbridge JW, Smith AJ, Barker SS, Robbie S, Henderson R, Balaggan K, Viswanathan A, Holder GE, Stockman A, Tyler N, Petersen-Jones S, Bhattacharya SS, Thrasher AJ, Fitzke FW, Carter BJ, Rubin GS, Moore AT, Ali RR (2008) Effect of gene therapy on visual function in Leber's congenital amaurosis. N Engl J Med 358:2231–2239

Bak RO, Hollensen AK, Primo MN, Sorensen CD, Mikkelsen JG (2013) Potent microRNA suppression by RNA Pol II-transcribed 'Tough Decoy' inhibitors. RNA 19:280–293

Balaggan KS, Ali RR (2012) Ocular gene delivery using lentiviral vectors. Gene Ther 19:145–153

Balaggan KS, Binley K, Esapa M, Iqball S, Askham Z, Kan O, Tschernutter M, Bainbridge JW, Naylor S, Ali RR (2006) Stable and efficient intraocular gene transfer using pseudotyped EIAV lentiviral vectors. J Gene Med 8:275–285

Bartel M, Schaffer D, Buning H (2011) Enhancing the clinical potential of AAV vectors by capsid engineering to evade pre-existing immunity. Front Microbiol 2:204

Bayer M, Kantor B, Cockrell A, Ma H, Zeithaml B, Li X, McCown T, Kafri T (2008) A large U3 deletion causes increased in vivo expression from a nonintegrating lentiviral vector. Mol Ther 16:1968–1976

Bennett J, Ashtari M, Wellman J, Marshall KA, Cyckowski LL, Chung DC, McCague S, Pierce EA, Chen Y, Bennicelli JL, Zhu X, Ying GS, Sun J, Wright JF, Auricchio A, Simonelli F, Shindler KS, Mingozzi F, High KA, Maguire AM (2012) AAV2 gene therapy readministration in three adults with congenital blindness. Sci Transl Med 4:120ra115

Bennicelli J, Wright JF, Komaromy A, Jacobs JB, Hauck B, Zelenaia O, Mingozzi F, Hui D, Chung D, Rex TS, Wei Z, Qu G, Zhou S, Zeiss C, Arruda VR, Acland GM, Dell'Osso LF, High KA, Maguire AM, Bennett J (2008) Reversal of blindness in animal models of leber congenital amaurosis using optimized AAV2-mediated gene transfer. Mol Ther 16:458–465

Booth C, Gaspar HB, Thrasher AJ (2011) Gene therapy for primary immunodeficiency. Curr Opin Pediatr 23:659–666

Boztug K, Schmidt M, Schwarzer A, Banerjee PP, Diez IA, Dewey RA, Bohm M, Nowrouzi A, Ball CR, Glimm H, Naundorf S, Kuhlcke K, Blasczyk R, Kondratenko I, Marodi L, Orange JS, von Kalle C, Klein C (2010) Stem-cell gene therapy for the Wiskott-Aldrich syndrome. N Engl J Med 363:1918–1927

Bramall AN, Wright AF, Jacobson SG, McInnes RR (2010) The genomic, biochemical, and cellular responses of the retina in inherited photoreceptor degenerations and prospects for the treatment of these disorders. Annu Rev Neurosci 33:441–472

Breckpot K, Escors D, Arce F, Lopes L, Karwacz K, Van Lint S, Keyaerts M, Collins M (2010) HIV-1 lentiviral vector immunogenicity is mediated by Toll-like receptor 3 (TLR3) and TLR7. J Virol 84:5627–5636

Brown BD, Venneri MA, Zingale A, Sergi Sergi L, Naldini L (2006) Endogenous microRNA regulation suppresses transgene expression in hematopoietic lineages and enables stable gene transfer. Nat Med 12:585–591.

Brown BD, Cantore A, Annoni A, Sergi LS, Lombardo A, Della Valle P, D'Angelo A, Naldini L (2007) A microRNA-regulated lentiviral vector mediates stable correction of hemophilia B mice. Blood 110:4144–4152

Burns JC, Friedmann T, Driever W, Burrascano M, Yee JK (1993) Vesicular stomatitis virus G glycoprotein pseudotyped retroviral vectors: concentration to very high titer and efficient gene transfer into mammalian and nonmammalian cells. Proc Natl Acad Sci U S A 90:8033–8037

Bushman FD, Miller MD (1997) Tethering human immunodeficiency virus type 1 preintegration complexes to target DNA promotes integration at nearby sites. J Virol 71:458–464

Calame M, Cachafeiro M, Philippe S, Schouwey K, Tekaya M, Wanner D, Sarkis C, Kostic C, Arsenijevic Y (2011) Retinal degeneration progression changes lentiviral vector cell targeting in the retina. PLoS ONE 6:e23782

Calcedo R, Vandenberghe LH, Gao G, Lin J, Wilson JM (2009) Worldwide epidemiology of neutralizing antibodies to adeno-associated viruses. J Infect Dis 199:381–390

Cartier N, Hacein-Bey-Abina S, Bartholomae CC, Veres G, Schmidt M, Kutschera I, Vidaud M, Abel U, Dal-Cortivo L, Caccavelli L, Mahlaoui N, Kiermer V, Mittelstaedt D, Bellesme C, Lahlou N, Lefrere F, Blanche S, Audit M, Payen E, Leboulch P, l'Homme B, Bougneres P, Von Kalle C, Fischer A, Cavazzana-Calvo M, Aubourg P (2009) Hematopoietic stem cell gene therapy with a lentiviral vector in X-linked adrenoleukodystrophy. Science 326:818–823

Cattoglio C, Facchini G, Sartori D, Antonelli A, Miccio A, Cassani B, Schmidt M, von Kalle C, Howe S, Thrasher AJ, Aiuti A, Ferrari G, Recchia A, Mavilio F (2007) Hot spots of retroviral integration in human $CD34^+$ hematopoietic cells. Blood 110:1770–1778

Cavazzana-Calvo M, Payen E, Negre O, Wang G, Hehir K, Fusil F, Down J, Denaro M, Brady T, Westerman K, Cavallesco R, Gillet-Legrand B, Caccavelli L, Sgarra R, Maouche-Chretien L, Bernaudin F, Girot R, Dorazio R, Mulder GJ, Polack A, Bank A, Soulier J, Larghero J, Kabbara N, Dalle B, Gourmel B, Socie G, Chretien S, Cartier N, Aubourg P, Fischer A, Cornetta K, Galacteros F, Beuzard Y, Gluckman E, Bushman F, Hacein-Bey-Abina S, Leboulch P (2010) Transfusion independence and HMGA2 activation after gene therapy of human beta-thalassaemia. Nature 467:318–322

Cepko CL, Vandenberghe LH (2013) Retinal gene therapy coming of age. Human Gene Ther 24:242–244

Cerullo V, Seiler MP, Mane V, Cela R, Clarke C, Kaufman RJ, Pipe SW, Lee B (2007) Correction of murine hemophilia A and immunological differences of factor VIII variants delivered by helper-dependent adenoviral vectors. Mol Ther 15:2080–2087

Chadderton N, Millington-Ward S, Palfi A, O'Reilly M, Tuohy G, Humphries MM, Li T, Humphries P, Kenna PF, Farrar GJ (2009) Improved retinal function in a mouse model of dominant retinitis pigmentosa following AAV-delivered gene therapy. Mol Ther 17:593–599

Cideciyan AV, Jacobson SG, Beltran WA, Sumaroka A, Swider M, Iwabe S, Roman AJ, Olivares MB, Schwartz SB, Komaromy AM, Hauswirth WW, Aguirre GD (2013) Human retinal gene therapy for Leber congenital amaurosis shows advancing retinal degeneration despite enduring visual improvement. Proc Natl Acad Sci U S A 110:E517–525

Colquitt JL, Jones J, Tan SC, Takeda A, Clegg AJ, Price A (2008) Ranibizumab and pegaptanib for the treatment of age-related macular degeneration: a systematic review and economic evaluation. Health Technol Assess 12:iii–iv, ix–201

Cornetta K, Nguyen N, Morgan RA, Muenchau DD, Hartley JW, Blaese RM, Anderson WF (1993) Infection of human cells with murine amphotropic replication-competent retroviruses. Hum Gene Ther 4:579–588

Cornu TI, Cathomen T (2007) Targeted genome modifications using integrase-deficient lentiviral vectors. Mol Ther 15:2107–2113

Coutant F, Frenkiel MP, Despres P, Charneau P (2008) Protective antiviral immunity conferred by a nonintegrative lentiviral vector-based vaccine. PLoS ONE 3:e3973

Dalkara D, Kolstad KD, Caporale N, Visel M, Klimczak RR, Schaffer DV, Flannery JG (2009) Inner limiting membrane barriers to AAV-mediated retinal transduction from the vitreous. Mol Ther 17:2096–2102

Dalkara D, Byrne LC, Lee T, Hoffmann NV, Schaffer DV, Flannery JG (2012) Enhanced gene delivery to the neonatal retina through systemic administration of tyrosine-mutated AAV9. Gene Ther 19:176–181

Darby SC, Kan SW, Spooner RJ, Giangrande PL, Hill FG, Hay CR, Lee CA, Ludlam CA, Williams M (2007) Mortality rates, life expectancy, and causes of death in people with hemophilia A or B in the United Kingdom who were not infected with HIV. Blood 110:815–825

De Palma M Venneri MA Roca C Naldini L (2003) Targeting exogenous genes to tumor angiogenesis by transplantation of genetically modified hematopoietic stem cells. Nat Med 9:789–795

DePolo NJ, Reed JD, Sheridan PL, Townsend K, Sauter SL, Jolly DJ, Dubensky TW, Jr. (2000) VSV-G pseudotyped lentiviral vector particles produced in human cells are inactivated by human serum. Mol Ther 2:218–222

Donahue RE, Kessler SW, Bodine D, McDonagh K, Dunbar C, Goodman S, Agricola B, Byrne E, Raffeld M, Moen R et al (1992) Helper virus induced T cell lymphoma in nonhuman primates after retroviral mediated gene transfer. J Exp Med 176:1125–1135

Dull T, Zufferey R, Kelly M, Mandel RJ, Nguyen M, Trono D, Naldini L (1998) A third-generation lentivirus vector with a conditional packaging system. J Virol 72:8463–8471

Emery DW, Yannaki E, Tubb J, Stamatoyannopoulos G (2000) A chromatin insulator protects retrovirus vectors from chromosomal position effects. Proc Natl Acad Sci U S A 97:9150–9155

Evans JT, Garcia JV (2000) Lentivirus vector mobilization and spread by human immunodeficiency virus. Hum Gene Ther 11:2331–2339

Ferrari FK, Samulski T, Shenk T, Samulski RJ (1996) Second-strand synthesis is a rate-limiting step for efficient transduction by recombinant adeno-associated virus vectors. J Virol 70:3227–3234

Fisher KJ, Gao GP, Weitzman MD, DeMatteo R, Burda JF, Wilson JM (1996) Transduction with recombinant adeno-associated virus for gene therapy is limited by leading-strand synthesis. J Virol 70:520–532

Galy A, Thrasher AJ (2011) Gene therapy for the Wiskott-Aldrich syndrome. Curr Opin Allergy Clin Immunol 11:545–550

Gao GP, Alvira MR, Wang L, Calcedo R, Johnston J, Wilson JM (2002) Novel adeno-associated viruses from rhesus monkeys as vectors for human gene therapy. Proc Natl Acad Sci U S A 99:11854–11859

Gaspar HB, Cooray S, Gilmour KC, Parsley KL, Adams S, Howe SJ, Al Ghonaium A, Bayford J, Brown L, Davies EG, Kinnon C, Thrasher AJ (2011) Long-term persistence of a polyclonal T cell repertoire after gene therapy for X-linked severe combined immunodeficiency. Sci Transl Med 3:97ra79

Gruter O, Kostic C, Crippa SV, Perez MT, Zografos L, Schorderet DF, Munier FL, Arsenijevic Y (2005) Lentiviral vector-mediated gene transfer in adult mouse photoreceptors is impaired by the presence of a physical barrier. Gene Ther 12:942–947

Hacein-Bey-Abina S, Von Kalle C, Schmidt M, McCormack MP, Wulffraat N, Leboulch P, Lim A, Osborne CS, Pawliuk R, Morillon E, Sorensen R, Forster A, Fraser P, Cohen JI, de Saint Basile G, Alexander I, Wintergerst U, Frebourg T, Aurias A, Stoppa-Lyonnet D, Romana S, Radford-Weiss I, Gross F, Valensi F, Delabesse E, Macintyre E, Sigaux F, Soulier J, Leiva LE, Wissler M, Prinz C, Rabbitts TH, Le Deist F, Fischer A, Cavazzana-Calvo M (2003) LMO2-associated clonal T cell proliferation in two patients after gene therapy for SCID-X1. Science 302:415–419

Hacein-Bey-Abina S, Hauer J, Lim A, Picard C, Wang GP, Berry CC, Martinache C, Rieux-Laucat F, Latour S, Belohradsky BH, Leiva L, Sorensen R, Debre M, Casanova JL, Blanche S, Durandy A, Bushman FD, Fischer A, Cavazzana-Calvo M (2010) Efficacy of gene therapy for X-linked severe combined immunodeficiency. N Engl J Med 363:355–364

Hansen TB, Jensen TI, Clausen BH, Bramsen JB, Finsen B, Damgaard CK, Kjems J (2013) Natural RNA circles function as efficient microRNA sponges. Nature 495:384–388

Harper AL, Sudol M, Katzman M (2003) An amino acid in the central catalytic domain of three retroviral integrases that affects target site selection in nonviral DNA. J Virol 77:3838–3845

Hartong DT, Berson EL, Dryja TP (2006) Retinitis pigmentosa. Lancet 368:1795–1809

Hassel D, Cheng P, White MP, Ivey KN, Kroll J, Augustin HG, Katus HA, Stainier DY, Srivastava D (2012) MicroRNA-10 regulates the angiogenic behavior of zebrafish and human endothelial cells by promoting vascular endothelial growth factor signaling. Circ Res 111:1421–1433

Helseth E, Kowalski M, Gabuzda D, Olshevsky U, Haseltine W, Sodroski J (1990) Rapid complementation assays measuring replicative potential of human immunodeficiency virus type 1 envelope glycoprotein mutants. J Virol 64:2416–2420

Hioki H, Kameda H, Nakamura H, Okunomiya T, Ohira K, Nakamura K, Kuroda M, Furuta T, Kaneko T (2007) Efficient gene transduction of neurons by lentivirus with enhanced neuron-specific promoters. Gene Ther 14:872–882

Hollensen AK, Bak RO, Haslund D, Mikkelsen JG (2013) Suppression of microRNAs by dual-targeting and clustered Tough Decoy inhibitors. RNA Biol 10(3):406-14

Ikeda Y, Yonemitsu Y, Miyazaki M, Kohno R, Murakami Y, Murata T, Tabata T, Ueda Y, Ono F, Suzuki T, Ageyama N, Terao K, Hasegawa M, Sueishi K, Ishibashi T (2009) Stable retinal gene expression in nonhuman primates via subretinal injection of SIVagm-based lentiviral vectors. Hum Gene Ther 20:573–579

Inagaki K, Piao C, Kotchey NM, Wu X, Nakai H (2008) Frequency and spectrum of genomic integration of recombinant adeno-associated virus serotype 8 vector in neonatal mouse liver. J Virol 82:9513–9524

Jacobson SG, Cideciyan AV, Ratnakaram R, Heon E, Schwartz SB, Roman AJ, Peden MC, Aleman TS, Boye SL, Sumaroka A, Conlon TJ, Calcedo R, Pang JJ, Erger KE, Olivares MB, Mullins CL, Swider M, Kaushal S, Feuer WJ, Iannaccone A, Fishman GA, Stone EM, Byrne BJ, Hauswirth WW (2012) Gene therapy for leber congenital amaurosis caused by RPE65 mutations: safety and efficacy in 15 children and adults followed up to 3 years. Arch Ophthalmol 130:9–24

Karali M, Manfredi A, Puppo A, Marrocco E, Gargiulo A, Allocca M, Corte MD, Rossi S, Giunti M, Bacci ML, Simonelli F, Surace EM, Banfi S and Auricchio A (2011) MicroRNA-restricted transgene expression in the retina. PLoS ONE 6:e22166

Kim VN, Mitrophanous K, Kingsman SM, Kingsman AJ (1998) Minimal requirement for a lentivirus vector based on human immunodeficiency virus type 1. J Virol 72:811–816

Kong J, Kim SR, Binley K, Pata I, Doi K, Mannik J, Zernant-Rajang J, Kan O, Iqball S, Naylor S, Sparrow JR, Gouras P, Allikmets R (2008) Correction of the disease phenotype in the mouse model of Stargardt disease by lentiviral gene therapy. Gene Ther 15:1311–1320

Kotin RM, Menninger JC, Ward DC, Berns KI (1991) Mapping and direct visualization of a region-specific viral DNA integration site on chromosome 19q13-qter. Genomics 10:831–834

Kumar-Singh R (2008) Barriers for retinal gene therapy: separating fact from fiction. Vision Res 48:1671–1680

Labrador M, Corces VG (2002) Setting the boundaries of chromatin domains and nuclear organization. Cell 111:151–154

Le Meur G Stieger K Smith AJ Weber M Deschamps JY Nivard D Mendes-Madeira A Provost N Pereon Y Cherel Y Ali RR Hamel C Moullier P Rolling F (2007) Restoration of vision in RPE65-deficient Briard dogs using an AAV serotype 4 vector that specifically targets the retinal pigmented epithelium. Gene Ther 14:292–303

Leavitt AD, Robles G, Alesandro N, Varmus HE (1996) Human immunodeficiency virus type 1 integrase mutants retain in vitro integrase activity yet fail to integrate viral DNA efficiently during infection. J Virol 70:721–728

Leung DW, Cachianes G, Kuang WJ, Goeddel DV, Ferrara N (1989) Vascular endothelial growth factor is a secreted angiogenic mitogen. Science 246:1306–1309

Lombardo A, Cesana D, Genovese P, Di Stefano B, Provasi E, Colombo DF, Neri M, Magnani Z, Cantore A, Lo Riso P, Damo M, Pello OM, Holmes MC, Gregory PD, Gritti A, Broccoli V, Bonini C, Naldini L (2011) Site-specific integration and tailoring of cassette design for sustainable gene transfer. Nat Methods 8:861–869

Lopes L, Dewannieux M, Gileadi U, Bailey R, Ikeda Y, Whittaker C, Collin MP, Cerundolo V, Tomihari M, Ariizumi K, Collins MK (2008) Immunization with a lentivector that targets tumor antigen expression to dendritic cells induces potent CD8+ and CD4 + T-cell responses. J Virol 82:86–95

Lukason M, DuFresne E, Rubin H, Pechan P, Li Q, Kim I, Kiss S, Flaxel C, Collins M, Miller J, Hauswirth W, Maclachlan T, Wadsworth S, Scaria A (2011) Inhibition of choroidal neovascularization in a nonhuman primate model by intravitreal administration of an AAV2 vector expressing a novel anti-VEGF molecule. Mol Ther 19:260–265

Maeda Y, Ikeda U, Shimpo M, Ueno S, Ogasawara Y, Urabe M, Kume A, Takizawa T, Saito T, Colosi P, Kurtzman G, Shimada K, Ozawa K (1998) Efficient gene transfer into cardiac myocytes using adeno-associated virus (AAV) vectors. J Mol Cell Cardiol 30:1341–1348

Maguire AM, Simonelli F, Pierce EA, Pugh EN, Jr., Mingozzi F, Bennicelli J, Banfi S, Marshall KA, Testa F, Surace EM, Rossi S, Lyubarsky A, Arruda VR, Konkle B, Stone E, Sun J, Jacobs J, Dell'Osso L, Hertle R, Ma JX, Redmond TM, Zhu X, Hauck B, Zelenaia O, Shindler KS, Maguire MG, Wright JF, Volpe NJ, McDonnell JW, Auricchio A, High KA, Bennett J (2008) Safety and efficacy of gene transfer for Leber's congenital amaurosis. N Engl J Med 358:2240–2248

Maguire AM, High KA, Auricchio A, Wright JF, Pierce EA, Testa F, Mingozzi F, Bennicelli JL, Ying GS, Rossi S, Fulton A, Marshall KA, Banfi S, Chung DC, Morgan JI, Hauck B, Zelenaia O, Zhu X, Raffini L, Coppieters F, De Baere E, Shindler KS, Volpe NJ, Surace EM, Acerra C, Lyubarsky A, Redmond TM, Stone E, Sun J, McDonnell JW, Leroy BP, Simonelli F, Bennett J (2009) Age-dependent effects of RPE65 gene therapy for Leber's congenital amaurosis: a phase 1 dose-escalation trial. Lancet 374:1597–1605

Maheshri N, Koerber JT, Kaspar BK, Schaffer DV (2006) Directed evolution of adeno-associated virus yields enhanced gene delivery vectors. Nat Biotechnol 24:198–204

Malhotra JD, Miao H, Zhang K, Wolfson A, Pennathur S, Pipe SW, Kaufman RJ (2008) Antioxidants reduce endoplasmic reticulum stress and improve protein secretion. Proc Natl Acad Sci U S A 105:18525–18530

Manno CS, Pierce GF, Arruda VR, Glader B, Ragni M, Rasko JJ, Ozelo MC, Hoots K, Blatt P, Konkle B, Dake M, Kaye R, Razavi M, Zajko A, Zehnder J, Rustagi PK, Nakai H, Chew A, Leonard D, Wright JF, Lessard RR, Sommer JM, Tigges M, Sabatino D, Luk A, Jiang H, Mingozzi F, Couto L, Ertl HC, High KA, Kay MA (2006) Successful transduction of liver in hemophilia by AAV-Factor IX and limitations imposed by the host immune response. Nat Med 12:342–347

Marshall E (1999) Gene therapy death prompts review of adenovirus vector. Science 286:2244–2245

McCarty DM, Monahan PE, Samulski RJ (2001) Self-complementary recombinant adeno-associated virus (scAAV) vectors promote efficient transduction independently of DNA synthesis. Gene Ther 8:1248–1254

McIntosh J, Lenting PJ, Rosales C, Lee D, Rabbanian S, Raj D, Patel N, Tuddenham EG, Christophe OD, McVey JH, Waddington S, Nienhuis AW, Gray JT, Fagone P, Mingozzi F, Zhou SZ, High KA, Cancio M, Ng CY, Zhou J, Morton CL, Davidoff AM, Nathwani AC (2013) Therapeutic levels of FVIII following a single peripheral vein administration of rAAV vector encoding a novel human factor VIII variant. Blood 121:3335–3344

Metais JY, Topp S, Doty RT, Borate B, Nguyen AD, Wolfsberg TG, Abkowitz JL, Dunbar CE (2010) Feline leukemia virus integrase and capsid packaging functions do not change the insertion profile of standard Moloney retroviral vectors. Gene Ther 17:799–804

Miao HZ, Sirachainan N, Palmer L, Kucab P, Cunningham MA, Kaufman RJ, Pipe SW (2004) Bioengineering of coagulation factor VIII for improved secretion. Blood 103:3412–3419

Miller DG, Rutledge EA, Russell DW (2002) Chromosomal effects of adeno-associated virus vector integration. Nat Genet 30:147–148

Mingozzi F, Chen Y, Edmonson SC, Zhou S, Thurlings RM, Tak PP, High KA, Vervoordeldonk MJ (2012a) Prevalence and pharmacological modulation of humoral immunity to AAV vectors in gene transfer to synovial tissue. Gene Ther 20(4):417-24.

Mingozzi F, Chen Y, Murphy SL, Edmonson SC, Tai A, Price SD, Metzger ME, Zhou S, Wright JF, Donahue RE, Dunbar CE, High KA (2012b) Pharmacological modulation of humoral immunity in a nonhuman primate model of AAV gene transfer for hemophilia B. Mol Ther 20:1410–1416

Miyoshi H, Takahashi M, Gage FH, Verma IM (1997) Stable and efficient gene transfer into the retina using an HIV-based lentiviral vector. Proc Natl Acad Sci U S A 94:10319–10323

Miyoshi H, Blomer U, Takahashi M, Gage FH, Verma IM (1998) Development of a self-inactivating lentivirus vector. J Virol 72:8150–8157

Moiseyev G, Chen Y, Takahashi Y, Wu BX, Ma JX (2005) RPE65 is the isomerohydrolase in the retinoid visual cycle. Proc Natl Acad Sci USA 102:12413–12418

Monteilhet V, Saheb S, Boutin S, Leborgne C, Veron P, Montus MF, Moullier P, Benveniste O, Masurier C (2011) A 10 patient case report on the impact of plasmapheresis upon neutralizing factors against adeno-associated virus (AAV) types 1, 2, 6, and 8. Mol Ther 19:2084–2091

Montini E, Cesana D, Schmidt M, Sanvito F, Ponzoni M, Bartholomae C, Sergi Sergi L, Benedicenti F, Ambrosi A, Di Serio C, Doglioni C, von Kalle C, Naldini L (2006) Hematopoietic stem cell gene transfer in a tumor-prone mouse model uncovers low genotoxicity of lentiviral vector integration. Nat Biotechnol 24:687–696

Murakami Y, Ikeda Y, Yonemitsu Y, Miyazaki M, Inoue M, Hasegawa M, Sueishi K, Ishibashi T (2010) Inhibition of choroidal neovascularization via brief subretinal exposure to a newly developed lentiviral vector pseudotyped with Sendai viral envelope proteins. Human Gene Ther 21:199–209

Mussolino C, Sanges D, Marrocco E, Bonetti C, Di Vicino U, Marigo V, Auricchio A, Meroni G, Surace EM (2011) Zinc-finger-based transcriptional repression of rhodopsin in a model of dominant retinitis pigmentosa. EMBO Mol Med 3:118–128

Naim C, Yerevanian A, Hajjar RJ (2013) Gene therapy for heart failure: where do we stand? Curr Cardiol Rep 15:333

Nakai H, Yant SR, Storm TA, Fuess S, Meuse L, Kay MA (2001) Extrachromosomal recombinant adeno-associated virus vector genomes are primarily responsible for stable liver transduction in vivo. J Virol 75:6969–6976

Nathwani AC, Tuddenham EG (1992) Epidemiology of coagulation disorders. Baillieres Clin Haematol 5:383–439

Nathwani AC, Gray JT, Ng CY, Zhou J, Spence Y, Waddington SN, Tuddenham EG, Kemball-Cook G, McIntosh J, Boon-Spijker M, Mertens K, Davidoff AM (2006) Self-complementary adeno-associated virus vectors containing a novel liver-specific human factor IX expression cassette enable highly efficient transduction of murine and nonhuman primate liver. Blood 107:2653–2661

Nathwani AC, Gray JT, McIntosh J, Ng CY, Zhou J, Spence Y, Cochrane M, Gray E, Tuddenham EG, Davidoff AM (2007) Safe and efficient transduction of the liver after peripheral vein infusion of self-complementary AAV vector results in stable therapeutic expression of human FIX in nonhuman primates. Blood 109:1414–1421

Nathwani AC, Tuddenham EG, Rangarajan S, Rosales C, McIntosh J, Linch DC, Chowdary P, Riddell A, Pie AJ, Harrington C, O'Beirne J, Smith K, Pasi J, Glader B, Rustagi P, Ng CY, Kay MA, Zhou J, Spence Y, Morton CL, Allay J, Coleman J, Sleep S, Cunningham JM, Srivastava D, Basner-Tschakarjan E, Mingozzi F, High KA, Gray JT, Reiss UM, Nienhuis AW, Davidoff AM (2011) Adenovirus-associated virus vector-mediated gene transfer in hemophilia B. N Engl J Med 365:2357–2365

Natkunarajah M, Trittibach P, McIntosh J, Duran Y, Barker SE, Smith AJ, Nathwani AC, Ali RR (2008) Assessment of ocular transduction using single-stranded and self-complementary recombinant adeno-associated virus serotype 2/8. Gene Ther 15:463–467

Ortiz R, Melguizo C, Prados J, Alvarez PJ, Caba O, Rodriguez-Serrano F, Hita F, Aranega A (2012) New gene therapy strategies for cancer treatment: a review of recent patents. Recent Pat Anticancer Drug Discov 7:297–312

Page KA, Landau NR, Littman DR (1990) Construction and use of a human immunodeficiency virus vector for analysis of virus infectivity. J Virol 64:5270–5276

Pang J, Cheng M, Haire SE, Barker E, Planelles V, Blanks JC (2006) Efficiency of lentiviral transduction during development in normal and rd mice. Mol Vis 12:756–767

Park F, Ohashi K, Kay MA (2003) The effect of age on hepatic gene transfer with self-inactivating lentiviral vectors in vivo. Mol Ther 8:314–323

Paskowitz DM, Greenberg KP, Yasumura D, Grimm D, Yang H, Duncan JL, Kay MA, Lavail MM, Flannery JG, Vollrath D (2007) Rapid and stable knockdown of an endogenous gene in retinal pigment epithelium. Hum Gene Ther 18:871–880

Perrault I, Rozet JM, Gerber S, Ghazi I, Leowski C, Ducroq D, Souied E, Dufier JL, Munnich A, Kaplan J (1999) Leber congenital amaurosis. Mol Genet Metab 68:200–208

Petrs-Silva H, Dinculescu A, Li Q, Min SH, Chiodo V, Pang JJ, Zhong L, Zolotukhin S, Srivastava A, Lewin AS, Hauswirth WW (2009) High-efficiency transduction of the mouse retina by tyrosine-mutant AAV serotype vectors. Mol Ther 17:463–471

Pihlmann M, Askou AL, Aagaard L, Bruun GH, Svalgaard JD, Holm-Nielsen MH, Dagnaes-Hansen F, Bek T, Mikkelsen JG, Jensen TG, Corydon TJ (2012) Adeno-associated virus-delivered polycistronic microRNA-clusters for knockdown of vascular endothelial growth factor in vivo. J Gene Med 14:328–338

Puppo A, Bello A, Manfredi A, Cesi G, Marrocco E, Della Corte M, Rossi S, Giunti M, Bacci ML, Simonelli F, Surace EM, Kobinger GP, Auricchio A (2013) Recombinant vectors based on porcine adeno-associated viral serotypes transduce the murine and pig retina. PLoS ONE 8:e59025

Rabinowitz JE, Rolling F, Li C, Conrath H, Xiao W, Xiao X, Samulski RJ (2002) Cross-packaging of a single adeno-associated virus (AAV) type 2 vector genome into multiple AAV serotypes enables transduction with broad specificity. J Virol 76:791–801

Radcliffe PA, Sion CJ, Wilkes FJ, Custard EJ, Beard GL, Kingsman SM, Mitrophanous KA (2008) Analysis of factor VIII mediated suppression of lentiviral vector titres. Gene Ther 15:289–297

Report (2002) Assessment of adenoviral vector safety and toxicity: report of the National Institutes of Health Recombinant DNA Advisory Committee. Hum Gene Ther 13:3–13

Rodriguez-Lebron E, Gouvion CM, Moore SA, Davidson BL, Paulson HL (2009) Allele-specific RNAi mitigates phenotypic progression in a transgenic model of Alzheimer's disease. Mol Ther 17:1563–1573

Sastry L, Cornetta K (2009) Detection of replication competent retrovirus and lentivirus. Methods Mol Biol 506:243–263

Schmeckpeper J, Ikeda Y, Kumar AH, Metharom P, Russell SJ, Caplice NM (2009) Lentiviral tracking of vascular differentiation in bone marrow progenitor cells. Differentiation 78:169–176

Schroder AR, Shinn P, Chen H, Berry C, Ecker JR, Bushman F (2002) HIV-1 integration in the human genome favors active genes and local hotspots. Cell 110:521–529

Seymour LW, Thrasher AJ (2012) Gene therapy matures in the clinic. Nat Biotechnol 30:588–593

Shibagaki Y, Chow SA (1997) Central core domain of retroviral integrase is responsible for target site selection. J Biol Chem 272:8361–8369

Singerman L (2009) Combination therapy using the small interfering RNA bevasiranib. Retina 29:S49–50

Sun L, Li J, Xiao X (2000) Overcoming adeno-associated virus vector size limitation through viral DNA heterodimerization. Nat Med 6:599–602

Surace EM, Auricchio A (2008) Versatility of AAV vectors for retinal gene transfer. Vision Res 48:353–359

Teramoto S, Bartlett JS, McCarty D, Xiao X, Samulski RJ, Boucher RC (1998) Factors influencing adeno-associated virus-mediated gene transfer to human cystic fibrosis airway epithelial cells: comparison with adenovirus vectors. J Virol 72:8904–8912

Testa F, Maguire AM, Rossi S, Pierce EA, Melillo P, Marshall K, Banfi S, Surace EM, Sun J, Acerra C, Wright JF, Wellman J, High KA, Auricchio A, Bennett J, Simonelli F (2013) Three-year follow-up after unilateral subretinal delivery of adeno-associated virus in patients with leber congenital amaurosis type 2. Ophthalmology 120(6):1283-91

Tschernutter M, Schlichtenbrede FC, Howe S, Balaggan KS, Munro PM, Bainbridge JW, Thrasher AJ, Smith AJ, Ali RR (2005) Long-term preservation of retinal function in the RCS rat model of retinitis pigmentosa following lentivirus-mediated gene therapy. Gene Ther 12:694–701

Vardas E, Kaleebu P, Bekker LG, Hoosen A, Chomba E, Johnson PR, Anklesaria P, Birungi J, Barin B, Boaz M, Cox J, Lehrman J, Stevens G, Gilmour J, Tarragona T, Hayes P, Lowenbein

S, Kizito E, Fast P, Heald AE, Schmidt C (2010) A phase 2 study to evaluate the safety and immunogenicity of a recombinant HIV type 1 vaccine based on adeno-associated virus. AIDS Res Hum Retroviruses 26:933–942

Wang Z, Ma HI, Li J, Sun L, Zhang J, Xiao X (2003) Rapid and highly efficient transduction by double-stranded adeno-associated virus vectors in vitro and in vivo. Gene Ther 10:2105–2111

Ward NJ, Buckley SM, Waddington SN, Vandendriessche T, Chuah MK, Nathwani AC, McIntosh J, Tuddenham EG, Kinnon C, Thrasher AJ, McVey JH (2011) Codon optimization of human factor VIII cDNAs leads to high-level expression. Blood 117:798–807

Weber M, Rabinowitz J, Provost N, Conrath H, Folliot S, Briot D, Cherel Y, Chenuaud P, Samulski J, Moullier P, Rolling F (2003) Recombinant adeno-associated virus serotype 4 mediates unique and exclusive long-term transduction of retinal pigmented epithelium in rat, dog, and nonhuman primate after subretinal delivery. Mol Ther 7:774–781

Xiao X, Li J, Samulski RJ (1998) Production of high-titer recombinant adeno-associated virus vectors in the absence of helper adenovirus. J Virol 72:2224–2232

Yang L, Bailey L, Baltimore D, Wang P (2006) Targeting lentiviral vectors to specific cell types in vivo. Proc Natl Acad Sci U S A 103:11479–11484

Yu SF, von Ruden T, Kantoff PW, Garber C, Seiberg M, Ruther U, Anderson WF, Wagner EF, Gilboa E (1986) Self-inactivating retroviral vectors designed for transfer of whole genes into mammalian cells. Proc Natl Acad Sci U S A 83:3194–3198

Yuan MK, Tao Y, Yu WZ, Kai W, Jiang YR (2010) Lentivirus-mediated RNA interference of vascular endothelial growth factor in monkey eyes with iris neovascularization. Mol Vis 16:1743–1753

Zennou V, Serguera C, Sarkis C, Colin P, Perret E, Mallet J, Charneau P (2001) The HIV-1 DNA flap stimulates HIV vector-mediated cell transduction in the brain. Nat Biotechnol 19:446–450

Zhong L, Li B, Mah CS, Govindasamy L, Agbandje-McKenna M, Cooper M, Herzog RW, Zolotukhin I, Warrington KH, Jr., Weigel-Van Aken KA, Hobbs JA, Zolotukhin S, Muzyczka N, Srivastava A (2008) Next generation of adeno-associated virus 2 vectors: point mutations in tyrosines lead to high-efficiency transduction at lower doses. Proc Natl Acad Sci U S A 105:7827–7832

Zincarelli C, Soltys S, Rengo G, Rabinowitz JE (2008) Analysis of AAV serotypes 1–9 mediated gene expression and tropism in mice after systemic injection. Mol Ther 16:1073–1080

Zufferey R, Nagy D, Mandel RJ, Naldini L, Trono D (1997) Multiply attenuated lentiviral vector achieves efficient gene delivery in vivo. Nat Biotechnol 15:871–875

Zufferey R, Donello JE, Trono D, Hope TJ (1999) Woodchuck hepatitis virus posttranscriptional regulatory element enhances expression of transgenes delivered by retroviral vectors. J Virol 73:2886–2892

Chapter 4
Nonviral Gene Therapy—The Challenge of Mobilizing DNA

Jacob Giehm Mikkelsen

4.1 Introduction

Through millions of years viruses have swept through populations. Devastating epidemics have roared and shaped the history of mankind and viruses have evolved and exploited the unique capacity to transport genetic material from individual to individual and from cell to cell. As a result, genetic cargo of ancient viruses has accumulated in the germline and shaped the human genome in ways that we are slowly beginning to appreciate (Kunarso et al. 2010). Along with the fundamental understanding of virus replication and spread, virus invasions have taught us the basic principles of gene transfer. Pioneering studies from the early 1980s showed that viruses could be engineered to carry heterologous genetic material. By providing the viral proteins necessary for production of retrovirus particles *in trans*, such helper-free viruses were able to incorporate heterologous RNA carrying the secondary RNA motifs for proper packaging into assembled virus particles. Such findings laid the foundation for the concept of therapeutic gene transfer (Friedmann and Roblin 1972; Anderson 1984; Blaese et al. 1995) and initiated a now 30-year long quest for development and refinement of safe and efficient vector systems for gene therapy treatment of genetic disorders. Years of research with ups and many downs are now finally bearing fruit, and the gene therapy community has recently celebrated successes with effective therapies for serious genetic conditions such as blindness caused by inherited retinal disorders (Bainbridge et al. 2008; Maguire et al. 2009), immunodeficiency syndromes (Aiuti et al. 2009; Cartier et al. 2009; Hacein-Bey-Abina et al. 2010), lipoprotein lipase deficiency (Gaudet et al. 2012), and hemophilia (Nathwani et al. 2011).

Viruses are structured, chemical entities consisting of nucleic acids wrapped in a core of viral proteins. A lipid envelope derived from the former host cell may for some viruses surround the protein core. This design is evolutionarily adapted

J. G. Mikkelsen (✉)
Department of Biomedicine, Faculty of Health, Aarhus University, Wilh. Meyers Allé 4, 8000 Aarhus C, Denmark
e-mail: giehm@biomed.au.dk

X.-Q. Li et al. (eds.), *Somatic Genome Manipulation*,
DOI 10.1007/978-1-4939-2389-2_4

to confronting the many barriers that a virus encounters on its way from the external environment to the nuclei of host cells. The viral package not only offers protection of the viral nucleic acids within the host, but also allows trafficking within extracellular matrices and facilitates host cell specificity due to interactions between proteins embedded in the cell membrane and proteins distributed on the surface of the virus. In addition, viruses traverse the first cellular barrier, the plasma membrane, though a variety of routes including endocytosis and fusion with the membrane, leading to internalization and release of the virus core (Klasse 2012). The next barrier, the highly viscous and crowded cytoplasm is bypassed by some viruses by active trafficking along the network of microtubules (Matarrese and Malorni 2005), whereas the third major barrier, the nuclear membrane, is overcome by access through the nuclear pores (Kobiler et al. 2012). Even inside the nucleus, viruses may possess mechanisms that allow genomic attraction of the viral DNA, facilitating the insertion of viral genetic cargo into transcriptionally active chromatin (Mitchell et al. 2004).

Such properties have promoted the development of efficient virus-derived gene vectors, and vectors, based on gamma-retroviruses, lentiviruses, adenoviruses, and adeno-associated viruses, are being heavily scrutinized for their applicability in the treatment of a wide variety of diseases. Chapter 3 in this book covers the progress and success of viral vector technologies. As the broad clinical use of viral vectors seems to emerge ahead of us, and viruses are turning into drugs, why should we bother pondering on alternative ways to achieve therapeutic gene transfer? Well, the truth is that viral vectors may have drawbacks that can be avoided with approaches that are not based on virus-based gene transportation. Most importantly is perhaps the risk of a potentially strong immune response induced by injection of a large dose of viral particles and for some applications the oncogenic potential due to insertional mutagenesis, but high production costs and challenges related to large-scale manufacturing of viral vectors are other main reasons that nonviral gene delivery strategies continue to draw attention in the gene therapy community.

In its most primitive form, a nonviral vector system consists of naked plasmid DNA carrying a promoter driving the expression of a therapeutic gene. Given the challenges that have been fought by viruses over millions of years, it does not come as a surprise that naked DNA transfer is rather inefficient and requires additional action to facilitate nuclear DNA uptake. Such actions may serve to increase cell permeability, for example by exposing target cells to an externally applied electrical field or high pressure. Other strategies aim at mimicking viral gene transfer by engineering synthetic particles carrying the DNA, whereas yet other strategies are based on the cellular uptake of DNA-conjugated nanoparticles. The term 'gene therapy' refers in the narrowest definition to the delivery of a therapeutic gene to cells in order to change their functional properties and treat disease. In this chapter, the current status of nonviral gene transfer is reviewed, with primary focus on gene-encoding DNA. Efforts to deliver DNA oligonucleotides, locked nucleic acids (LNA), and synthetic RNAs, like small interfering RNA, will not be covered here, although such methods may certainly fit within a broader definition of gene therapy. Rather, based on recent findings, the focus will be primarily on the DNA itself and a description of how DNA, as the source of the therapeutic gene in nonviral delivery approaches, undergoes changes that may optimize future performance in terms of safety and efficiency. Also, with focus on the adaptation of DNA transposable elements for genomic gene insertion, emphasis will be put on novel gene-insertion technologies, which have successfully addressed the lack of gene expression persistency, typically considered one of the major drawbacks of nonviral gene therapy. A short glimpse into current clinical trials involving nonviral gene delivery is given towards the end of the chapter.

4.2 Gene Delivery by Naked Plasmid DNA—as Primitive as It Gets

'At the risk of adding to this list, I propose *plasmid* as a generic term for any extrachrosomal hereditary determinant' is the quote that Joshua Lederberg used when he in 1952 introduced the term "plasmid", referring to a long list of terms that had previously been used to describe extra-chromosomal genetic particles and the concept of cytoplasmic heredity (Lederberg 1952). The definition of plasmid DNA was later refined to describe ring-shaped, double-stranded DNA molecules that replicate independently of chromosomal DNA. Obviously, such circular DNA molecules, replicated in and purified from bacteria, have played a fundamental role in the development of molecular cloning and serve on a daily basis as an invaluable tool in the hands of students and researchers world-wide. Plasmid DNA is the backbone of today's highly advanced methods of genetic engineering and represents the simple key constituent of most nonviral gene vehicles.

4.2.1 Plasmids—the Nonviral Gene Vehicle

Plasmids are capable of replicating autonomously in bacteria. Each plasmid molecule carries an origin of replication which constitutes the initiation site for replication of the plasmid and serves to regulate the number of plasmids within each bacterial cell. In a natural setting, plasmids may typically provide bacteria a selective advantage. Such plasmid-encoded bacterial resistance factors, originally referred to as R factors (Datta and Kontomichalou 1965), are transferable between bacteria and include proteins conferring resistance to antibacterial drugs like neomycin, kanamycin, and penicillins including ampicillin (Sutcliffe 1978). By including such antibiotic resistance genes, like the ampicillin resistance gene (Amp^R), in the plasmid backbone, a single species of plasmid DNA may be selectively amplified in bacteria, usually *Escherichia coli*, grown in ampicillin-containing growth medium, resulting in high total yields of plasmid DNA. Plasmids are harvested from lysed bacteria by selective alkaline denaturation of chromosomal DNA (Birnboim and Doly 1979), leading upon neutralization with sodium acetate to DNA renaturation and formation of insoluble aggregates of high molecular weight DNA. Complexes of protein and SDS as well as high-molecular weight RNAs are also precipitated under these conditions, whereas renatured covalently closed plasmid DNA stays in solution and may be harvested subsequently by ethanol precipitation or column purification. Lipopolysaccharides (LPS), often referred to as endotoxins, are present in high numbers in the outer bacterial membrane and are shed to the surrounding supernatant during growth and lysis. Due to the size, negative charge, and chemical structure of LPS, these molecules are co-purified with plasmid DNA and are frequent contaminants of plasmid preparations. It is well-known that LPS trigger an inflammatory response and therefore must be removed prior to the use of plasmids in sensitive experimentation in cultured cells or for in vivo applications like gene therapy. Modern purification methods therefore include a removal buffer that prevents LPS from binding to the resin in plasmid DNA purification columns, leading to preparations of DNA with almost negligible levels of LPS.

Plasmid DNA is a macromolecule with a relative high molecular weight. Despite its defined size in terms of the number of basepairs (typically in the range of 5–10 kb dependent on the specific application), even a pure preparation of plasmid DNA may be heterologous due to different conformations of the plasmid. Covalently closed, supercoiled DNA is the most condensed form in which the DNA double helix coil is further coiled to form a twisted conformation of the plasmid (Fig. 4.1a). Excessive alkaline treatment during plasmid DNA preparation may cause increased denaturation of supercoiled DNA, which is usually not desired. Also, during plasmid extraction and preparation, single-stranded nicks may occur, leading to unwinding of plasmid and formation of a relaxed open circular conformation (Fig. 4.1b). Alternatively, plasmids may adapt a relaxed conformation by enzymatic removal of supercoils leading to an unwound circular form with both DNA strands uncut (Fig. 4.1c). Double-stranded DNA breaks may also occur, leading to a linear conformation without supercoiling (Fig. 4.1d).

Plasmid DNA can be engineered to express any gene of interest and even several genes if this is relevant for the specific application. In its most primitive form, a gene-carrying plasmid contains a simple expression cassette consisting of a ubiquitously expressed promoter flanked downstream by an intron-devoid open reading frame, derived from cDNA, and a polyadenylation sequence (Fig. 4.1, upper right). However, the sky is the limit when it comes to the creativity of designing gene-expressing plasmids, and it is beyond the scope of this chapter to review all possibilities. Nevertheless, an almost infinite list of inducible and cell- and tissue-regulatable expression strategies has been employed to obtain optimized expression strength, increased RNA stability and processing, spatiotemporal control and so on. Based on successful plasmid construction and purification, the researcher is faced with the challenges of mobilizing plasmid DNA in cultured cells and perhaps even in live individuals. Some of the main barriers for in vivo delivery of plasmid DNA are discussed here (see Fig. 4.2 for a schematic representation).

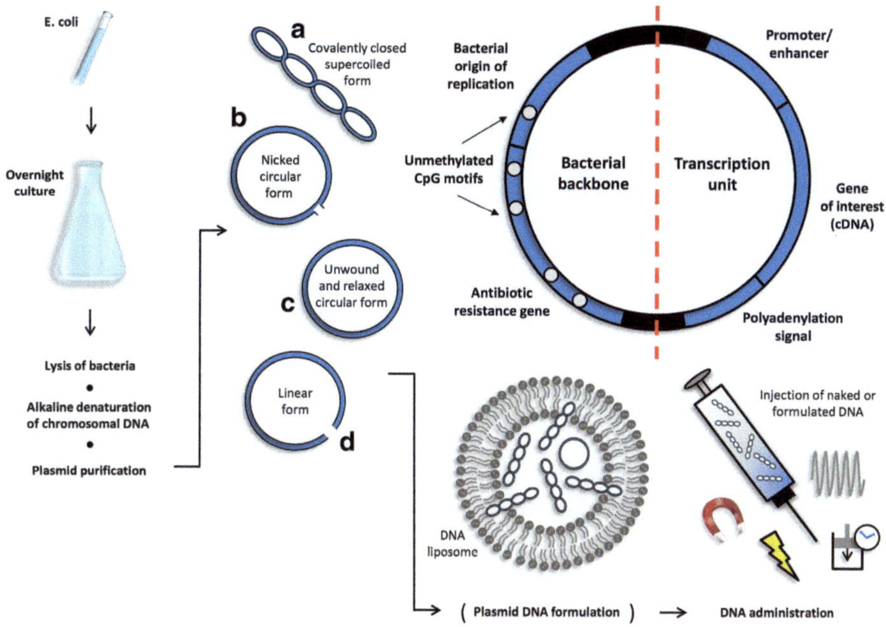

Fig. 4.1 Plasmid DNA—the nonviral gene vehicle. The figure gives a schematic representation of some of the most important features of plasmid DNA from basics of production (*left*) to different structural forms (*middle*, **a–d**), and typical content of a standard plasmid utilized for gene delivery (*upper right*). Plasmid DNA can be formulated (a DNA-containing liposome is shown) and administered as either naked or formulated DNA. DNA uptake can be facilitated by treating target tissues by magnetic force (magnetofection), ultrasound (sonoporation), electric pulses (electroporation), or high-pressure (hydrodynamic injection), as indicated by the symbols (*lower right*)

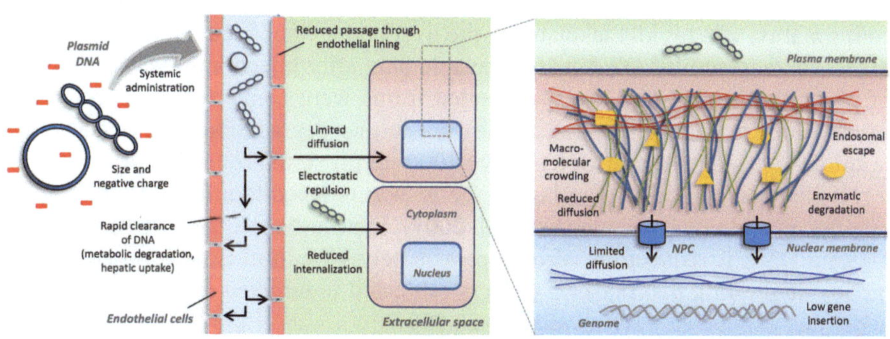

Fig. 4.2 The major barriers to plasmid-based gene transfer. The barriers to systemically administered plasmid DNA are depicted. Primary challenges to the in vivo mobilization of plasmid DNA include barriers within the capillaries (*left*), within tissues and extracellular space (*middle*), and within cells (*right*). The right panel presents the concept of macromolecular crowding caused by the dense network of microtubules (*blue lines*), intermediate filaments (*green lines*), and actin filaments (*red lines*). The nuclear matrix within the nucleus is represented by thin blue lines

4.2.2 *Getting Plasmids To the Nucleus—Barriers On the Way to the Checkered Flag*

Mobilization of plasmid DNA in a naked form is complicated by the fact that DNA is negatively charged and that molecules of plasmid DNA therefore represent a strong anionic charge. Due to the negative charge on the surface of cells, plasmid DNA is naturally repelled from the cell surface by electrostatic repulsion leading to limited DNA uptake. Also, following conventional intravenous administration in mice, unformulated plasmid DNA is rapidly cleared from the blood stream, at least partially due to metabolic degradation due to hepatic uptake and clearance (Kawabata et al. 1995) (Fig. 4.2, left). Such clearance is based on the uptake of plasmid DNA by nonparenchymal cells, like endothelial and Kuppfer cells, by mechanisms involving the recognition of anions by scavenger receptor-like surface proteins. Plasmid DNA entering the systemic circulation has direct access to capillary endothelial cells and needs to traverse the endothelial lining to get access to tissues. In many tissues, the layer of endothelial cells and underlying structures of the capillaries do not allow passage of macromolecules like plasmid DNA, whereas a few tissues, such as the liver, contain capillaries that are characterized by a permeable endothelial lining with gaps that allow passage of DNA molecules into the tissue. Hence, plasmid DNA uptake is largely restricted to endothelial cells, circulating blood cells, and cells in the liver and spleen. The resulting degradation of plasmid DNA, for example in Kuppfer cells in the liver, results in only negligible gene expression following a standard intravenous injection of naked plasmid DNA (Liu et al. 1999). Hence, the overall pharmacokinetic properties are limited by the size and charge of plasmid DNA, rendering the transport and access to tissues of interest rather inefficient. Notably, in a study based on intravenous plasmid injection in rats, it was observed that supercoiled plasmid, relative to the other plasmid conformations, was rapidly cleared from the circulation, suggesting that this more compact conformation was rapidly taken up by the liver and metabolized (Houk et al. 2001). Moreover, supercoiled DNA was detectable only after injection of very high doses of plasmid DNA (2 mg), and supercoiled forms were converted to open circular forms which were then linearized, allowing further degradation. In a related study in mice, levels of plasmid DNA above 1000 ng/ml serum, obtained 30 s after intravenous injection of 50 μg plasmid DNA, was reduced to a level of about 3 ng/ml after an additional 30 min (Oh et al. 2001). In accordance, the half-life of naked plasmid DNA is reduced in vitro (approximately 10 min in isolated mouse whole blood), suggesting that plasmid DNA is subjected to targeted degradation by both endo- and exonucleases (Kawabata et al. 1995).

The difficulties for plasmid DNA on the journey into cells are far from over in case the DNA should in fact live long enough to encounter and attach to a cell. In fact, the challenges may have just begun, since the plasmid DNA must now face the cellular barriers in shape of the cellular and nuclear membranes and not least the crowded cytoplasm (Fig. 4.2, right panel). Cellular internalization of naked DNA across the cell membrane is inefficient but may occur through endocytosis and subsequent escape from endosomes into the cytoplasm. However, diffusion of DNA is highly restricted in the cytoplasm. Whereas mobility of small DNA fragments (<250 bp) are only mildly hindered, larger plasmid-size DNA molecules (>2000 bp) are increasingly impeded with increasing size (Lukacs et al. 2000). Studies based on measurements of the cytoplasmic mobilization of fluorophore-labelled DNA molecules suggested that size-dependent reduction of diffusion is caused by macromolecular crowding in the cytoplasm and that mainly fixed structures, such as the highly cross-linked actin cytoskeleton, trap plasmid DNA and, hence, represent a major barrier to mobilization through the cytoplasm from the endosome to nuclear pores (Dauty and Verkman 2005). In fact, it seems unlikely that plasmid DNA is able to move through the cytoplasm by diffusion, and directed trafficking along the cytoskeleton is a more likely scenario (Vaughan et al. 2006). It is well-known that viruses travel through the cytoplasm along the network of microtubules (McDonald et al. 2002). In experiments based on microinjection of GFP-expressing plasmids into cells in culture, plasmid trafficking, and resulting expression of GFP, was found to rely on the microtubule network. Hence, the expression of GFP decreased in cells in which the network was disrupted (Vaughan and Dean 2006). Moreover, the molecular motor protein dynein was identified as a player in facilitating DNA movement. Also, cytoplasmic DNA transport was found to benefit from the association of plasmid DNA with proteins containing a nuclear localization signal (NLS) (Mesika et al. 2005). This interaction not only facilitates nuclear entry but also mediates transport of protein-DNA complexes along microtubules in a dynein-dependent manner and in a fashion that seems to involve NLS-directed recognition of the microtubule-interacting importins (Mesika et al. 2005). Interestingly, recent evidence suggests that transcription factor binding sites within plasmids may support their cytoplasmic trafficking. Hence, binding of cyclic AMP response-element binding protein (CREB) to sites that are present in the commonly used cytomegalovirus (CMV) promoter was crucial for interaction with microtubules and enhanced movement of plasmid DNA through the cytoplasm (Badding et al. 2012). A proteomics approach used to identify components of plasmid-trafficking complexes identified microtubule-associated motor proteins, including dynein and importin β1, as key players in plasmid transport through the cytoplasm (Badding et al. 2013), supporting the notion that nonviral gene delivery relies on active intracellular DNA trafficking facilitated by sequence-directed complex formation between plasmid DNA and cytoplasmic proteins (Lam and Dean 2010). Despite active transport through the cytoplasm, plasmid DNA is still sensitive to enzymatic degradation in the cytoplasm, resulting in a relatively short half-life, which by estimations range from 1 to 5 h (reviewed in (Vaughan et al. 2006)) with suspected variations between plasmids and cell types.

The nuclear membrane is more or less impermeable to plasmid DNA and represents the final solid barrier for incoming plasmid DNA. Hence, a large percentage of the plasmid DNA is stuck in the cytoplasm and never gains access to the nucleus (Vaughan et al. 2006). It is well-known that the nuclear uptake of DNA is increased in actively dividing cells due to the disassembly of the nuclear membrane. Hence, plasmid DNA may gain easier access to and transiently accumulate in relatively high numbers in the nuclei of cells in a dividing cell population. Still, exogenous plasmid DNA may be able to enter the nucleus in non-dividing, terminally differentiated cells. This transport across the nuclear membrane is facilitated by nuclear pore complexes (NPCs), which are membrane-penetrating, aqueous channels consisting of multiple copies of approximately thirty different proteins. The structure of those large channel complexes is largely conserved, but heterogeneity in composition and function is evident in different cell types and tissues (see (Raices and D'Angelo 2012) for review). Due to the size, intruding plasmid DNA is not likely to diffuse through NPCs in a signal-independent fashion and will rely on facilitated import through the pore. To be ferried across the nuclear membrane, larger molecules must interact with ferry proteins—exportins or importins—that pull the cargo through the membrane. For import, importin-α recognizes a specific amino acid sequence—the nuclear localization signal (NLS)—on the cargo protein and allows also importin-β to become part of the complex that is subsequently carried through the pore. Once inside the nucleus, the complex is disassembled, leading to release of the cargo molecule, by the actions of a small GTPase called Ran. RanGTP facilitates the displacement of importins from the complex and forms RanGTP-importin complexes that are subsequently shuttled back to the cytoplasm. Upon arrival in the cytoplasm, RanGTP is hydrolyzed to RanGDP leading to release of the importins.

Importin-based nuclear import of plasmid DNA requires association of the DNA with NLS-containing proteins that may facilitate the binding to importin-α. In the cytoplasm, plasmid DNA becomes complexed with cellular proteins (e.g. transcription factors) that may play a role in the attraction towards NPCs. Obviously, such interactions will depend on sequences in the plasmid, and several sequences, including the frequently used SV40 enhancer (reviewed in (Vaughan et al. 2006), are known to facilitate trafficking through the pore most likely by binding NLS-containing transcription factors.

4.2.3 Expressing Genes from Naked Plasmid DNA—Crossing the Barriers with Force

Despite the many barriers to delivery of genes carried on naked plasmid DNA, successful gene transfer has indeed been achieved in vivo with naked DNA. However, as an overall trend, this kind of gene delivery seems to depend on the application of extra physical force to the target cells. As indicated in Fig. 4.1, such force can be achieved by subjecting cells or tissues to an externally applied electrical field (electroporation), sound of ultrasonic frequencies (sonoporation), or a magnetic field (magnetofection). Also, the application of high pressure by a controlled hydrodynamic pressure in capillaries (hydrodynamic gene delivery) facilitates increased permeability of the endothelial lining and parenchymal cells.

The first evidence of in vivo gene expression achieved by administration of naked plasmid DNA was provided by Wolff and coworkers who injected plasmid DNA into skeletal muscle and observed subsequent expression of marker genes in transfected myofibers (Wolff et al. 1990). Gene expression was achieved in later studies by injection of plasmid DNA into other tissues, including liver (Wolff 1997), heart (Acsadi et al. 1991; Buttrick et al. 1992), skin (Hengge et al. 1996), and the thyroid gland (Sikes et al. 1994). However, the efficiency of this method was relatively low resulting at best in short-term gene expression without any clear clinical perspective. Budker and coworkers subsequently reported substantial levels of gene expression in skeletal muscle after rapid injection of a plasmid DNA solution into the femoral artery and simultaneous occlusion of blood vessels leading into and out of the hindlimb (Budker et al. 1998). This finding laid the foundation for a gene delivery strategy based on applying a controlled hydrodynamic pressure in the capillaries. Soon after, reports by Liu et al. (1999) and Zhang et al. (1999) established the hydrodynamics-based procedure for delivery of plasmid DNA to mouse liver. By rapid injection of a large volume of plasmid DNA (2 ml, equivalent to 8–10 % of the body weight) through the tail vein, a high level of gene expression could be measured in the liver, resulting in a peak of expression approximately 8 h after injection. Although gene expression would subsequently decline, the absence of cell division among normal hepatocytes allowed gene expression to prevail at a reasonable level due to the persistence of plasmid DNA in the transfected cells. Also, peak levels of expression could be re-established by repeated injection of plasmid DNA. By transferring plasmid encoding β-galactosidase, it was estimated in the early studies that as many as 40 % of the hepatocytes were transfected by this method (Liu et al. 1999). However, this transfection rate may certainly depend on the amount of plasmid, its size and sequence, the mouse strain, and the experience and skills of the researcher performing the injection. Nevertheless, this approach for systemic administration of plasmid DNA established the mouse liver as a preferred model system for nonviral gene delivery, and numerous studies have benefitted from this technique allowing in vivo insight into the uptake of plasmid DNA and the mobilization of DNA inside cells. As we shall see later, the establishment of DNA transposon-based gene vehicles was one of numerous technologies that developed through work in hydrodynamically injected mice. Importantly, in several cases hydrodynamic plasmid injections were found leading to long-term expression of genes with therapeutic applicability. Hence, injected naked plasmid DNA encoding α1 antitrypsin and human factor IX resulted in levels of protein that reached therapeutic levels (Miao et al. 2000).

Hydrodynamic gene delivery exploits the close association between the capillary endothelium and parenchymal cells. This allows direct access of the DNA to surrounding tissues during rupture of the endothelial barrier in thin-layered capillaries. High pressure does not only induce breakage of the endothelial lining, but also causes transient hyperpermeability of the surrounding cells (Suda and Liu 2007) supporting a mechanism that involves hydroporation (Zhang et al. 2004). The structure and architecture of the capillaries and the surrounding tissue have major impact on the efficiency of the process. In liver, it is believed that the large volume injected into the mouse in a period of 5–7 s enters directly into the inferior vena cava and thereby stretches heart muscle fibers to an extent that leads to cardiac congestion. This causes the accumulation of the injected solution in the inferior vena cava, and an elevated intravascular pressure drives the DNA solution into the liver in retrograde via the hepatic vein while the existing blood in the liver is pushed backward toward the portal vein. Inside the liver, the elevated pressure expands the sinusoids and concomitantly causes membrane defects in hepatocytes allowing subsequent DNA uptake. The procedure—carefully reviewed and modeled in Suda and Liu (2007) and Bonamassa et al. (2011)—causes temporary cardiac dysfunction as well as liver toxicity due to liver expansion and structural deformation. The enlarged liver returns to its original size approximately 30 min after injection, and hepatocyte membrane defects are reportedly 're-sealed' a few minutes after the pressure impact (Bonamassa et al. 2011).

Hydrodynamic plasmid DNA delivery has been performed in a variety of species and tissues using different routes of administration. It is beyond the scope here to review these studies, some of which are listed in a review by Suda and Liu (Suda and Liu 2007). Nevertheless, although hydrodynamic DNA injection was for some time considered a convenient—but not necessarily a clinically relevant—model, researchers in the field have strived to advance the technology for use in larger animals and humans. By injection of plasmid DNA encoding the lacZ transgene into the femoral artery of pigs, Danialou and colleagues were able to balance the pressure in the skeletal muscle providing optimized conditions for the injection. As a result, more than 20 % of the muscle fibers in the pig hind limb were transfected by this procedure (Danialou et al. 2005). Encouraged by such findings, new devices are being developed to assist reproducible DNA delivery. Automatic computer-controlled hydrodynamic was achieved with a new tool, designated the HydroJector-EM, that allows reproducible effect in mouse liver (Yokoo et al. 2013). A related injection device promises to mimic the effects of conventional, manual methods and was found to assist delivery of plasmid DNA encoding the luciferase reporter gene to pig liver through the lateral hepatic vein in combination with image-guided catheterization and complete occlusion (Suda et al. 2008). Previous studies suggested that only about 2 % of the pig hepatocytes were transfected by this type of procedure (Kamimura et al. 2009). However, by adapting the method for gene transfer to skeletal muscle of pig, Kamimura and coworkers were able to transfect 70–90 % of the muscle groups in the hind limb leading to persistent reporter gene expression (Kamimura et al. 2010). Hydrodynamic injection into the limb vein was established also for forced delivery of naked plasmid to the myofibers of nonhuman primates (Hegge et al. 2010). By a single hydrodynamic limb vein injection, Wooddell and coworkers were able to transfect an average of 10 % of all myofibers in the targeted muscle of the arms and legs (Wooddell et al. 2011), providing hope that this method may have clinical relevance. Despite such promising findings, it is still the overall impression that hydrodynamic injection may need further optimization for effective and safe use in humans. In attempts to scale-up the technology, several groups have experienced disappointing results, and it has been suggested that the impulse, rather than the pressure, is a critical determinant for the potency of hydrodynamic delivery (Hackett et al. 2011).

The application of an electrical field to cells and tissues represents an alternative strategy for facilitating forced DNA delivery into cells. Short and intense electrical pulses that may last from a few microseconds to a millisecond disturb the phospholipid bilayer of the cell membrane and trigger a transient increase in the permeability of the cell membrane. Along with the formation of temporary pores, the electrical potential across the membrane assists in mobilizing the negatively charged plasmid DNA, allowing the influx of plasmid DNA from the surroundings as well as enhanced intracellular mobilization and transport across the nuclear membrane. Plasmid DNA is thought to enter the cell by a two-step mechanism by which the pulse facilitates a direct interaction between the DNA molecules and the destabilized membrane prior to passage of plasmid DNA across the membrane (Escoffre et al. 2009).

Electroporation of cultured cells and bacteria has been used as a routine technique for many years and is now playing an increasingly dominant role in in vivo nonviral gene transfer in relation to DNA-based vaccination and gene therapies. So far, the majority of clinical trials based on DNA delivery by electrotransfer addresses either anticancer applications by DNA delivery into tumors or the use of DNA as antiviral vaccines (Bodles-Brakhop et al. 2009). However, for gene therapy purposes, electrotransfer-mediated gene delivery to muscle seems particularly promising (see Hojman (2010) for review). The application of electric pulses to muscles of mice injected with naked plasmid DNA was first shown by Aihara and Miyazaki to promote enhanced transgene expression (Aihara and Miyazaki 1998). By optimizing the regimens of pulses, gene transfer to muscle has been further refined allowing only minimal perturbation of cell physiology (Hojman et al. 2008). Using mouse muscle as a target tissue, therapeutic delivery of a gene expression cassette encoding erythropoietin (Epo) was achieved by a single electrotransfer-supported injection of only 500 ng of plasmid DNA (Hojman et al. 2007). Although promising, these findings were not directly mirrored by studies of electrotransfer of plasmid DNA to the gluteal muscle in the pig (Spanggaard et al. 2012). Hence, using luciferase as a reporter, gene electrotransfer was found to facilitate localized gene expression only in an area delineated by the electrodes. Nevertheless, the impact on cells facilitated by an electrical field is an example of how physical force may aid the uptake of naked DNA with a substantial therapeutic potential for local gene delivery, for example into tumors (Andre et al. 2008).

4.3 Delivery of Plasmid DNA Wrapped in Synthetic Carrier Materials—Bypassing Barriers

Although physical force may in some selected cases be applicable for in vivo delivery of naked plasmid DNA, alternative strategies may be needed to ferry foreign genetic material to a specific tissue and cellular target. With inspiration from the world of viruses, many attempts have been made to wrap negatively charged high molecular weight plasmid DNA in synthetic, chemical coats that may to some degree mimic the composition of viruses. Many different types of liposomal and polymeric DNA delivery systems have been explored for transfection of cells both in vitro and in vivo. Such nonviral delivery platforms are efficient in many culture systems, and the development of effective DNA transfection reagents has become a profitable business based on the daily need of potent DNA transfection in laboratories world-wide. For the same reason, the exact chemistry of such commercial lipid- and cationic polymer-based reagents is often a well-kept secret. Nevertheless, the chemistry related to in vivo delivery of nucleic acids is a growing independent research area which has been thoroughly covered in previous reviews (Ewert et al. 2010; Zhu and Mahato 2010; Kang et al. 2012). With reference to these reviews, some of the basic principles of lipid- and polymer-based delivery strategies are mentioned here.

Nonviral DNA formulations are based on the electrostatic interactions between the cationic carrier material and the anionic plasmid DNA. In the chemical complex formed around the plasmid DNA as a scaffold, the negatively charged DNA is masked and either encapsulated or condensed (Xu et al. 1999), allowing easier in vivo mobilization due to charge changes and a significant reduction in molecular volume. Also, encapsulated plasmid DNA is protected against nucleolytic degradation. Lipids used for the DNA complex formation consist of a cationic head group, which is linked by a spacer to a tail of hydrophobic hydrocarbons (e.g fatty acids or cholesterol derivatives). See for example Nguyen and Szoka (2012) for graphic representation. The head group is responsible for electrostatic interactions with the DNA and the cell membrane, whereas the spacer contributes to stability and biodegradability of the DNA-complexed macromolecule. Plasmid DNA complexed with the cationic lipid GL67A is currently being used in the only active clinical trial for cystic fibrosis (Alton et al. 2013). These positively charged fatty liposomes are relatively stable and able to withstand the aerosolisation process that drives the delivery of the nonviral vector to the lung epithelium of cystic fibrosis patients (Griesenbach and Alton 2013).

Like lipids, polycationic polymers electrostatically complex with negatively charged plasmid DNA. The reduced size and net cationic surface charge of DNA-polymer complexes facilitate cellular entry. Furthermore, high flexibility of polymers supports strong molecular interactions with the plasmid and provides efficient protection against enzymatic degradation. However, in vivo use of such complexes may be restricted by toxic effects caused by polymer-induced aggregation of erythrocytes and complex formation with anionic blood proteins. The most frequently used polymers include polylysine and polyethyleneimine (PEI) which are both available in linear and branched versions (Kadlecova et al. 2013). Highly branched polymers provide increased flexibility and stable complex formation but are thought also to display higher cytotoxic effects due to their increased affinity for the cell membrane leading to damage of the membrane (Fischer et al. 2003; Kadlecova et al. 2012).

The mechanisms of lipid- and polymer-based gene transfer are vaguely understood, but lipoplexes and polyplexes are thought to enter cells mainly through endocytosis. During this process, part of the cell membrane engulfs the macromolecular DNA complex bound on the cell surface. The intracellular vesicle—often designated the endocytic vesicle—serves as an intracellular distribution center, allowing distribution to various intracellular destinations. Escape from endocytic vesicles and early endosomes is a crucial factor for further transport of the DNA towards the nucleus. Means of promoting escape from endosomal entraptment by the inclusion of endosomal escape devices include fusogenic lipids and membrane-disruptive peptides and polymers. These and related strategies are carefully reviewed in El-Sayed et al. (2009). The fate of plasmid DNA that enters cells through endocytic uptake may depend on details of the uptake mechanism. Hence, different endocytic pathways are known and classified according to the regions of the membrane that become part of the endocytic vesicle. Some areas of the cell membrane, lipid raft domains, are characterized by an ordered structure composed of organized clusters of lipids. However, the membrane is not uniform, and lipid raft domains exist in a generally quite disordered milieu of lipids. Endocytosis outside the lipid raft areas may occur through chathrin-mediated endocytosis, phagocytosis, or macropinocytosis, whereas several different mechanisms may facilitate DNA uptake in lipid raft domains, as reviewed in El-Sayed and Harashima (2013). Although the importance of these different pathways for intracellular distribution of plasmid DNA is far from fully understood, it seems reasonable to believe that the intracellular fate of plasmid DNA can be further controlled in new vector designs capitalizing on increased understanding of the processes of endocytosis.

4.4 Safer and More Efficient Gene Transfer by Minicircle DNA—Getting Rid of Bacterial DNA

Until recently, plasmid DNA has been the key component of vectors for nonviral gene delivery, but this status may be changing. As the growth and amplification of plasmid DNA in bacteria requires bacterial genes and elements situated in the plasmid backbone, such sequence stowaways are automatically ferried into plasmid-treated cells and may potentially compromise the safety of plasmid-based gene delivery. Risks include an unintended immune response evoked by the expression of antibiotic resistance genes and potential transmission of antibiotic resistance genes to host bacteria via horizontal gene transfer. For these reasons, the regulatory agencies recommend avoiding the use of antibiotic resistance markers (Mayrhofer et al. 2009; Vandermeulen et al. 2011). Moreover, unmethylated CpG motifs are frequently present in the plasmid backbone and may activate the innate immune system through recognition by the Toll-like receptor 9 (TLR9), the receptor that recognizes immunostimulatory CpG motifs, in antigen-presenting cells (Krieg et al. 1995; Hemmi et al. 2000). Indeed, such inflammatory response was reduced in TLR9-deficient transgenic mice (Zhao et al. 2004). CpG-triggered immune responses may result in the loss of cells carrying the plasmid (Yew et al. 1999; Hodges et al. 2004; Yew and Cheng 2004), and plasmids with a reduced number of CpG motifs are less immunogenic and confer long-term gene expression with less toxicity (Yew et al. 2002; Hyde et al. 2008).

Pioneering work by Chen and co-workers demonstrated that linear DNA devoid of the bacterial backbone induced higher levels of gene expression in mouse liver than was obtained with ordinary plasmid DNA (Chen et al. 2001). In a systematic study of a panel of expression cassettes, the general pattern was that animals receiving the purified expression cassette expressed persistently higher levels of the transgene than animals treated with plasmid DNA (Chen et al. 2004). These authors also observed that the linear DNA concatamerized after mouse liver transfection by hydrodynamic injection through the tail vein and that some of the linear fragments formed small circles upon uptake in the liver (Chen et al. 2001). These findings gave a first indication that the linkage between the expression cassette and sequences in the backbone could have an inhibitory effect on expression and that optimized and longer-term gene expression could potentially be achieved with circles devoid of the backbone. Chen *et al.* then moved on to engineer a technology based on the recombination activity of the phage PhiC31 integrase for production of such 'minicircles' excluding the bacterial backbone and carrying only the transgene expression cassette (Chen et al. 2003). The minicircles were produced by transforming bacteria with a plasmid from which expression of the PhiC31 integrase was induced by incubation of the bacteria in L-arabinose. By flanking the expression cassette with the *attB* and *attP* PhiC31 recognition sites, PhiC31-directed recombination between the two sites could facilitate the release of a minicircle from the original plasmid, allowing subsequent purification on a CsCl gradient. Such minicircles were found to express transgenes in liver up to more than 500-fold more efficiently than standard plasmid (Chen et al. 2003). Notably, this difference was seemingly not caused by increased transfection of minicircle DNA, as the vector copy numbers in mouse liver after minicircle and plasmid injection were comparable (Chen et al. 2001; Chen et al. 2003; Riu et al. 2007). Interestingly, several heterochromatin markers (e.g. di- and trimethylation of lysine 9 on histone 3 and di- and tri-methylation of lysine 20 on histone 4) were found by chromatin immunoprecipitation to be associated with plasmid DNA, but not with minicircle DNA, analyzed about a month after DNA transfection of the mouse liver (Riu et al. 2007). In particular, plasmid sequences have been observed to be dramatically enriched with trimethylated lysine 27 on histone 3 (Gracey Maniar et al. 2013). In contrast, higher amounts of euchromatin markers (e.g. acetylated histones 3 and 4, di-methylation of lysine 79 on histone 3, and acetylation of lysine 9 on histone 3) were identified on minicircle DNA, suggesting that the persistence of gene expression from minicircles is correlated with an open chromatin state (Riu et al. 2007). Perhaps surprisingly, transcriptional silencing in the mouse liver seemed not to be associated with the content of CpG motifs or the presence or absence of methylation in those motifs (Chen et al. 2008). However, recent findings suggest that preferential silencing of plasmid DNA occurs inside the nucleus on a stage that precedes the export of mRNA to the cytoplasm (Gracey Maniar et al. 2013). This indicates that plasmid silencing is not due to aberrant mRNA processing and that transcriptional blockage is enforced through plasmid nucleosome formation and subsequent heterochromatin formation. Interestingly, injections into mouse liver of minicircles with various lengths of extragenic spacer sequences have indicated that the length and not the sequence or

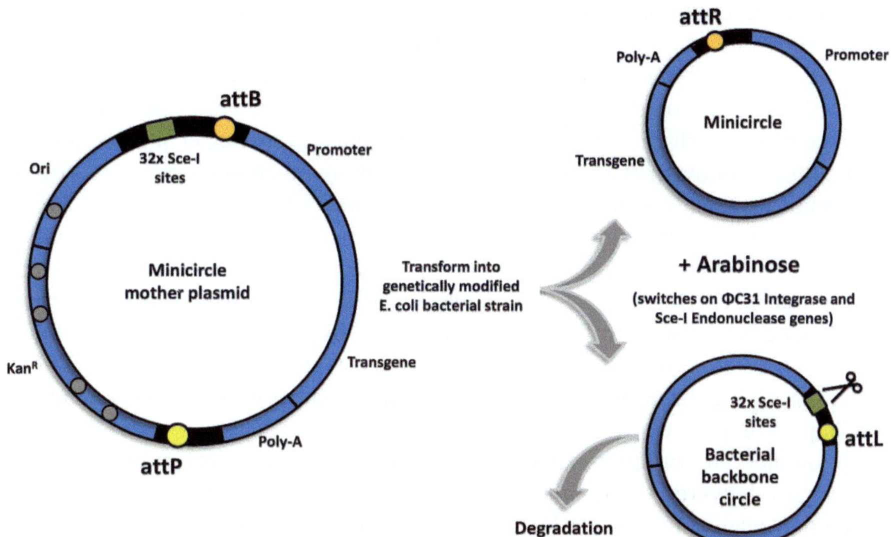

Fig. 4.3 Basic concepts in the production of minicircles (see text for details)

origin of this extragenic spacer is the key determinant, at least in this tissue, for transcriptional silencing of episomal transgene expression cassettes (Lu et al. 2012).

To allow efficient large-scale production of minicircles, Kay and colleagues produced a genetically modified *E. coli* producer strain containing the components for potent minicircle production (Kay et al. 2010). This bacterial strain contains expression cassettes for L-arabinose-inducible production of the PhiC31 integrase and the endonuclease I-SceI. After overnight culturing of this strain transformed with the mother plasmid, incubation of the bacteria in L-arabinose facilitates PhiC31-catalyzed release of the minicircle (Fig. 4.3). Simultaneous degradation of plasmid DNA and backbone circles is facilitated by nucleolytic processing by the ISceI endonuclease which cuts at one or more of the numerous ISceI restriction sites situated in the plasmid backbone. Large amounts of pure minicircles can now be isolated on conventional plasmid purification columns. Alternative methodologies also include recombinase-based release of minicircles but generally suffer from high contamination rates, low yield, or the need for time-consuming purification processes (Darquet et al. 1997; Bigger et al. 2001; Chen et al. 2003; Mayrhofer et al. 2008; Kobelt et al. 2013). With the method developed by Kay et al. minicircle production resembles conventional plasmid production and has become considerably less labor-intensive. Hence, it is to be expected that minicircles will become standard tools in biomedical experimentation and may rapidly rise to become key carriers of genetic cargo in nonviral gene therapies.

Preclinical evidence of minicircle-based gene delivery is now emerging in different corners of the gene therapy field. Liver-directed minicircle administration was explored also for delivery of the α-L-Iduronidase (IDUA) gene for treatment of mice with mucopolysaccharidosis type 1, a lysosomal storage disease (Osborn et al. 2011). Previous attempts to express the IDUA gene from plasmid DNA in mouse liver have resulted in only short-term expression in immunocompetent mice, leading to the assumption that transgene-expressing cells were eradicated by a robust immune reaction triggered by the transgenic protein (Aronovich et al. 2007; Osborn et al. 2008; Aronovich et al. 2009). Although copy number analyses in mouse liver have indicated that neither immune-based clearance of liver cells nor the loss of vector genomes explains the loss of IDUA production, expression from plasmids as well as from minicircles was found to be compromised in immune-competent mice (Osborn et al. 2011). Interestingly, however, minicircles carrying a CMV-driven IDUA gene were found to facilitate significantly higher, persistent levels of gene expression relative to full-length plasmid DNA in immunodeficient NOD/SCID mice, supporting the notion that the bacterial backbone also in this case has a negative impact on gene expression (Osborn et al. 2011).

In another line of study, Huang et al. demonstrated efficient expression from minicircles delivered to the mouse heart (resulting in 5-fold higher levels of gene expression relative to plasmid DNA) and found that transgene expression was considerably more persistent with minicircles than with regular plasmid DNA (Huang et al. 2009a). The authors also managed to treat myocardial infarction by intramyocardial injection of minicircles encoding hypoxia-inducible factor-1α (HIF-1α). Notably, the levels of HIF-1α protein were significantly higher in minicircle-treated hearts relative to hearts treated with standard plasmid, and cardiac function was more efficiently restored with minicircles (Huang et al. 2009a). HIF-1α protein has a short half-life due to rapid proteosomal degradation mediated by hydroxylating enzymes. In accordance, minicircle-encoded small hairpin RNAs (shRNAs) directed toward HIF-1α-degrading enzymes were found to induce an increased level of HIF-1α protein in intramyocardially injected mouse hearts (Huang et al. 2011). In this case, minicircle gene therapy enhanced the mobilization of stem cells allowing increased myocardial angiogenesis and improved cardiac function after induced myocardial infarction. Studies in mouse heart with minicircles containing the human vascular endothelial growth factor (hVEGF) gene driven by a CMV promoter did not show an obvious difference after injection with equal molar amounts of minicircle and plasmid DNA (Stenler et al. 2009). This particular minicircle was more potent than the plasmid counterpart only when the same dose in milligrams was used, indicating that transgene expression in this case was affected primarily by the copy number and to a lesser degree by differential transcriptional silencing of minicircles and plasmid DNA. Using similar VEGF-encoding minicircles (with expression driven by a β-actin promoter) Kwon et al. treated skin wounds of diabetic mice by subcutaneous injection of minicircles complexed with a cationic dendrimer carrier. However, in this study, only dendrimer-complexed minicircles, and not naked minicircles, resulted in VEGF production after injection, leading to a marked increase in the healing rate of wounds in mice with streptozotocin-induced diabetes (Kwon et al. 2012).

Systemic administration by intravenous injection of minicircles is an attractive delivery scenario for many nonviral gene therapies. Delivery of liposome-complexed plasmid DNA elicits a type I interferon-mediated inflammatory response

(Sellins et al. 2005), which may potentially be reduced by the use of minicircles. Although systemically administered minicircle liposomes carrying the gene encoding manganese superoxide dismutase (MnSOD) were found to protect mice from total body irradiation, it is currently unclear if minicircles for this application are more suitable carriers than plasmid DNA (Zhang et al. 2008).

Recent intradermal injections of minicircles provide the first evidence that minicircles may also function as an optimized platform for new vaccines (Dietz et al. 2013). Dietz and coworkers used a tattooing approach to deliver plasmid DNA and the minicircle counterpart, and found that the immunogenicity caused by transgene-encoded proteins was significantly higher with injected minicircles than with plasmid DNA. With this approach, the authors obtained protection against bacterial infection, which was the result of the higher and more prolonged gene expression after injection of minicircles (Dietz et al. 2013). In a similar fashion, minicircles transferred to mouse muscle by electroporation resulted in higher levels of gene expression relative to plasmid DNA (Chabot et al. 2013).

Minicircles are likely to be both safer and more efficient carriers of transgenes than plasmid DNA. These circles free of bacterial DNA recently moved into the world of stem cells; hence, 'reprogramming' minicircles, carrying all four 'Yamanaka' factors, were utilized for generation of induced pluripotent stem cells (Jia et al. 2010; Narsinh et al. 2011). Although companies are now offering minicircles, the relative cheap and easy production of minicircles will allow general use in laboratories worldwide. The reduced size and improved safety will certainly fuel their use in nonviral gene therapies.

4.5 Gene Insertion by Integrating Nonviral Vectors—From Short-Term to Persistent Gene Expression

Upon reaching the final destination of the journey—the transcriptional machinery within the nucleus—transfected plasmid or minicircle DNA molecules will be living a life primarily as nuclear episomes. Hence, only a minor and insignificant proportion of the molecules will randomly insert into the genome of the transfected cell, for example by jumping into double-stranded DNA breaks. In nondividing or slowly dividing cells, like muscle and liver cells, such episomes may persist in the nucleus for a long time allowing prolonged expression of the transgene. However, gene expression may eventually cease due to the loss of episomes during cell replication and/or transcriptional silencing facilitated by the bacterial backbone or sequences in the promoter. In proliferating cells, nonintegrated plasmid or minicircle DNA is rapidly diluted and lost as a result of cell division, leading to only short-term gene expression. Such transient expression of the transgene was for long an Achilles Heel of nonviral gene therapies, and several strategies have been explored to secure stable gene expression. One of these is based on the inclusion of scaffold matrix attachment regions (also referred to as S/MAR elements) in the vector DNA. S/MARs are genomic DNA sequences that facilitate the interaction between chromatin and the nuclear matrix. See for example review Argyros et al. (2011b).

Introduction of S/MAR sequences in plasmids has been shown to facilitate episomal replication and maintenance even in dividing cells (Piechaczek et al. 1999; Baiker et al. 2000), although there is still some controversy associated with the capacity of S/MARs to guide plasmid replication in in vivo models, like mouse liver (Argyros et al. 2008; Argyros et al. 2011a; Argyros et al. 2011b). Whereas nuclear maintenance of episomal DNA remains a promising but also challenging approach, enzymatically catalyzed insertion of the transgene cassette into the genome ensures transmission of the transgene to daughter cells, allowing stable gene expression in a population of dividing cells.

With the reconstructed *Sleeping Beauty* (*SB*) DNA transposon as the locomotive, transposable DNA elements have emerged during the last decade as a primary tool for genomically inserting genes carried by nonviral vectors (Fig. 4.4). DNA trans-

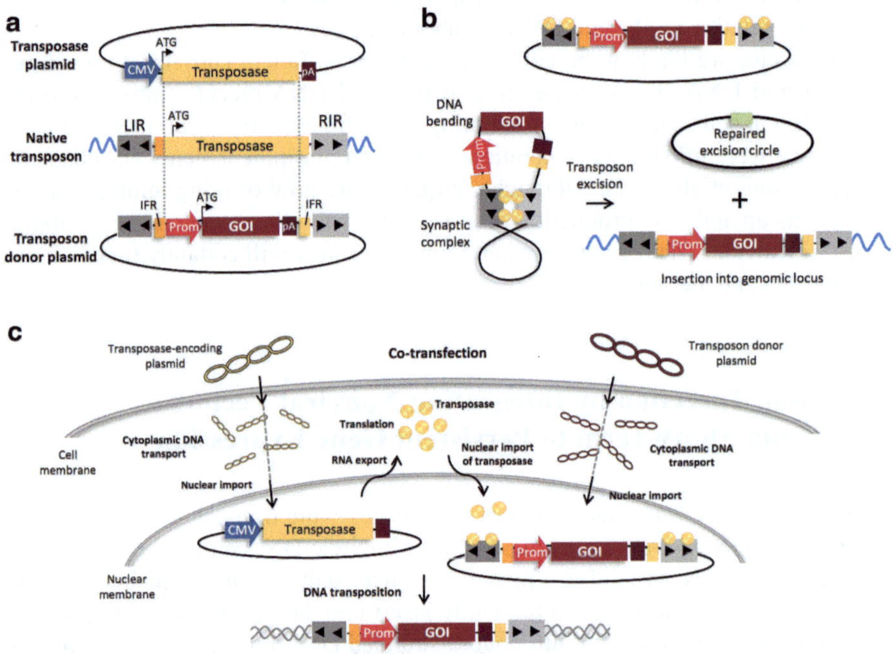

Fig. 4.4 Gene insertion by DNA transposition. The figure illustrates the components and mechanisms of *Sleeping Beauty* DNA transposition. **a** Development of the two-component transposition system from a native transposon. Based on the revived DNA transposable element, one plasmid (*top*) is designed to express the transposase, whereas another plasmid (*below*) carries the transposon with the gene of interest (GOI) expression cassette. The transposon is flanked by the left and right inverted repeats (LIR and RIR) which both carry two direct repeats (indicated by black triangles). The inner flanking regions (IFRs) were defined and described in (Moldt et al. 2007). 'CMV' indicates the promoter derived from cytomegalovirus, whereas 'Prom' indicates any promoter driving the expression of the GOI. **b** Mobilization of the *SB* transposon vector from plasmid DNA by DNA transposition involving the formation of the synaptic transposition complex. Excision circles are generated by cellular repair of the excision site (indicated by the small green box). **c** A schematic overview of the cellular actions leading to gene insertion by *SB* DNA transposition after co-transfection of transposase-encoding plasmid (*left*) and transposon-containing plasmid DNA (*right*)

posons are primitive genetic elements which have colonized living organisms from plants to bacteria and mammals. Such elements are originally mobile with the ability to jump from one position to another within the genome in a reaction catalyzed by the transposase protein encoded by the element itself. However, through evolution, most DNA transposons have accumulated inactivating mutations and have been left behind as fossil relics without any obvious function in the host cell or organism. Based on detailed insight into the phylogenetics of DNA transposon elements in the genome of salmonid fish, Ivics and colleagues were able to 're-awaken' the *SB* element from ancient Tc1/*mariner* elements through a series of mutagenesis steps (Ivics et al. 1997). These efforts primarily focused on re-establishing the function of the transposase protein which was in a final, revived version able to catalyze the mobilization of a transposon element, containing a transgene expression cassette, from transfected plasmid DNA into the genome of human cells. These findings marked the birth of a simple two-component vector technology consisting of (i) a plasmid encoding the transposase and (ii) a plasmid containing the transposon vector consisting of two inverted repeat (IR) structures (that were originally derived from an element present in the genome of *Tanichthys albonubes*) flanking the transgene cassette (Fig. 4.4a). With this technique, the transgene could be mobilized the last step into the genome providing essentially a nonviral alternative to the integrating viral vectors.

The 340 amino acid long *SB* transposase is composed of two major domains; the N-terminal DNA-binding domain and the C-terminal catalytic domain of the enzyme containing the characteristic DDE domain found in many evolutionary related proteins, including the retroviral integrases. (Izsvak et al. 2002). A nuclear localisation signal (NLS) is located between the two domains and is important for nuclear import of the protein. The outer borders of the *SB* transposon are defined by the two 225-bp IR sequences (left and right IR) each containing two 30-bp direct repeats (DRs) (Cui et al. 2002). Binding of a single transposase subunit to each DRs in the two IRs results in the generation of a synaptic complex in which the ends of the transposon are held together by a tetramer of transposases (Ivics et al. 1997; Geurts et al. 2003) (Fig. 4.4b). The transposon is then excised from the donor locus by double-strand DNA breaks which generate 3-bp 3′-overhangs at the transposon ends and at the excision site. The released transposon is re-inserted into a TA-dinucleotide target sequence which is duplicated upon transposon insertion. Together with the TA target-site duplication, repair of the 3′-overhang at the excision site creates a characteristic 5-bp *SB* "footprint". This footprint can be identified in repaired plasmid molecules ("excision circles") that have served as a donor for DNA transposition. A schematic overview of cellular actions related to the insertion of transgenes by the DNA transposition system is provided in Fig. 4.4c.

Yant and co-workers were the first to demonstrate the potency of DNA transposon-based gene delivery in vivo (Yant et al. 2000). By hydrodynamically co-injecting two plasmids, one expressing the *SB* transposase from a CMV promoter and one carrying a *SB* transposon with the human factor IX gene (driven by an elongation factor 1α promoter), the authors demonstrated high stability of transgene expression in the mouse liver triggered by mobilization of the transposon from the plasmid and into the genome. Notably, transgene expression in transposon-treated livers remained stable even after cell proliferation induced by partial hepatectomy, demonstrating that the transgene cassette had indeed been inserted into the genome of the mouse hepatocytes. Using this technique, a mouse model of hemophilia B was successfully treated by a single injection with the two plasmids (Yant et al. 2000). Since then, numerous studies have been carried out to understand and further develop DNA transposon elements as gene carriers in mammalian cells. Moreover, the battery of DNA transposon elements with a capacity to jump in mammalian cells has slowly been expanded. In addition to *SB*, the *piggyBac* (*PB*) element, isolated from the cabbage looper moth *Trichoplusia ni* (Elick et al. 1996; Fraser et al. 1996), is showing efficient DNA transposition in human cells (Adelman et al. 2002; Ding et al. 2005; Wilson et al. 2007) and is currently intensively studied for gene delivery. Also, *Tol2* derived from the genome of the Japanese medaka fish *Oryzias latipes* (Kawakami et al. 1998; Kawakami et al. 2000; Balciunas et al. 2006), *Frog Prince* (*FP*) isolated from the genome of the leopard frog *Rana pipiens* (Miskey et al. 2003), *Passport* derived from the fish *Pleuronectes platessa* (Clark et al. 2009), and *Himar1* from the hornfly *Haematobia irritans* (Keravala et al. 2006) are elements with robust transposition activity in mammalian cells. Newly identified elements include *piggyBat* isolated from the bat *Myotis lucifungus* (Mitra et al. 2013) and *TcBuster* from the red flour beetle *Tribolium castaneum* (Woodard et al. 2012; Li et al. 2013).

Although these distinct elements may have different properties in different cell types, they seem to share the common trait that mobilization does not require species-specific host factors. However, host factors like DNA-bending proteins may indeed support the transposition process, as has been seen for *SB* and *FP* (Zayed et al. 2003). Despite the potential benefits of choosing among a variety of DNA transposon systems, *SB* and *PB* are currently the most effective systems in mammalian cells and now together represent the forefront in transposon-based gene therapy.

DNA transposition and gene delivery by *SB* and *PB* has been reviewed on numerous occasions (Izsvak and Ivics 2004; Hackett et al. 2005; Ivics and Izsvak 2006; Claeys Bouuaert and Chalmers 2009; VandenDriessche et al. 2009; Izsvak et al. 2010; Ammar et al. 2012b; Di Matteo et al. 2012; Swierczek et al. 2012), and it is not the intention in this chapter to review detailed properties and compare the two elements. However, it should be noted, that both these elements have been further developed since their original discovery. Mutagenesis of the SB transposase sequence has produced a series of hyperactive transposase mutants with enhanced transposition activities (Yant et al. 2004; Zayed et al. 2004; Baus et al. 2005). Most recently, a high-throughput PCR-based DNA-shuffling strategy led to the development of the hyperactive transposase SB100X, which was initially demonstrated to be 100-fold more active than the original *SB* transposase (originally referred to as SB10) (Mates et al. 2009). Improved *PB* transposition has been achieved by codon optimization of the transposase gene (Cadinanos and Bradley 2007; Lacoste et al. 2009). In addition, a hyperactive PB transposase (hyPBase) was recently developed (Yusa et al. 2011).

DNA transposon vectors facilitate active insertion of the transgene into the genome and thereby facilitate long-term gene expression. The other side of the coin is that insertion of the gene generates a risk of disrupting or activating cellular genes. For this reason, the integration profiles of integrating vectors—viral and nonviral—have been attracting increased attention. Interestingly, both *SB* and *PB* transposons preferentially integrate into DNA sequences that are rich in thymidines and adenines (Liu et al. 2005; Yant et al. 2005; Geurts et al. 2006; Wilson et al. 2007), but differ in global integration site preferences. Mapping of *SB* insertion sites from mouse liver (Yant et al. 2005), human primary T cells (Huang et al. 2010), HeLa cells (Vigdal et al. 2002), NIH 3T3 mouse fibroblasts (Yant et al. 2005), and mouse embryonic stem cells (Liang et al. 2009) reveals a near random integration profile with less than 40% integration within active genes. The *PB* transposon, in contrast, seems to have a bias towards intragenic regions as demonstrated by studies in HEK293 and HeLa cells (Wilson et al. 2007; Meir et al. 2011) and in primary human T cells (Galvan et al. 2009; Huang et al. 2010). It is currently not known why the integration profiles of the two systems differ, but the more random profile of *SB* has sometimes been used to back up the claim that *SB* is a safer vector system than *PB*. Nevertheless, even randomly inserting vectors may hit cellular genes, and attempts have been made to target the *SB* transposon to predetermined genomic loci by fusing the transposase with the adeno-associated virus Rep protein (Ammar et al. 2012a) or zinc-finger proteins (Yant et al. 2007; Voigt et al. 2012). A similar strategy was utilized to pursue site-directed *PB* insertion (Kettlun et al. 2011). These studies show the feasibility of targeting DNA transposons to predefined loci but also give indications that directed insertion into the genome will remain a very challenging goal for future therapeutic use of DNA transposons.

The potency of transposon-based vectors in preclinical settings has been demonstrated in mouse liver on several occasions (Yant et al. 2000; Mikkelsen et al. 2003; Yant and Kay 2003; Ehrhardt et al. 2005; Ohlfest et al. 2005b; Aronovich et al. 2007). Hyperactive transposase variants SB100X and HyPBase have been shown to increase in vivo efficacy even further (Mates et al. 2009; Doherty et al. 2012) and currently represent the prime choices for therapeutic mobilization of transgenes. However, although high efficiency is certainly attractive in hard-to-transfect cell types or tissues that are difficult to reach with plasmid DNA, it should be kept in mind that the most efficacious transposases hold the potential to insert several transposons into the same cell (Sharma et al. 2012), leading in many cases to an increased risk of insertional mutagenesis. Therefore, the amount of transfected plasmid (transposon donor as well as transposase-encoding plasmid) should be optimized for any given application. Early experimentation in the mouse liver also revealed that the *SB* system, in particular, seems to be vulnerable to an inhibitory response caused by overproduction of the transposase. Hence, whereas injection of 1 µg plasmid DNA encoding the SB10 transposase from a CMV promoter resulted in efficient DNA transposition in mouse liver, transposition was not detectable with an increased plasmid dose (25 µg plasmid DNA) (Yant et al. 2000). This effect was initially thought to be caused by some evolutionarily conserved mechanism that protects host organisms from accelerating transposition, for example by forming aggregates of transposase proteins. Although this is still a possible explanation, analyses of liver samples by Southern blotting indicated that plasmid DNA was diminished from livers treated with high amounts of transposase-encoding plasmid (Yant, Mikkelsen, Kay, unpublished observations). In accordance, later studies showed that overexpressed transposases had a negative impact on the growth of cells (Walisko et al. 2006; Galla et al. 2011). It is likely, therefore, that hepatocytes with boosted expression of the transposase died upon hydrodynamic gene delivery resulting in low levels of transposition and reduced gene expression, which has been consistently observed with high amounts of transposase.

In attempts to optimize delivery of the *SB* transposon system, both the transposon and the transposase expression cassette were included on the same plasmid (Mikkelsen et al. 2003). Since the two components were hence delivered in a 1:1 ratio on a single plasmid, it was necessary to balance the expression of the transposase by utilizing a promoter that was relatively weak in mouse liver. Using such helper-independent transposon-transposase vectors (HITT, later most often referred to as 'cis' vectors), optimized DNA transposition in mouse liver was achieved and a haemophilia B mouse model treated (Mikkelsen et al. 2003). To further improve the safety profile and the efficacy of the *SB* system, the *SB* transposon vector technology was recently combined with the minicircle DNA technology (Sharma et al. 2013). By exploiting minicircles as donors for *SB* transposons, efficient transposition and, at low transposase dosages, stable transfection rates were achieved that were improved relative to the standard plasmid donor. These findings demonstrated that minicircle-based *SB* transposon vectors represent an efficient alternative to plasmid DNA as both donors of *SB* transposons and carriers of *SB* transposase expression cassettes, and suggest that circular DNA, devoid of the bacterial backbone, should be utilized in future clinical applications.

Today, more than 17 years after the initial report on *SB*, long-term gene expression after in vivo treatment with the *SB* system has been reported for a range of tissues. These include lung (Belur et al. 2003; Liu et al. 2004; Liu et al. 2006a; Liu et al. 2006c; Belur et al. 2007; Lin et al. 2011), skin (Ortiz-Urda et al. 2003), and brain (Ohlfest et al. 2004; Ohlfest et al. 2005a) of adult mice injected with plasmid DNA. However, the versatility of this approach in cultured cells may probably have the most immediate impact in developing therapies. Gene insertion in $CD34^+$ hematopoietic stem cells (Hollis et al. 2006; Staunstrup et al. 2009; Sumiyoshi et al. 2009; Xue et al. 2009), primary T cells (Huang et al. 2008; Singh et al. 2008; Huang et al. 2009b; Jin et al. 2011), and human embryonic stem cells (Wilber et al. 2007; Orban et al. 2009) is creating hope for the use of DNA transposon-based gene mobilization for genetic engineering of such cell types. Indeed, the first clinical trial utilizing *SB*-directed gene insertion has focus on gene insertion in primary T cells (Singh et al. 2008) and was recently initiated for treatment of patients with B-lymphoid malignancies with adoptive immunotherapy (Williams 2008; Kebriaei et al. 2012; Torikai et al. 2012).

During the time of development of *SB*-based vector technologies, several recombinases, including the phage PhiC31 integrase, have attracted attention due to their potential for facilitating site-directed gene insertion (Thyagarajan et al. 2001). By hydrodynamic injection of plasmids carrying a recombination substrate and encoding the PhiC31 integrase, respectively, site-specific gene insertion was found to facilitate therapeutic gene expression in mouse liver (Olivares et al. 2002). Despite the enormous potential of this technique, the interest in PhiC31-based gene transfer has declined substantially due to the risk of causing interchromosomal translocations in treated cells (Ehrhardt et al. 2005; Ehrhardt et al. 2006; Liu et al. 2006b; Liu et al. 2009). With the invention of designer nucleases, like zinc-finger nucleases (ZFNs) and transcription activator-like nucleases (TALENs) (Clark et al. 2011; McMahon et al. 2012), new winds are blowing in the field of genetic engineering. Such nucleases generate site-directed double-stranded DNA breaks and facilitate gene insertion by homologous recombination (Porteus and Carroll 2005; Urnov et al. 2005; Mahfouz et al. 2011) and may in the future allow site-directed gene insertion and repair of disease mutations in stem cells and in vivo.

4.6 Nonviral Gene Therapy Clinical Trials—Hopes and Challenges

By understanding and surpassing the major barriers to delivery of plasmid DNA to organs and cells, research in plasmid production and administration has brought nonviral vectors much closer to successful clinical use. Although it seems unlikely that plasmid- or minicircle-based gene delivery techniques will replace viral vector technologies, which have also been further optimized and made safer in recent years, the ease of production of plasmid/minicircle DNA, the growing expertise and reproducibility in production of chemical carrier materials, and not least the high level of safety and low immunogenicity will make nonviral vectors the vehicles of choice in selected therapeutic applications.

Since the first-in-human plasmid delivery treatment was carried out by intratumoral injection of plasmid-liposome complexes (encoding a foreign major histocompability complex protein, HLA-B7) back in 1992 (Nabel et al. 1993), a total of 112 trials using lipids and liposomes as carrier of plasmid DNA have been initialized according to available online sources ('Gene Therapy Clinical Trials worldwide' provided by the Journal of Gene Medicine at www.wiley.co.uk/genmed/clinical). Trials involving the use of naked plasmid delivery constitute a total of 346 trials. The fact that only less than 3 % of the nonviral gene therapy trials have been phase II/III or III trials (counting only trials based on naked or lipid-aided plasmid delivery) bears witness to the obvious challenges of obtaining therapeutic gene expression. However, past trials have laid the foundation for technological improvements, and twelve new trials were initiated in 2012. Among these, five trials are aimed at treating various forms of cancer, two of them by intratumoral injection of naked plasmid DNA. Three trials target cardiovascular diseases including critical limp ischemia, diabetic peripheral neuropathy, and heart failure, whereas two trials explore the continued potential of intramuscular plasmid delivery in DNA-based vaccines.

One of the trials that were initiated in 2012 is focused on the treatment of cystic fibrosis (CF) by delivery of plasmid DNA embedded in cationic lipids to the lung. This trial is run by The UK Cystic Fibrosis Gene Therapy Consortium (UK-CFGTC), which consists of three research groups with a combined staff of more than 50 individuals working solely with the aim of developing gene therapies for CF. In many ways, the hopes and challenges of bringing nonviral gene therapy to clinical reality are well represented by the past and current efforts to establish a genetic medicine for treatment of CF. CF is a common inherited disease caused by mutations within the *CFTR* gene encoding the CF conductance regulator which is involved in ion transport in the lung epithelium. Defective ion transport causes formation of viscous secretions in the airways leading to inflammation, infections, and lung damage. The requirement for repeated treatment (due to the limited life span of respiratory epithelial cells) as well as the marked immune recognition of viral vectors upon repeated vector administration (Alton et al. 2013) has put attention on the use of nonviral gene delivery for treatment of CF.

The ongoing UKCFGTC trial is a representative example of how the improved understanding of gene expression, plasmid-directed inflammation, and chemical formulations may be utilized to develop an optimized clinical protocol (Alton et al. 2013). Delivery of naked plasmid DNA to the airways is very inefficient and aerosolization of plasmid DNA to lungs requires that the DNA is 'packaged' in a liposome formulation. Early clinical studies using pulmonary and nasal administration of a standard plasmid with CMV-driven expression of the *CFTR* gene demonstrated partial correction of the chloride transport abnormality but also signs of an inflammatory response (Alton et al. 1999). This trial utilized a formulation based on the cationic lipid GL67, which was generated by chemical optimization of a cationic cholesterol derivative mixed with dimethylaminoethane-carbamoyl (DC-Chol) (Lee et al. 1996). An improved formulation was created by adding the neutral lipid DOPE, thought to mediate improved escape of the complex from endosomes, along with the lipid DMPE-PEG5000 which stabilizes the liposome for in vivo delivery in lung epithelium (Eastman et al. 1997). This formulation was found to be successfully aerosolized in the lungs of CF patients (Ruiz et al. 2001), and more recent, comparative studies have not been able to identify alternative formulations that work more efficiently for aerosol delivery of plasmid DNA (McLachlan et al. 2011). The plasmid expressing CFTR has undergone several changes since the first trials in the 1990s. Production of the CFTR protein was improved by expressing a codon-optimized version of the *CFTR* gene from the human EF1α promoter (supported by a CMV enhancer), which was shown by preclinical experimentation to provide sustained expression (Hyde et al. 2008). Moreover, the plasmid (now referred to as pGM169) was engineered to become entirely free of CpG dinucleotides, which are known to induce inflammation and compromise long-term gene expression in epithelial lung tissue (Hyde et al. 2008). The trial based on GL67A-formulated pGM169 is currently carried out as a blinded, placebo-controlled treatment and is planned to enroll a total of 130 patients who will each receive twelve doses of the nonviral vector over a period of about a year (Alton et al. 2013).

4.7 Conclusions

The pGM169 plasmid is—according to the UKCFGTC web site—the most abundant plasmid to be produced so far for human clinical use. The fact alone stands as a clear indication that nonviral gene therapies have come a long way and have moved from studies of plasmid design and chemical wrapping to high-profile clinical trials including a large cohort of patients. The near future will demonstrate not only if CF is treated by repeated application of nonviral vectors but also if the many efforts to overcome the obstacles to plasmid-based gene delivery will pay off and result in improved and successful treatment of inherited diseases and infections as well as of cancers and cardiovascular diseases. In recent years, the gene therapy field has created success stories with a constantly increasing pace. With the discovery of nonviral integrating vector technologies (e.g. DNA transposons), novel genome editing tools (e.g. designer nucleases), and minicircles devoid of bacterial DNA, there is no doubt that nonviral techniques will become established for in vitro applications including perhaps most importantly the genetic modifications of stem cells. In the years to come, we will learn whether plasmid-based gene delivery can grow to compete with viral vector technologies for in vivo applications and therapy. Come join the challenge of mobilizing DNA.

Acknowledgments Studies of nonviral vector technogies in the laboratory of JGM have been carried out through support from the Lundbeck Foundation, the Novo Nordisk Foundation, Aase og Ejnar Danielsens Fond, Agnes og Poul Friis Fond, Kgl. Hofbuntmager Aage Bangs Fond, Grosserer A. V. Lykfeldt og Hustrus Legat, Else og Mogens Wedell-Wedellsborgs Fond, Fonden af 17-12-1981, Kong Christian den Tiendes Fond, Civilingeniør Frode V. Nyegaard og hustrus Fond, Clara Hansens Mindelegat, Kirsten Anthonius Mindelegat, Snedkermester Sophus Jacobsen and Hustru Astrid Jacobsens Fond og Frits, Georg og Marie Cecilie Gluds Legat. JGM is the director of *Gene Therapy Initiative Aarhus* (GTI-Aarhus) funded by the Lundbeck Foundation and a member of *the Aarhus Research Center for Innate Immunology* (ARCII) established through funding by the AU-Ideas program at Aarhus University.

References

Acsadi G, Jiao SS, Jani A, Duke D, Williams P, Chong W, Wolff JA (1991) Direct gene transfer and expression into rat heart in vivo. New Biol 3:71–81

Adelman ZN, Jasinskiene N, James AA (2002) Development and applications of transgenesis in the yellow fever mosquito, *Aedes aegypti*. Mol Biochem Parasitol 121:1–10

Aihara H, Miyazaki J (1998) Gene transfer into muscle by electroporation in vivo. Nat Biotechnol 16:867–870

Aiuti A, Cattaneo F, Galimberti S, Benninghoff U, Cassani B, Callegaro L, Scaramuzza S, Andolfi G, Mirolo M, Brigida I, Tabucchi A, Carlucci F, Eibl M, Aker M, Slavin S, Al-Mousa H, Al Ghonaium A, Ferster A, Duppenthaler A, Notarangelo L, Wintergerst U, Buckley RH, Bregni M, Marktel S, Valsecchi MG, Rossi P, Ciceri F, Miniero R, Bordignon C, Roncarolo MG (2009) Gene therapy for immunodeficiency due to adenosine deaminase deficiency. N Engl J Med 360:447–458

Alton EW, Stern M, Farley R, Jaffe A, Chadwick SL, Phillips J, Davies J, Smith SN, Browning J, Davies MG, Hodson ME, Durham SR, Li D, Jeffery PK, Scallan M, Balfour R, Eastman SJ, Cheng SH, Smith AE, Meeker D, Geddes DM (1999) Cationic lipid-mediated CFTR gene transfer to the lungs and nose of patients with cystic fibrosis: a double-blind placebo-controlled trial. Lancet 353:947–954

Alton EW, Boyd AC, Cheng SH, Cunningham S, Davies JC, Gill DR, Griesenbach U, Higgins T, Hyde SC, Innes JA, Murray GD, Porteous DJ (2013) A randomised, double-blind, placebo-controlled phase IIB clinical trial of repeated application of gene therapy in patients with cystic fibrosis. Thorax 68:1075–1077

Ammar I, Gogol-Doring A, Miskey C, Chen W, Cathomen T, Izsvak Z, Ivics Z (2012a) Retargeting transposon insertions by the adeno-associated virus Rep protein. Nucleic Acids Res 40:6693–6712

Ammar I, Izsvak Z, Ivics Z (2012b) The Sleeping Beauty transposon toolbox. Methods Mol Biol 859:229–240

Anderson WF (1984) Prospects for human gene therapy. Science 226:401–409

Andre FM, Gehl J, Sersa G, Preat V, Hojman P, Eriksen J, Golzio M, Cemazar M, Pavselj N, Rols MP, Miklavcic D, Neumann E, Teissie J, Mir LM (2008) Efficiency of high- and low-voltage pulse combinations for gene electrotransfer in muscle, liver, tumor, and skin. Hum Gene Ther 19:1261–1271

Argyros O, Wong SP, Niceta M, Waddington SN, Howe SJ, Coutelle C, Miller AD, Harbottle RP (2008) Persistent episomal transgene expression in liver following delivery of a scaffold/matrix attachment region containing non-viral vector. Gene Ther 15:1593–1605

Argyros O, Wong SP, Fedonidis C, Tolmachov O, Waddington SN, Howe SJ, Niceta M, Coutelle C, Harbottle RP (2011a) Development of S/MAR minicircles for enhanced and persistent transgene expression in the mouse liver. J Mol Med (Berl) 89:515–529

Argyros O, Wong SP, Harbottle RP (2011b) Non-viral episomal modification of cells using S/MAR elements. Expert Opin Biol Ther 11:1177–1191

Aronovich EL, Bell JB, Belur LR, Gunther R, Koniar B, Erickson DC, Schachern PA, Matise I, McIvor RS, Whitley CB, Hackett PB (2007) Prolonged expression of a lysosomal enzyme in mouse liver after Sleeping Beauty transposon-mediated gene delivery: implications for non-viral gene therapy of mucopolysaccharidoses. J Gene Med 9:403–415

Aronovich EL, Bell JB, Khan SA, Belur LR, Gunther R, Koniar B, Schachern PA, Parker JB, Carlson CS, Whitley CB, McIvor RS, Gupta P, Hackett PB (2009) Systemic correction of storage disease in MPS I NOD/SCID mice using the Sleeping Beauty transposon system. Mol Ther 17:1136–1144

Badding MA, Vaughan EE, Dean DA (2012) Transcription factor plasmid binding modulates microtubule interactions and intracellular trafficking during gene transfer. Gene Ther 19:338–346

Badding MA, Lapek JD, Friedman AE, Dean DA (2013) Proteomic and functional analyses of protein-DNA complexes during gene transfer. Mol Ther 21:775–785

Baiker A, Maercker C, Piechaczek C, Schmidt SB, Bode J, Benham C, Lipps HJ (2000) Mitotic stability of an episomal vector containing a human scaffold/matrix-attached region is provided by association with nuclear matrix. Nat Cell Biol 2:182–184

Bainbridge JW, Smith AJ, Barker SS, Robbie S, Henderson R, Balaggan K, Viswanathan A, Holder GE, Stockman A, Tyler N, Petersen-Jones S, Bhattacharya SS, Thrasher AJ, Fitzke FW, Carter BJ, Rubin GS, Moore AT, Ali RR (2008) Effect of gene therapy on visual function in Leber's congenital amaurosis. N Engl J Med 358:2231–2239

Balciunas D, Wangensteen KJ, Wilber A, Bell J, Geurts A, Sivasubbu S, Wang X, Hackett PB, Largaespada DA, McIvor RS, Ekker SC (2006) Harnessing a high cargo-capacity transposon for genetic applications in vertebrates. PLoS Genet 2:e169

Baus J, Liu L, Heggestad AD, Sanz S, Fletcher BS (2005) Hyperactive transposase mutants of the Sleeping Beauty transposon. Mol Ther 12:1148–1156

Belur LR, Frandsen JL, Dupuy AJ, Ingbar DH, Largaespada DA, Hackett PB, Scott McIvor R (2003) Gene insertion and long-term expression in lung mediated by the Sleeping Beauty transposon system. Mol Ther 8:501–507

Belur LR, Podetz-Pedersen K, Frandsen J, McIvor RS (2007) Lung-directed gene therapy in mice using the nonviral Sleeping Beauty transposon system. Nat Protoc 2:3146–3152
Bigger BW, Tolmachov O, Collombet JM, Fragkos M, Palaszewski I, Coutelle C (2001) An araC-controlled bacterial cre expression system to produce DNA minicircle vectors for nuclear and mitochondrial gene therapy. J Biol Chem 276:23018–23027
Birnboim HC, Doly J (1979) A rapid alkaline extraction procedure for screening recombinant plasmid DNA. Nucleic Acids Res 7:1513–1523
Blaese RM, Culver KW, Miller AD, Carter CS, Fleisher T, Clerici M, Shearer G, Chang L, Chiang Y, Tolstoshev P, Greenblatt JJ, Rosenberg SA, Klein H, Berger M, Mullen CA, Ramsey WJ, Muul L, Morgan RA, Anderson WF (1995) T lymphocyte-directed gene therapy for ADA-SCID: initial trial results after 4 years. Science 270:475–480
Bodles-Brakhop AM, Heller R, Draghia-Akli R (2009) Electroporation for the delivery of DNA-based vaccines and immunotherapeutics: current clinical developments. Mol Ther 17:585–592
Bonamassa B, Hai L, Liu D (2011) Hydrodynamic gene delivery and its applications in pharmaceutical research. Pharm Res 28:694–701
Budker V, Zhang G, Danko I, Williams P, Wolff J (1998) The efficient expression of intravascularly delivered DNA in rat muscle. Gene Ther 5:272–276
Buttrick PM, Kass A, Kitsis RN, Kaplan ML, Leinwand LA (1992) Behavior of genes directly injected into the rat heart in vivo. Circ Res 70:193–198
Cadinanos J, Bradley A (2007) Generation of an inducible and optimized piggyBac transposon system. Nucleic Acids Res 35:e87
Cartier N, Hacein-Bey-Abina S, Bartholomae CC, Veres G, Schmidt M, Kutschera I, Vidaud M, Abel U, Dal-Cortivo L, Caccavelli L, Mahlaoui N, Kiermer V, Mittelstaedt D, Bellesme C, Lahlou N, Lefrere F, Blanche S, Audit M, Payen E, Leboulch P, l'Homme B, Bougneres P, Von Kalle C, Fischer A, Cavazzana-Calvo M, Aubourg P (2009) Hematopoietic stem cell gene therapy with a lentiviral vector in X-linked adrenoleukodystrophy. Science 326:818–823
Chabot S, Orio J, Schmeer M, Schleef M, Golzio M, Teissie J (2013) Minicircle DNA electrotransfer for efficient tissue-targeted gene delivery. Gene Ther 20:62–68
Chen ZY, Yant SR, He CY, Meuse L, Shen S, Kay MA (2001) Linear DNAs concatemerize in vivo and result in sustained transgene expression in mouse liver. Mol Ther 3:403–410
Chen ZY, He CY, Ehrhardt A, Kay MA (2003) Minicircle DNA vectors devoid of bacterial DNA result in persistent and high-level transgene expression in vivo. Mol Ther 8:495–500
Chen ZY, He CY, Meuse L, Kay MA (2004) Silencing of episomal transgene expression by plasmid bacterial DNA elements in vivo. Gene Ther 11:856–864
Chen ZY, Riu E, He CY, Xu H, Kay MA (2008) Silencing of episomal transgene expression in liver by plasmid bacterial backbone DNA is independent of CpG methylation. Mol Ther 16:548–556
Claeys Bouuaert C, Chalmers RM (2009) Gene therapy vectors: the prospects and potentials of the cut-and-paste transposons. Genetica 138:473–484
Clark KJ, Carlson DF, Leaver MJ, Foster LK, Fahrenkrug SC (2009) Passport, a native Tc1 transposon from flatfish, is functionally active in vertebrate cells. Nucleic Acids Res 37:1239–1247
Clark KJ, Voytas DF, Ekker SC (2011) A TALE of two nucleases: gene targeting for the masses? Zebrafish 8:147–149
Cui Z, Geurts AM, Liu G, Kaufman CD, Hackett PB (2002) Structure-function analysis of the inverted terminal repeats of the Sleeping Beauty transposon. J Mol Biol 318:1221–1235
Danialou G, Comtois AS, Matecki S, Nalbantoglu J, Karpati G, Gilbert R, Geoffroy P, Gilligan S, Tanguay JF, Petrof BJ (2005) Optimization of regional intraarterial naked DNA-mediated transgene delivery to skeletal muscles in a large animal model. Mol Ther 11:257–266
Darquet AM, Cameron B, Wils P, Scherman D, Crouzet J (1997) A new DNA vehicle for nonviral gene delivery: supercoiled minicircle. Gene Ther 4:1341–1349
Datta N, Kontomichalou P (1965) Penicillinase synthesis controlled by infectious R factors in Enterobacteriaceae. Nature 208:239–241
Dauty E, Verkman AS (2005) Actin cytoskeleton as the principal determinant of size-dependent DNA mobility in cytoplasm: a new barrier for non-viral gene delivery. J Biol Chem 280:7823–7828

Di Matteo M, Belay E, Chuah MK, Vandendriessche T (2012) Recent developments in transposon-mediated gene therapy. Expert Opin Biol Ther 12:841–858

Dietz WM, Skinner NE, Hamilton SE, Jund MD, Heitfeld SM, Litterman AJ, Hwu P, Chen ZY, Salazar AM, Ohlfest JR, Blazar BR, Pennell CA, Osborn MJ (2013) Minicircle DNA is superior to plasmid DNA in eliciting antigen-specific CD8 T-cell responses. Mol Ther 21:1526–1535

Ding S, Wu X, Li G, Han M, Zhuang Y, Xu T (2005) Efficient transposition of the piggyBac (PB) transposon in mammalian cells and mice. Cell 122:473–483

Doherty JE, Huye LE, Yusa K, Zhou L, Craig NL, Wilson MH (2012) Hyperactive piggyBac gene transfer in human cells and in vivo. Hum Gene Ther 23:311–320

Eastman SJ, Lukason MJ, Tousignant JD, Murray H, Lane MD, George JA St, Akita GY, Cherry M, Cheng SH, Scheule RK (1997) A concentrated and stable aerosol formulation of cationic lipid: DNA complexes giving high-level gene expression in mouse lung. Hum Gene Ther 8:765–773

Ehrhardt A, Xu H, Huang Z, Engler JA, Kay MA (2005) A direct comparison of two nonviral gene therapy vectors for somatic integration: in vivo evaluation of the bacteriophage integrase phiC31 and the Sleeping Beauty transposase. Mol Ther 11:695–706

Ehrhardt A, Engler JA, Xu H, Cherry AM, Kay MA (2006) Molecular analysis of chromosomal rearrangements in mammalian cells after phiC31-mediated integration. Hum Gene Ther 17:1077–1094

El-Sayed A, Harashima H (2013) Endocytosis of gene delivery vectors: from clathrin-dependent to lipid raft-mediated endocytosis. Mol Ther 21:1118–1130

El-Sayed A, Futaki S, Harashima H (2009) Delivery of macromolecules using arginine-rich cell-penetrating peptides: ways to overcome endosomal entrapment. AAPS J 11:13–22

Elick TA, Bauser CA, Fraser MJ (1996) Excision of the piggyBac transposable element in vitro is a precise event that is enhanced by the expression of its encoded transposase. Genetica 98:33–41

Escoffre JM, Portet T, Wasungu L, Teissie J, Dean D, Rols MP (2009) What is (still not) known of the mechanism by which electroporation mediates gene transfer and expression in cells and tissues. Mol Biotechnol 41:286–295

Ewert KK, Zidovska A, Ahmad A, Bouxsein NF, Evans HM, McAllister CS, Samuel CE, Safinya CR (2010) Cationic liposome-nucleic acid complexes for gene delivery and silencing: pathways and mechanisms for plasmid DNA and siRNA. Top Curr Chem 296:191–226

Fischer D, Li Y, Ahlemeyer B, Krieglstein J, Kissel T (2003) In vitro cytotoxicity testing of polycations: influence of polymer structure on cell viability and hemolysis. Biomaterials 24:1121–1131

Fraser MJ, Ciszczon T, Elick T, Bauser C (1996) Precise excision of TTAA-specific lepidopteran transposons piggyBac (IFP2) and tagalong (TFP3) from the baculovirus genome in cell lines from two species of Lepidoptera. Insect Mol Biol 5:141–151

Friedmann T, Roblin R (1972) Gene therapy for human genetic disease? Science 175:949–955

Galla M, Schambach A, Falk CS, Maetzig T, Kuehle J, Lange K, Zychlinski D, Heinz N, Brugman MH, Gohring G, Izsvak Z, Ivics Z, Baum C (2011) Avoiding cytotoxicity of transposases by dose-controlled mRNA delivery. Nucleic Acids Res 39:7147–7160

Galvan DL, Nakazawa Y, Kaja A, Kettlun C, Cooper LJ, Rooney CM, Wilson MH (2009) Genome-wide mapping of piggyBac transposon integrations in primary human T cells. J Immunother 32:837–844

Gaudet D, Methot J, Dery S, Brisson D, Essiembre C, Tremblay G, Tremblay K, de Wal J, Twisk J, van den Bulk N, Sier-Ferreira V, van Deventer S (2012) Efficacy and long-term safety of alipogene tiparvovec (AAV1-LPL(S447X)) gene therapy for lipoprotein lipase deficiency: an open-label trial. Gene Ther 20:361–369

Geurts AM, Yang Y, Clark KJ, Liu G, Cui Z, Dupuy AJ, Bell JB, Largaespada DA, Hackett PB (2003) Gene transfer into genomes of human cells by the Sleeping Beauty transposon system. Mol Ther 8:108–117

Geurts AM, Hackett CS, Bell JB, Bergemann TL, Collier LS, Carlson CM, Largaespada DA, Hackett PB (2006) Structure-based prediction of insertion-site preferences of transposons into chromosomes. Nucleic Acids Res 34:2803–2811

Gracey Maniar LE, Maniar JM, Chen ZY, Lu J, Fire AZ, Kay MA (2013) Minicircle DNA vectors achieve sustained expression reflected by active chromatin and transcriptional level. Mol Ther 21:131–138

Griesenbach U, Alton EW (2013) Expert opinion in biological therapy: update on developments in lung gene transfer. Expert Opin Biol Ther 13:345–360

Hacein-Bey-Abina S, Hauer J, Lim A, Picard C, Wang GP, Berry CC, Martinache C, Rieux-Laucat F, Latour S, Belohradsky BH, Leiva L, Sorensen R, Debre M, Casanova JL, Blanche S, Durandy A, Bushman FD, Fischer A, Cavazzana-Calvo M (2010) Efficacy of gene therapy for X-linked severe combined immunodeficiency. N Engl J Med 363:355–364

Hackett PB, Ekker SC, Largaespada DA, McIvor RS (2005) Sleeping Beauty transposon-mediated gene therapy for prolonged expression. Adv Genet 54:189–232

Hackett PB Jr, Aronovich EL, Hunter D, Urness M, Bell JB, Kass SJ, Cooper LJ, McIvor S (2011) Efficacy and safety of Sleeping Beauty transposon-mediated gene transfer in preclinical animal studies. Curr Gene Ther 11:341–349

Hegge JO, Wooddell CI, Zhang G, Hagstrom JE, Braun S, Huss T, Sebestyen MG, Emborg ME, Wolff JA (2010) Evaluation of hydrodynamic limb vein injections in nonhuman primates. Hum Gene Ther 21:829–842

Hemmi H, Takeuchi O, Kawai T, Kaisho T, Sato S, Sanjo H, Matsumoto M, Hoshino K, Wagner H, Takeda K, Akira S (2000) A Toll-like receptor recognizes bacterial DNA. Nature 408:740–745

Hengge UR, Walker PS, Vogel JC (1996) Expression of naked DNA in human, pig, and mouse skin. J Clin Invest 97:2911–2916

Hodges BL, Taylor KM, Joseph MF, Bourgeois SA, Scheule RK (2004) Long-term transgene expression from plasmid DNA gene therapy vectors is negatively affected by CpG dinucleotides. Mol Ther 10:269–278

Hojman P (2010) Basic principles and clinical advancements of muscle electrotransfer. Curr Gene Ther 10:128–138

Hojman P, Gissel H, Gehl J (2007) Sensitive and precise regulation of haemoglobin after gene transfer of erythropoietin to muscle tissue using electroporation. Gene Ther 14:950–959

Hojman P, Gissel H, Andre FM, Cournil-Henrionnet C, Eriksen J, Gehl J, Mir LM (2008) Physiological effects of high- and low-voltage pulse combinations for gene electrotransfer in muscle. Hum Gene Ther 19:1249–1260

Hollis RP, Nightingale SJ, Wang X, Pepper KA, Yu XJ, Barsky L, Crooks GM, Kohn DB (2006) Stable gene transfer to human CD34(+) hematopoietic cells using the Sleeping Beauty transposon. Exp Hematol 34:1333–1343

Houk BE, Martin R, Hochhaus G, Hughes JA (2001) Pharmacokinetics of plasmid DNA in the rat. Pharm Res 18:67–74

Huang X, Guo H, Kang J, Choi S, Zhou TC, Tammana S, Lees CJ, Li ZZ, Milone M, Levine BL, Tolar J, June CH, Scott McIvorR, Wagner JE, Blazar BR, Zhou X (2008) Sleeping Beauty transposon-mediated engineering of human primary T cells for therapy of CD19+ lymphoid malignancies. Mol Ther 16:580–589

Huang M, Chen Z, Hu S, Jia F, Li Z, Hoyt G, Robbins RC, Kay MA, Wu JC (2009a) Novel minicircle vector for gene therapy in murine myocardial infarction. Circulation 120:S230–237

Huang X, Wilber A, McIvor RS, Zhou X (2009b) DNA transposons for modification of human primary T lymphocytes. Methods Mol Biol 506:115–126

Huang X, Guo H, Tammana S, Jung YC, Mellgren E, Bassi P, Cao Q, Tu ZJ, Kim YC, Ekker SC, Wu X, Wang SM, Zhou X (2010) Gene transfer efficiency and genome-wide integration profiling of Sleeping Beauty, Tol2, and piggyBac transposons in human primary T cells. Mol Ther 18:1803–1813

Huang M, Nguyen P, Jia F, Hu S, Gong Y, de Almeida PE, Wang L, Nag D, Kay MA, Giaccia AJ, Robbins RC, Wu JC (2011) Double knockdown of prolyl hydroxylase and factor-inhibiting hypoxia-inducible factor with nonviral minicircle gene therapy enhances stem cell mobilization and angiogenesis after myocardial infarction. Circulation 124:S46–54

Hyde SC, Pringle IA, Abdullah S, Lawton AE, Davies LA, Varathalingam A, Nunez-Alonso G, Green AM, Bazzani RP, Sumner-Jones SG, Chan M, Li H, Yew NS, Cheng SH, Boyd AC,

Davies JC, Griesenbach U, Porteous DJ, Sheppard DN, Munkonge FM, Alton EW, Gill DR (2008) CpG-free plasmids confer reduced inflammation and sustained pulmonary gene expression. Nat Biotechnol 26:549–551

Ivics Z, Izsvak Z (2006) Transposons for gene therapy! Curr Gene Ther 6:593–607

Ivics Z, Hackett PB, Plasterk RH, Izsvak Z (1997) Molecular reconstruction of Sleeping Beauty, a Tc1-like transposon from fish, and its transposition in human cells. Cell 91:501–510

Izsvak Z, Ivics Z (2004) Sleeping Beauty transposition: biology and applications for molecular therapy. Mol Ther 9:147–156

Izsvak Z, Khare D, Behlke J, Heinemann U, Plasterk RH, Ivics Z (2002) Involvement of a bifunctional, paired-like DNA-binding domain and a transpositional enhancer in Sleeping Beauty transposition. J Biol Chem 277:34581–34588

Izsvak Z, Hackett PB, Cooper LJ, Ivics Z (2010) Translating Sleeping Beauty transposition into cellular therapies: victories and challenges. Bioessays 32:756–767

Jia F, Wilson KD, Sun N, Gupta DM, Huang M, Li Z, Panetta NJ, Chen ZY, Robbins RC, Kay MA, Longaker MT, Wu JC (2010) A nonviral minicircle vector for deriving human iPS cells. Nat Methods 7:197–199

Jin Z, Maiti S, Huls H, Singh H, Olivares S, Mates L, Izsvak Z, Ivics Z, Lee DA, Champlin RE, Cooper LJ (2011) The hyperactive Sleeping Beauty transposase SB100X improves the genetic modification of T cells to express a chimeric antigen receptor. Gene Ther 18:849–856

Kadlecova Z, Baldi L, Hacker D, Wurm FM, Klok HA (2012) Comparative study on the in vitro cytotoxicity of linear, dendritic, and hyperbranched polylysine analogues. Biomacromolecules 13:3127–3137

Kadlecova Z, Rajendra Y, Matasci M, Baldi L, Hacker DL, Wurm FM, Klok HA (2013) DNA delivery with hyperbranched polylysine: a comparative study with linear and dendritic polylysine. J Control Release 169:276–288

Kamimura K, Suda T, Xu W, Zhang G, Liu D (2009) Image-guided, lobe-specific hydrodynamic gene delivery to swine liver. Mol Ther 17:491–499

Kamimura K, Zhang G, Liu D (2010) Image-guided, intravascular hydrodynamic gene delivery to skeletal muscle in pigs. Mol Ther 18:93–100

Kang HC, Huh KM, Bae YH (2012) Polymeric nucleic acid carriers: current issues and novel design approaches. J Control Release 164:256–264

Kawabata K, Takakura Y, Hashida M (1995) The fate of plasmid DNA after intravenous injection in mice: involvement of scavenger receptors in its hepatic uptake. Pharm Res 12:825–830

Kawakami K, Koga A, Hori H, Shima A (1998) Excision of the Tol2 transposable element of the medaka fish, *Oryzias latipes*, in zebrafish, *Danio rerio*. Gene 225:17–22

Kawakami K, Shima A, Kawakami N (2000) Identification of a functional transposase of the Tol2 element, an Ac-like element from the Japanese medaka fish, and its transposition in the zebrafish germ lineage. Proc Natl Acad Sci U S A 97:11403–11408

Kay MA, He CY, Chen ZY (2010) A robust system for production of minicircle DNA vectors. Nat Biotechnol 28:1287–1289

Kebriaei P, Huls H, Jena B, Munsell M, Jackson R, Lee DA, Hackett PB, Rondon G, Shpall E, Champlin RE, Cooper LJ (2012) Infusing CD19-directed T cells to augment disease control in patients undergoing autologous hematopoietic stem-cell transplantation for advanced B-lymphoid malignancies. Hum Gene Ther 23:444–450

Keravala A, Liu D, Lechman ER, Wolfe D, Nash JA, Lampe DJ, Robbins PD (2006) Hyperactive Himar1 transposase mediates transposition in cell culture and enhances gene expression in vivo. Hum Gene Ther 17:1006–1018

Kettlun C, Galvan DL, George AL Jr, Kaja A, Wilson MH (2011) Manipulating piggyBac transposon chromosomal integration site selection in human cells. Mol Ther 19:1636–1644

Klasse PJ (2012) The molecular basis of HIV entry. Cell Microbiol 14:1183–1192

Kobelt D, Schleef M, Schmeer M, Aumann J, Schlag PM, Walther W (2013) Performance of high quality minicircle DNA for in vitro and in vivo gene transfer. Mol Biotechnol 53:80–89

Kobiler O, Drayman N, Butin-Israeli V, Oppenheim A (2012) Virus strategies for passing the nuclear envelope barrier. Nucleus 3:526–539

Krieg AM, Yi AK, Matson S, Waldschmidt TJ, Bishop GA, Teasdale R, Koretzky GA, Klinman DM (1995) CpG motifs in bacterial DNA trigger direct B-cell activation. Nature 374:546–549
Kunarso G, Chia NY, Jeyakani J, Hwang C, Lu X, Chan YS, Ng HH, Bourque G (2010) Transposable elements have rewired the core regulatory network of human embryonic stem cells. Nat Genet 42:631–634
Kwon MJ, An S, Choi S, Nam K, Jung HS, Yoon CS, Ko JH, Jun HJ, Kim TK, Jung SJ, Park JH, Lee Y, Park JS (2012) Effective healing of diabetic skin wounds by using nonviral gene therapy based on minicircle vascular endothelial growth factor DNA and a cationic dendrimer. J Gene Med 14:272–278
Lacoste A, Berenshteyn F, Brivanlou AH (2009) An efficient and reversible transposable system for gene delivery and lineage-specific differentiation in human embryonic stem cells. Cell Stem Cell 5:332–342
Lam AP, Dean DA (2010) Progress and prospects: nuclear import of nonviral vectors. Gene Ther 17:439–447
Lederberg J (1952) Cell genetics and hereditary symbiosis. Physiol Rev 32:403–430
Lee ER, Marshall J, Siegel CS, Jiang C, Yew NS, Nichols MR, Nietupski JB, Ziegler RJ, Lane MB, Wang KX, Wan NC, Scheule RK, Harris DJ, Smith AE, Cheng SH (1996) Detailed analysis of structures and formulations of cationic lipids for efficient gene transfer to the lung. Hum Gene Ther 7:1701–1717
Li X, Ewis H, Hice RH, Malani N, Parker N, Zhou L, Feschotte C, Bushman FD, Atkinson PW, Craig NL (2013) A resurrected mammalian hAT transposable element and a closely related insect element are highly active in human cell culture. Proc Natl Acad Sci U S A 110:E478–E487
Liang Q, Kong J, Stalker J, Bradley A (2009) Chromosomal mobilization and reintegration of Sleeping Beauty and piggyBac transposons. Genesis 47:404–408
Lin EH, Keramidas M, Rome C, Chiu WT, Wu CW, Coll JL, Deng WP (2011) Lifelong reporter gene imaging in the lungs of mice following polyethyleneimine-mediated sleeping-beauty transposon delivery. Biomaterials 32:1978–1985
Liu F, Song Y, Liu D (1999) Hydrodynamics-based transfection in animals by systemic administration of plasmid DNA. Gene Ther 6:1258–1266
Liu L, Sanz S, Heggestad AD, Antharam V, Notterpek L, Fletcher BS (2004) Endothelial targeting of the Sleeping Beauty transposon within lung. Mol Ther 10:97–105
Liu G, Geurts AM, Yae K, Srinivasan AR, Fahrenkrug SC, Largaespada DA, Takeda J, Horie K, Olson WK, Hackett PB (2005) Target-site preferences of Sleeping Beauty transposons. J Mol Biol 346:161–173
Liu H, Liu L, Fletcher BS, Visner GA (2006a) Sleeping Beauty-based gene therapy with indoleamine 2,3-dioxygenase inhibits lung allograft fibrosis. Faseb J 20:2384–2386
Liu J, Jeppesen I, Nielsen K, Jensen TG (2006b) Phi c31 integrase induces chromosomal aberrations in primary human fibroblasts. Gene Ther 13:1188–1190
Liu L, Mah C, Fletcher BS (2006c) Sustained FVIII expression and phenotypic correction of hemophilia A in neonatal mice using an endothelial-targeted Sleeping Beauty transposon. Mol Ther 13:1006–1015
Liu J, Skjorringe T, Gjetting T, Jensen TG (2009) PhiC31 integrase induces a DNA damage response and chromosomal rearrangements in human adult fibroblasts. BMC Biotechnol 9:31
Lu J, Zhang F, Xu S, Fire AZ, Kay MA (2012) The extragenic spacer length between the 5′ and 3′ ends of the transgene expression cassette affects transgene silencing from plasmid-based vectors. Mol Ther 20:2111–2119
Lukacs GL, Haggie P, Seksek O, Lechardeur D, Freedman N, Verkman AS (2000) Size-dependent DNA mobility in cytoplasm and nucleus. J Biol Chem 275:1625–1629
Maguire AM, High KA, Auricchio A, Wright JF, Pierce EA, Testa F, Mingozzi F, Bennicelli JL, Ying GS, Rossi S, Fulton A, Marshall KA, Banfi S, Chung DC, Morgan JI, Hauck B, Zelenaia O, Zhu X, Raffini L, Coppieters F, De Baere E, Shindler KS, Volpe NJ, Surace EM, Acerra C, Lyubarsky A, Redmond TM, Stone E, Sun J, McDonnell JW, Leroy BP, Simonelli F, Bennett J (2009) Age-dependent effects of RPE65 gene therapy for Leber's congenital amaurosis: a phase 1 dose-escalation trial. Lancet 374:1597–1605

Mahfouz MM, Li L, Shamimuzzaman M, Wibowo A, Fang X, Zhu JK (2011) De novo-engineered transcription activator-like effector (TALE) hybrid nuclease with novel DNA binding specificity creates double-strand breaks. Proc Natl Acad Sci U S A 108:2623–2628

Matarrese P, Malorni W (2005) Human immunodeficiency virus (HIV)-1 proteins and cytoskeleton: partners in viral life and host cell death. Cell Death Differ 12(Suppl 1):932–941

Mates L, Chuah MK, Belay E, Jerchow B, Manoj N, Acosta-Sanchez A, Grzela DP, Schmitt A, Becker K, Matrai J, Ma L, Samara-Kuko E, Gysemans C, Pryputniewicz D, Miskey C, Fletcher B, Vandendriessche T, Ivics Z, Izsvak Z (2009) Molecular evolution of a novel hyperactive Sleeping Beauty transposase enables robust stable gene transfer in vertebrates. Nat Genet 41:753–761

Mayrhofer P, Blaesen M, Schleef M, Jechlinger W (2008) Minicircle-DNA production by site specific recombination and protein-DNA interaction chromatography. J Gene Med 10:1253–1269

Mayrhofer P, Schleef M, Jechlinger W (2009) Use of minicircle plasmids for gene therapy. Methods Mol Biol 542:87–104

McDonald D, Vodicka MA, Lucero G, Svitkina TM, Borisy GG, Emerman M, Hope TJ (2002) Visualization of the intracellular behavior of HIV in living cells. J Cell Biol 159:441–452

McLachlan G, Davidson H, Holder E, Davies LA, Pringle IA, Sumner-Jones SG, Baker A, Tennant P, Gordon C, Vrettou C, Blundell R, Hyndman L, Stevenson B, Wilson A, Doherty A, Shaw DJ, Coles RL, Painter H, Cheng SH, Scheule RK, Davies JC, Innes JA, Hyde SC, Griesenbach U, Alton EW, Boyd AC, Porteous DJ, Gill DR, Collie DD (2011) Pre-clinical evaluation of three non-viral gene transfer agents for cystic fibrosis after aerosol delivery to the ovine lung. Gene Ther 18:996–1005

McMahon MA, Rahdar M, Porteus M (2012) Gene editing: not just for translation anymore. Nat Methods 9:28–31

Meir YJ, Weirauch MT, Yang HS, Chung PC, Yu RK, Wu SC (2011) Genome-wide target profiling of piggyBac and Tol2 in HEK 293: pros and cons for gene discovery and gene therapy. BMC Biotechnol 11:28

Mesika A, Kiss V, Brumfeld V, Ghosh G, Reich Z (2005) Enhanced intracellular mobility and nuclear accumulation of DNA plasmids associated with a karyophilic protein. Hum Gene Ther 16:200–208

Miao CH, Ohashi K, Patijn GA, Meuse L, Ye X, Thompson AR, Kay MA (2000) Inclusion of the hepatic locus control region, an intron, and untranslated region increases and stabilizes hepatic factor IX gene expression in vivo but not in vitro. Mol Ther 1:522–532

Mikkelsen JG, Yant SR, Meuse L, Huang Z, Xu H, Kay MA (2003) Helper-Independent Sleeping Beauty transposon-transposase vectors for efficient nonviral gene delivery and persistent gene expression in vivo. Mol Ther 8:654–665

Miskey C, Izsvak Z, Plasterk RH, Ivics Z (2003) The Frog Prince: a reconstructed transposon from *Rana pipiens* with high transpositional activity in vertebrate cells. Nucleic Acids Res 31:6873–6881

Mitchell RS, Beitzel BF, Schroder AR, Shinn P, Chen H, Berry CC, Ecker JR, Bushman FD (2004) Retroviral DNA integration: ASLV, HIV, and MLV show distinct target site preferences. PLoS Biol 2:E234

Mitra R, Li X, Kapusta A, Mayhew D, Mitra RD, Feschotte C, Craig NL (2013) Functional characterization of piggyBat from the bat *Myotis lucifugus* unveils an active mammalian DNA transposon. Proc Natl Acad Sci U S A 110:234–239

Moldt B, Yant SR, Andersen PR, Kay MA, Mikkelsen JG (2007) Cis-acting gene regulatory activities in the terminal regions of Sleeping Beauty DNA transposon-based vectors. Hum Gene Ther 18:1193–1204

Nabel GJ, Nabel EG, Yang ZY, Fox BA, Plautz GE, Gao X, Huang L, Shu S, Gordon D, Chang AE (1993) Direct gene transfer with DNA-liposome complexes in melanoma: expression, biologic activity, and lack of toxicity in humans. Proc Natl Acad Sci U S A 90:11307–11311

Narsinh KH, Jia F, Robbins RC, Kay MA, Longaker MT, Wu JC (2011) Generation of adult human induced pluripotent stem cells using nonviral minicircle DNA vectors. Nat Protoc 6:78–88

Nathwani AC, Tuddenham EG, Rangarajan S, Rosales C, McIntosh J, Linch DC, Chowdary P, Riddell A, Pie AJ, Harrington C, O'Beirne J, Smith K, Pasi J, Glader B, Rustagi P, Ng CY, Kay MA, Zhou J, Spence Y, Morton CL, Allay J, Coleman J, Sleep S, Cunningham JM, Srivastava D, Basner-Tschakarjan E, Mingozzi F, High KA, Gray JT, Reiss UM, Nienhuis AW, Davidoff AM (2011) Adenovirus-associated virus vector-mediated gene transfer in hemophilia B. N Engl J Med 365:2357–2365

Nguyen J, Szoka FC (2012) Nucleic acid delivery: the missing pieces of the puzzle? Acc Chem Res 45:1153–1162

Oh YK, Kim JP, Yoon H, Kim JM, Yang JS, Kim CK (2001) Prolonged organ retention and safety of plasmid DNA administered in polyethylenimine complexes. Gene Ther 8:1587–1592

Ohlfest JR, Lobitz PD, Perkinson SG, Largaespada DA (2004) Integration and long-term expression in xenografted human glioblastoma cells using a plasmid-based transposon system. Mol Ther 10:260–268

Ohlfest JR, Demorest ZL, Motooka Y, Vengco I, Oh S, Chen E, Scappaticci FA, Saplis RJ, Ekker SC, Low WC, Freese AB, Largaespada DA (2005a) Combinatorial antiangiogenic gene therapy by nonviral gene transfer using the Sleeping Beauty transposon causes tumor regression and improves survival in mice bearing intracranial human glioblastoma. Mol Ther 12:778–788

Ohlfest JR, Frandsen JL, Fritz S, Lobitz PD, Perkinson SG, Clark KJ, Nelsestuen G, Key NS, McIvor RS, Hackett PB, Largaespada DA (2005b) Phenotypic correction and long-term expression of factor VIII in hemophilic mice by immunotolerization and nonviral gene transfer using the Sleeping Beauty transposon system. Blood 105:2691–2698

Olivares EC, Hollis RP, Chalberg TW, Meuse L, Kay MA, Calos MP (2002) Site-specific genomic integration produces therapeutic Factor IX levels in mice. Nat Biotechnol 20:1124–1128

Orban TI, Apati A, Nemeth A, Varga N, Krizsik V, Schamberger A, Szebenyi K, Erdei Z, Varady G, Karaszi E, Homolya L, Nemet K, Gocza E, Miskey C, Mates L, Ivics Z, Izsvak Z, Sarkadi B (2009) Applying a "double-feature" promoter to identify cardiomyocytes differentiated from human embryonic stem cells following transposon-based gene delivery. Stem Cells 27:1077–1087

Ortiz-Urda S, Lin Q, Yant SR, Keene D, Kay MA, Khavari PA (2003) Sustainable correction of junctional epidermolysis bullosa via transposon-mediated nonviral gene transfer. Gene Ther 10:1099–1104

Osborn MJ, McElmurry RT, Peacock B, Tolar J, Blazar BR (2008) Targeting of the CNS in MPS-IH using a nonviral transferrin-alpha-L-iduronidase fusion gene product. Mol Ther 16:1459–1466

Osborn MJ, McElmurry RT, Lees CJ, DeFeo AP, Chen ZY, Kay MA, Naldini L, Freeman G, Tolar J, Blazar BR (2011) Minicircle DNA-based gene therapy coupled with immune modulation permits long-term expression of alpha-L-iduronidase in mice with mucopolysaccharidosis type I. Mol Ther 19:450–460

Piechaczek C, Fetzer C, Baiker A, Bode J, Lipps HJ (1999) A vector based on the SV40 origin of replication and chromosomal S/MARs replicates episomally in CHO cells. Nucleic Acids Res 27:426–428

Porteus MH, Carroll D (2005) Gene targeting using zinc finger nucleases. Nat Biotechnol 23:967–973

Raices M, D'Angelo MA (2012) Nuclear pore complex composition: a new regulator of tissue-specific and developmental functions. Nat Rev Mol Cell Biol 13:687–699

Riu E, Chen ZY, Xu H, He CY, Kay MA (2007) Histone modifications are associated with the persistence or silencing of vector-mediated transgene expression in vivo. Mol Ther 15:1348–1355

Ruiz FE, Clancy JP, Perricone MA, Bebok Z, Hong JS, Cheng SH, Meeker DP, Young KR, Schoumacher RA, Weatherly MR, Wing L, Morris JE, Sindel L, Rosenberg M, van Ginkel FW, McGhee JR, Kelly D, Lyrene RK, Sorscher EJ (2001) A clinical inflammatory syndrome attributable to aerosolized lipid-DNA administration in cystic fibrosis. Hum Gene Ther 12:751–761

Sellins K, Fradkin L, Liggitt D, Dow S (2005) Type I interferons potently suppress gene expression following gene delivery using liposome(-)DNA complexes. Mol Ther 12:451–459

Sharma N, Hollensen AK, Bak RO, Staunstrup NH, Schroder LD, Mikkelsen JG (2012) The impact of cHS4 insulators on DNA transposon vector mobilization and silencing in retinal pigment epithelium cells. PLoS ONE 7:e48421
Sharma N, Cai Y, Bak RO, Jakobsen MR, Schroder LD, Mikkelsen JG (2013) Efficient Sleeping Beauty DNA transposition from DNA minicircles. Mol Ther Nucleic Acids 2:e74
Sikes ML, O'Malley BW Jr, Finegold MJ, Ledley FD (1994) In vivo gene transfer into rabbit thyroid follicular cells by direct DNA injection. Hum Gene Ther 5:837–844
Singh H, Manuri PR, Olivares S, Dara N, Dawson MJ, Huls H, Hackett PB, Kohn DB, Shpall EJ, Champlin RE, Cooper LJ (2008) Redirecting specificity of T-cell populations for CD19 using the Sleeping Beauty system. Cancer Res 68:2961–2971
Spanggaard I, Corydon T, Hojman P, Gissel H, Dagnaes-Hansen F, Jensen TG, Gehl J (2012) Spatial distribution of transgenic protein after gene electrotransfer to porcine muscle. Hum Gene Ther Methods 23:387–392
Staunstrup NH, Moldt B, Mates L, Villesen P, Jakobsen M, Ivics Z, Izsvak Z, Mikkelsen JG (2009) Hybrid lentivirus-transposon vectors with a random integration profile in human cells. Mol Ther 17:1205–1214
Stenler S, Andersson A, Simonson OE, Lundin KE, Chen ZY, Kay MA, Smith CI, Sylven C, Blomberg P (2009) Gene transfer to mouse heart and skeletal muscles using a minicircle expressing human vascular endothelial growth factor. J Cardiovasc Pharmacol 53:18–23
Suda T, Liu D (2007) Hydrodynamic gene delivery: its principles and applications. Mol Ther 15:2063–2069
Suda T, Suda K, Liu D (2008) Computer-assisted hydrodynamic gene delivery. Mol Ther 16:1098–1104
Sumiyoshi T, Holt NG, Hollis RP, Ge S, Cannon PM, Crooks GM, Kohn DB (2009) Stable transgene expression in primitive human CD34+ hematopoietic stem/progenitor cells, using the Sleeping Beauty transposon system. Hum Gene Ther 20:1607–1626
Sutcliffe JG (1978) Nucleotide sequence of the ampicillin resistance gene of *Escherichia coli* plasmid pBR322. Proc Natl Acad Sci U S A 75:3737–3741
Swierczek M, Izsvak Z, Ivics Z (2012) The Sleeping Beauty transposon system for clinical applications. Expert Opin Biol Ther 12:139–153
Thyagarajan B, Olivares EC, Hollis RP, Ginsburg DS, Calos MP (2001) Site-specific genomic integration in mammalian cells mediated by phage phiC31 integrase. Mol Cell Biol 21:3926–3934
Torikai H, Reik A, Liu PQ, Zhou Y, Zhang L, Maiti S, Huls H, Miller JC, Kebriaei P, Rabinovitch B, Lee DA, Champlin RE, Bonini C, Naldini L, Rebar EJ, Gregory PD, Holmes MC, Cooper LJ (2012) A foundation for universal T-cell based immunotherapy: T cells engineered to express a CD19-specific chimeric-antigen-receptor and eliminate expression of endogenous TCR. Blood 119:5697–5705
Urnov FD, Miller JC, Lee YL, Beausejour CM, Rock JM, Augustus S, Jamieson AC, Porteus MH, Gregory PD, Holmes MC (2005) Highly efficient endogenous human gene correction using designed zinc-finger nucleases. Nature 435:646–651
VandenDriessche T, Ivics Z, Izsvak Z, Chuah MK (2009) Emerging potential of transposons for gene therapy and generation of induced pluripotent stem cells. Blood 114:1461–1468
Vandermeulen G, Marie C, Scherman D, Preat V (2011) New generation of plasmid backbones devoid of antibiotic resistance marker for gene therapy trials. Mol Ther 19:1942–1949
Vaughan EE, Dean DA (2006) Intracellular trafficking of plasmids during transfection is mediated by microtubules. Mol Ther 13:422–428
Vaughan EE, DeGiulio JV, Dean DA (2006) Intracellular trafficking of plasmids for gene therapy: mechanisms of cytoplasmic movement and nuclear import. Curr Gene Ther 6:671–681
Vigdal TJ, Kaufman CD, Izsvak Z, Voytas DF, Ivics Z (2002) Common physical properties of DNA affecting target site selection of Sleeping Beauty and other Tc1/mariner transposable elements. J Mol Biol 323:441–452
Voigt K, Gogol-Doring A, Miskey C, Chen W, Cathomen T, Izsvak Z, Ivics Z (2012) Retargeting Sleeping Beauty transposon insertions by engineered zinc finger DNA-binding domains. Mol Ther 20:1852–1862

Walisko O, Izsvak Z, Szabo K, Kaufman CD, Herold S, Ivics Z (2006) Sleeping Beauty transposase modulates cell-cycle progression through interaction with Miz-1. Proc Natl Acad Sci U S A 103:4062–4067

Wilber A, Linehan JL, Tian X, Woll PS, Morris JK, Belur LR, McIvor RS, Kaufman DS (2007) Efficient and stable transgene expression in human embryonic stem cells using transposon-mediated gene transfer. Stem Cells 25:2919–2927

Williams DA (2008) Sleeping Beauty vector system moves toward human trials in the United States. Mol Ther 16:1515–1516

Wilson MH, Coates CJ, George AL Jr (2007) PiggyBac transposon-mediated gene transfer in human cells. Mol Ther 15:139–145

Wolff JA (1997) Naked DNA transport and expression in mammalian cells. Neuromuscul Disord 7:314–318

Wolff JA, Malone RW, Williams P, Chong W, Acsadi G, Jani A, Felgner PL (1990) Direct gene transfer into mouse muscle in vivo. Science 247:1465–1468

Woodard LE, Li X, Malani N, Kaja A, Hice RH, Atkinson PW, Bushman FD, Craig NL, Wilson MH (2012) Comparative analysis of the recently discovered hAT transposon TcBuster in human cells. PLoS ONE 7:e42666

Wooddell CI, Hegge JO, Zhang G, Sebestyen MG, Noble M, Griffin JB, Pfannes LV, Herweijer H, Hagstrom JE, Braun S, Huss T, Wolff JA (2011) Dose response in rodents and nonhuman primates after hydrodynamic limb vein delivery of naked plasmid DNA. Hum Gene Ther 22:889–903

Xu Y, Hui SW, Frederik P, Szoka FC Jr (1999) Physicochemical characterization and purification of cationic lipoplexes. Biophys J 77:341–353

Xue X, Huang X, Nodland SE, Mates L, Ma L, Izsvak Z, Ivics Z, LeBien TW, McIvor RS, Wagner JE, Zhou X (2009) Stable gene transfer and expression in cord blood-derived CD34+ hematopoietic stem and progenitor cells by a hyperactive Sleeping Beauty transposon system. Blood 114:1319–1330

Yant SR, Kay MA (2003) Nonhomologous-end-joining factors regulate DNA repair fidelity during Sleeping Beauty element transposition in mammalian cells. Mol Cell Biol 23:8505–8518

Yant SR, Meuse L, Chiu W, Ivics Z, Izsvak Z, Kay MA (2000) Somatic integration and long-term transgene expression in normal and haemophilic mice using a DNA transposon system. Nat Genet 25:35–41

Yant SR, Park J, Huang Y, Mikkelsen JG, Kay MA (2004) Mutational analysis of the N-terminal DNA-binding domain of Sleeping Beauty transposase: critical residues for DNA binding and hyperactivity in mammalian cells. Mol Cell Biol 24:9239–9247

Yant SR, Wu X, Huang Y, Garrison B, Burgess SM, Kay MA (2005) High-resolution genome-wide mapping of transposon integration in mammals. Mol Cell Biol 25:2085–2094

Yant SR, Huang Y, Akache B, Kay MA (2007) Site-directed transposon integration in human cells. Nucleic Acids Res 35:e50

Yew NS, Cheng SH (2004) Reducing the immunostimulatory activity of CpG-containing plasmid DNA vectors for non-viral gene therapy. Expert Opin Drug Deliv 1:115–125

Yew NS, Wang KX, Przybylska M, Bagley RG, Stedman M, Marshall J, Scheule RK, Cheng SH (1999) Contribution of plasmid DNA to inflammation in the lung after administration of cationic lipid: pDNA complexes. Hum Gene Ther 10:223–234

Yew NS, Zhao H, Przybylska M, Wu IH, Tousignant JD, Scheule RK, Cheng SH (2002) CpG-depleted plasmid DNA vectors with enhanced safety and long-term gene expression in vivo. Mol Ther 5:731–738

Yokoo T, Kamimura K, Suda T, Kanefuji T, Oda M, Zhang G, Liu D, Aoyagi Y (2013) Novel electric power-driven hydrodynamic injection system for gene delivery: safety and efficacy of human factor IX delivery in rats. Gene Ther 20:816–823

Yusa K, Zhou L, Li MA, Bradley A, Craig NL (2011) A hyperactive piggyBac transposase for mammalian applications. Proc Natl Acad Sci U S A 108:1531–1536

Zayed H, Izsvak Z, Khare D, Heinemann U, Ivics Z (2003) The DNA-bending protein HMGB1 is a cellular cofactor of Sleeping Beauty transposition. Nucleic Acids Res 31:2313–2322

Zayed H, Izsvak Z, Walisko O, Ivics Z (2004) Development of hyperactive Sleeping Beauty transposon vectors by mutational analysis. Mol Ther 9:292–304
Zhang G, Budker V, Wolff JA (1999) High levels of foreign gene expression in hepatocytes after tail vein injections of naked plasmid DNA. Hum Gene Ther 10:1735–1737
Zhang G, Gao X, Song YK, Vollmer R, Stolz DB, Gasiorowski JZ, Dean DA, Liu D (2004) Hydroporation as the mechanism of hydrodynamic delivery. Gene Ther 11:675–682
Zhang X, Epperly MW, Kay MA, Chen ZY, Dixon T, Franicola D, Greenberger BA, Komanduri P, Greenberger JS (2008) Radioprotection in vitro and in vivo by minicircle plasmid carrying the human manganese superoxide dismutase transgene. Hum Gene Ther 19:820–826
Zhao H, Hemmi H, Akira S, Cheng SH, Scheule RK, Yew NS (2004) Contribution of Toll-like receptor 9 signaling to the acute inflammatory response to nonviral vectors. Mol Ther 9:241–248
Zhu L, Mahato RI (2010) Lipid and polymeric carrier-mediated nucleic acid delivery. Expert Opin Drug Deliv 7:1209–1226

Chapter 5
Human Stromal Stem Cell Therapy Using Gene-Modified Cells

Walid Zaher and Moustapha Kassem

List of Abbreviations

Runx2	Runt-related transcription factor 2
BMB2	Bone morphogenetic protein 2
HIF-1α	Hypoxia-inducible factor 1, α subunit
Osx	Osterix
IGF-1	Insulin-like growth factor 1
VEGF	Vascular endothelial growth factor
Akt	Protein kinase B
ANG1	Angiopoietin 1
BCL2	B cell lymphoma 2
SDF1	Stromal-derived factor 1
TNF	Tumor necrosis factor
CXCR4	Chemokine receptor type 4
NICD	Notch intracellular domain
GLP-1	Glucagon-like peptide 1

M. Kassem (✉) · W. Zaher
Department of Endocrinology and Metabolism, Endocrine Research Laboratory (KMEB), Odense University Hospital, University of Southern Denmark, Kloevervaenget 6, 4th floor, 5000 Odense, Denmark
e-mail: mkassem@health.sdu.dk

M. Kassem
DanStem (Danish stem cell center), Panum Institute, University of Copenhagen, Copenhagen, Denmark

W. Zaher · M. Kassem
Stem Cell Unit, Department of Anatomy, College of Medicine, King Saud University, Riyadh, Saudi Arabia

X.-Q. Li et al. (eds.), *Somatic Genome Manipulation*,
DOI 10.1007/978-1-4939-2389-2_5

IFN-α	Interferon-α
IFN-β	Interferon-β
IL-2	Interleukin 2
IL12	Interleukin 12
CCr1	Chemokine receptor type 1

5.1 Introduction

There is an increasing need to develop novel approaches for the treatment of a large number of chronic degenerative diseases affecting primarily the aging population and where there is currently no effective therapy. These diseases include a variety of conditions including Parkinson's disease, liver failure, diabetes, osteoarthritis, and osteoporosis. Cell therapy or cellular therapeutics is being introduced into clinical practice with the aim of restoring defective organ functions through transplanting healthy and functional cells. Stem cells, because of their nature, are currently considered the most suitable cells for cell therapy. Combining gene therapy with stem cell therapy can potentially enhance the beneficial characteristics of stem cells and prevent possible side effects.

5.2 Stem Cells: Definition and Types

Stem cells are defined as cells that can self-renew and differentiate into mature cells (Keller 2005). Based on their differentiation potential, stem cells can be classified as pluripotent where stem cells can differentiate potentially into all somatic cell types and exemplified by embryonic stem cells (ESC), or multipotent, bipotent, or monopotent where stem cells can differentiate into a limited number of cell types and this usually the case of tissue-specific (adult) stem cells. ESCs are isolated from the inner cell mass of mammalian blastocysts and the first human ESC (hESC) line was derived in 1998 by Thomson and colleagues from supernumerary in vitro fertilization (IVF) clinic embryos (Thomson et al. 1998). Tissue-specific stem cells (or adult stem cells) are isolated from different tissues using either a physicochemical procedure, e.g., adherence to plastic, density gradient centrifugation, or through immune cell sorting using antibodies against specific cell membrane-associated antigens.

A novel "manufactured" stem cell type that has received enormous attention in recent years due to its potential use in therapy is the induced pluripotent stem cells (iPSC). iPSC cells are pluripotent, ESC-like cells, established via reprogramming of somatic cells using specific factors called "reprogramming factors" (Takahashi et al. 2007). The most widely used set of reprogramming factors are Oct4, Sox2, Klf4, and c-Myc (Takahashi and Yamanaka 2006). However, combinations of other genes, proteins, microRNA (miRNAs), or chemicals have been reported to create iPS cells with varying degree of efficiency (Hanna et al. 2009; Plath and Lowry 2011).

5.3 Mesenchymal (Stromal) Stem Cells (MSC)

MSC are a multipotent, adult stem cell type not only identified in bone marrow stroma but also present in the stroma of other organs. We focus on MSC in this chapter since it is being introduced into clinical medicine in an increasing number of clinical trials. MSC are fusiform, fibroblast-like cells that are negative for hematopoietic surface markers: CD34, CD45, CD14, CD31, CD133, and positive for a number of markers: Stro-1, CD90, CD73, CD105, CD166, and CD44. Traditionally, human MSC have been isolated from low-density mononuclear cell populations of bone marrow, based on their selective adherence to plastic surfaces, compared to hematopoietic cells (Friedenstein et al. 1970; Luria et al. 1971; Kassem et al. 1993; Rickard et al. 1996). Single-cell cloning of MSC has demonstrated that only 30 % of putative MSC clones are multipotential and thus true MSC (Kuznetsov et al. 1997). MSC-like cells have been isolated from other sources including peripheral blood (Kuznetsov et al. 2001), umbilical cord blood (Rosada et al. 2003), synovial membranes (De Bari et al. 2001), deciduous teeth (Miura et al. 2003), and amniotic fluid (De Coppi et al. 2007). These various MSC populations share some common properties and cell surface phenotype but differ in their differentiation potential and their gene expression profile (Al-Nbaheen et al. 2012). Bone marrow-derived MSC are characterized by their ability to differentiate into mesoderm-like cells, e.g., osteoblasts (Kassem et al. 1993), adipocytes (Abdallah et al. 2005), and chondrocytes (Johnstone et al. 1998) in vitro and to form bone and bone marrow organ that support hematopoiesis upon implantation in an open system (subcutaneous implantation) in immune-deficient mice (Abdallah et al. 2008).

5.4 Stem Cell Therapy

Two approaches employing stem cells are being examined in a number of preclinical and clinical (mostly phase I) trials. The first is to transplant stem cells that have been differentiated into a specific lineage to tissues with the aim of replacing the diseased or dysfunctional tissues (called replacement therapy), e.g., transplantation of stem-cell-differentiated neuronal cells in preclinical animal models of stroke or spinal cord injury (Salazar et al. 2010; Jensen et al. 2011). The second approach makes use of the fact that stem cells (and in particular MSC) can secrete a large number of regeneration-enhancing growth factors and thus when transplanted into injured tissues, the stem cells enhance tissue regeneration. This form of therapy has been termed stem-cell-based humoral therapy. An example is the implantation of MSC into damaged myocardium in order to enhance its regeneration (Grauss et al. 2007).

5.5 Use of Genetically Modified Stem Cells in Preclinical and Clinical Therapy

Targeting stem cells for gene therapy offers unique advantages over direct gene transfer into the body since it makes use of the possible long half-life of stem cells within the tissues as well as the natural ability of stem cells for integration within the tissues (Abdallah and Kassem 2009). In addition, gene therapy can act to confer new characteristic on stem cells, e.g., stem cells can act as carriers for therapeutic genes or proteins into organs or tissues. A number of recent studies have employed gene-modified MSC in preclinical and clinical settings and some of the results provide basis for human clinical translation. Some of these preclinical trials in different medical specialties are summarized in Table 5.1 and discussed in the following paragraphs.

Table 5.1 Fields of regenerative medicine 21-1: examples of preclinical studies using gene-modified human mesenchymal stem cells in different disease models

System	Disease model	Cell model	Gene altered	Rationale and outcome	Reference
Bone	Critical defect or fractures	Rat BMMSC	Runx2	Accelerate healing of critical-sized defects	(Wojtowicz et al. 2010)
		Human adipose-derived MSC	BMB2	Enhance bone formation	(Peterson et al. 2005)
		Rat BMMSC	HIF-1α	Enhance angiogenesis and bone formation	(Zou et al. 2011)
		Mouse BMMSC	Osx	Enhance bone formation	(Tu et al. 2007)
		Mouse cloned BMMSC	IGF-1	Enhance homing and bone formation	(Shen et al. 2002)
		Mouse BMMSC	BMB 2 and VEGF	Enhance vascularization and bone formation	(Kumar et al. 2010b)
	Osteoporosis	Murine BMMSC	BMB2 and α4 integrin	Improve homing to bone marrow and bone mass	(Kumar et al. 2010a)

Table 5.1 (continued)

System	Disease model	Cell model	Gene altered	Rationale and outcome	Reference
Cardio-vascular	Acute Myocardial Infarction	Rat BMMSC	Akt and ANG1	Prevent cell apoptosis, enhance angiogenesis	(Shujia et al. 2008)
		Rat BMMSC	BCL2	Prevent apoptosis and enhance survival	(Li et al. 2007)
		Rat BMMSC	SDF1 and VEGF	Improve homing and enhance survival	(Tang et al. 2010)
		Rat BMMSC	IGF-1	Improve survival, and engraftment, promote stem cell recruitment	(Haider et al. 2008)
	Ischemic heart disease	Mouse BMMSC	CXCR4	Enhance binding to SDF1, Improve homing and engraftment	(Zhang et al. 2008)
Neuro-logical	Controlled cortical impact (CCI)	Human telomerized bone marrow stromal cells	GLP-1	Reduction of hippocampal cell loss as well as attenuation of cortical neuronal and glial abnormalities	(Heile et al. 2009)
	Parkinson's disease	Rat and human bone marrow MSCs	NICD	Functional improvement in rotational behavior and adjusting step and paw-reaching tests	(Dezawa et al. 2004)
Cancer	Melanoma	Mouse BMMSC	IFN–α	Increase survival, anti-tumorigenic effects	(Ren et al. 2008a)
	Prostate cancer	Mouse BMMSC	IFN-β	Increase survival, anti-tumorigenic effects	(Ren et al. 2008b)
	Melanoma and Cervical cancer	Mouse BMMSC	IL12	Increase survival, anti-tumorigenic effects	(Seo et al. 2011)
Homing	AMI	Mouse BMMSC	CCr1	Enhance viability and migration	(Huang et al. 2010)
	AMI	Rat BMMSC	CXCR4	Enhance homing and migration	(Cheng et al. 2008)

BMMSC bone marrow mesenchymal stem cell, *AMI* acute myocardial infarction, *MSC* mesenchymal stem cells

5.6 Use of Gene-Modified MSC in Skeletal Repair

Manufactured bone tissue is needed for replacement in an increasing number of common disorders, e.g., bone defects resulting from complicated fractures caused by trauma or surgically induced amputation, reconstruction procedures carried out by cranial/oro-maxillo-facial surgeons for genetic or acquired diseases and in situations of nonunion fracture healing caused by defective bone formation. Historically, the paradigm treatment for these conditions has been the use of the patient's own bone (autograft) harvested from the iliac crest or fibula. However, there are significant limitations for the use of autografts, including the risk of severe or long-term discomfort at the bone graft donor site, functional or cosmetic deficit, and complications such as infection and nerve damage (Ahlmann et al. 2002). Allografts obtained from cadaveric donors or from patients undergoing surgery have been employed as an alternative for autografts (Emms et al. 2009), However, the use of fresh allografts carries risk of pathogen transmission.

Transplantation of MSC that are capable of bone generation in vivo is therefore an attractive alternative approach to autograft and allograft techniques. Genetic modification of MSC has been tried to enhance bone-forming capacity and efficacy of bone formation through a number of approaches. Overexpression of growth factors known to enhance bone formation has been tried, e.g., overexpression of members of bone morphogenetic proteins (BMP) (Chen et al. 2004) or vascular endothelial growth factor (VEGF) (Johnson et al. 1992; Galloway et al. 2000; Carano and Filvaroff 2003; Kneser et al. 2006; Ho and Kuo 2007; Kodach et al. 2008; Kanczler et al. 2010). Another approach is to enhance osteoblast differentiation capacity and hence bone formation through overexpression of osteoblast master transcription factors, e.g., Runx2 (Zhao et al. 2005) and Osterix (Tu et al. 2007). Alternatively, enhancing bone formation through augmentation of vascularization ability of MSC has been tried (see below). The functionality of gene-modified MSC have been tested by delivering the cells at the injury site with or without an osteoconductive scaffold.

A number of studies have examined the effect of BMP-overexpressing MSC on fracture healing. In a recent publication, human adipose tissue-derived MSC overexpressing BMP-2 were loaded on a collagen–ceramic carrier and implanted into 6 mm femorals with critical size defect in adult rats. After 8 weeks, transgenic cells resulted in full healing of bone defects whereas transplanting control (non-transgenic) cells did not elicit significant bone formation (Peterson et al. 2005). Similarly, murine bone marrow-derived MSC overexpressing BMP-2 and VEGF were systemically administered in a mouse segmental bone defect in the tibia and led to enhanced bone formation due to increased number of osteoblastic cells as well as enhanced tissue vascularity (Kumar et al. 2010b). Murine bone marrow MSC overexpressing BMP-2 and α-4 integrin were shown to home to bone marrow following systemic infusion and to increase bone mass in ovariectomy-induced osteoporotic mice (Kumar et al. 2010a). Other bone formation enhancing growth factors have been employed. Transgenic IGF-1-overexpressing murine bone marrow MSC were systemically administered into mice with stabilized mid-femur closed fracture and resulted in an improved healing with an increased volume of mineralized callus compared with mice treated with control non-transfected cells (Shen et al. 2002).

The possibility of increasing bone formation through increased mature osteoblast formation of MSC has also been tested. Runx2, the master transcription factor of osteoblastic phenotype, was overexpressed in rat bone marrow MSC and the cells implanted with a scaffold in critical-sized segmental defects in rats and the transgenic cells resulted in accelerated healing of bone defect compared to control unmodified MSC or defects treated with cell-free scaffold (Wojtowicz et al. 2010). Another study tested the effects of overexpression of osterix (Osx), another known osteoblast-specific transcription factor, in murine MSC on their ability for bone regeneration in vivo. Bone marrow-derived MSC overexpressing osterix were implanted in mouse calvarial critical-sized bone defects and resulted in healing of bone defects. The total amount of newly formed bone was five times greater in the osterix-treated group compared to control group implanted with non-transgenic cells (Tu et al. 2007). Furthermore, in this study, the Osx-transduced cells were identified in areas of newly formed bone (Tu et al. 2007).

One of the limitations of successful in vivo bone formation by the transplanted MSC is their limited ability to enhance vascular formation. Genetic modification of MSC has tried to enhance their vascularization potential. For example, rat bone marrow MSC were transfected with VEGF, cultured on nano-hydroxyapatite/collagen (NHAC) scaffold, and implanted locally in mice. VEGF-overexpressing MSC exhibited higher levels of alkaline phosphatase activity and a larger amount of ectopic bone formation associated with enhanced vascularization compared to control non-transfected cells. Also, hypoxia-inducible factor-1α (HIF-1α) has recently been identified as a major regulator of angiogenic–osteogenic coupling and a number of osteogenesis- and vasculogenesis-associated genes are the direct targets of HIF-1a including VEGF and stromal-derived factor 1 (SDF1, (Sarkar et al. 2009). In a recent study, rat bone marrow-derived MSC were transduced by lentivirus containing either wild-type or constitutively active HIF-1α (cHIF). HIF-1α-overexpressing MSC exhibited an enhanced osteogenic differentiation in vitro and enhanced the repair of critical-sized calvarial defects in rats in vivo with evidence of increased newly formed bone volume and blood vessel volume (Zou et al. 2011).

The above-cited studies demonstrate a proof of concept of the possibility of enhancing the osteoblast differentiation and in vivo bone formation capacity of MSC through genetic manipulation. However, clinical translation will require confirmatory studies in large animal models as well as consideration of the safety issues discussed below.

5.7 Use of Gene-Modified MSC in Cardiovascular Diseases

Cardiovascular diseases including heart failure due to mechanical impairment of myocardium is one of the major causes of mortality worldwide (Fuster et al. 2011) and MSC have been tested for their ability to enhance myocardial tissue regeneration following acute myocardial infarction (AMI) or chronic ischemic heart failure (CHF). The beneficial effects of MSC in these conditions are most probably related to "humoral" factors secreted by MSC with regeneration enhancing effects. Genetic modification of MSC to enhance production of a particular growth factor or cytokine has been tested for their usefulness in enhancing cardiac regeneration. For example, rat MSC overexpressing IGF-1 were injected intra-myocardially in a rat model of AMI. The injected cells resulted in increased local production of stromal-derived factor-1 (SDF-1) a chemotactic factor associated with increased recruitment of c-kit(+), MDR1(+), CD31(+), and CD34(+) vascular stem cells into the infarcted myocardium resulting in reduced infarction size and increased ejection fraction (Haider et al. 2008). In another study, rat bone marrow MSC overexpressing SDF-1 and VEGF were implanted locally into infarcted rat myocardium and led to reduced infarct size and postinfarction fibrosis as well as increased vascularization in the damaged myocardium (Tang et al. 2010).

A member of the chemokine family is the C–C chemokine receptor type 1 (CCR1) that is overexpressed in mouse bone marrow MSC. These transgenic cells were injected locally 1 h after coronary artery ligation in a mouse model of acute myocardial infarction. The CCR1-overexpressing injected mice showed reduced cardiomyocyte apoptosis, increased capillary density and restoration of cardiac function, and reduction in the infarct size (Huang et al. 2010).

Limited survival of transplanted stem cells in the myocardium has been one of the factors compromising the therapeutic benefits of transplantation therapy. An example to overcome this limitation is through overexpression of survival-enhancing factors. VEGF (Tao et al. 2011) and Akt (Shiojima and Walsh 2006) are the major angiogenic factors being investigated for the treatment of myocardial infarction (MI). Rat MSC overexpressing survival protein B-cell lymphoma 2 (Bcl-2) were locally injected into a rat model of AMI via intracardiac route. The modified cells exhibited long-term survival at the infarction location and resulted in 17% reduction in infarct size compared to controls. Interestingly, the beneficial effects of Bcl-2 overexpression were mediated through an increased production of VEGF (Li et al. 2007). In another study, the survival of injected rat MSC was improved through overexpression of Akt protein (also known as protein kinase B). Transplanting Akt-overexpressing cells in a rat model of AMI significantly reduced infarct size and improved left ventricular function (Mangi et al. 2003). Similarly, intramyocardial injection of rat MSC over-expressing both angiopoietin 1 (Ang-1) and Akt in rat AMI model resulted in a better functional recovery of myocardium (Shujia et al. 2008).

5.8 Use of Gene-Modified MSC in Neurological Diseases

Stroke, spinal cord injuries, Parkinson's disease, and Alzheimer disease are major causes of disability and death and bone marrow stromal cells are one of the most promising candidates for cellular therapy of these conditions. MSC possess neuron markers such as glial fibrillary acidic protein (GFAP), neuron-specific enolase, nestin (Khoo et al. 2008). While MSC have been reported to differentiate into neuronal-like cells (Wislet-Gendebien et al. 2012), this differentiation capacity was not reproducible by other investigators (Bertani et al. 2005; Krabbe et al. 2005) and most probably represents an in vitro phenomenon and not a physiological mechanisms taking place in vivo.

Some investigators tried to induce a neuronal-like phenotype in MSC through genetic manipulation and in vitro "reprogramming" with neuro-associated differentiation factors. For example, rat and human bone marrow MSC transfected with the intracellular domain of Notch, a key regulator of the terminal differentiation of neurons and glial cells, treated in vitro with basic fibroblast growth factor (bFGF), forskolin, and ciliary neurotrophic factor followed by treatment with glial cell-line-derived neurotrophic factor (GDNF) produced a proportion of cells that were tyrosine hydroxylase positive. Local (intra-striatal) implantation of these genetically modified and reprogrammed rat MSC cells into a rat model of Parkinson's led to improvements in apomorphine-induced rotational behavior and adjusting step and paw-reaching tests (Dezawa et al. 2004). However, these results need confirmation by other investigators.

The transplantation of genetically modified MSC as vehicle for the production of important therapeutic proteins has been tested in a number of preclinical studies. glucagon-like peptide 1 (GLP-1) is a 30-amino-acid peptide hormone produced in the gut epithelial endocrine L cells and it exerts neuro-protective properties (Holst 2004; Perry and Greig 2004). Heile et al. tested the efficacy of human bone marrow-derived MSC transduced with the human telomerase reverse transcriptase gene (hTERT), human MSC-telomerase reverse transcriptase gene (hMSC-TERT) (Simonsen et al. 2002), and overexpressing GLP-1 in a controlled cortical impact (CCI) rat model. Results demonstrated that transplanted hMSC-TERT-GLP1 cells resulted in the reduction of hippocampal cell loss as well as attenuation of cortical neuronal and glial abnormalities, as measured by MAP-2 and GFAP expression (Heile et al. 2009). In another study, the intraventricular implantation of encapsulated hMSC-TERT-GLP1 cells resulted in decreased iglial fibrillary acidic protein and microglial integrin α M (CD11b) immunoreactivity in the frontal lobes suggesting anti-inflammatory and neuro-protective properties of the transplanted cells (Klinge et al. 2011).

5.9 Use of Gene-Modified MSC in Cancer Therapy

Due to their immune-modulatory effects and tumor tropism, MSC have been considered as an ideal delivery vehicle for cancer gene therapies (Seo et al. 2011). MSC possess different molecular mechanisms involving chemokine receptors and adhesion molecules that facilitate MSC migration toward tumor microenvironment (Menon et al. 2007; Chamberlain et al. 2008; McGrail et al. 2012). Genetic modification of MSC can further enhance migration potential of MSC to specific tumors

(Menon et al. 2008). This approach method holds great promise for the treatment of cancers that are otherwise resistant to conventional therapies.

Interferon-α (IFN-α) is one of the most frequently used adjuvant therapies to eradicate micro-metastatic deposits in patients with a high risk of systemic recurrence in many cancer types (Omori et al. 2012) and it is one of the cancer frontline treatment modalities in ovarian cancer (Ingersoll et al. 2012) and renal cell carcinoma (Kruck et al. 2012). Intravenous infusion of murine MSC overexpressing IFN-α homed to mouse melanoma lung metastasis and was associated with reduced tumor growth and significantly increased survival of the treated animals (Ren et al. 2008a). In addition, the same group reported that intravenous infusion of MSC overexpressing IFN-β reduced tumor growth in a mouse model of prostate cancer with lung metastasis due to increased metastatic tumor cell apoptosis and decreased tumor cell proliferation and vasculature (Ren et al. 2008b).

Interleukins are another group of cytokines that are essential members of the immune system. Rat MSC overexpressing human interleukin 2 (IL-2) injected directly into either the glioma growth in rat brain or the contralateral hemisphere, resulted in inhibition of tumor growth in both experiments as evidenced by magnetic resonance imaging (MRI) scanning, and was associated with increased survival of treated animals (Nakamura et al. 2004).

Recently, the best route for administering genetically modified MSC as antitumour therapy was tested. Rat MSC overexpressing interleukin 12 (IL12) were injected into mice bearing solid and metastatic B16F10 melanoma via intra-tumor, subcutaneous, or intravenous route. Animals receiving local (intra-tumor) injection exhibited stronger tumor-specific T-cell responses and antitumor effects as well as more sustained expressions of IL-12 and interferon (IFN)-γ at the tumor site. On the other hand, subcutaneous injection of IL-12-MSC at the contralateral site exhibited similar levels of serum IL-12 and IFN-γ but much weaker antitumor effects. Intravenous injection elicited earlier peaks in serum levels of IL12 and IFN-γ but induced weaker tumor-specific T-cell responses and antitumor effects. These findings suggested that intra-tumor injection of MSC is a more efficient approach for therapy (Seo et al. 2011).

5.10 Use of Gene-Modified MSC to Enhance Homing to Injured Tissues

The inefficient homing of systemically injected MSC to noninjured tissues (Bentzon et al. 2005) remains one of the major obstacles facing cell therapy. Homing to injured tissues has been studied extensively in leucocytes and it is a complex and highly regulated multistep process in which circulating cells loosely roll/adhere upon sinusoidal endothelial cells followed by firm adhesion that prevents their movement back to circulation and facilitate their transmigration to their destined tissues under the influence of cytokines and adhesion molecules (Karp and Leng Teo 2009). Following injury, damaged cells secrete a number of chemokines that act as attractants to cells participating in tissue repair (Wu and Zhao 2012). Several groups have demonstrated successful but rather limited homing after systemic delivery to ischemic, irradiated or otherwise injured skeletal tissues in which only a small number of the transplanted cells were found in the target tissue.

One of the hypothesized reasons for poor homing capacity of MSC is inadequate expression of homing associated chemokines and adhesion molecules. For example, CXCR4, a known homing signaling molecule, is expressed at low levels in MSC (Wynn et al. 2004; Cheng et al. 2008; Sackstein et al. 2008). Interestingly, in a number of in vivo studies, homing of transplanted MSC to myocardium (Zhang et al. 2008) and bone (Devine et al. 2002) has been improved by overexpressing CXCR4 in MSC. The functional effects of increased homing to injured sites have also been reported. MSC overexpressing CXCR4 infused intravenously 24 h after coronary occlusion in a rat model of AMI where the cells homed to the infarcted myocardium and resulted in better recovery of left ventricular function (Cheng et al. 2008). Similarly, a murine bone marrow cell line overexpressing CXCR4 was intravenously administrated in immune-competent glucocorticoid-induced osteoporotic mice and better homed to bone marrow and increased bone mass (Lien et al. 2009).

5.11 Challenges for Clinical Use of Genetically Modified Stem Cells

The safety record of non-genetically modified adult stem cells is excellent (Lepperdinger et al. 2008). However, there are concerns that genetic manipulation of stem cells may lead to genetic instability and increase the risk for tumor transformation upon transplantation. Currently, two clinical trials are being conducted using genetically modified neuronal stem cells in patients with stroke and safety data from these trials will provide important insight into the safety of clinical use of genetically modified stem cells in therapy. The first trial is conducted by the CellMed company in collaboration with International Neuroscience Institute, Hanover, Germany, where telomerized MSC (hMSC-TERT) (Simonsen et al. 2002) were genetically engineered to overexpress GLP-1 protein known to enhance brain regeneration following injury (see above). It is a phase I clinical trial in patients with cerebral hemorrhage that require surgery. After removing of the bleed, the GLP-1-overexpressing hMSC-TERT cells encapsulated in alginate beads are transplanted within a retrievable mesh device (has been described in the popular press as "tea-bag") and removed after the treatment period of 14 days to carry out histological analysis of the cells within the mesh to ensure safety as well as quality control related to the viability of the cells during this procedure (http://www.biocompatibles.com/media/press-releases/first-ever-treatment-of-stroke-patient-with-stem-cell-therapy-product). The second trial is also a phase I safety study and is conducted by ReNeuron where a neuronal stem cell product, ReN001 is being administered in ascending doses to a total of 12 stroke patients suffering from ischaemic stroke (http://clinicaltrials.gov/ct2/show/NCT01151124?term=stem+cells+and+stroke&rank=6)

The cells used in this trial are conditionally immortalized neuronal cells with a gene-enhancing cell proliferation: c-myc under tamoxifen-inducible promoter (Pollock et al. 2006). In the presence of tamoxifen, the cells proliferate and can be expanded in vitro to generate sufficient number of cells needed for clinical trials and when the drug is removed, the cells behave as "normal" cells.

Additional concerns related to use of genetically modified MSC in therapy is related to immunological rejection. The two abovementioned trials are based on using "off-the-shelf" allogeneic stem cells that facilitate clinical use of stem cell therapeutics. However, the use of the "universal" stem cells for therapy creates concerns regarding immunological rejection. MSC is generally considered as "immunologically tolerant" due to their immune suppressive effects and thus are being tested in allogenic transplantation protocols. However, immunological consequences following allogenic genetically modified stem cell transplantation will require further investigation during any planned clinical trials.

5.12 Conclusion and Future Directions

Combining gene therapy and stem cell therapy through the creation of the clinical-grade gene-modified stem cells has enormous potential for treating and ameliorating the conditions of a large number of patients suffering from chronic diseases. It also allows the creation of "off-the-shelf" stem cell products ready for use in therapy and thus facilitates the use of stem cell-based products in clinical trials. The clinical use of gene-modified stem cells requires careful monitoring for long-term side effects in the form of tumor formation or undesired immunological responses. Understanding the biology of gene-modified stem cells and the development of in vitro assays predictive for their in vivo behavior and genetic stability will allow wider use of gene-modified stem cells in clinical protocols.

Acknowledgments We acknowledge the support of grants from the Lundbeck foundation (Denmark), NovoNordisk foundation (Denmark), and KACST (Project Code: 10-BIO1308–02) (KSA).

References

Abdallah BM, Kassem M (2009) The use of mesenchymal (skeletal) stem cells for treatment of degenerative diseases: current status and future perspectives. J Cell Physiol 218:9–12

Abdallah BM, Haack-Sorensen M, Burns JS, Elsnab B, Jakob F, Hokland P, Kassem M (2005) Maintenance of differentiation potential of human bone marrow mesenchymal stem cells immortalized by human telomerase reverse transcriptase gene despite [corrected] extensive proliferation. Biochem Biophys Res Commun 326:527–538

Abdallah BM, Ditzel N, Kassem M (2008) Assessment of bone formation capacity using in vivo transplantation assays: procedure and tissue analysis. Methods Mol Biol 455:89–100

Ahlmann E, Patzakis M, Roidis N, Shepherd L, Holtom P (2002) Comparison of anterior and posterior iliac crest bone grafts in terms of harvest-site morbidity and functional outcomes. J Bone Joint Surg Am 84-A:716–720

Al-Nbaheen M, Vishnubalaji R, Ali D, Bouslimi A, Al-Jassir F, Megges M, Prigione A, Adjaye J, Kassem M, Aldahmash A (2012) Human stromal (mesenchymal) stem cells from bone marrow, adipose tissue and skin exhibit differences in molecular phenotype and differentiation potential. Stem Cell Rev 9(1):32–43

Bentzon JF, Stenderup K, Hansen FD, Schroder HD, Abdallah BM, Jensen TG and Kassem M (2005) Tissue distribution and engraftment of human mesenchymal stem cells immortalized by human telomerase reverse transcriptase gene. Biochem Biophys Res Commun 330:633–640

Bertani N, Malatesta P, Volpi G, Sonego P, Perris R (2005) Neurogenic potential of human mesenchymal stem cells revisited: analysis by immunostaining, time-lapse video and microarray. J Cell Sci 118:3925–3936

Carano RA, Filvaroff EH (2003) Angiogenesis and bone repair. Drug DiscovToday 8:980–989

Chamberlain G, Wright K, Rot A, Ashton B, Middleton J (2008) Murine mesenchymal stem cells exhibit a restricted repertoire of functional chemokine receptors: comparison with human. PLoS One 3:e2934

Chen D, Zhao M, Mundy GR (2004) Bone morphogenetic proteins. Growth Factors 22:233–241

Cheng Z, Ou L, Zhou X, Li F, Jia X, Zhang Y, Liu X, Li Y, Ward CA, Melo LG, Kong D (2008) Targeted migration of mesenchymal stem cells modified with CXCR4 gene to infarcted myocardium improves cardiac performance. Mol Ther 16:571–579

De Bari C Dell'Accio F Tylzanowski P Luyten FP (2001) Multipotent mesenchymal stem cells from adult human synovial membrane. Arthritis Rheum 44:1928–1942

De Coppi P Bartsch G Jr. Siddiqui MM Xu T Santos CC Perin L Mostoslavsky G Serre AC Snyder EY Yoo JJ Furth ME Soker S Atala A (2007) Isolation of amniotic stem cell lines with potential for therapy. Nat Biotechnol 25:100–106

Devine MJ, Mierisch CM, Jang E, Anderson PC, Balian G (2002) Transplanted bone marrow cells localize to fracture callus in a mouse model. J Orthop Res 20:1232–1239

Dezawa M, Kanno H, Hoshino M, Cho H, Matsumoto N, Itokazu Y, Tajima N, Yamada H, Sawada H, Ishikawa H, Mimura T, Kitada M, Suzuki Y, Ide C (2004) Specific induction of neuronal cells from bone marrow stromal cells and application for autologous transplantation. J Clin Invest 113:1701–1710

Emms NW, Buckley SC, Stockley I, Hamer AJ, Kerry RM (2009) Mid- to long-term results of irradiated allograft in acetabular reconstruction: a follow-up report. J Bone Joint Surg Br 91:1419–1423

Friedenstein AJ, Chailakhjan RK, Lalykina KS (1970) The development of fibroblast colonies in monolayer cultures of guinea-pig bone marrow and spleen cells. Cell Tissue Kinet 3:393–403

Fuster V, Kelly BB, Vedanthan R (2011) Global cardiovascular health: urgent need for an intersectoral approach. J Am Coll Cardiol 58:1208–1210

Galloway SM, McNatty KP, Cambridge LM, Laitinen MP, Juengel JL, Jokiranta TS, McLaren RJ, Luiro K, Dodds KG, Montgomery GW, Beattie AE, Davis GH, Ritvos O (2000) Mutations in an oocyte-derived growth factor gene (BMP15) cause increased ovulation rate and infertility in a dosage-sensitive manner. Nat Genet 25:279–283

Grauss RW, Winter EM, van TJ, Pijnappels DA, Steijn RV, Hogers B, van der Geest RJ, de Vries AA, Steendijk P, van der Laarse A, Gittenberger-de Groot AC, Schalij MJ, Atsma DE (2007) Mesenchymal stem cells from ischemic heart disease patients improve left ventricular function after acute myocardial infarction. Am J Physiol Heart Circ Physiol 293:H2438–H2447

Haider HK, Jiang S, Idris NM, Ashraf M (2008) IGF-1-overexpressing mesenchymal stem cells accelerate bone marrow stem cell mobilization via paracrine activation of SDF-1alpha/CXCR4 signaling to promote myocardial repair. Circ Res 103:1300–1308

Hanna J, Saha K, Pando B, van ZJ, Lengner CJ, Creyghton MP, van OA, Jaenisch R (2009) Direct cell reprogramming is a stochastic process amenable to acceleration. Nature 462:595–601
Heile AM, Wallrapp C, Klinge PM, Samii A, Kassem M, Silverberg G, Brinker T (2009) Cerebral transplantation of encapsulated mesenchymal stem cells improves cellular pathology after experimental traumatic brain injury. Neurosci Lett 463:176–181
Ho QT, Kuo CJ (2007) Vascular endothelial growth factor: biology and therapeutic applications. Int J Biochem Cell Biol 39:1349–1357
Holst JJ (2004) On the physiology of GIP and GLP-1. Horm Metab Res 36:747–754
Huang J, Zhang Z, Guo J, Ni A, Deb A, Zhang L, Mirotsou M, Pratt RE, Dzau VJ (2010) Genetic modification of mesenchymal stem cells overexpressing CCR1 increases cell viability, migration, engraftment, and capillary density in the injured myocardium. Circ Res 106:1753–1762
Ingersoll SB, Ahmad S, Finkler NJ, Edwards JR, Holloway RW (2012) Cellular therapy for ovarian cancer: experimental and clinical perspectives. Curr Med Chem 19(22):3787–3793
Jensen MB, Yan H, Krishnaney-Davison R, Al SA, Zhang SC (2011) Survival and differentiation of transplanted neural stem cells derived from human induced pluripotent stem cells in a rat stroke model. J Stroke Cerebrovasc Dis 22(4):304–308
Johnson EE, Urist MR, Finerman GA (1992) Resistant nonunions and partial or complete segmental defects of long bones. Treatment with implants of a composite of human bone morphogenetic protein (BMP) and autolyzed, antigen-extracted, allogeneic (AAA) bone. Clin Orthop Relat Res:229–237
Johnstone B, Hering TM, Caplan AI, Goldberg VM, Yoo JU (1998) In vitro chondrogenesis of bone marrow-derived mesenchymal progenitor cells 43. Exp Cell Res 238:265–272
Kanczler JM, Ginty PJ, White L, Clarke NM, Howdle SM, Shakesheff KM and Oreffo RO (2010) The effect of the delivery of vascular endothelial growth factor and bone morphogenic protein-2 to osteoprogenitor cell populations on bone formation. Biomaterials 31:1242–1250
Karp JM, Leng Teo GS (2009) Mesenchymal stem cell homing: the devil is in the details. Cell Stem Cell 4:206–216
Kassem M, Mosekilde L, Eriksen EF (1993) 1,25-dihydroxyvitamin D3 potentiates fluoride-stimulated collagen type I production in cultures of human bone marrow stromal osteoblast-like cells. J Bone Miner Res 8:1453–1458
Keller G (2005) Embryonic stem cell differentiation: emergence of a new era in biology and medicine. Genes Dev 19:1129–1155
Khoo ML, Shen B, Tao H, Ma DD (2008) Long-term serial passage and neuronal differentiation capability of human bone marrow mesenchymal stem cells. Stem Cells Dev 17:883–896
Klinge PM, Harmening K, Miller MC, Heile A, Wallrapp C, Geigle P, Brinker T (2011) Encapsulated native and glucagon-like peptide-1 transfected human mesenchymal stem cells in a transgenic mouse model of Alzheimer's disease. Neurosci Lett 497:6–10
Kneser U, Schaefer DJ, Polykandriotis E, Horch RE (2006) Tissue engineering of bone: the reconstructive surgeon's point of view. J Cell Mol Med 10:7–19
Kodach LL, Wiercinska E, de Miranda NF, Bleuming SA, Musler AR, Peppelenbosch MP, Dekker E, van den Brink GR, van Noesel CJ, Morreau H, Hommes DW, Ten DP, Offerhaus GJ, Hardwick JC (2008) The bone morphogenetic protein pathway is inactivated in the majority of sporadic colorectal cancers. Gastroenterology 134:1332–1341
Krabbe C, Zimmer J, Meyer M (2005) Neural transdifferentiation of mesenchymal stem cells—a critical review. *APMIS* 113:831–844.
Kruck S, Bedke J, Kuczyk MA, Merseburger AS (2012) Second-line systemic therapy for the treatment of metastatic renal cell cancer. Expert Rev Anticancer Ther 12:777–785
Kumar S, Nagy TR, Ponnazhagan S (2010a) Therapeutic potential of genetically modified adult stem cells for osteopenia. *Gene Ther* 17:105–116
Kumar S, Wan C, Ramaswamy G, Clemens TL, Ponnazhagan S (2010b) Mesenchymal stem cells expressing osteogenic and angiogenic factors synergistically enhance bone formation in a mouse model of segmental bone defect. Mol Ther 18:1026–1034.

Kuznetsov SA, Krebsbach PH, Satomura K, Kerr J, Riminucci M, Benayahu D, Robey PG (1997) Single-colony derived strains of human marrow stromal fibroblasts form bone after transplantation in vivo. J Bone Miner Res 12:1335–1347

Kuznetsov SA, Mankani MH, Gronthos S, Satomura K, Bianco P, Robey PG (2001) Circulating skeletal stem cells. J Cell Biol 153:1133–1140

Lepperdinger G, Brunauer R, Jamnig A, Laschober G, Kassem M (2008) Controversial issue: is it safe to employ mesenchymal stem cells in cell-based therapies? Exp Gerontol 43:1018–1023

Li W, Ma N, Ong LL, Nesselmann C, Klopsch C, Ladilov Y, Furlani D, Piechaczek C, Moebius JM, Lutzow K, Lendlein A, Stamm C, Li RK, Steinhoff G (2007) Bcl-2 engineered MSCs inhibited apoptosis and improved heart function. Stem Cells 25:2118–2127

Lien CY, Chih-Yuan HK, Lee OK, Blunn GW, Su Y (2009) Restoration of bone mass and strength in glucocorticoid-treated mice by systemic transplantation of CXCR4 and cbfa-1 co-expressing mesenchymal stem cells. J Bone Miner Res 24:837–848

Luria EA, Panasyuk AF, Friedenstein AY (1971) Fibroblast colony formation from monolayer cultures of blood cells. Transfusion 11:345–349

Mangi AA, Noiseux N, Kong D, He H, Rezvani M, Ingwall JS, Dzau VJ (2003) Mesenchymal stem cells modified with Akt prevent remodeling and restore performance of infarcted hearts. Nat Med 9:1195–1201

McGrail DJ, Ghosh D, Quach ND, Dawson MR (2012) Differential mechanical response of mesenchymal stem cells and fibroblasts to tumor-secreted soluble factors. PLoS One 7:e33248

Menon LG, Picinich S, Koneru R, Gao H, Lin SY, Koneru M, Mayer-Kuckuk P, Glod J, Banerjee D (2007) Differential gene expression associated with migration of mesenchymal stem cells to conditioned medium from tumor cells or bone marrow cells. Stem Cells 25:520–528

Menon LG, Shi VJ, Carroll RS (2008) Mesenchymal stromal cells as a drug delivery system. Harvard Stem Cell Institute, Cambridge

Miura M, Gronthos S, Zhao M, Lu B, Fisher LW, Robey PG, Shi S (2003) SHED: stem cells from human exfoliated deciduous teeth. Proc Natl Acad Sci USA 100:5807–5812

Nakamura K, Ito Y, Kawano Y, Kurozumi K, Kobune M, Tsuda H, Bizen A, Honmou O, Niitsu Y, Hamada H (2004) Antitumor effect of genetically engineered mesenchymal stem cells in a rat glioma model. Gene Ther 11:1155–1164

Omori R, Eguchi J, Hiroishi K, Ishii S, Hiraide A, Sakaki M, Doi H, Kajiwara A, Ito T, Kogo M, Imawari M (2012) Effects of interferon-alpha-transduced tumor cell vaccines and blockade of programmed cell death-1 on the growth of established tumors. Cancer Gene Ther 19(9):637–643

Perry TA, Greig NH (2004) A new Alzheimer's disease interventive strategy: GLP-1. Curr Drug Targets 5:565–571

Peterson B, Zhang J, Iglesias R, Kabo M, Hedrick M, Benhaim P, Lieberman JR (2005) Healing of critically sized femoral defects, using genetically modified mesenchymal stem cells from human adipose tissue. Tissue Eng 11:120–129

Plath K, Lowry WE (2011) Progress in understanding reprogramming to the induced pluripotent state. Nat Rev Genet 12:253–265

Pollock K, Stroemer P, Patel S, Stevanato L, Hope A, Miljan E, Dong Z, Hodges H, Price J, Sinden JD (2006) A conditionally immortal clonal stem cell line from human cortical neuroepithelium for the treatment of ischemic stroke. Exp Neurol 199:143–155

Ren C, Kumar S, Chanda D, Chen J, Mountz JD, Ponnazhagan S (2008a) Therapeutic potential of mesenchymal stem cells producing interferon-alpha in a mouse melanoma lung metastasis model. Stem Cells 26:2332–2338

Ren C, Kumar S, Chanda D, Kallman L, Chen J, Mountz JD, Ponnazhagan S (2008b) Cancer gene therapy using mesenchymal stem cells expressing interferon-beta in a mouse prostate cancer lung metastasis model. Gene Ther 15:1446–1453

Rickard DJ, Kassem M, Hefferan TE, Sarkar G, Spelsberg TC, Riggs BL (1996) Isolation and characterization of osteoblast precursor cells from human bone marrow. J Bone Miner Res 11:312–324

Rosada C, Justesen J, Melsvik D, Ebbesen P, Kassem M (2003) The human umbilical cord blood: a potential source for osteoblast progenitor cells. Calcif Tissue Int 72:135–142

Sackstein R, Merzaban JS, Cain DW, Dagia NM, Spencer JA, Lin CP, Wohlgemuth R (2008) Ex vivo glycan engineering of CD44 programs human multipotent mesenchymal stromal cell trafficking to bone. Nat Med 14:181–187

Salazar DL, Uchida N, Hamers FP, Cummings BJ, Anderson AJ (2010) Human neural stem cells differentiate and promote locomotor recovery in an early chronic spinal cord injury NOD-scid mouse model. PLoS One 5:e12272

Sarkar K, Fox-Talbot K, Steenbergen C, Bosch-Marce M, Semenza GL (2009) Adenoviral transfer of HIF-1alpha enhances vascular responses to critical limb ischemia in diabetic mice. Proc Natl Acad Sci USA 106:18769–18774.

Seo SH, Kim KS, Park SH, Suh YS, Kim SJ, Jeun SS, Sung YC (2011) The effects of mesenchymal stem cells injected via different routes on modified IL-12-mediated antitumor activity. Gene Ther 18:488–495

Shen FH, Visger JM, Balian G, Hurwitz SR, Diduch DR (2002) Systemically administered mesenchymal stromal cells transduced with insulin-like growth factor-I localize to a fracture site and potentiate healing. J Orthop Trauma 16:651–659

Shiojima I, Walsh K (2006) Regulation of cardiac growth and coronary angiogenesis by the Akt/PKB signaling pathway. Genes Dev 20:3347–3365

Shujia J, Haider HK, Idris NM, Lu G, Ashraf M (2008) Stable therapeutic effects of mesenchymal stem cell-based multiple gene delivery for cardiac repair. Cardiovasc Res 77:525–533

Simonsen JL, Rosada C, Serakinci N, Justesen J, Stenderup K, Rattan SI, Jensen TG, Kassem M (2002) Telomerase expression extends the proliferative life-span and maintains the osteogenic potential of human bone marrow stromal cells. Nat Biotechnol 20:592–596

Takahashi K, Yamanaka S (2006) Induction of pluripotent stem cells from mouse embryonic and adult fibroblast cultures by defined factors. Cell 126:663–676

Takahashi K, Tanabe K, Ohnuki M, Narita M, Ichisaka T, Tomoda K, Yamanaka S (2007) Induction of pluripotent stem cells from adult human fibroblasts by defined factors. Cell 131:861–872

Tang J, Wang J, Zheng F, Kong X, Guo L, Yang J, Zhang L, Huang Y (2010) Combination of chemokine and angiogenic factor genes and mesenchymal stem cells could enhance angiogenesis and improve cardiac function after acute myocardial infarction in rats. Mol Cell Biochem 339:107–118

Tao Z, Chen B, Tan X, Zhao Y, Wang L, Zhu T, Cao K, Yang Z, Kan YW, Su H (2011) Coexpression of VEGF and angiopoietin-1 promotes angiogenesis and cardiomyocyte proliferation reduces apoptosis in porcine myocardial infarction (MI) heart. Proc Natl Acad Sci U S A 108:2064–2069

Thomson JA, Itskovitz-Eldor J, Shapiro SS, Waknitz MA, Swiergiel JJ, Marshall VS, Jones JM (1998) Embryonic stem cell lines derived from human blastocysts. Science 282:1145–1147

Tu Q, Valverde P, Li S, Zhang J, Yang P, Chen J (2007) Osterix overexpression in mesenchymal stem cells stimulates healing of critical-sized defects in murine calvarial bone. Tissue Eng 13:2431–2440

Wislet-Gendebien S, Laudet E, Neirinckx V, Alix P, Leprince P, Glejzer A, Poulet C, Hennuy B, Sommer L, Shakhova O, Rogister B (2012) Mesenchymal stem cells and neural crest stem cells from adult bone marrow: characterization of their surprising similarities and differences. Cell Mol Life Sci 69:2593–2608

Wojtowicz AM, Templeman KL, Hutmacher DW, Guldberg RE, Garcia AJ (2010) Runx2 overexpression in bone marrow stromal cells accelerates bone formation in critical-sized femoral defects. Tissue Eng Part A 16:2795–2808

Wu Y, Zhao RC (2012) The role of chemokines in mesenchymal stem cell homing to myocardium. Stem Cell Rev 8:243–250

Wynn RF, Hart CA, Corradi-Perini C, O'Neill L, Evans CA, Wraith JE, Fairbairn LJ, Bellantuono I (2004) A small proportion of mesenchymal stem cells strongly expresses functionally active CXCR4 receptor capable of promoting migration to bone marrow. Blood 104:2643–2645

Zhang D, Fan GC, Zhou X, Zhao T, Pasha Z, Xu M, Zhu Y, Ashraf M, Wang Y (2008) Overexpression of CXCR4 on mesenchymal stem cells augments myoangiogenesis in the infarcted myocardium. J Mol Cell Cardiol 44:281–292

Zhao Z, Zhao M, Xiao G, Franceschi RT (2005) Gene transfer of the Runx2 transcription factor enhances osteogenic activity of bone marrow stromal cells in vitro and in vivo. Mol Ther 12:247–253

Zou D, Zhang Z, Ye D, Tang A, Deng L, Han W, Zhao J, Wang S, Zhang W, Zhu C, Zhou J, He J, Wang Y, Xu F, Huang Y, Jiang X (2011) Repair of critical-sized rat calvarial defects using genetically engineered bone marrow-derived mesenchymal stem cells overexpressing hypoxia-inducible factor-1alpha. Stem Cells 29:1380–1390

Chapter 6
Somatic Cell Nuclear Transfer and the Creation of Transgenic Large Animal Models

Naomi Dicks, Luis B. Agellon and Vilceu Bordignon

6.1 Introduction

The concept of nuclear transfer and the ability to clone individuals is for many a recent development marked by the creation of the cloned sheep, Dolly, in 1996. In fact, the idea was first introduced in the late 1930s, but only successfully applied 20 years later in amphibians (Edwards et al. 2003). In 1952, Briggs and King first reported successful cloning in the frog (Briggs and King 1952), followed by the mouse in 1981 by Illmensee and Hoppe, and many other species in the late 1980s including rabbit, sheep, cattle and pigs (reviewed in Edwards et al. 2003). These pioneering studies created cloned animals by taking donor nuclei, from *totipotent* or *pluripotent* embryonic cells, and transferring them into enucleated oocytes. Effectively, individuals were cloned at a very early stage in life—the embryo. The creation of Dolly in 1996 was a hallmark scientific discovery, as it was the first successful cloning of an *adult* animal (Wilmut et al. 1997). Adult *differentiated* cells were used as the donor nuclei as opposed to totipotent embryonic cells, a feat which many scientists initially considered impossible. Somatic cell nuclear transfer (SCNT) was hence born and applied to clone adult animals of many other species

V. Bordignon (✉)
Department of Animal Science, McGill University, 21,111 Lakeshore Road, Ste-Anne-de-Bellevue, QC H9X 3V9, Canada
e-mail: vilceu.bordignon@mcgill.ca

L. B. Agellon (✉)
School of Dietetics and Human Nutrition, McGill University, 21,111 Lakeshore Road, Ste-Anne-de-Bellevue, QC H9X 3V9, Canada
e-mail: luis.agellon@mcgill.ca

N. Dicks
Department of Animal Science, McGill University, Ste-Anne-de-Bellevue, QC, Canada

X.-Q. Li et al. (eds.), *Somatic Genome Manipulation*,
DOI 10.1007/978-1-4939-2389-2_6

including mice, cattle, pigs, goats, cats, rabbits, rats, dogs, horses, mules and deer (Edwards et al. 2003; Campbell et al. 2005; Jang et al. 2010).

Throughout the years, the SCNT technique has largely remained unchanged, with only minor alterations to the general procedure. There has been increasing development, however, in the use of specialized gene-editing techniques to modify donor cell DNA prior to SCNT. This has greatly facilitated the creation of transgenic cloned animals harbouring targeted genetic alterations. Clones expressing novel non-native genes (knockin), reduced gene expression (gene attenuation or knockdown) and even the creation of animals with non-functional genes (gene disruption or knockout), are examples. The ability to create engineered and cloned animals is essentially becoming as simple as modifying the genome within a cell. The purpose of this chapter is to review the various techniques used for transgenic animal production, with focus on the advantages of SCNT in the creation of large animal models. Furthermore, it will highlight current gene-editing technologies and how they can be applied to create unique transgenic cloned large animal strains for use in agriculture, research and other biomedical applications. Transgenic large animals that have been successfully produced to date will also be discussed.

6.2 Importance of Transgenic Large Animals

There are many applications for transgenic animals, particularly in the development of biomedical models for the study of disease. Genetically modified mice have been used for years in research for the study of biological processes fundamental to the pathogenesis of complex diseases. While studies in mice have been instrumental in understanding basic cellular processes and disease pathophysiology, there are a number of limitations with this model species. Small size, short lifespan, as well as poor anatomical and physiological comparison with humans are examples. These hindrances limit the ability to translate findings in mice research directly to humans. The development of transgenic models in larger species is needed to bridge this translational gap, given their longer lifespan and larger size. In particular, the porcine model is gaining favour because the pig shares similarities in anatomy and physiology with humans. Furthermore, when compared to previously engineered mouse models, pig models exhibit greater similarity of phenotypes to those seen in human disease (Rogers et al. 2008). In addition, these similarities make pigs excellent potential organ donors for xenotransplantation (Ibrahim et al. 2006).

While transgenic large animals, particularly pigs, are important for the development of more suitable biomedical models, they are also important for other uses. Animal "pharming", or the use of transgenic animals for the production of protein-based pharmaceuticals, is more practical in larger animals due to the greater amount of product that can be harvested, as well as the ease of collection methods. Furthermore, transgenic large animals have direct applications for agriculture through the development of animals with increased growth, decreased disease susceptibility and improved animal product characteristics (Maksimenko et al 2013).

6.3 Techniques Used for the Production of Transgenic Animals

Various techniques to produce genetically altered animals have been developed throughout the years including pronuclear microinjection, sperm-mediated gene transfer (SMGT), oocyte transduction and finally, SCNT. All methods involve the integration of a specific transgene within the genome of the target cell. The transgene is composed of a promoter sequence and a structural gene sequence. The promoter sequence defines the control of transgene expression, and it can be designed to express in all cells (constitutive promoter) or in specific cells (cell-specific promoter). This component of the transgene can be exploited to create transgenic animals that express the transgene in all cells of the body (pan expression), versus within a specific tissue (tissue-specific expression). The structural gene sequence encodes the protein that ultimately will be produced within the cell. This can be a natural or synthetic gene, and either native or non-native to the target species. Furthermore, the gene does not necessarily need to encode a physiologically active protein, but one that can interfere with protein translation (RNA interference) or disrupt other genetic sequences (DNA-specific nucleases) to create decreased or loss of gene function.

Specialized transgenes can be designed based on the intention for the transgenic animal, such as the creation of an animal that expresses a novel protein in its milk, or an animal that lacks a particular protein. Once constructed, the transgene must somehow be integrated within the genome of the target cell or embryo. The different methods of transgenic animal production are discussed and the efficiency of transgene integration compared.

6.3.1 Pronuclear Microinjection

In 1981, Gordon and Ruddle developed pronuclear microinjection, a technique whereby foreign DNA is injected into the pronucleus of a developing zygote (Gordon and Ruddle 1981). This technology resulted in the production of various transgenic animals by the end of the 1980s, including rabbits, sheep and pigs (Wall 2001). The major disadvantage of pronuclear injection, while quite simple to perform, is the low efficiency of production of transgenic offspring (Luo et al. 2012). This is because there is no control over which zygotes will incorporate the injected transgene, nor is there control as to the location where the transgene will integrate within the genome. Effectively, once the transgene is injected, it is left to chance whether it will be integrated into the genome to produce a transgenic animal. As a result, only 5–10 % of offspring born through this technique will harbour the transgene (Wall 2001; Luo et al. 2012). Some may not even express it appropriately, especially if it has been incorporated in areas of gene silencing (Wall 2001; Luo et al. 2012). In fact, it has been shown that only 50 % of transgenic animals will express the transgene at useful levels (Wall 2001). Furthermore, only 70 % of the offspring will be able to pass the genetic alteration onto the next generation (Wall 2001; Luo et al. 2012). Using this technique in mice may not seem as daunting, given that completion of gestation and sexual maturity can be attained within 2 months, as well as the large litter size obtained. Furthermore, housing and caring for mice are practical and costs are relatively low, so reduced efficiency in the creation of transgenic founders may not pose a huge financial setback. In the context of large animal species, however, the low efficiency of pronuclear injection greatly hinders the usefulness of the technique. Large animal species have much longer times to sexual maturity, longer gestation periods and significantly smaller litter sizes. This, coupled with the higher cost for basic husbandry makes pronuclear injection extremely unpractical. For example, with the low efficiency of transgenic offspring production, a maximum of one per ten live piglets born (10 %) would be expected to be transgenic. Furthermore, there is only a 50 % chance this animal will express the transgene and a 70 % chance it will pass the gene on to further generations with breeding. In cattle, sheep or goats, the numbers would be even more dismal given their smaller litter sizes. While easy to perform and potentially useful for small animals with high reproductive potential, pronuclear injection is simply not efficient enough to justify its use in larger animal models.

6.3.2 Sperm-Mediated Gene Transfer

In 1989, Lavitrano and colleagues introduced SMGT, a technique that is based on the natural ability of sperm to adsorb foreign DNA. There is evidence that exogenous DNA (e.g. transgenes) can bind to sperm cells which can then be used in standard in vitro fertilization (IVF) or artificial insemination (AI) to produce transgenic animals (Lavitrano et al. 2003). In 2003, eight separate experiments conducted by the Lavitrano laboratory using SMGT resulted in an average of 57 % transgenic offspring, a stark improvement compared to reported values using pronuclear microinjection. Unfortunately, the high efficiency reported by these studies has not been reliably repeated or confirmed in other species. Indeed, the efficiency of SMGT is still variable since there is no reliable way to select the transfected sperm prior to IVF or AI. Adaptations to the procedure have been tested, including the utilization of intracytoplasmic sperm injection-mediated gene transfer (ICSI-MGT) directly into oocytes. Different research groups have used this procedure with variable success in developing live offspring (Lai et al. 2001; Pereyra-Bonnet et al. 2008; Watanabe et al. 2012). Live transgenic piglets were born using this method by Watanabe and colleagues in 2012; however, their efficiency of producing transgenic offspring remained low at 12–18 %. While this efficiency may be slightly better compared to pronuclear injection using these methods, the technique has addressed neither the issue of nonintegration nor the random location of transgene insertion within the genome.

6.3.3 Oocyte Transduction

Oocyte transduction is another method used for developing transgenic animals and involves the injection of a transgene encoded by retroviral vectors into the oocyte prior to IVF. In 2001, Cabot and colleagues successfully introduced the enhanced green fluorescent protein gene into pigs using this method (Cabot et al. 2001). Unfortunately, much like pronuclear injection, SMGT and ICSI-MGT, there is no control in transgene integration in oocyte transduction, resulting in low efficiency of transgenic animal production. Furthermore, while very large transgenes can be injected in pronuclear injection and ICSI-MGT, the size of transgene utilized in oocyte transduction is limited by the size of the viral vector (Prather et al. 2013).

6.3.4 Somatic Cell Nuclear Transfer

As described in the introduction, nuclear transfer and cloning have been successfully applied in mammals since the 1980s, albeit using totipotent or pluripotent donor cells. The further development of using *somatic* donor cell nuclei in the late 1990s has enabled the technology to be easily applied to many species where embryonic stem cell populations are not available. This improvement has particularly facilitated the production of transgenics in large animal species. The SCNT technique essentially involves the introduction of a somatic cell nucleus (the donor) into the cytoplasm of an enucleated oocyte (the recipient), prior to the oocyte being activated and developing into an embryo (Fig. 6.1). Several methods are available for enucleating oocytes. The most commonly used method involves a specialized micromanipulation system, in which one micropipette is used to gently hold the oocyte while another micropipette is used to aspirate the nucleus (traditional cloning method; Niemann 2012). Alternatively, enucleation can also be accomplished by manually cutting a portion of the oocyte that includes the DNA using a sharp blade (hand-made cloning method; Vajta 2007). The location of the oocyte DNA is normally identified by the polar body in matured meiosis II (MII) stage oocytes (Campbell et al. 2007). Once the mature oocyte is enucleated, it is fused with the donor cell through a direct current (DC) electrical pulse (Campbell et al. 2007). While termed the "nuclear donor", the cell is actually fused in its entirety with the enucleated oocyte, therefore fusing both the donor nucleus and cytoplasm with the recipient. After fusion, the reconstructed embryo must be activated to allow further development. This can be accomplished via another DC electrical pulse or chemical treatments such as ionomycin or strontium chloride. The basis for the activation is to mimic calcium concentration oscillations within the embryo, which occur during normal spermatic fertilization. Embryos are cultured in vitro until various stages of development, depending on the species, followed by embryo transfer to the surrogate female for implantation and continued development (Campbell et al. 2007; Bordignon 2011) (Fig. 6.1).

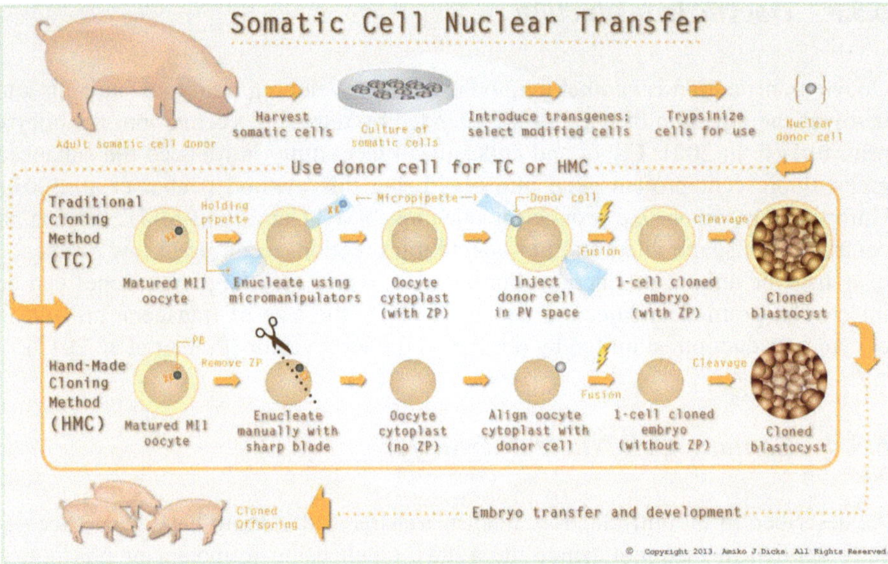

Fig. 6.1 Somatic cell nuclear transfer (*SCNT*). Diagram depicting the SCNT procedure with comparison of the traditional and hand-made cloning methods. Transgenes can be introduced into cultured donor somatic cells and then preferentially selected for use in the SCNT procedure and creation of transgenic cloned offspring. *PB* polar body, *ZP* zona pellucida, *PV* perivitteline; A.J. Dicks ©

The major advantage imparted by SCNT over the aforementioned techniques, is the ability to genetically alter donor cells and select for the cells with desired modifications prior to nuclear transfer (Fig. 6.2). By selecting only transgenic cells

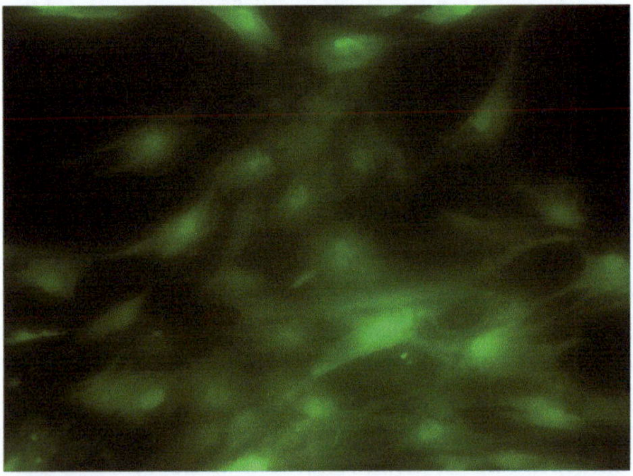

Fig. 6.2 Porcine fetal fibroblasts transfected with apolipoprotein E RNA interference transgenes (shRNA), which also contain the green fluorescent protein marker. These transfected cells may be preferentially selected and used for SCNT and generation of transgenic offspring. *SCNT* somatic cell nuclear transfer, *shRNA* short hairpin RNA

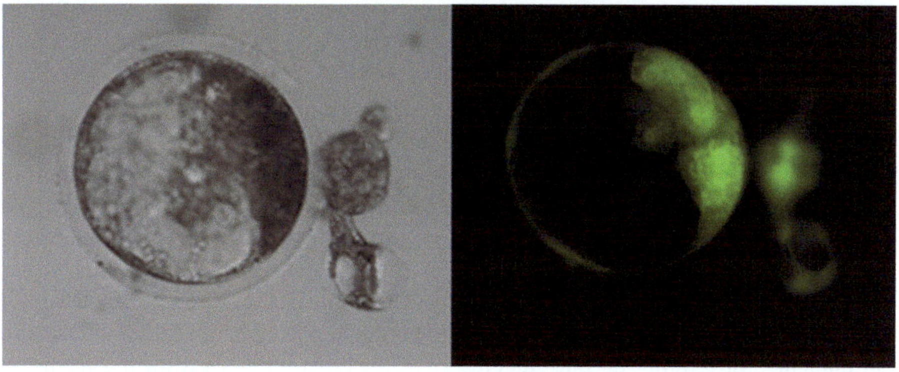

Fig. 6.3 Porcine blastocyst which has been created through SCNT using porcine fetal fibroblasts transfected with apolipoprotein E shRNA. Phase contrast image (*left*) compared to fluorescence (*right*). This transgenic blastocyst can be transferred to a surrogate female for further development. *SCNT* somatic cell nuclear transfer, *shRNA* short hairpin RNA

for the nuclear transfer and embryo development (Fig. 6.3), the efficiency of producing a transgenic animal rises to 100% which is not achievable by pronuclear microinjection, SMGT, ICSI-MGT or oocyte transduction. Furthermore, SCNT is more amenable to the application of gene-editing techniques to create targeted alterations to the genome, including gene inactivation. While transgenic animals expressing novel genes or attenuated expression of specific endogenous genes can be created by the aforementioned techniques, it is much more difficult to create targeted gene-disrupted animals using these methods. The reason for this is primarily how the genetic alterations are accomplished. In pronuclear microinjection, SMGT, ICSI-MGT and oocyte transduction, exogenous DNA is introduced. This exogenous DNA can be in the form of a novel gene not inherently expressed in that species or a synthetic gene that encodes an interfering molecule (e.g. short hairpin RNA, shRNA) which blocks the expression of an endogenous gene.

The development of targeted gene-disrupted animals was, until recently, only accomplished using homologous recombination, which is laborious and time consuming. Pronuclear microinjection, SMGT, ICSI-MGT and oocyte transduction methods have not produced gene-disrupted animals, as it requires direct removal or alteration of the gene sequence. However, this may change with the advent of new gene-editing techniques that can introduce specific gene mutations through expression of transgenes encoding specialized nucleases. Regardless, tools for direct identification and selection of specific embryos that have incorporated the desired alterations before implantation are not yet available. As a result, the creation of gene-disrupted transgenic animals using these methods remains inefficient at this time.

Despite the 100% transgenic efficiency in large animal species afforded by SCNT, the cloning efficiency, which refers to the number of live animals obtained for every reconstructed embryo transferred into the surrogate female, is low at the present time. The cloning efficiency varies depending on the species and the donor cell population, but on average is less than 5% (Bordignon 2011). Mortality of cloned embryos is seen throughout various stages of pregnancy, although it occurs largely during the preimplantation period (Edwards et al. 2003). Among those that survive to delivery, there is a higher incidence of perinatal death in cloned animals compared to naturally conceived animals (Bordignon 2011). Cloned animals sometime exhibit no abnormalities on gross or histopathological examination, while others may be born exhibiting signs of "large offspring syndrome" (Edwards et al. 2003). This condition encompasses a multitude of defects such as oversized birth weight, respiratory difficulties, cardiac disease, contracted tendons, or higher disease susceptibility, among other defects (Niemann 2012). Reasons for the high embryonic loss during development and the perinatal period is not fully understood, but nuclear reprogramming deficiencies as well as disturbances resulting from extensive in vitro manipulations are suspected to play a large role (Bordignon 2011). The inappropriate, misdirected or lack of reprogramming within the donor somatic cell nucleus likely results in abnormal gene regulation and development, resulting in pre- and post-implantation death (Campbell et al. 2007). Refinements in this area would greatly improve the yield of live transgenic animals, and subsequently decrease the time and labour associated with successive cloning procedures (Edwards et al. 2003). Fortunately, most studies indicate that compared to naturally conceived animals, cloned large animals that survive past the weaning period have normal survivability (Niemann 2012) as well as growth, meat/milk production and reproductive efficiency (Bordignon 2011). Further research into how embryonic reprogramming occurs and how it can be facilitated will increase our understanding of development and thereby enhance the cloning efficiency as well as improve the health of cloned offspring.

6.4 Gene-Editing Techniques Used to Create Transgenic Animals

The ability to select the modified genome of choice prior to nuclear transfer permits the utilization of virtually any gene-editing technique available to create the genetically altered donor cells. This is particularly useful in the creation of gene-modified animals, where the gene must be altered to either delete or activate gene function. In mice, the creation of gene-inactivated animals has been facilitated by embryonic stem cells and homologous recombination (see below). In large animal species, the lack of embryonic stem cells that can be used for this purpose forces the application of other techniques to create strains with loss-of-function mutations. Fortunately, the exponential development of gene-editing tools along with SCNT has significantly enhanced the ease of gene modifications in cells, facilitating the introduction of missense mutations or gene deletions. In particular, it might be possible to use the RNA-guided gene-editing method (see below) to create gain-of-function mutations by allowing site-specific changes to be made in the genome, such as in promoter regions to enable constitutive expression, or release from repression. These new technologies are making it plausible to simply and effectively create "designer" cloned animals for use in research and other biomedical applications.

6.4.1 Homologous Recombination

There are numerous methods for altering DNA sequences, including the natural process of homologous recombination. Homologous recombination has been extensively used in the mouse for two reasons: the availability of pluripotent embryonic stem cells that can be maintained and manipulated in cell culture, and the high reproductive potential of the murine species. The latter allows rapid generation of progeny leading to the production of strains that are homozygous for the disrupted allele, in the case of nonlethal loss-of-function mutations. The mutation is delivered to the cell using a targeting vector carrying the desired genetic modification. A segment of the targeting vector is integrated into the genome via recombination events designed to occur at regions flanking the altered DNA sequence, thus effectively replacing a DNA segment at the locus of interest. The addition of genetic elements belonging to specific recombination systems, such as the Cre-Lox, flippase-flippase recognition target (FLP-FRT) systems (Mallo 2006), is an improvement to the original system in that they permit temporal or tissue- and cell-type specific disruption of specific genes. This method has been successfully used in murine embryonic stem cells, which exhibit a 100-fold greater rate of recombination compared to somatic cells (Wang and Zhou 2003). Furthermore, the ability to screen for altered cells using positive and negative selection pressures (e.g. resistance or susceptibility to antibiotics) allows for a 1000-fold enrichment of cells with recombined genomes (Wang and Zhou 2003). The lack of pluripotent embryonic stem cells in large animal species known to be fit for this application is seen as the major limitation in adopting this method for the development of transgenic animals. While inefficient in somatic cells, homologous recombination has been used to create targeted gene-disrupted somatic cells in large animal species. These have subsequently been utilized in SCNT to generate gene-ablated animal clones. Such is the case used to create pig strains that are deficient in α-1,3-galactosyltransferase (GGTA; Phelps et al. 2003), cystic fibrosis transmembrane conductance regulator (Rogers et al. 2008) and coagulation factor VIII (Kashiwakura et al. 2012).

6.4.2 RNA Interference

RNA interference represents another genetic engineering tool, which was first discovered in *Caenorhabitis elegans* (Hannon and Rossi 2004; Fire et al.1998). This method utilizes a natural host defence mechanism designed to elicit an immune response to exogenous RNA, namely RNA viruses, and results in decreased translation of the viral mRNA as well as an inflammatory response (Hannon and Rossi 2004). When double-stranded RNA (dsRNA) is introduced into the cell, it is pro-

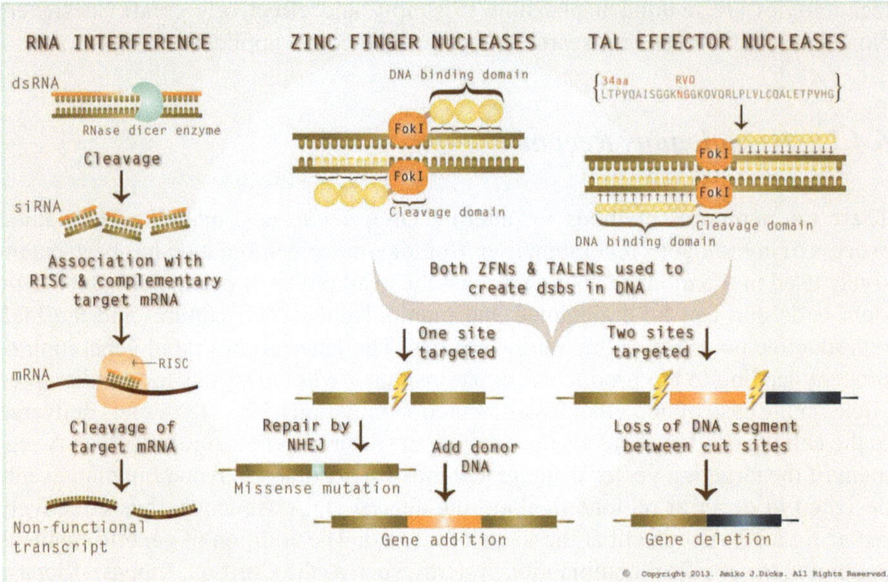

Fig. 6.4 Comparison of gene-editing techniques: RNA interference (*left*), zinc finger nucleases (ZFNs; *middle*) and transcription activator-like (*TAL*) effector nucleases (*right*). ZFNs differ compared to TALENs with respect to the DNA-binding domain. The ZFN depicted shows three zinc finger proteins (*ZFP*), each recognizing three nucleotides of the DNA sequence. The TALEN depicted is composed of 12 tandem arrays of repeating protein sequences of 34 amino acids long. Each repeating sequence varies at the 12th–13th amino acid (RVD) and this confers the recognition of a single nucleotide of the DNA sequence. ZFNs and TALENs can be exploited to create missense mutations, gene additions and gene deletions. *RISC* RNA induced silencing complex, *FokI* FokI endonuclease, *RVD* repeating variable diresidue, *dsbs* double stranded breaks, *NHEJ* non-homologous end-joining; A.J. Dicks ©

cessed by an RNase dicer enzyme into small-interfering RNA (siRNA), which then incorporates into the RNA-induced silencing complex (RISC) (Fig. 6.4). The RISC subsequently binds to mRNA complementary to the bound siRNA and impedes protein production through mRNA cleavage (Fig. 6.4). This process has been exploited by introducing specifically designed siRNA complementary to endogenous genes to attenuate their expression, resulting in gene knockdown within cells. Since its discovery, RNA interference has been widely used in gene function studies to selectively reduce target gene expression in cells and transgenic animals. For RNA interference to be useful in transgenic animal production, modifications to the technique had to be made to ensure consistent production of siRNA throughout the life of the cell and the animal. Direct introduction of dsRNA into cells results in only transient gene knockdown (Paddison et al. 2002). In contrast, integration of a transgene encoding a shRNA molecule into the host DNA, ensures long-lasting production of dsRNA/siRNA which permits continual repression of target mRNA translation (Paddison et al. 2002). Cells transfected with these targeted shRNA sequences can subsequently be used as donor cells for SCNT and the production of cloned

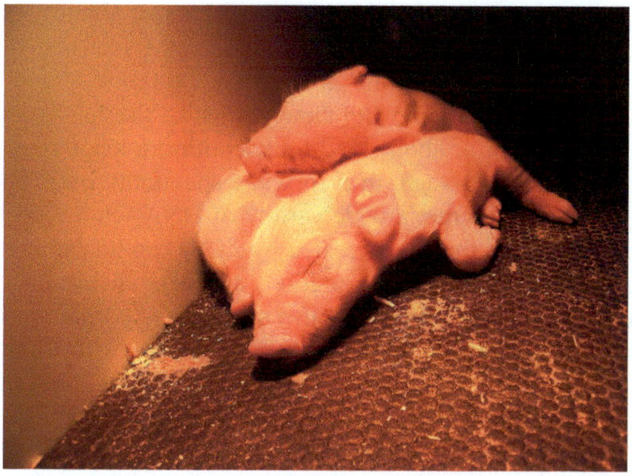

Fig. 6.5 Gene-attenuated (*knockdown*) ApoE transgenic piglets produced via SCNT and RNA interference technology. *SCNT* somatic cell nuclear transfer. (Bordignon et al. 2013)

transgenic animals. This approach has already been successfully used to produce engineered pig strains with attenuated gene expression, including transgenic piglets expressing apolipoprotein E shRNA (Bordignon et al. 2013; Fig. 6.5) and porcine endogenous retrovirus shRNA (Ramsoondar et al. 2009). The former transgenic animal was created as a potential model for atherosclerosis and the latter to create an improved tissue donor for xenotransplantation studies.

6.4.3 DNA-Specific Nucleases

6.4.3.1 Zinc Finger Nucleases

While RNA interference has been used successfully to decrease the expression of target genes, there is variable efficacy in gene expression attenuation. Sequences selected for this purpose must be validated for specificity as well as efficiency. Fortunately, transgene expression can also be exploited to enable gene inactivation. The development of DNA sequence-specific nucleases has revolutionized the ability to create targeted genetic alterations. Like RNA interference, this method also relies on naturally occurring processes. Zinc finger nucleases (ZFN) are composed of the bacterial FokI endonuclease, which facilitates strand breakage, coupled with engineered DNA-specific binding proteins, termed zinc finger proteins (ZFP), which have been engineered to bind to certain DNA sequences (Fig. 6.4). The ZFPs actually derive from natural transcription binding factors discovered in *Xenopus* in 1985 (Miller et al. 1985). Each ZFP is traditionally composed of three to four Cys2–His2 zinc fingers arranged in tandem arrays, each of which recognizes and binds a specific 3-nucleotide DNA sequence (Urnov et al. 2010; Fig. 6.4). Thus, ZFPs allow the assembly of modules with DNA binding specificity of about 9–12 nucleotides for a single ZFN. This functionality is enhanced by the fact that the engineered FokI endonuclease associated with the ZFN will not catalyze strand breakage unless dimerized (Urnov et al. 2010). The dimerization of the ZFNs results in a specific DNA-binding domain of about 18–24 nucleotides, a length that significantly limits potential off-target cleavage. In theory, the specificity of ZFNs is attractive as they can be designed to target a wide variety of DNA sequences. Furthermore, the double-strand breaks in the DNA created by ZFNs can be exploited to produce other genetic alterations than simple mutations and gene disruption. The addition of donor DNA along with the ZFNs can also allow the incorporation of novel sequences at the target site, such as small DNA fragments to restore a non-functional target gene, or a complete gene, or even a series of genes (Fig. 6.4). Thus, the ZFN technology can be used not only for gene inactivation but also for DNA insertions as well as gene repair (Urnov et al. 2010).

An additional advantage of ZFN technology is that the rate of biallelic modifications to the DNA is approximately 1–2%, compared to 0% for homologous recombination (Whyte and Prather 2012). Albeit the low frequency, the ability to obtain homozygous transgenic animals at the first founder generation is an attractive characteristic, as it limits cost and time associated with breeding to obtain the same results. The ease and specificity of targeted alterations to the genome using ZFNs has facilitated the production of transgenic animals through SCNT. An example includes the generation of bilallelic knockout pigs targeting the GGTA gene by Hauschild and colleagues in 2011 (Hauschild et al. 2011). This transgenic strain was created among many others as a biomedical model for xenotransplantation studies.

6.4.3.2 Transcription Activator-Like Effector Nucleases

Some drawbacks of ZFNs are cost and the complexity of design to enable targeting of some gene sequences. Transcription activator-like effector nucleases (TALENs) are a new technology that is similar to ZFNs but are less costly to produce and simpler to design. Much like the ZFNs, the TALENs also comprise two domains—the FokI endonuclease domain and a specific DNA-binding domain. The two engineered nucleases differ in the origin of the DNA-binding domain. The basis of the TALEN DNA-binding domain is the transcription activator-like effector proteins discovered in the plant pathogen *Xanthomonas* (Mussolino et al. 2011). The domain is made up of conserved repeating protein sequences of about 33–35 amino acids long, which are arranged in 12–30 tandem arrays (Mussolino et al. 2011). Each array of conserved repeating sequences varies at the 12th–13th amino acid, and this variability confers specificity with one of the four possible DNA nucleotides (Li et al. 2011). The specificity of the repeating variable diresidues enables arrays to be arranged and designed to recognize specific DNA sequences (Fig. 6.4). The ability to design TALENs for virtually any sequence of DNA is highly attractive and studies have indicated it is as effective as ZFNs at targeted gene modification (Mussolino et al. 2011). Furthermore, there is evidence that TALENs have less toxicity or off-target effects than ZFNs (Mussolino et al. 2011). TALEN technology has already been used to create genetically altered donor cells used for SCNT to produce low-density lipoprotein receptor gene-disrupted pigs (Carlson et al. 2012). These animals were created as a large animal model for human familial hypercholesterolaemia.

6.4.3.3 RNA-Guided Gene Editing

While ZFNs and TALENs are currently the most widely used gene-editing tools, new technologies are constantly being discovered and developed. Most recently, prokaryotic immune response mechanisms have been exploited to generate RNA-guided gene-editing techniques that have been shown to be effective in mammalian cells (Cong et al. 2013). The technology is based on the clustered regularly interspersed short palindromic repeats (CRISPR) or CRISPR-associated (Cas) system, which normally protects bacteria and archaea from foreign viruses and plasmids through RNA-guided gene silencing (Chang et al. 2013). The system is composed of two non-coding RNAs, the CRISPR RNA (crRNA) and the *trans*-activating CRISPR RNA (tracrRNA), along with a Cas9 nuclease. A 19–23 base pair portion of the crRNA provides the DNA-binding specificity of the system. The design of this DNA-binding domain is virtually unrestricted, except for an NGG sequence that must be present on the 3' end of the target cleavage site. This NGG sequence, otherwise known as the protospacer adjacent motif, is needed to direct cleavage of the DNA by the Cas9 nuclease just to the 5' end of the motif. The advantage this system provides over ZFNs or TALENs, is that only the specific crRNA needs to be tailored for the target sequence of choice. The remainder of the system remains identical. With only a small 19–23 base pair RNA sequence to design, this method can provide a simple and effective means at gene targeting (Chang et al. 2013). It has been shown effective in zebrafish as well as mammalian cells, with similar or increased efficiency compared to TALENs (Cong et al. 2013). Moreover, this method was successfully used to generate mice with multiple gene mutations by injecting zygotes with the Cas9 nuclease and two different crRNA guides to direct DNA-specific cleavage (Wang et al. 2013). The genetic alteration of DNA from larger animal species has also been of interest using this method (Sato et al. 2014; Niu et al. 2014). Whitworth and colleagues successfully developed transgenic piglets with biallelic mutation of multiple cluster of differentiation (CD) genes using the CRISPR-associated system and SCNT in 2014. While further studies need to be accomplished to determine whether toxicity from off-target effects are a concern, this new technology is definitely promising for future applications in the development of transgenic animals via SCNT.

6.5 The Creation of Transgenic Animals Using SCNT and Their Applications

Given the advantages of SCNT, and the concurrent development of exciting new gene-editing technologies, it is not surprising there has been substantial development in the area of large animal transgenesis using these approaches.

6.5.1 Transgenic Animals in Agriculture

In agriculture, transgenic animals have the ability to address many issues including poor production animal growth, susceptibility to disease and poor or inadequate nutrient composition of final products. Many transgenic animals either have been developed or are in development, using SCNT and gene-editing techniques. Given potential human health benefits related to high ω-3 fatty acid composition in the diet, multiple transgenic farm animals have been created to improve the content within food products. Examples include the production of transgenic cows and pigs expressing ω-3 fatty acid desaturase, the enzyme required to convert ω-6 fatty acids to ω-3 fatty acids, thus resulting in higher levels of ω-3 fatty acid in their milk and meat, respectively (Lai et al. 2006; Pan et al. 2010; Wu et al. 2012). Furthermore, the problem of cow's milk allergies related to β-lactoglobulin, a protein that is not present in human milk, requires resolution. Transgenic cows created through RNA interference and SCNT have been shown to produce lower levels of β-lactoglobulin and higher levels of casein (Jabed et al. 2012). This cow strain not only has the potential to address allergenic disease but also has the added benefit of higher calcium content and cheese yield due to the higher casein content of the milk obtained (Jabed et al. 2012)

6.5.2 Transgenic Animals and Therapeutic Cloning

Therapeutic cloning is another potential application of SCNT and animal transgenesis. It involves the harvesting of pluripotent cells from the developing reconstructed embryo for use to treat traumatic injuries as well as degenerative or genetic diseases. The major advantage of therapeutic cloning is that the pluripotent cells that are harvested are autologous, as they are created from somatic cells of the patient. This addresses major issues regarding tissue graft rejections. The potential uses for these pluripotent cells are enormous, including repopulation of blood cells, corneal injury repair, regeneration of nervous tissue, treatment of diabetes and more (Watt and Driskell 2010). In addition, potential curative strategies for inherited diseases can be developed by modifying donor somatic cells, reconstructing them into embryos, and the resultant "cured" pluripotent stem cells returned to the host. Furthermore, this approach can also be utilized to confer resistance genes in farm animals to make them resistant to various diseases. While therapeutic cloning has remarkable potential to address both animal and human disease, studies in humans are hampered by ethical and legal concerns relating to embryo development (Watt and Driskell 2010). Another avenue is the use of induced pluripotent stem cells, which were discovered by Yamanaka and Takahashi (2006) who showed that the addition of factors Oct3/4, Sox2, c-Myc and Klf4 transforms adult differentiated cells into cells that exhibit similar properties to embryonic stem cells.

6.5.3 *Transgenic Animals and Animal Pharming*

Since the discovery of passive immunization, there has been progressive development in the production of antibodies and antitoxins for therapeutic use (Redwan 2009). With this came the concept of harvesting these pharmaceutical products from laboratory and farm animals—otherwise termed animal pharming. Production of antibodies in farm animals was favoured given their large size and higher blood volume. Furthermore, the possibility of generating pharmaceutical products within the milk was exciting. The concept was driven by the demand for products other than antibodies, the need to increase harvest yield and also the need to facilitate product collection. This method simplifies and improves the efficiency of product purification by directing high-level secretion of the product into the milk (Redwan 2009). An assortment of high-value pharmaceutical products has been engineered to be expressed in transgenic animals, including albumin, antithrombin, peptide hormones, coagulation factors, as well as other agents to treat toxicities and bacterial infections (Maksimenko et al. 2013). Wall and colleagues (2005) created transgenic cows producing lysostaphin in their milk via this method. Lysostaphin has been used as an aid in treating methicillin-resistant bacterial infections by breaking down biofilms present in the body (Wall et al. 2005). Similarly, recombinant growth hormone has been produced as a component of the milk of genetically engineered cattle and goats (Lee et al. 2006; Salamone et al. 2006); while albumin has been isolated from the milk of transgenic cows (Echelard et al. 2009). In general, gene-editing techniques can be used to introduce novel genes into donor cells under the control of a promoter associated with milk production and then use the modified cells as donors for SCNT to generate transgenic animals which express the novel gene of interest only in the milk. The potential for further development in this area is great, especially using SCNT and novel gene-editing techniques. These technologies particularly facilitate the production of transgenic large animals, which have the potential to create higher product yields compared to small animals or basic cell cultures.

6.5.4 *Transgenic Animals and Xenotransplantation*

Xenotransplantation is a topic that has garnered much research interest over the years. The demand for organ donors far outweighs the supply and the use of porcine organs as alternatives is a potentially viable option. The pig model for xenotransplantation was selected given the similar anatomy and physiology shared with humans in multiple organ systems (Ibrahim et al. 2006). However, organ transplantations are affected by complications associated with graft–host rejection, even among human-to-human transplantations. There are three main rejection phases that occur when a non-autologous organ is transferred into a patient: the hyperacute rejection, acute humoral xenograft rejection and immune cell-mediated rejection. An approach taken to quell these rejection reactions is by developing porcine transgenic models that exhibit altered cell-surface markers and express various immunomodulatory proteins (Luo et al. 2012). The hyperacute rejection involves immediate graft cell destruction by the human immune system via recognition of porcine cell surface markers (Takahagi et al. 2005). The primary xenoantigen, α-Gal epitope, which is not present on human cells, elicits complement activation and subsequently cell death (Takahagi et al. 2005). In 2002 and 2003, respectively, Lai and Phelps groups (Lai et al. 2002; Phelps et al. 2003) developed SCNT knockout pigs for GGTA, the enzyme needed to create the α-Gal epitope. Similarly, in 2005, a pig model exhibiting the GGTA-KO phenotype, as well as expression of human decay-accelerating factor (DAF) and human *N*-acetylglucosaminyltransferase III (GnT-III) genes, was produced. Human DAF and hGnT-III, inhibit complement activation and reduce xenoantigenicity, respectively. This pig model was also produced using the SCNT technique (Takahagi et al. 2005).

Acute humoral xenograft rejection is initiated by non-Gal epitope antibodies, which activate porcine endothelial cells resulting in thrombosis and inflammation (Cho et al. 2011). Two groups attempted to address this rejection reaction by creating SCNT models expressing the human A20 gene and a soluble human tumour necrosis factor (TNF) receptor I-Fc (Oropeza et al. 2009; Cho et al. 2011). Human A20 protein protects against TNF-mediated apoptosis and inflammation while human TNF receptor I-Fc scavenges TNF-α and limits its pro-inflammatory actions (Oropeza et al. 2009; Cho et al. 2011).

The immune cell-mediated rejection (ICMR) that occurs in xenotransplantation is mediated by T cell host responses that are naturally inhibited by co-stimulation of the cytotoxic T cell lymphocyte-associated antigen 4, CTLA4, receptor, which is expressed on CD4+T cells (Phelps et al. 2009). The ICMR can be downregulated by stimulating the CTLA4 receptor with an anti-CTLA4 immunoglobulin. This process was exploited in transgenic pigs expressing anti-CTLA4 immunoglobulin as a potential solution to this phase of the xenograft rejection reaction (Phelps et al. 2009).

Finally, aside from host–graft rejection reactions, porcine organ transplantation has the additional concern of viral disease transmission. While less risky compared to xenotransplantation using non-human primate donors, pigs harbour endogenous retroviruses within their genome, which can potentially be transmitted to humans. While there is no evidence that these porcine endogenous retroviruses (PERV) cause disease in humans, there is still potential, given possible genetic recombination causing viral reactivation, as well as increased patient susceptibility from immunosuppression to prevent graft rejection (Ramsoondar et al. 2009). To address this concern, Ramsoondar and colleagues used a vector with tandem RNA interference transgenes to create somatic cells with reduced mRNA of two primary PERV genes. These cells were subsequently used for SCNT to create PERV knockdown pig models (Ramsoondar et al. 2009).

Xenotransplantation rejection reactions are complex and require further study; however, the current development of transgenic models is promising. Multi-transgenic pigs are likely needed to provide the best success for long-term xenograft survival, and SCNT offers an ideal approach given the possibility to perform multiple genetic mutations and cell selection prior to the cloning procedure.

6.5.5 Transgenic Animals as Biomedical Models

The creation of other biomedical models for research is another important application of SCNT and large animal transgenesis. A prime example is the creation of a cystic fibrosis transmembrane conductance regulator (CFTR) knockout pig model of human cystic fibrosis in 2008 by Rogers and colleagues. This transgenic animal, produced by homologous recombination and SCNT, resulted in a more representative model of the pathophysiology and clinical signs seen in humans, as compared to that seen in the mouse CFTR knockout (Rogers et al. 2008). Porcine models are gaining interest given their similar anatomy, lifespan and organ physiology compared to humans, making them potentially more representative of humans (Ibrahim et al. 2006). This has led to development of genetically modified pigs models carrying specific alterations that are known to cause hyperlipidaemia or atherosclerosis in humans, such as transgenic pigs expressing reduced levels of LDL receptor (Carlson et al. 2012), and apolipoprotein E (Bordignon et al. 2013), as well as overexpressing human apolipoprotein CIII (Wei et al. 2012).

Endocrine disease such as diabetes is another target illness that researchers hope to gain more information about through biomedical models. In 2009, a porcine model of mature-onset diabetes of the young, a hereditary form of diabetes caused by a mutation in an autosomal dominant gene, was produced (Umeyama et al. 2009). This model was created by targeting the hepatocyte nuclear factor 1-α (HNF-1α) gene through sequential SMGT and SCNT cloning techniques (Umeyama et al. 2009). Transgenic piglets exhibited elevated glucose levels, poor insulin secretion, glucose intolerance and stunted Langerhans islets. Most of these piglets did not survive past weaning, with only four piglets living up to 1–6 months of age (Umeyama et al. 2009). Renner and colleagues have also been developing models of insulin-independent diabetes (Renner et al. 2010) as well as permanent neonatal diabetes (Renner et al. 2013). A suitable model for type II diabetes remains to be developed and the ongoing effort into the creation of models for this particular disease will be indispensable due to the increasing incidence of this disorder worldwide.

The list of porcine transgenic biomedical models that have been developed so far is extensive. There have been advances in models for ophthalmic diseases such as retinitis pigmentosa and macular dystrophy, neurodegenerative disorders like

Alzheimer's and Huntington's diseases, as well as, various cancer models (Luo et al. 2012). It is clear that the potential for future development in this area is endless, given the advantages afforded by SCNT coupled with current gene-editing techniques.

6.6 Concluding Remarks and Perspectives

The demand for genetically altered large animal strains is increasing because of numerous applications including the creation of higher-quality production animals, therapeutic cloning, animal pharming, the creation of organ donors for xenotransplantation, and the creation of large animal models for biomedical research. These animal strains have the potential to improve agriculture, generate stem cell therapies, as well as facilitate research into disease pathogenesis and treatment, which can be eventually translated to humans. Since the isolation and culture of totipotent stem cells from large animal species useful for creating transgenics remains to be accomplished, it is currently extremely difficult to apply traditional methods used to generate gene deletions in these species. Genetic engineering technologies used to alter the cell genomes continue to improve and are becoming easier to adopt. With the advent of ZFNs, the evolution of TALENs and the emergence of RNA-guided gene editing, the list of gene-editing strategies now available in our toolbox make it simpler to introduce highly specific modifications into the genome. SCNT offers a facile method for creating cloned animals from engineered cells. While additional refinements are needed to further improve efficiency, SCNT technology is currently the best method available to generate transgenic large animal clones.

References

Bordignon V (2011) Animal systems animal cloning: state of the art and applications. In: Murray M-Y (ed) Comprehensive biotechnology, vol 4. 4441–4456

Bordignon V, El-Beirouthi N, Gasperin BG, Albornoz MS, Martinez-Diaz MA, Schneider C, Laurin D, Zadworny D, Agellon LB (2013) Production of cloned pigs with targeted attenuation of gene expression. PLoS ONE 8:e64613

Briggs R, King TJ (1952) Transplantation of living nuclei from blastula cells into enucleated frogs' eggs. Proc Natl Acad Sci U S A 38:455–463

Cabot RA, Kuhholzer B, Chan AW, Lai L, Park KW, Chong KY, Schatten G, Murphy CN, Abeydeera LR, Day BN, Prather RS (2001) Transgenic pigs produced using in vitro matured oocytes infected with a retroviral vector. Anim Biotechnol 12:205–214

Campbell KH, Alberio R, Choi I, Fisher P, Kelly RD, Lee JH, Maalouf W (2005) Cloning: eight years after Dolly. Reprod Domes Anim 40:256–268

Campbell KH, Fisher P, Chen WC, Choi I, Kelly RD, Lee JH, Xhu J (2007). Somatic cell nuclear transfer: past, present and future perspectives. Theriogenology 68(Suppl 1):214–31

Carlson DF, Tan W, Lillico SG, Stverakova D, Proudfoot C, Christian M, Voytas DF, Long CR, Whitelaw CB, Fahrenkrug SC (2012) Efficient TALEN-mediated gene knockout in livestock. Proc Natl Acad Sci U S A 109:17382–17387

Chang N, Sun C, Gao L, Zhu D, Xu X, Zhu X, Xiong JW, Xi JJ (2013) Genome editing with RNA-guided Cas9 nuclease in Zebrafish embryos. Cell Res 23:465–472

Cho B, Koo OJ, Hwang JI, Kim H, Lee EM, Hurh S, Park SJ, Ro H, Yang J, Surh CD, D'Apice AJ, Lee BC, Ahn C (2011) Generation of soluble human tumor necrosis factor-alpha receptor 1-Fc transgenic pig. Transplantation 92:139–147

Cong L, Ran FA, Cox D, Lin S, Barretto R, Habib N, Hsu PD, Wu X, Jiang W, Marraffini LA, Zhang F (2013) Multiplex genome engineering using CRISPR/Cas systems. Science 339:819–823

Echelard Y, Williams JL, Destrempes MM, Koster JA, Overton SA, Pollock DP, Rapiejko KT, Behboodi E, Masiello NC, Gavin WG, Pommer J, Van Patten SM, Faber DC, Cibelli JB, Meade HM (2009) Production of recombinant albumin by a herd of cloned transgenic cattle. Transgenic Res 18:361–376

Edwards JL, Schrick FN, McCracken MD, van Amstel SR, Hopkins FM, Welborn MG, Davies CJ (2003) Cloning adult farm animals: a review of the possibilities and problems associated with somatic cell nuclear transfer. Am J Reprod Immunol 50:113–123

Fire A, Xu S, Montgomery MK, Kostas SA, Driver SE, Mello CC (1998) Potent and specific genetic interference by double-stranded RNA in. Caenorhabditis elegans Nature 391:806–811

Gordon JW, Ruddle FH (1981) Integration and stable germ line transmission of genes injected into mouse pronuclei. Science 214:1244–1246

Hannon GJ, Rossi JJ (2004) Unlocking the potential of the human genome with RNA interference. Nature 431:371–378

Hauschild J, Petersen B, Santiago Y, Queisser AL, Carnwath JW, Lucas-Hahn A, Zhang L, Meng X, Gregory PD, Schwinzer R, Cost GJ, Niemann H (2011) Efficient generation of a biallelic knockout in pigs using zinc-finger nucleases. Proc Natl Acad Sci U S A 108:12013–12017

Ibrahim Z, Busch J, Awwad M, Wagner R, Wells K, Cooper DK (2006) Selected physiologic compatibilities and incompatibilities between human and porcine organ systems. Xenotransplantation 13:488–499

Illmensee K, Hoppe PC (1981) Nuclear transplantation in Mus musculus: developmental potential of nuclei from preimplantation embryos. Cell 23:9–18

Jabed A, Wagner S, McCracken J, Wells DN, Laible G (2012) Targeted microRNA expression in dairy cattle directs production of beta-lactoglobulin-free, high-casein milk. Proc Natl Acad Sci U S A 109:16811–16816

Jang G, Kim MK, Lee BC (2010) Current status and applications of somatic cell nuclear transfer in dogs. Theriogenology 74:1311–1320

Kashiwakura Y, Mimuro J, Onishi A, Iwamoto M, Madoiwa S, Fuchimoto D, Suzuki S, Suzuki M, Sembon S, Ishiwata A, Yasumoto A, Sakata A, Ohmori T, Hashimoto M, Yazaki S, Sakata Y (2012) Porcine model of hemophilia A. PLoS ONE 7:e49450

Lai L, Sun Q, Wu G, Murphy CN, Kuhholzer B, Park KW, Bonk AJ, Day BN, Prather RS (2001) Development of porcine embryos and offspring after intracytoplasmic sperm injection with liposome transfected or non-transfected sperm into in vitro matured oocytes. Zygote 9:339–346

Lai L, Kolber-Simonds D, Park KW, Cheong HT, Greenstein JL, Im GS, Samuel M, Bonk A, Rieke A, Day BN, Murphy CN, Carter DB, Hawley RJ, Prather RS (2002) Production of alpha-1,3-galactosyltransferase knockout pigs by nuclear transfer cloning. Science 295:1089–1092

Lai L, Kang JX, Li R, Wang J, Witt WT, Yong HY, Hao Y, Wax DM, Murphy CN, Rieke A, Samuel M, Linville ML, Korte SW, Evans RW, Starzl TE, Prather RS, Dai Y (2006) Generation of cloned transgenic pigs rich in omega-3 fatty acids. Nat Biotechnol 24:435–436

Lavitrano M, Forni M, Bacci ML, Di Stefano C, Varzi V, Wang H, Seren E (2003) Sperm mediated gene transfer in pig: Selection of donor boars and optimization of DNA uptake. Mol Reprod Dev 64:284–291

Lee C-S, Lee DS, Fang N, Oh KB, Shin ST, Lee K-K (2006). Integration and expression of goat β-casein/hGH hybrid gene in a transgenic goat. Reprod Dev Biol 30:293–299.

Li T, Huang S, Jiang WZ, Wright D, Spalding MH, Weeks DP, Yang B (2011) TAL nucleases (TALNs): hybrid proteins composed of TAL effectors and FokI DNA-cleavage domain. Nucleic Acids Res 39:359–372

Luo Y, Lin L, Bolund L, Jensen TG, Sorensen CB (2012) Genetically modified pigs for biomedical research. J Inherit Metab Dis 35:695–713

Maksimenko OG, Deykin AV, Khodarovich YM, Georgiev PG (2013) Use of transgenic animals in biotechnology: prospects and problems. Acta Naturae 5:33–46

Mallo M (2006) Controlled gene activation and inactivation in the mouse. Front Biosci 11:313–327

Miller J, McLachlan AD, Klug A (1985) Repetitive zinc-binding domains in the protein transcription factor IIIA from Xenopus oocytes. EMBO J 4:1609–1614

Mussolino C, Morbitzer R, Lutge F, Dannemann N, Lahaye T, Cathomen T (2011) A novel TALE nuclease scaffold enables high genome editing activity in combination with low toxicity. Nucleic Acids Res 39:9283–9293

Niemann H, Lucas-Hahn A (2012) Somatic cell nuclear transfer cloning: practical applications and current legislation. Reprod Domes Anim 47(Suppl 5):2–10

Niu Y, Shen B, Cui Y, Chen Y, Wang J, Wang L, Kang Y, Zhao X, Si W, Li W, Xiang AP, Zhou J, Guo X, Bi Y, Si C, Hu B, Dong G, Wang H, Zhou Z, Li T, Tan T, Pu X, Wang F, Ji S, Zhou Q, Huang X, Ji W, Sha J (2014). Generation of gene-modified cynomolgus monkey via Cas9/RNA-mediated gene targeting in one-cell embryos. Cell 156:836–43

Oropeza M, Petersen B, Carnwath JW, Lucas-Hahn A, Lemme E, Hassel P, Herrmann D, Barg-Kues B, Holler S, Queisser AL, Schwinzer R, Hinkel R, Kupatt C, Niemann H (2009). Transgenic expression of the human A20 gene in cloned pigs provides protection against apoptotic and inflammatory stimuli. Xenotransplantation 16:522–34

Paddison PJ, Caudy AA, Bernstein E, Hannon GJ, Conklin DS (2002) Short hairpin RNAs (shRNAs) induce sequence-specific silencing in mammalian cells. Genes Dev 16:948–958

Pan D, Zhang L, Zhou Y, Feng C, Long C, Liu X, Wan R, Zhang J, Lin A, Dong E, Wang S, Xu H, Chen H (2010) Efficient production of omega-3 fatty acid desaturase (sFat-1)-transgenic pigs by somatic cell nuclear transfer. Sci China Life Sci 53:517–523

Pereyra-Bonnet F, Fernandez-Martin R, Olivera R, Jarazo J, Vichera G, Gibbons A, Salamone D (2008) A unique method to produce transgenic embryos in ovine, porcine, feline, bovine and equine species. Reprod Fertil Dev 20:741–749

Phelps CJ, Koike C, Vaught TD, Boone J, Wells KD, Chen SH, Ball S, Specht SM, Polejaeva IA, Monahan JA, Jobst PM, Sharma SB, Lamborn AE, Garst AS, Moore M, Demetris AJ, Rudert WA, Bottino R, Bertera S, Trucco M, Starzl TE, Dai Y, Ayares DL (2003) Production of alpha 1,3-galactosyltransferase-deficient pigs. Science 299:411–414

Phelps CJ, Ball SF, Vaught TD, Vance AM, Mendicino M, Monahan JA, Walters AH, Wells KD, Dandro AS, Ramsoondar JJ, Cooper DK, Ayares DL (2009) Production and characterization of transgenic pigs expressing porcine CTLA4-Ig. Xenotransplantation 16:477–485

Prather RS, Lorson M, Ross JW, Whyte JJ, Walters E (2013) Genetically engineered pig models for human diseases. Annu Rev Anim Biosci 1:203–219

Ramsoondar J, Vaught T, Ball S, Mendicino M, Monahan J, Jobst P, Vance A, Duncan J, Wells K, Ayares D (2009) Production of transgenic pigs that express porcine endogenous retrovirus small interfering RNAs. Xenotransplantation 16:164–180

Redwan el RM (2009) Animal-derived pharmaceutical proteins. J Immunoass Immunochem 30:262–290

Renner S, Fehlings C, Herbach N, Hofmann A, von Waldthausen DC, Kessler B, Ulrichs K, Chodnevskaja I, Moskalenko V, Amselgruber W, Goke B, Pfeifer A, Wanke R, Wolf E (2010) Glucose intolerance and reduced proliferation of pancreatic beta-cells in transgenic pigs with impaired glucose-dependent insulinotropic polypeptide function. Diabetes 59:1228–1238

Renner S, Braun-Reichhart C, Blutke A, Herbach N, Emrich D, Streckel E, Wunsch A, Kessler B, Kurome M, Bahr A, Klymiuk N, Krebs S, Puk O, Nagashima H, Graw J, Blum H, Wanke R, Wolf E (2013) Permanent neonatal diabetes in INSC94Y transgenic pigs. Diabetes 62:1505–1511

Rogers CS, Stoltz DA, Meyerholz DK, Ostedgaard LS, Rokhlina T, Taft PJ, Rogan MP, Pezzulo AA, Karp PH, Itani OA, Kabel AC, Wohlford-Lenane CL, Davis GJ, Hanfland RA, Smith TL, Samuel M, Wax D, Murphy CN, Rieke A, Whitworth K, Uc A, Starner TD, Brogden KA, Shi-

lyansky J, McCray PB Jr, Zabner J, Prather RS, Welsh MJ (2008) Disruption of the CFTR gene produces a model of cystic fibrosis in newborn pigs. Science 321:1837–1841
Salamone D, Baranao L, Santos C, Bussmann L, Artuso J, Werning C, Prync A, Carbonetto C, Dabsys S, Munar C, Salaberry R, Berra G, Berra I, Fernandez N, Papouchado M, Foti M, Judewicz N, Mujica I, Munoz L, Alvarez SF, Gonzalez E, Zimmermann J, Criscuolo M, Melo C (2006) High level expression of bioactive recombinant human growth hormone in the milk of a cloned transgenic cow. J Biotechnol 124:469–472
Sato M, Miyoshi K, Nagao Y, Nishi Y, Ohtsuka M, Nakamura S, Sakurai T, Watanabe S (2014) The combinational use of CRISPR/Cas9-based gene editing and targeted toxin technology enables efficient biallelic knockout of the α-1,3-galactosyltransferase gene in porcine embryonic fibroblasts. Xenotransplantation 21:291–300
Takahagi Y, Fujimura T, Miyagawa S, Nagashima H, Shigehisa T, Shirakura R, Murakami H (2005) Production of alpha 1,3-galactosyltransferase gene knockout pigs expressing both human decay-accelerating factor and N-acetylglucosaminyltransferase III. Mol Reprod Dev 71:331–338
Umeyama K, Watanabe M, Saito H, Kurome M, Tohi S, Matsunari H, Miki K, Nagashima H (2009) Dominant-negative mutant hepatocyte nuclear factor 1alpha induces diabetes in transgenic-cloned pigs. Transgenic Res 18:697–706
Urnov FD, Rebar EJ, Holmes MC, Zhang HS, Gregory PD (2010) Genome editing with engineered zinc finger nucleases. Nat Rev Genet 11:636–646
Vajta G (2007) Handmade cloning: the future way of nuclear transfer? Trends Biotechnol 25:250–253
Wall RJ (2001) Pronuclear microinjection. Cloning Stem Cells 3:209–220
Wall RJ, Powell AM, Paape MJ, Kerr DE, Bannerman DD, Pursel VG, Wells KD, Talbot N, Hawk HW (2005) Genetically enhanced cows resist intramammary *Staphylococcus aureus* infection. Nat Biotechnol 23:445–451
Wang B, Zhou J (2003) Specific genetic modifications of domestic animals by gene targeting and animal cloning. Reprod Biol Endocrinol 103:1–8
Wang H, Yang H, Shivalila CS, Dawlaty MM, Cheng AW, Zhang F, Jaenisch R (2013) One-step generation of mice carrying mutations in multiple genes by CRISPR/Cas-mediated genome engineering. Cell 153:910–918
Watanabe M, Kurome M, Matsunari H, Nakano K, Umeyema K, Shiota A, Nakauchi H, Nagashima H (2012) The creation of transgenic pigs expressing human proteins using BAC-derived, full-length genes and intracytoplasmic sperm injection-mediated gene transfer. Transgenic Res 21:605–618
Watt FM, Driskell RR (2010) The therapeutic potential of stem cells. Philos Trans R Soc Lond B Biol Sci 365:155–163
Wei J, Ouyang H, Wang Y, Pang D, Cong NX, Wang T, Leng B, Li D, Li X, Wu R, Ding Y, Gao F, Deng Y, Liu B, Li Z, Lai L, Feng H, Liu G, Deng X (2012) Characterization of a hypertriglyceridemic transgenic miniature pig model expressing human apolipoprotein CIII. FEBS J 279:91–99
Whitworth KM, Lee K, Benne JA, Beaton BP, Spate LD, Murphy SL, Samuel MS, Mao J, O'Gorman C, Walters EM, Murphy CN, Driver JP, Mileham A, McLaren D, Wells KD, Prather RS (2014). Use of the CRISPR/Cas9 system to produce genetically engineered pigs from in vitro-derived oocytes and embryos. Biol Reprod 91(3):78. pii: biolreprod.114.121723
Whyte JJ, Prather RS (2012) Cell biology symposium: Zinc finger nucleases to create custom-designed modifications in the swine (Sus scrofa) genome. J Anim Sci 90:1111–1117
Wilmut I, Schnieke AE, McWhir J, Kind AJ, Campbell KH (1997) Viable offspring derived from fetal and adult mammalian cells. Nature 385:810–813
Wu X, Ouyang H, Duan B, Pang D, Zhang L, Yuan T, Xue L, Ni D, Cheng L, Dong S, Wei Z, Li L, Yu M, Sun QY, Chen DY, Lai L, Dai Y, Li GP (2012) Production of cloned transgenic cow expressing omega-3 fatty acids. Transgenic Res 21:537–543
Yamanaka S, Takahashi K (2006) Induction of pluripotent stem cells from mouse embryonic and adult fibroblast cultures by defined factors. Cell 126:663–676

Part II
Plants

Chapter 7
Apomixis: The Asexual Formation of Seed

Ross Bicknell and Andrew Catanach

7.1 Introduction

Apomixis is the clonal propagation of a plant through seed, avoiding the processes of meiosis and fertilization. In common with other natural forms of clonal reproduction, apomixis leads to the formation of uniform populations, and these can extend over large areas of land and persist over long periods of time (Horandl 2008). Several comprehensive reviews have been written on apomixis detailing the variation observed in known natural systems of apomixis (Crane 2001; Nogler 1984a), the genetics of the trait (Ozias-Akins and van Dijk 2007), its evolutionary and ecological implications (van Dijk 2003; van Dijk et al. 2009), the role of temporal and epigenetic factors (Koltunow and Grossniklaus 2003) and the potential that apomixis offers for increasing the rate of plant breeding for seed-propagated crops (Savidan 2000a). This chapter is intended to inform the reader of the potential of apomixis and of progress being made in apomixis research, focusing on more recent findings and contemporary approaches. A general description of the trait is given to provide context to the latter discussion, but this is deliberately broad in scope.

7.2 Why Study Apomixis?

Apomixis is not, in itself, a form of somatic manipulation. It is, however, a potentially invaluable tool for applying advances in somatic manipulation to seed-propagated crops. Most of the world's key crops, as measured by traded value, land under cultivation, and/or nutritional importance, are propagated by seed. Examples include wheat, rice, soy, maize, cotton and most timber, forage, oil and fibre crops. The reasons for using seed are compelling. Typically, seed is small, robust, contains

R. Bicknell (✉) · A. Catanach
Plant and Food Research, Private Bag 4704, Christchurch, New Zealand
e-mail: ross.bicknell@plantandfood.co.nz

X.-Q. Li et al. (eds.), *Somatic Genome Manipulation*,
DOI 10.1007/978-1-4939-2389-2_7

storage reserves to support germination and establishment, and (in most cases) is easily produced, harvested, cleaned, transported, stored and sown. However, seed is usually also the product of sexual reproduction, specifically of the combined actions of meiosis and fertilization, and is therefore genetically diverse. The degree of this diversity varies between breeding systems. Self-fertile systems, such as wheat and rice, can be inbred to form uniform (true breeding) lines; however, they are produced at the cost of lost heterosis (hybrid vigour). On the other hand, outcrossing species tend to retain heterosis, but the progeny are variable, compromising commercial utility. Again, a breeder can reduce variation through inbreeding but at the cost of lost heterosis. In the special case of F_1 hybrid breeding, variation is reduced and heterosis is obtained, but this is achieved at a high economic cost. Cloning allows the multiplication of elite forms, but it is typically expensive and restricted to plants of high unit value such as fruit trees. The obvious exceptions to this are plants that already have an efficient form of cloning which can be used for production, such as the tubers of potato and the runners of strawberry.

As apomixis is cloning through seed, it offers the uniformity and predictability of clonal reproduction along with the economic advantages of seed production. In agriculture, it could be used to generate new hybrid varieties quickly and to perpetuate them cheaply and indefinitely through seed. As some forms of apomixis can occur without pollination, it could be used to avoid difficulties associated with cross compatibility and pollinator availability. Finally, even in vegetatively propagated plants like potatoes, apomixis may be useful in avoiding viral transfer, as very few viruses are able to transmit through true seed (Hanna 1995; Jefferson and Bicknell 1995; Koltunow et al. 1995; Savidan 2000a, b). For potato production in temperate regions, this would need to be managed as part of a high health programme as it is only possible to reach an acceptable tuber size in a short temperate growing season by using tubers as the planting material.

7.3 The Key Components of Apomixis

Apomixis is currently recorded in more than 400 taxa of flowering plants, involving 32 plant families, with a pattern of distribution that indicates it has arisen many times amongst the flowering plants (Carman 1997). It is generally agreed that in most if not all of these cases, apomixis has evolved through the modification of sexual reproduction. Evidence to support this conclusion comes from comparative studies of the structures, processes and gene expression patterns of sexual and apomictic plants (see below). Before describing the processes that underlie apomixis therefore, we have included a brief review of the current knowledge of sexual reproduction, focusing on developmental elements that either have remained constant in apomicts or have been the key steps of change permitting the evolution of apomictic development.

Most importantly, the two defining events of sexual reproduction that are not involved in apomixis are meiosis and fertilization. Clearly these are the events that give rise to the unique genotype of the sexually derived seedling and both, therefore, are avoided during the development of asexually derived embryos. That said, incomplete meiosis is a common feature of apomicts, in which the progression of meiosis is interrupted either in the first cycle (first division restitution) or in the second cycle (second division restitution) of meiotic division. Similarly, fertilization of the egg is not seen in apomictic development, but fertilization of the central cell, giving rise to the endosperm, is commonly seen (see below).

Within an angiosperm flower, the megagametophyte generation is represented by an embryo sac, typically comprising of eight meiotically reduced nuclei and seven cells. The embryo sac is surrounded by the tissues of the ovule and that in turn is encased within an ovary. The structures of the ovary, ovule and even the embryo sac differ significantly between species. Four key developmental features, however, remain constant amongst sexual flowering plant species. These features are also known to be common to all forms of apomictic seed development. First, only a single cell within the embryo sac differentiates into an egg cell and that is the only cell within the embryo sac that is fully totipotent. Many forms of apomixis involve the embryo sac (see below) and in all of these it is the egg cell that gives rise to the new embryo. Similarly, only the central cell can give rise to the endosperm and this is true in apomicts as well. Other cells within the somatic (sporophytic) plant body can be totipotent, which forms the basis of the sporophytic mechanisms of apomixis discussed below.

Second, in sexual plants the development of the angiosperm embryo requires the presence of a nutritive endosperm tissue, and that tissue is the product of fertilization. There are some exceptions to this rule. Orchids and members of the Podostemaceae lack an endosperm and both have evolved other mechanisms for nourishing the young embryo. Only two known apomicts lack an endosperm. Both are orchids (Teppner and Klein 1993; Huang et al. 2009). In all other apomicts, an endosperm is required. The endosperm tissue in apomicts may arise following fertilization (pseudogamy) or it may arise spontaneously without fertilization (autonomous endospermy), but, as in sexual species, it is always the derivative of the central cell.

Third, in flowering plants the endosperm and embryo typically have different ploidy states. Most commonly, this is 2n:3n for the embryo and endosperm respectively. Other ratios are known to occur in nature, particularly amongst apomicts, and plants are known to differ in their tolerances to different ploidy ratios (Scott et al. 1998, Berger 1999). In grasses, however, a 2n:3n ploidy balance appears to be particularly critical. Apomixis is a common trait amongst the grasses and most of the developmental mechanisms seen amongst apomicts in this group result in a 2n:3n ploidy balance (Crane 2001).

Finally, in sexual plants, embryogenesis and endosperm formation take place within a wider developmental programme of reproduction including flowering, gametogenesis, seed maturation and fruit formation. These events are tightly and coordinately regulated. As all forms of apomixis also employ the structures of the flower, seed and fruit, coordinated developmental timing is a critical parameter in the success of these mechanisms.

7.4 The Mechanisms of Apomixis

The known mechanisms of apomixis are subdivided into two primary groupings (sporophytic and gametophytic) based on the tissue involved in the formation of the clonal embryo. In the case of gametophytic types, they are then further subdivided based on the cell type giving rise to the embryo sac. As described above, mechanisms are also separately categorized with respect to whether the endosperm tissue arises spontaneously or as the result of fertilization.

7.4.1 Sporophytic Apomixis

In sporophytic apomixis, the clonal embryo arises directly from a somatic cell (or cells) of the ovule (Fig. 7.1). Citrus and mango provide good examples of this kind of apomixis (Wakana and Uemoto 1987, 1988; Litz et al. 1995). In these plants, the sexual reproductive pathway proceeds normally with the fertilization of meiotically derived egg and central cells. Somatic embryo formation occurs in the surrounding ovule tissue, normally coinciding with the early development of the sexually derived embryo. As the endosperm arises normally from the union of a 2n central cell nucleus and a 1n sperm nucleus, it has a 3n ploidy state. Both the sexually derived embryo and any clonally derived somatic embryos within the ovule have a ploidy of 2n, so a 2n:3n ploidy balance is achieved in this system. Within the developing seed of these plants, it is common for there to be more than one embryo developing (polyembryony), including one sexually derived embryo and one or more asexually derived embryos. Only one endosperm tissue is present, however, so polyembryony in sporophytic apomicts leads to competition for the limited reserves of the ovule and in particular for the reserves of the endosperm. Paradoxically, therefore, although polyembryony is a reflection of the highly embryogenic state of the ovule tissue, it can often lead to reduced seed viability.

As sporophytic apomicts generate seedlings derived both sexually and asexually, it is not uncommon for seedling populations to consist of a mixture of genetically identical (and maternal) and genetically unique (sexually derived) individuals. This combination of progeny types is referred to as "facultative apomixis." It is particularly typical of sporophytic apomicts because a sexually derived embryo and endosperm must form in these species, but it is also seen amongst gametophytic apomicts as well (see below). It is theorized that facultative apomixis may confer some evolutionary advantages, as it provides apomicts with an opportunity to evolve in response to environmental changes while still expending most of their reproductive effort on the perpetuation of a successful clonal lineage (van Dijk 2003). To breeders of crops where sporophytic apomixis is common, however, such as mango and citrus, it can pose quite a challenge, as progeny testing is needed to determine which seedlings are clonal and which are not (López-Valenzuela et al. 1997; de Oliveira et al. 2002).

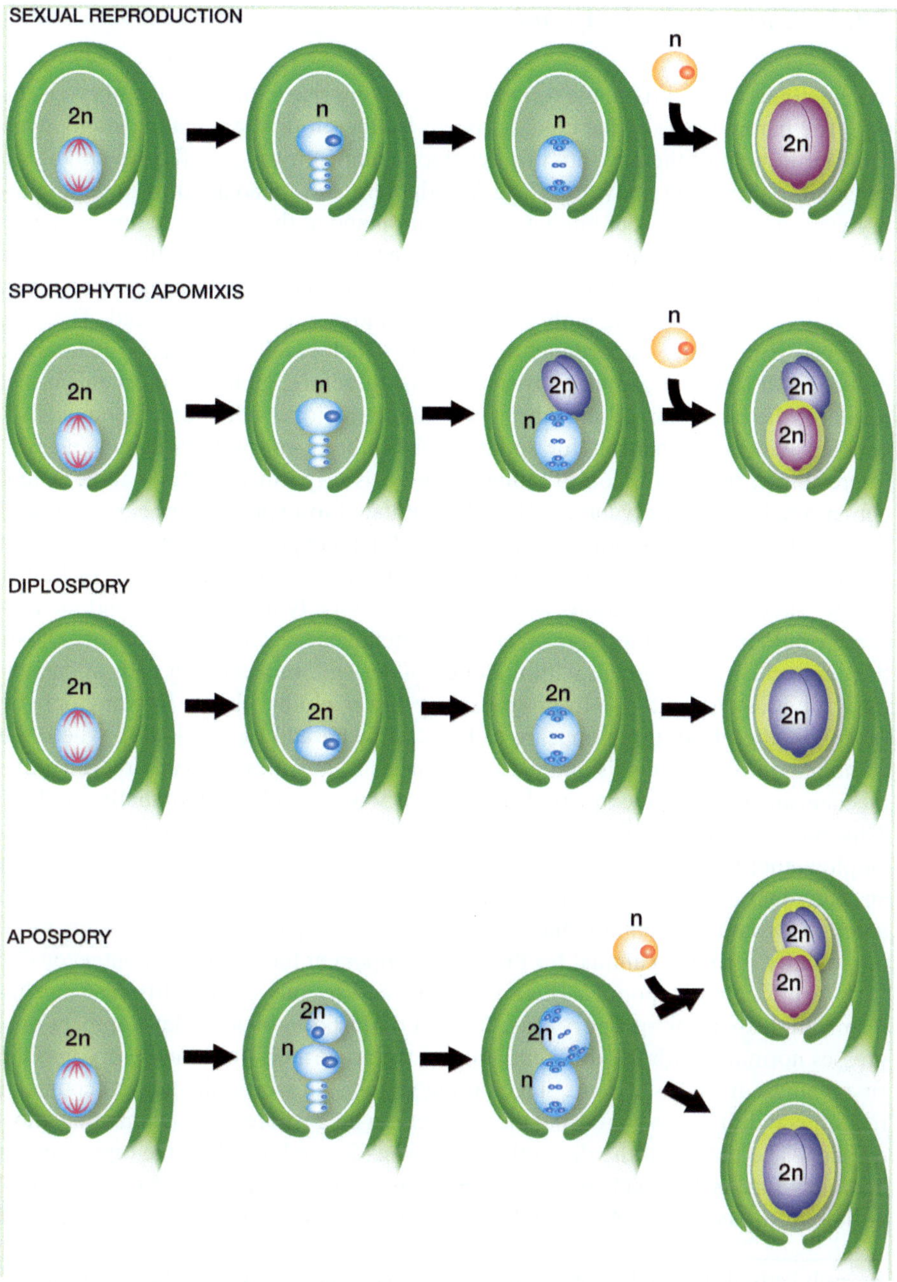

Fig. 7.1 Sexual reproduction compared with three forms of apomixis in flowering plants at four key stages of development: the initiation of meiosis, following megaspore differentiation, the differentiation of a mature embryo sac and early embryogenesis. For simplicity, only the pathway to embryo formation is depicted. Endosperm origin is not detailed but the endosperm tissue is depicted in *light green*. Sperm cells involved in the fertilization of the egg are depicted in *red*. During sporophytic apomixis and apospory, fertilization can lead to the formation of both hybrid and

7.4.2 Gametophytic Apomixis

In gametophytic apomixis, the totipotent cell giving rise to the clonal embryo is the egg. The parent tissue of the clonal progeny is, therefore, the female gametophyte. The recruitment of the egg into this developmental pathway satisfies several of the constraints discussed above: the egg is totipotent and preadapted to undergo embryogenesis, the egg occurs in the right location and at the right time during development for it to divide in concordance with the broader programme of reproductive development and finally, as the egg is the sole totipotent cell within the embryo sac, it will give rise to a single embryo for each endosperm tissue produced. The egg, however, is normally the product of meiosis, which results in it being genetically reduced. All forms of gametophytic apomixis, therefore, include a mechanism that results in the formation of an embryo sac without meiotic reduction ("apomeiosis") coupled with a mechanism for spontaneous embryogenesis ("parthenogenesis"). Interestingly, the expression of either apomeiosis or parthenogenesis alone would confer a selective disadvantage. If a plant mutated to express apomeiosis but still required fertilization to form an embryo (i.e., did not express parthenogenesis), then unreduced eggs would combine with sperm nuclei, resulting in plants of higher ploidy. Successive generations would have successively higher ploidy and the outcome would be untenable. Conversely, if parthenogenesis arose without apomeiosis the opposite would happen, with meiotically derived, reduced egg cells spontaneously dividing to give rise to progressively lower ploidy plants. In gametophytic apomixis, therefore, there appears to be a "chicken and egg" paradox where the combination of apomeiosis and parthenogenesis establishes a tenable reproductive mechanism, but either expressed on its own would lead to reproductive failure. This may explain why almost all gametophytic apomicts are perennials and many employ other forms of vegetative reproduction, such as stolons, to sustain the life of an individual genet. In a perennial, a mutation conferring one part of gametophytic apomixis would not be immediately deleterious to the parent plant, providing more opportunity than in an annual for the other component then to arise spontaneously.

In gametophytic apomicts, meiosis is avoided by one of two different mechanisms, diplospory or apospory (Fig. 7.1). In diplospory, meiotic development progresses normally with the differentiation of the archesporial cell and entry into the first meiotic division. At some stage during meiosis, however, the normal course of reductional division is halted and the product is a megaspore with the maternal ploidy state. The timing of the disruption of meiosis is used to subdivide diplosporous apomicts further by mechanism, the details of which are well covered by several reviews (Nogler 1984a; Asker and Jerling 1992; Crane 2001). Once an unreduced

asexually derived progeny. In sporophytic apomixis, only a meiotically derived egg is present and that is the egg that is fertilized. In apospory, although both reduced and unreduced eggs are present, the sperm nucleus typically just fertilizes the meiotically derived reduced egg. During development in aposporous apomicts, however, meiotically derived structures are often lost. In this case, only asexually derived embryo sacs are present and only asexually derived embryos form. The nuclear state (n or 2n) of the cells or structures involved is noted. Where an embryo is clonal in origin it is depicted in *blue* and where it is hybrid in origin it is depicted in *purple* megaspore is formed, it undergoes gametogenesis to give rise to an unreduced embryo sac, within which an unreduced egg and unreduced polar nuclei differentiate. The spontaneous division of the egg initiates embryogenesis without fertilization (parthenogenesis). The development of the endosperm in diplosporous apomicts may be either pseudogamous or autonomous, as described previously. Sexually derived progeny can still arise in this system, and they do so when female meiosis is successfully achieved. Meiotically derived seedlings, however, are typically rare in diplosporous apomicts. For this discussion, the main feature of note in diplospory is that both the megaspore and the egg are directly involved in the formation of the clonal embryo. Only a single embryo arises in each seed, and its development is typically in synchrony with related events of seed and fruit development. Unlike sporophytic apomicts, therefore, polyembryony is uncommon in these plants.

In apospory, one or more somatic cell(s) of the ovule (or occasionally of the ovary) differentiate into a structure that is similar in form and function to the megaspore. These cells are called "aposporous initials". As they are somatically derived, aposporous initials have the maternal genotype. Once formed, aposporous initials divide and differentiate into an embryo sac, within which an unreduced egg and unreduced polar nuclei form. The egg then undergoes parthenogenesis to initiate embryogenesis. As with diplospory, both pseudogamous and autonomous endospermy are seen amongst aposporous apomicts. In aposporous apomicts, both the meiotic pathway and the asexual pathway occur in parallel and both can contribute to the production of seedlings. Typically, one pathway predominates at the other's expense; however, polyembryonic seeds are common in these plants. Multiple embryos can arise either from the survival of both sexually and asexually derived progeny, and/or from multiple asexually derived embryos arising from multiple aposporous initials.

7.4.3 Gametophytic Apomixis and Polyploidy

There is a highly significant association between the expression of gametophytic apomixis and the occurrence of polyploidy in plants naturally expressing the trait. Of the hundreds of taxa known to include gametophytic apomicts (Carman 1997), naturally occurring diploid apomicts have only been clearly identified in *Boechera* (Kantama et al. 2007). Müntzing (1928) reported diploid apomicts in the genus *Potentilla,* but the reproductive biology of this material is now contested (Holm and Ghatnekar 1996). Even in *Boechera,* the majority of apomictic field collections are reported to be polyploids (Schranz et al. 2005). The reason for the close association between gametophytic apomixis and polyploidy is unknown. Amongst sporophytic apomicts diploidy is common, so it does not appear that the reproductive phenomenon of apomixis induces or requires polyploidy. Diploid apomicts have been identified as weak individuals in experimental populations of some gametophytic species (Nogler 1982; Bicknell 1997; Kojima and Nagato 1997) indicating that polyploidy is not a prerequisite for the expression of gametophytic apomixis. Similarly, polyploidy has been induced in many species, yet this has very seldom resulted in apomixis of any type. Instead, it has been proposed that polyploidy may be necessary for the transmission of the trait via gametes. If true, this would, over evolutionary time, lead to its near exclusive expression in polyploid races. Nogler (1984a) working in *Ranunculus auricomus,* noted that apomixis would not transmit via a haploid gamete, yet sexual reproduction would. He proposed that a genetic element involved in apomixis was associated with a recessive gamete lethal factor, which became exposed in haploid gametes. There is now evidence of transmission bias against apomixis in *Tripsacum* (Grimanelli et al., 1998a) and *Pennisetum* (Roche et al. 2001; Jessup et al. 2002), suggesting that the ability to transmit the trait may be an important factor limiting native gametophytic apomixis to polyploid genotypes.

The role of polyploidy in the expression of apomixis is of particular interest because it represents a potential impediment to the installation of apomixis into target crop species. Many of our crop species are diploids, including some that have been explicitly identified as potential candidates for apomixis to facilitate breeding progress. Cultivated rice, in particular, is a diploid inbreeding species which is known to express significant hybrid vigour (Yu et al. 1997). Apomixis would have a profound impact on this important crop but this would ideally not require a change in the ploidy of the cultivated varieties.

7.4.4 Can Apomixis be Installed into Crop Species, and What Form of Apomixis Would Best Meet Our Needs?

Apomixis is already known to occur in a small number of crop species. Kentucky blue grass (*Poa pratensis*) is an important temperate turf and fodder grass that reproduces by gametophytic apomixis (Naumova et al. 1992). Similarly, *Brachiaria decumbens, Panicum maximum* and *Tripsacum dactyloides* are tropical forage grasses that are gametophytic apomicts (Naumova et al. 1999; Savidan 1980; Leblanc et al. 1995a). As mentioned above, sporophytic apomixis is common in citrus and in some tropical fruits such as mango and mangosteen (*Garcinia mangostana*). In commerce, however, these trees are typically reproduced by vegetative means and apomixis is not employed in the husbandry of the crop as it is in the grasses. In other cases, apomixis is not known in a crop species, but does occur in a closely related species. Attempts to transfer it by conventional introgression through crossing have been reported in maize (from *T. dactyloides*; Leblanc et al. 1995b; 1997), millet (from *Pennisetum squamulatum*; Dujardin and Hanna 1983; Roche et al. 2001), wheat (from *Elymus rectisetus*; Peel et al. 1997) and beet (from *Beta lomatogona*; Cleij et al. 1976). In each case, the trait was best expressed in polyploid F_1 progeny, and its expression fell as the percentage of the apomictic parent was reduced in subsequent backcross generations. The reason for this remains unknown.

For most crops, however, an apomictic relative is unavailable, so conventional introgression cannot be used to introduce the trait into commercially adapted germplasm. Even in cases where introgression is an option, the type of apomixis available may not be ideal for the intended purpose. Most current research, therefore, is directed at understanding the mechanisms that underlie the expression of apomixis in model species and using this information to direct a synthetic approach to engineering the trait. As apomixis appears to have arisen as a modified form of sexual reproduction, work is progressing on both sexual and apomictic model plants. Before reviewing our current understanding of the genetics of apomixis, it is valuable to consider what form of apomixis would best meet our future needs. To have maximum utility installed, apomixis should address the demands of plant breeders, seed producers, farmers and consumers alike. Seed multiplication, commercial production and downstream processing would all benefit from asexual uniformity, so apomixis would need to be highly expressed during multiplication and commercial production. Plant breeding, however, combines phases where variation is encouraged, with phases in the breeding cycle where uniformity is valued. It would, therefore, be highly advantageous for apomixis to be available in an inducible format. A plant breeder would then be able to switch it off to allow sexual reproduction to generate variation, and then switch it on to permit the clonal multiplication of desirable genotypes.

As mentioned above, many of the natural forms of apomixis lead to polyembryony, the formation of multiple embryos in a seed. This tends to reduce the viability of the seed and it can also increase the variability of the seedlings produced if some of the embryos are sexually produced while others are clonal. An ideal form of apomixis would only result in a single embryo in each seed. Of the known natural forms of apomixis, diplospory would best suit this requirement, as it involves the conversion of the meiotic apparatus to a structure that produces a single maternally derived seedling.

Finally, the formation of the endosperm is a prerequisite for seed formation in most flowering plants and this must also be addressed in any engineered mechanism of apomixis. Grasses are widely cited as potential targets for engineered apomixis, and the endosperm is a particularly prominent tissue in the grass seed. Critically, it is also a key nutritional component of cereals such as wheat, rice and maize, so the success of endosperm formation also dictates the food value of these crops. As mentioned above, the embryo/endosperm ploidy balance is a factor in the normal development of many seeds and it is one that is particularly critical in the grasses. Fortunately, apomixis occurs in many grass species, so asexual mechanisms are already known that enable endosperm formation and appropriate ploidy balance within the grass seed.

7.5 The Genetics of Apomixis

7.5.1 Mutations of Genes in Sexual Plants with Features of Apomixis

The model plant *Arabidopsis thaliana* has been mutagenized extensively and a number of mutants have been reported that appear to display aspects of apomixis.

Mutations reflecting apomeiosis include the *dyad* allele of *SWITCH1*. Plants homozygous for the recessive allele *dyad* form a high proportion of unreduced egg cells that can participate in fertilization (Ravi et al. 2008). Male meiosis is not affected, so triploid progeny are often observed when these plants are allowed to self-fertilize. Similarly, d'Erfurth et al. (2009) identified a gene, *omission of second division* (*OSD1*) that controls the entry of cells into the second meiotic division. Plants with *osdl,* a mutant form of this gene, underwent second division restitution (SDR), producing unreduced gametes.

Several other mutations appear to reflect parthenogenesis. The *LEAFY COTYLEDON* genes *LEC1, LEC2* and *FUSCA3* regulate embryogenesis, specifically the identity of the suspensor, cotyledons and other structures arising late in embryo development (Meinke et al. 1994; West et al. 1994; Lotan et al. 1998; Stone et al. 2001). The ectopic expression of *LEC1* and *LEC2* in vegetative cells induces the expression of embryo-specific genes and also leads to the development of embryo-like structures (Lotan et al. 1998; Stone et al. 2001). The chromatin-remodelling factor *PICKLE* (*PKL*) is a master regulator of *LEC1, LEC2* and *FUSCA3* (Ogas et al. 1999; Rider et al. 2003). In *pkl* mutants, embryo-like structures arise on the roots, indicating that this gene acts to repress embryogenic development throughout the body of the plant (Henderson et al. 2004). Polycomb group proteins participate in the stable repression of specific genes through chromatin remodelling. The fertilization-independent seed (FIS) genes of *Arabidopsis* are members of this group which are involved in endosperm formation and in establishing early parent-of-origin effects (Koltunow and Grossniklaus 2003). *BABY BOOM* (*BBM*) is an AP2 domain transcription factor, the ectopic expression of which induced embryo-like structures to form on the cotyledons and leaves of *Arabidopsis* (Boutilier et al. 2002). *Somatic embryogenesis receptor kinase* (SERK) was found as an upregulated transcript in carrot (*Daucus carota)* cultures induced to form somatic embryos (Schmidt et al. 1997). Interestingly, a similar transcript (PpSERK) was found to be differentially expressed between apomictic and sexual forms of the grass *Poa pratensis* (Albertini et al., 2005). Mutations that demonstrate spontaneous endosperm formation include the *fis* mutants described above, in which the endosperm over-proliferates but is typically poorly cellularized (Kinoshita et al. 1999; Sorensen 2001). *MEDICIS* is a member of this group (Guitton et al. 2004), the mutation of which leads to the formation of rudimentary embryo and endosperm-like structures in the absence of fertilization.

7.5.2 Forward Genetics with Native Apomicts

Research into the genetics of apomixis of native apomicts has largely been approached through mapping or expression analysis, followed by the targeted sequencing of associated genomic bacterial artificial chromosomes (BACs). Apomicts are often cross-compatible to sexual relatives, permitting the production of segregating populations, and apomixis loci are dominant and are often in simplex, so they are amenable to map-based cloning.

For sporophytic apomixis, *Citrus* is the best-studied system. García et al. (1999) reported the detection of six quantitative trait loci (QTL) associated with the extent of somatic embryo formation in a cross between *Poncirus trifoliate* and *Citrus volkameriana*. Recently, Nakano et al. (2008, 2012) provided an analysis of 380 kb of genomic sequence from *Citrus* that is believed to be strongly associated with the trait. The region contains 70 predicted open-reading frames with a diverse range of putative functions.

For gametophytic apomicts, a notable development was the identification of 12 markers tightly linked to the apospory-specific genomic region (ASGR) of *Pennisetum* (Ozias-Akins et al. 1998). No meiotic recombination was seen to occur between the markers and apospory. BACs carrying those markers were physically mapped to the ASGR using fluorescence *in situ* hybridization enabling the size of the region to be estimated at approximately 50 Mb. This is a very large region of nonrecombinant DNA, comparable to the human Y chromosome (59 Mb). Apospory in other grasses also maps to single dominant loci, for example *Brachiaria* (Pessino et al. 1997), *Paspalum* (Martínez et al. 2001) and *Tripsacum* (Grimanelli et al. 1998b).

Given difficulties associated with mapping genes at large nonrecombinant loci, Catanach et al. (2006) used deletion mutagenesis to identify mutations in apomixis loci in the aposporous apomict *Hieracium caespitosum*. Two loci were clearly identified, one associated with apomeiosis (the *LOA* locus) and one associated with parthenogenesis (the *LOP* locus). Amplified fragment length polymorphisms (AFLP) markers were used to form molecular maps of these loci and to confirm that they act as independently inherited, dominant, simplex factors in this plant. Inheritance studies have been reported in two other members of the Asteraceae, the diplosporous apomicts *Taraxacum* (van Dijk et al. 1999) and in *Erigeron* (Noyes and Rieseberg 2000). As in *Hieracium,* two loci appear to act independently, one controlling apomeiosis and the other controlling both parthenogenesis and autonomous endospermy. Recent evidence in *Hieracium* indicates that autonomous endospermy and parthenogenesis can be separated, suggesting that the factors involved are tightly linked rather than pleiotropic in this system (Ogawa et al. 2013).

Until recently, no native apomict was sequenced, nor were there any comprehensive resources such as transfer-DNA (T-DNA) mutant panels available. The current sequencing of *Boechera holboellii* is expected to assist in this effort and may have significant effects on the landscape of apomixis research. Apomixis in *B. holboellii* is typically diplosporous with pseudogamy, although autonomous endospermy is also observed (Naumova et al. 2001). *Boechera* is a relative of *Arabidopsis,* so it is hoped that the extensive resources associated with *Arabidopsis* will be transferable and can be used to identify factors influencing apomixis in this plant.

7.5.3 Genes at Loci That Control Apomixis: What Are the Critical Determinants?

In some apomicts, loci associated with the trait have been cloned and sequenced, revealing candidate regulatory sequences. Calderini et al. (2006) sequenced a BAC clone of approximately 129 kb linked to apomixis from *Paspalum simplex*. They found it contained two genes along with evidence of significant loss of coding capacity due to mutation and transposable element (TE) insertion. The large and nonrecombinant ASGR of *Pennisetum* has both gene-rich and gene-poor regions (Conner et al. 2008) and potentially a vast array of candidate genes. Shotgun sequencing of approximately 2.7 Mb of the ASGR from *Cenchrus ciliaris* and *Pennisetum squamulatum* revealed a total of 74 putative protein-coding regions (PPCRs). Of most interest to the authors were two distinct copies of *ASGR-BABY BOOM* (*BBM*)*-like* genes, orthologous to the *BBM* genes mentioned above. The polyembryony locus in citrus has been fully sequenced and out of a total of 70 open-reading frames, seven were found with similarity to transcription factors (Nakano et al. 2012). The authors propose that the gene or genes that control polyembryony are likely to be transcription factors, and of most interest was a MADS homolog, because of the role that MADS-box transcription factors have in development including embryogenesis.

Whilst some candidate genes for apomixis have been revealed, determining which of them are critical has remained elusive. Only a small part of the ASGR has been sequenced and the authors state that if the whole of the ASGR was sequenced, it would contain too many genes to be analyzed using a gene-by-gene approach (Conner et al. 2008). The polyembryony locus of *Citrus,* while significantly smaller, still contains a significant number of candidates. Approaches towards identifying critical genes include microsynteny, where DNA sequence is compared across related apomicts to identify highly conserved sequences, or between apomicts and their sexual relatives (Gualtieri et al. 2006), to identify key differences such as gene translocations, alterations in regulatory sequences or epigenetic features unique to the apomict. These approaches are available because of the likely single phylogenetic origins of apomixis within related species.

Reports of genes identified at apomixis loci of different species are yet to reveal any common candidate genes. For example, genes linked to apospory of *Paspalum simplex* (Calderini et al. 2006) and the ASGR (Gualtieri et al. 2006) show synteny with different regions of the rice genome, implying that different routes of evolution were taken resulting in positionally different genetic mechanisms of control. Similarly, overlap between apomixis-associated gene expression profiles of more distantly related species is yet to be reported. On the other hand, apomixis loci of apomicts of common lineage often do reflect common apomictic ancestors. The ASGR of the genera *Pennisetum* and *Cenchrus,* and the polyembryony haplotype among a range of *Citrus* species, are both notable examples.

A small number of studies have aimed to determine variations in gene expression seen in apomicts compared with their sexual counterparts. The techniques used have included differential display (Leblanc et al. 1997; Rodrigues et al. 2003) and complementary DNA (cDNA)-AFLP (Albertini et al. 2004; Polegri et al. 2010), and more recently transcriptomics using next-generation sequencing. While expression differences between related sexuals and apomicts have been revealed, it is likely that most, if not all of these differences reflect events downstream from critical signals. Findings from these studies have yet to yield expression differences that clearly reflect regulatory mechanisms underlying the trait. An interesting approach taken by Zeng et al. (2011) found only one transcript from the ASGR that was expressed in apomictic ovules, despite a significant number of genes present in this region (Conner et al. 2008). This result, however, underscores an important advance in expression analysis, where the transcriptomes of critical tissues at critical developmental time points, are used to analyze the role of genes identified at apomixis loci. Further work like this, where gene expression and gene mapping converge, is likely to continue into the foreseeable future.

7.5.4 Evolution of Apomixis and of Apomixis Loci

Models of the evolution of apomixis are currently subjects of considerable discussion in the literature. Determining the evolutionary paths taken for plants to acquire apomixis is not only of academic interest but may also assist in the identification of regulating factors and processes of apomixis, and may even provide opportunities to induce apomixis in sexual species. We have already mentioned the conundrum faced when considering how apomixis evolved, given that the modifications of sexual reproduction, apomeiosis and parthenogenesis (with or without autonomous endospermy) need to be present together for apomixis to be a viable form of reproduction. In most cases, two unlinked loci have been reported, one controlling apomeiosis and another controlling parthenogenesis (Ozias Akins and van Dijk 2007). In apomictic grasses, however, there is often a single locus that carries both apomeiosis and parthenogenesis, which may be a more evolutionarily advanced adaptation, given that both apomeiosis and parthenogenesis need to be present together. Also previously mentioned is how apomixis is often found in perennial plants that already have means of clonal reproduction (for example, through stolons), which may offer a greater opportunity for a plant to acquire all modifications of sexual reproduction separately.

While apomixis has been reported in 32 different plant families (Carman 1997), it is not uniformly distributed. Three families, the Poaceae, Asteraceae and to a lesser extent, Rosaceae, contain most-known apomictic species. This bias suggests the existence of predisposing factor(s) that made members of these families more likely to evolve apomixis. The nature of any predisposing factors is unknown but a genetic component could be speculated. For example, certain combinations of linked genes may provide suitable ancestral loci for genetic modifications for apomixis that took place in sexual ancestors. Physiological predisposing factors could include developmental plasticity such as tolerance to egg/endosperm dosage imbalances, a feature of *Hieracium,* or to the presence of heterochronies involving the timing of gene expression in key events involved in reproduction (Carman 2007). This will be covered in more detail below. It should not be overlooked that the skewed distribution of apomixis amongst a small number of families may also reflect, to some extent, a familial predisposition to ecological and evolutionary success with asexuality, which has manifested in greater representation of apomicts seen today. This raises an interesting view that apomixis or rudimentary forms of apomixis may occur or "evolve" repeatedly, only to be maladaptive to the species and the ecological niche in which it resides.

At the level of genomic structure, apomixis loci tend to have very low rates of meiotic recombination and show "scars" of their genomic isolation with gene mutation, typified by the presence of pseudogenes (Matzk et al. 2003; Calderini et al. 2006), allele sequence divergence (see van Dijk 2009 for review) and retro-element accumulation (Matzk et al. 2003; Akiyama et al. 2004; Okada et al. 2011). The ASGR has little or no recombination occurring across an estimated 50 Mb, and, as a likely consequence, appears diverged to the point of hemizygosity, while the rest of the genome is more able to recombine.

Despite the genomic isolation, the large numbers of functionally conserved predicted gene sequences at apomixis loci is striking. If apomixis utilizes or "exploits" genes normally used in sexual reproduction, the gene complement of apomixis loci might only be a small number of high-level regulators and perhaps lower-level effectors. Apomixis is almost always found in polyploids which have higher gene redundancy, and should therefore be more tolerant of mutation of genes that have no critical dosage requirements. Under these circumstances, any gene at an apomixis locus that is not required for the trait itself, may be prone to eventual loss-of-function mutation. However, counter to this argument, the roles of an apomixis locus may extend beyond the induction of apomeiosis and/or parthenogenesis; there may be genes that have roles in the maintenance of an apomixis locus through evolutionary time. Given that the success of a locus depends on that of the clones in which it resides, it may provide further redundancy, or even plasticity or functionality to ensure the fitness of the clone.

7.5.5 Apomixis as a Product of Hybridization

There is growing support for the hypothesis that apomixis may be induced by hybridization and polyploidization. Most, if not all, natural apomicts show features of being hybrids; in particular, they are usually polyploid and highly heterozygous. The potential role of hybridization in the establishment of apomixis was first postulated by Ernst (1918). Carman (2007) extended this hybridization model by postulating that apomixis is caused by the asynchronous expression of duplicated genes derived from each parent. Mistimed regulation of developmental checkpoints controlling megasporogenesis, embryo sac, embryo and endosperm formation, could lead to unreduced egg cell formation and spontaneous embryogenesis. Typically, when a wide hybrid forms between sexual species, any resulting asynchronous gene expression is under negative selective pressure and synchronization is re-established within a small number of sexual generations. However, in "fledgling apomicts" where the synchronizing fails because of epigenetic modifications of the polyploid genome, apomixis can establish then stabilize. Alternatively, sexual reproduction can be reinstated through successive generations with meiotic recombination disturbing duplicate gene asynchronies (Carman 2007). Fledgling apomicts have received little research interest, probably because of low expression of apomixis. They may, however, be valuable tools for detecting the most fundamental alterations required for apomixis to evolve, as these differences are obscured by stabilizing evolution in established apomictic species.

Evidence of asynchrony of gene expression was reported in ovules of *B. holboellii,* in which heterochrony, gene duplication and "parent of origin" effects were seen in the transcriptomic profiles associated with apomeiosis. In addition, the observation that large numbers of genes commonly occur in nonrecombinant apomixis-associated loci is consistent with the hybridization model. They presumably represent remnant linkage blocks inherited following a wide hybridization event, in which several genes of diverse functions cluster around a smaller number of key regulatory sequences that influence the expression of trait.

7.5.6 Epigenetic Regulation

Epigenetic regulation, including the role of transposable elements, is an expanding frontier of research into many biological processes, and apomixis is no exception. Given the conundrum that apomixis inhibits genetic change yet the expression of apomixis, and the life histories of apomictic species are often highly adapted, it appears that some form of genetic change is occurring at a rate that allows evolution to proceed. Epigenetic modifications can occur without meiosis or fertilization and they can lead to rapid heritable adaptations. Epigenetic modification is also particularly prevalent during hybridization and polyploidization, as discussed above.

Significant research is being undertaken on the regulatory roles of non-coding RNA in a wide range of processes, and roles in aspects of reproduction are now well established. The *Arabidopsis* protein AGONAUTE 9 (AGO9) appears to restrict female gamete formation while repressing transposable elements. Post-meiotic ovules of *ago9* mutants carry additional unreduced somatic cells that have the identity of functional megaspores, which resembles apospory (Olmedo-Monfil et al. 2010). A preliminary investigation into the roles of microRNA expression in apomixis has shown many conserved microRNAs expressed in both sexual and apomictic ovules. One microRNA target, the transcription factor squamosa promoter binding protein like SPL11, is significantly upregulated in *Boechera* stage 2 ovules at the time of megaspore mother cell (MMC) development (Amitey et al. 2011).

7.6 Synthetic Forms of Apomixis

Do we already know enough to attempt the synthesis of apomixis in a model system? As noted above, Ravi et al. (2008) reported that plants homozygous for *dyad* produced a high proportion of unreduced egg cells which could be fertilized. Similarly, d'Erfurth et al. (2009) noted that plants homozygous for *osd1* underwent second division restitution (SDR), producing unreduced gametes. Chromosome recombination and chromatid segregation, however, occur in the first meiotic division, so the gametes formed on their *osd1/osd1* plants did not share the maternal genotype. However, by combining *osd1*with mutations blocking recombination (*Atrec8*) and chromatid segregation (*Atspo11*), they were able to generate a genotype that produced unreduced maternal gametes, effectively synthesizing a form of apomeiosis in *Arabidopsis* (d'Erfurth et al. 2009). Plants with this genotype were named mitosis instead of meiosis (*MiMe*).

Ravi and Chan et al. (2010) constructed a chimeric sequence combining the core of the *Arabidopsis* centromeric histone protein CENH3, the amino terminal tail domain of histone H3.3 and the green fluorescent protein (GFP) reporter sequence. This was referred to as the GFP-tailswap construct. When introduced into a plant homozygous for *cenH3,* a recessive mutant allele of CENH3, none of the nuclear DNA from that modified plant transferred to its progeny. Due to problems with male fertility, this cross was conducted with the modified plant as the maternal parent. Seedlings formed following crosses with wild-type male pollen parents were, therefore, predominantly haploid and entirely paternal in origin. This demonstrated uniparental transmission through the egg—an outcome analogous to parthenogenesis. In this case, it was the male contribution that was conducted and the female genome that was lost. Male parthenogenesis is known to occur very rarely in native systems, the most well-studied example being the formation of patrogenic progeny in the conifer *Cupressus dupreziana* (Pichot et al. 2008).

Combining these findings, Marimuthu et al. (2011) recently demonstrated that when *Arabidopsis* plants with either the *dyad* or *the MiMe* triple mutation were crossed to plants with the GFP-tailswap construct and *cenH3* mutation, a significant proportion of the progeny were strictly maternal in origin. The fertility of these crosses was low and both parents needed to have specific mutant backgrounds, but this result does provide an example of how synthetic apomixis might be achieved.

7.7 Conclusion

Apomixis offers the potential to significantly increase the rate of genetic gain achievable in seed-propagated plants by allowing a breeder to produce true breeding lines with unique, selected, genetic combinations. The desirable genotype might include somatically derived mutations, transgenes and/or hybrid vigour. Research into apomixis is progressing on several fronts. The developmental genetics of *Arabidopsis* has contributed significantly and it is now clear that the regulation of apomixis is very similar to that of sexual reproduction. It appears in most cases to represent a derived form of sex, and this transition has occurred on several occasions in flowering plants. Advances in sequencing, expression analysis, histological imaging and bioinformatics are all contributing to our understanding, most notably in native apomictic systems, which are now gaining in prominence in the apomixis literature. Revealing the determinants of apomixis, however, remains elusive, and research in the future will focus on the convergence of genetic mapping, genomic sequencing and transcriptomics, and also on the role of epigenetic regulation. Investigating the evolution of apomixis is of great interest, and "fledgling apomicts", those that show apomixis or aspects of apomixis following recent hybridization and/or polyploidization events, may provide new clues on the key events that characterize the transition of plants from sexuality to apomixis. Alternatively, the convergence of this knowledge may result in a synthetic form of the trait which may, or may not, reflect the native systems currently known. In closing, therefore, it now seems reasonable to predict that apomixis will be part of the tool kit available to future breeders and propagators of our crop plants and that this will have a significant impact on the genetic advance and practice of agriculture in the coming decades.

Acknowledgments The authors would like to thank Philippa Barrell and Anne Gunson for their help with the chapter and Donna Gibson for providing the illustration.

References

Akiyama Y, Conner JA, Goel S, Morishige DT, Mullet JE, Hanna WW, Ozias-Akins P (2004) High-resolution physical mapping in *Pennisetum squamulatum* reveals extensive chromosomal heteromorphism of the genomic region associated with apomixis. Plant Physiol 134:1733–1741

Albertini E, Marconi G, Barcaccia G, Raggi L, Falcinelli M (2004) Isolation of candidate genes for apomixis in *Poa pratensis* L. Plant Mol Biol 56:879–894

Albertini E, Marconi G, Reale L, Barcaccia G, Porceddu A, Ferranti F, Falcinelli M (2005) SERK and APOSTART. Candidate genes for apomixis in *Poa pratensis*. Plant Physiol 138:2185–2199

Amitey S et al (2011) Analysis of conserved microRNAs in floral tissues of sexual and apomictic Boechera species. BMC Genomics 12(1):500

Asker SE, Jerling L (1992) Apomixis in plants. CRC, Boca Raton

Berger F (1999) Endosperm development. Curr Opin Plant Biol 2:28–32

Bicknell RA (1997) Isolation of a diploid, apomictic plant of *Hieracium aurantiacum*. Sex Plant Reprod 10:168–172

Boutilier K, Offringa R, Sharma VK, Kieft H, Ouellet T, Zhang L, Hattori J, Liu CM, van Lammeren AA, Miki BL et al (2002) Ectopic expression of BABY BOOM triggers a conversion from vegetative to embryonic growth. Plant Cell 14:1737–1749

Calderini O, Chang S, Jong H, Busti A, Paolocci F, Arcioni S, Vries S, Abma-Henkens MC, Lankhorst RK, Donnison I et al (2006) Molecular cytogenetics and DNA sequence analysis of an apomixis-linked BAC in *Paspalum simplex* reveal a non pericentromere location and partial microcolinearity with rice. Theor Appl Genet 112:1179–1191

Carman JG (1997) Asynchronous expression of duplicate genes in angiosperms may cause apomixis, bispory, tetraspory, and polyembryony. Biol J Linn Soc 61:51–94

Carman JG (2007) Do duplicate genes cause apomixis? In: Horandl E, Grossniklaus U, Van Dijk P, Sharbel T (eds) Apomixis: evolution, mechanisms and perspectives: A.R.G. Gantner Verlag, Rugell, pp 63–91

Catanach AS, Erasmuson SK, Podivinsky E, Jordan BR, Bicknell R (2006) Deletion mapping of genetic regions associated with apomixis in *Hieracium*. Proc Natl Acad Sci U S A 103:18650–18655

Cleij G, de Bock TSM, Lekkerkerker B (1976) Crosses between *Beta vulgais* L. and *Beta lomatogona* F. et M. Euphytica 25:539–547

Conner JA, Goel S, Gunawan G, Cordonnier-Pratt M-M, Johnson VE, Liang C, Wang H, Pratt LH, Mullet JE, DeBarry J et al (2008) Sequence analysis of bacterial artificial chromosome clones from the apospory-specific genomic region of *Pennisetum* and *Cenchrus*. Plant Physiol 147:1396–1411

Crane CF (2001) Classification of apomictic mechanisms. In: Savidan Y, Carman JG, Dresselhaus T (eds). Flowering of apomixis: from mechanisms to genetic engineering. CIMMYT, IRD, European Commission DG VI, Mexico, pp 24–34

de Oliveira AC Garcia AN Cristofani M Machado MA (2002) Identification of citrus hybrids through the combination of leaf apex morphology and SSR markers. Euphytica 128:397–403

d'Erfurth I, Jolivet S, Froger N, Catrice O, Novatchkova M, Mercier R (2009) Turning meiosis into mitosis. PLoS Biol 7:1–10.

Dujardin M, Hanna WW (1983) Apomictic and sexual pearl millet X *Pennisetum squamulatum* hybrids. J Hered 74:277–279

Ernst A (1918) Bastardierung als Ursache der Apogamie im Pflanzenreich. Fischer, Jena

García R, Asíns MJ, Forner J, Carbonell EA (1999) Genetic analysis of apomixis in *Citrus* and *Poncirus* by molecular markers. Theor Appl Genet 99:511–518

Grimanelli D, Leblanc O, Espinosa E, Perotti E, Gonzalez de Leon D, Savidan Y (1998a) Non-Mendelian transmission of apomixis in maize-*Tripsacum* hybrids caused by a transmission ratio distortion. Heredity 80:40–47

Grimanelli D, Leblanc O, Espinosa E, Perotti E, Gonzalez-de-Léon D, Savidan Y (1998b) Mapping diplosporous apomixis in tetraploid *Tripsacum*: one gene or several genes? Heredity 80:33–39

Gualtieri G, Conner JA, Morishige DT, Moore LD, Mullet JE, Ozias-Akins P (2006) A segment of the apospory-specific genomic region is highly microsyntenic not only between the apomicts *Pennisetum squamulatum* and buffelgrass, but also with a rice chromosome 11 centromeric-proximal genomic region. Plant Physiol 140:963–971

Guitton AE, Page DR, Chambrier P, Lionnet C, Faure JE, Grossniklaus U, Berger F (2004) Identification of new members of Fertilisation Independent Seed Polycomb Group pathway involved in the control of seed development in *Arabidopsis thaliana*. Development 131:2971–2981

Hanna WW (1995) Use of apomixis in cultivar development. Adv Agron 54:333–350

Henderson JT, Li HC, Rider SD, Mordhorst AP, Romero-Severson J, Cheng JC, Robey J, Sung ZR, de Vries SC, Ogas J (2004) *PICKLE* acts throughout the plant to repress expression of embryonic traits and may play a role in gibberellin-dependent responses. Plant Physiol 134:995–1005

Holm S, Ghatnekar L (1996) Sexuality and no apomixis found in crossing experiments with diploid *Potentilla argentea*. Hereditas 125:77–82

Horandl E (2008) Understanding the geographic distributions of apomictic plants: a case for a pluralistic approach. Plant Ecol Divers 1:309–320

Huang S-Q, Lu Y, Chen Y-Z, Luo YB, Delph LF (2009) Parthenogenesis Maintains Male Sterility in a Gynodioecious Orchid. Amer Nat 174:578–584

Jefferson RA, Bicknell RA (1995) The potential impacts of apomixis: a molecular genetics approach. In: Sobral BWS (ed) The impact of plant molecular genetics. Birklhäuser, Boston pp 87–101

Jessup RW, Burson BL, Burow GB, Wang YW, Chang C, Li Z, Paterson AH, Hussey MA (2002) Disomic inheritance, suppressed recombination, and allelic interactions govern apospory in buffelgrass as revealed by genome mapping. Crop Sci 42:1688–1694

Kantama L, Sharbel TF, Schranz ME, Mitchell-Olds T, de Vries S, de Jong H (2007) Diploid apomicts of the *Boechera holboellii* complex display large-scale chromosome substitutions and aberrant chromosomes. Proc Natl Acad Sci U S A 104:14026–14031

Kinoshita T, Yadegari R, Harada JJ, Goldberg RB, Fischer RL (1999) Imprinting of the *MEDEA* polycomb gene in *Arabidopsis* endosperm. Plant Cell 11:1945–1952

Kojima A, Nagato Y (1997) Discovery of highly apomictic and highly amphimictic dihaploids in *Allium tuberosum*. Sex Plant Reprod 10:8–12

Koltunow AM, Grossniklaus U (2003) Apomixis: a developmental perspective. Annu Rev Plant Biol 54:547–574

Koltunow AM, Bicknell RA, Chaudhury AM (1995) Apomixis: molecular strategies for the generation of genetically identical seeds without fertilization. Plant Physiol 108:1345–1352

Leblanc O, Grimanelli D, Gonzalez de Leon D, Savidan Y (1995a) Detection of the apomictic mode of reproduction in maize-*Tripsacum* hybrids using maize RFLP markers. Theor Appl Genet 90:1198–1203

Leblanc O, Peel MD, Carman JG, Savidan Y (1995b) Megasporogenesis and megagametogenesis in several *Tripsacum* species (Poaceae). Am J Bot 81:57–63

Leblanc O, Armstead I, Pessino S, Ortiz JPA, Evans C, do Valle C, Hayward MD (1997) Non-radioactive mRNA fingerprinting to visualise gene expression in mature ovaries of *Brachiaria* hybrids derived from *B. brizantha*, an apomictic tropical forage. Plant Sci 126:49–58

Litz RE, Moon PA, Mathews H, Jayasankar S, Monsalud MJ., Pliego-Alfaro F (1995) Somatic embryogenesis in mango (Mangifera indica L.). In: Jain SM, Gupta PK, Newton RJ (eds) Somatic embryogenesis in woody plants. Vol 2– Andiosperms On line: Springer Link. pp 341–356

López-Valenzuela JA, Martínez O, Paredes-López O (1997) Geographic differentiation and embryo type identification in *Mangifera indica* L. cultivars using RAPD markers. Hort Sci 32:1105–1108

Lotan T, Ohto M, Matsudaira Yee K, West MAL, Lo R, Kwong RW, Yamagishi K, Fischer RL, Goldberg RB, Harada JJ (1998) *Arabidopsis LEAFY COTYLEDON1* is sufficient to induce embryo development in vegetative cells. Cell 93:1195–1205

Marimuthu MPA, Jolivet S, Ravi M, Pereira L, Davda JN, Cromer L, Wang L, Nogué F, Chan SWL, Siddiqi I et al (2011) Synthetic clonal reproduction through seeds. Science 331:876

Martínez EJ, Urbani, MH, Quarin CL, Ortiz JPA (2001) Inheritance of apospory in bahiagrass, *Paspalum notatum*. Hereditas 135:19–25

Matzk F, Hammer K, Schubert I (2003) Coevolution of apomixis and genome size within the genus *Hypericum*. Sex Plant Reprod 16:51–58

Meinke DW, Franzmann LH, Nickle TC, Yeung EC (1994) Leafy cotyledon mutants of *Arabidopsis*. Plant Cell 6:1049–1064

Müntzing A (1928) Pseudogamie in der Gattung *Potentilla*. Hereditas 11:267–283

Nakano M, Shimizu T, Fujii H, Shimada T, Endo T, Nesumi H, Kuniga K, Omura M (2008) Marker enrichment and construction of haplotype-specific BAC contigs for the polyembryony genomic region in *Citrus*. Breeding Sci 58:375–383

Nakano M, Shimada T, Endo T, Fujii H, Nesumi H, Kita M, Ebina M, Shimizu T, Omura M (2012) Characterization of genomic sequence showing strong association with polyembryony among diverse *Citrus* species and cultivars, and its synteny with *Vitis* and *Populus*. Plant Sci 183:131–142

Naumova TN, den Nijs JCM, Willemse MTM (1992) Quantitative analysis of aposporous parthenogenesis in *Poa pratensis* genotypes. Acta Bot Neerl 43:1–14

Naumova TN, Hayward MD, Wagenvoort M (1999) Apomixis and sexuality in diploid and tetraploid accessions of *Brachiaria decumbens*. Sex Plant Reprod 12:43–52

Naumova T, van der Laak J, Osadtchiy J, Matzk F, Kravtchenko A, Bergervoet J, Ramulu K, Boutilier K (2001) Reproductive development in apomictic populations of *Arabis holboellii* (Brassicaceae). Sex Plant Reprod 14:195–200

Nogler GA (1982) How to obtain diploid apomictic *Ranunculus auricomus* plants not found in the wild state. Bot Helv 92:13–22

Nogler GA (1984a) Gametophytic apomixis. In: Johri BM (ed) Embryology of angiosperms. Springer-Verlag, Berlin p 475–518

Nogler GA (1984b) Genetics of apospory in apomictic *Ranunculus auricomus*. V. conclusion. Bot Helv 94:411–422

Noyes RD, Rieseberg LH (2000) Two independent loci control agamospermy (apomixis) in the triploid flowering plant *Erigeron annuus*. Genetics 155:379–390

Ogas J, Kaufmann S, Henderson J, Somerville C (1999) *PICKLE* is a CHD3 chromatin-remodeling factor that regulates the transition from embryonic to vegetative development in *Arabidopsis*. Proc Natl Acad Sci U S A 96:13839–13844

Ogawa D, Johnson S, Henderson S, Koltunow AG (2013) Genetic separation of autonomous endosperm formation (AutE) from the two other components of apomixis in *Hieracium*. Plant Reprod:1–11

Okada T, Ito K, Johnson SD, Oelkers K, Suzuki G, Houben A, Mukai Y, Koltunow AM (2011) Chromosomes carrying meiotic avoidance loci in three apomictic eudicot *Hieracium* subgenus *Pilosella* species share structural features with two monocot apomicts. Plant Physiol 157:1327–1341

Olmedo-Monfil V, Durán-Figueroa N, Arteaga-Vázquez M, Demesa-Arévalo E, Autran D, Grimanelli D, Slotkin RK, Martienssen RA, Vielle-Calzada JP (2010) Control of female gamete formation by a small RNA pathway in Arabidopsis. Nature 464:628–632

Ozias-Akins P, van Dijk PJ (2007) Mendelian genetics of apomixis in plants. Annu Rev Genet 41:509–537

Ozias-Akins P, Roche D, Hanna WW (1998) Tight clustering and hemizygosity of apomixis-linked molecular markers in *Pennisetum squamulatum* implies genetic control of apospory by a divergent locus that may have no allelic form in sexual genotypes. Proc Natl Acad Sci U S A 95:5127–5132

Peel MD, Carman JG, Leblanc O (1997) Megasporocyte callose in apomictic buffelgrass, Kentucky bluegrass, *Pennisetum squamulatum* Fresen, *Tripsacum* L., and weeping lovegrass. Crop Sci 37:724–732

Pessino SC, Oritz JPA, Leblanc O, do Valle CB, Evans C, Hayward MD (1997) Identification of a maize linkage group related to apomixis in *Brachiaria*. Theor Appl Genet 94:439–444

Pichot C, Liens B, Rivera Nava JL, Bachelier JB, Maâtaoui EM (2008) Cypress surrogate mother produces haploid progeny from Alien Pollen. Genetics 178:379–383

Polegri L, Calderini O, Arcioni S, Pupilli F (2010) Specific expression of apomixis-linked alleles revealed by comparative transcriptomic analysis of sexual and apomictic *Paspalum simplex* Morong flowers. J Exp Bot: erq054

Ravi M, Chan SWL (2010) Haploid plants produced by centromere-mediated genome elimination. Nature 464:615–619

Ravi M, Marimuthu MPA, Siddiqi I (2008) Gamete formation without meiosis in Arabidopsis. Nature 451:1121–1124

Rider SDJ, Henderson JT, Jerome RE, Edenberg HJ, Romero-Severson J, Ogas J (2003) Coordinate repression of regulators of embryonic identity by *PICKLE* during germination in *Arabidopsis*. Plant J 35:33–43

Roche D, Chen ZB, Hanna WW, Ozias-Akins P (2001) Non-Mendelian transmission of an apospory-specific genomic region in a reciprocal cross between sexual pearl millet (*Pennisetum glaucum*) and an apomictic F1 (*P-glaucum* x *P-squamulatum*). Sex Plant Reprod 13:217–223

Rodrigues JCM, Cabral GB, Dusi DMA, de Mello LV, Rigden DJ, Carneiro VTC (2003) Identification of differentially expressed cDNA sequences in ovaries of sexual and apomictic plants of *Brachiaria brizantha*. Plant Mol Biol 53:745–757

Savidan Y (1980) Chromosomal and embryological analyses in sexual x apomictic hybrids of *Panicum maximum* Jacq. Theor Appl Genet 57:153–156

Savidan Y (2000a) Apomixis, the way of cloning seeds. Biofutur 198:38–43

Savidan Y (2000b) Apomixis: genetics and breeding. In: Janick J (ed) Plant breed. Rev. Wiley, New York, pp 13–86

Schmidt ED, Guzzo F, Toonen MA, de Vries SC (1997) A leucine-rich repeat containing receptor-like kinase marks somatic plant cells competent to form embryos. Development 124:2049–2062

Schranz ME, Dobe C, Koch MA, Mitchell-Olds T (2005) Sexual reproduction, hybridization, apomixis, and polyploidization in the genus *Boechera* (Brassicaceae). Amer J Bot 92:1797–1810

Scott RJ, Spielman M, Bailey J, Dickinson HG (1998) Parent-of-origin effects on seed development in *Arabidopsis thaliana*. Development 125:3329–3341

Sorensen MB, Chaudhury AM, Robert H, Bancharel E, Berger F (2001) Polycomb group genes control pattern formation in plant seed. Curr Biol 11:277–281

Stone SL, Kwong LW, Yee KM, Pelletier J, Lepiniec L, Fischer RL, Goldberg RB, Harada JJ (2001) *LEAFY COTYLEDON2* encodes a B3 domain transcription factor that induces embryo development. Proc Natl Acad Sci U S A 98:11806–11811

Teppner H, Klein E (1993) *Nigritella gabasiana* spec. nova, *N. nigra* subsp. *iberica* subsp. nova *(Orchidaceae—Orchideae)* und deren Embryologie. Phyton 33:179–209

van Dijk PJ (2003) Ecological and evolutionary opportunities of apomixis: insights from *Taraxacum* and *Chondrilla*. Philos T Roy Soc B 358:1113–1121

van Dijk PJ Tas ICQ Falque M Bakx-Schotman T (1999) Crosses between sexual and apomictic dandelions (*Taraxacum*). II. The breakdown of apomixis. Heredity 83:715–721

van Dijk P de Jong HKV Biere A (2009) An apomixis-gene's view on dandelions. In: Schön I, Martens K, van Dijk P (eds) Lost sex. Springer, Dordrecht p 475–494

Wakana A, Unemoto S (1987) Adventive embryogenesis in Citrus. I. The occurrence of adventive embryos without pollination or fertilization. Amer J Bot 74:517–530

Wakana A, Uemoto S (1988) Adventive embryogenesis in *Citrus* (Rutaceae) II. Postfertilization development. Amer J Bot 75:1033–1047

West M, Yee KM, Danao J, Zimmerman JL, Fischer RL, Goldberg RB, Harada JJ (1994) *LEAFY COTYLEDON1* is an essential regulator of late embryogenesis and cotyledon identity in *Arabidopsis*. Plant Cell 6:1731–1745

Yu SB, Li JX, Xu CG, Tan YF, Gao YJ, Li XH, Zhang Q, Maroof MAS (1997) Importance of epistasis as the genetic basis of heterosis in an elite rice hybrid. Proc Natl Acad Sci USA 94:9226–9231

Zeng Y, Conner J, Ozias-Akins P (2011) Identification of ovule transcripts from the Apospory-Specific Genomic Region (ASGR)-carrier chromosome. BMC Genomics 12:206

Chapter 8
Somatic Embryogenesis for Potato (*Solanum tuberosum* L.) Improvement

A. M. K. Nassar, S. Kubow and D. J. Donnelly

8.1 Importance of Potato

Potato (*Solanum tuberosum* L.) is the world's most important vegetable crop. It can be grown under a wide range of altitudes, latitudes, and climatic environments; from sea level to >4000 m in altitude and from the equator to >40° north and south (Harris 1992). Potato stem tubers are an indispensable dietary item for much of the world's population. Global consumption of potato is increasing in developing countries, particularly China and India (FAO 2013). The annual consumption for 2008 in Africa, Asia/Oceania, Europe, Latin America, North America, and the world average was 15.60, 26.60, 85.90, 28.20, 53.60, and 33.00 kg capita^{-1} year^{-1}, respectively (FAO 2013). Potato is a rich source of nutrients including carbohydrate, protein, vitamins, minerals, and phytonutrients (Davies 2002; Brown 2008; Camire et al. 2009). For example, in 2007 potato provided 0.60, 1.10, 3.90, 2.50, 2.10, 1.70, and 1.40 g capita^{-1} day^{-1} of protein supply for Africa, Asia, Europe, North America, Oceania, South America, and the world average, respectively (FAO 2013).

D. J. Donnelly (✉) · A. M. K. Nassar
Plant Science Department, McGill University, 21,111 Lakeshore Rd., Sainte Anne de Bellevue, QC H9X 3V9, Canada
e-mail: danielle.donnelly@mcgill.ca

A. M. K. Nassar · S. Kubow
School of Dietetics & Human Nutrition, McGill University, 21,111 Lakeshore Rd., Sainte Anne de Bellevue, QC H9X 3V9, Canada

A. M. K. Nassar
Faculty of Agriculture, Pest Control and Environmental Protection Department, Damanhour University, Al-Gomhouria Street, PO Box 22516 Damanhour, Albeheira, Egypt

X.-Q. Li et al. (eds.), *Somatic Genome Manipulation*,
DOI 10.1007/978-1-4939-2389-2_8

8.2 Techniques for Potato Improvement

Potato breeders and researchers seek to improve potato qualities, including yield, storability and processing, nutritional features including protein concentration, and disease resistance towards a range of pathogens (Beukema and Zaag 1990; Mullins et al. 2006; Nassar et al. 2011). Potato improvement is important to cope with the increasing world population and food scarcity, especially in developing countries (Cotes et al. 2002; IYP 2008). Many potato cultivars are sterile due to incompatibility after crossing or failure of seeds to complete development after fertilization (Walker 1955; Hawkes 1990). Cultivated potatoes are most commonly tetraploids with high variability in phenotype, including yield from season to season, and low seed viability (Gardner and Snustad 1986). About 235 species of the genus *Solanum* are tuber bearing. Only the tuberosum specie is cultivated (Hawkes 1990). This means a narrow genetic base is present in most cultivated potato, which is problematic for potato improvement (Bradshaw and Mackay 1994).

The first report of a potato improvement programme was in the early nineteenth century, in the UK, by Knight (1814). Knight suggested that early-season genotypes with increased tuber yield could be attained (Knight 1814; Bradshaw and Mackay 1994). Potato yield increased about sixfold per unit growing area in the USA since the 1920s. This was due to improved cultural practices (fertilization, irrigation, pest management, etc.), a shift in production to the western USA, and through the introduction of improved cultivars through public breeding efforts (Lucier et al. 1990; Douches et al. 1996). However, improving yield, a multi-gene-controlled trait (Cassells et al. 1983) with high variability (Neele et al. 1988; Jones and Cassells 1995), is not an easy task.

During the twentieth century, many new techniques for potato improvement were developed, including expansion of breeding methods though the use of wild species, haploids, and *neo*tuberosum, molecular techniques such as marker-assisted selection (MAS), genetic engineering, and use of a wide range of tissue culture methods including calliclones and somatic hybridization. Exploration within wild species started with the production of a population of *Phureja/Stenotomum* adapted to long days (Carroll 1982). Natural seeds were used and both seedling and tuber populations were produced over a 2-year cycle. After several cycles, the population became adapted to long-day conditions and yield improved; mainly through increased tuber size and reduced tuber numbers. Direct hybridization of members of this improved diploid *S. phureja* × *S. stenotomum* population with tetraploid *S. tuberosum* cultivars resulted in tetraploid hybrids superior in both marketable and total yield, with more tubers per plant and slightly lesser mean tuber weights (Carroll and De Maine 1989). Crossing of cultivated tetraploid 'Norgleam' × wild diploid specie (W5279–4) was accomplished, and resulted in the release of the now popular 'Yukon Gold' (Johnston and Rowberry 1981) and identification of other superior clones with lighter chip colour and generally good tuber appearance (Lauer et al. 1988; Darmo and Peloquin 1991). Tetraploid cultivars were also produced by crossing diploid × dihaploid species of *S. tuberosum* followed by recurrent phenotypic mass selection of the tetraploid hybrids (Bradshaw and Mackay 1994).

Haploid breeding through ploidy manipulation led to enhancement of genetic diversity of germplasm holdings through introduction of genes of desirable traits from wild diploid species into cultivated tetraploid potato (Jansky et al. 1990; Carputo et al. 1997). Use of multiple bridge crosses permitted the transfer of the resistance genes (R-genes) from *S. bulbocastanum* to *S. tuberosum* (Hermsen and Ramanna 1973). Multiple bridge crossing of *S. bulbocastanum* with *S. acaule* and *S. phureja* and then backcrossing of hybrids with *S. tuberosum* resulted in the release of 'Toluca' (Jacobsen and Schouten 2008) and 'Biogold' (Visser 2009). These cultivars were released around 2005, described as resistant to *P. infestans,* and are apparently in use for organic farming in the Netherlands (Visser 2009).

Haploids are plants with half the chromosome number (equal to the gametophytic chromosome number) of their parents (Tai 2005). They have uses in gene mapping, identification of major and quantitative trait loci (QTL), genetic transformation, somatic fusion, and MAS in breeding for new cultivars. Haploids were produced by androgenesis (paternal haploids) and/or gynogenesis (maternal haploids; Tai 2005). Maternal haploids can be produced parthenogenetically by interspecific hybridization of cultivated tetraploid potato cultivars (4x) (maternal; pistillate) using *S. phureja* (Juz. and Bukasov) pollen (2x) (pollinator; parent) (Peloquin et al. 1996; reviewed by Ortiz 1998). Parthenogenesis can occur when a paternal nucleus fertilizes the polar nuclei but the egg cell is not fertilized. Parental haploids are obtained using male gametes via anther or microspore culture (Uhrig and Salamini 1987; Calleberg and Johansson 1993; Rokka et al. 1998). De Maine (1995) examined effects of a second haploid generation on inbreeding of somatically chromosome-doubled (homozygous dihaploid) potato from crosses with *S. phureja* clones. The dihaploids had greater seed yield than the original haploid generation, but no other apparent differences occurred, including similar tuber yield. While haploids have use in plant improvement programmes, including potato, no reports have occurred describing potato cultivar release based on this method.

*Neo*tuberosum is a breeding technique that involves identification of new-*Tuberosum* species from *Andigena*-based populations. Several authors from different countries attempted to pursue this technique, for example, in 1959 in the UK (Simmonds 1969), in 1963 in the USA on material supplied by Simmonds (Plaisted 1987), in 1968 in Canada (Glendinning 1987), and in 1976 in the Netherlands (Maris 1989). Large numbers of seedlings from non-adapted *Andigena* accessions were grown in the field and selection was applied for greatest yield with acceptable tuber size, shape, and colour. Moreover, berries were collected to produce sexual generations. Selection for resistance to late blight (*Phytophthora infestans*) was improved by imposing severe late blight infection in the field (Bradshaw and Mackay 1994). Field results of clones selected after crossing clones of *S. andigena* with *S. tuberosum* showed tuber uniformity and yield heterosis (Tarn and Tai 1983; Plaisted 1987). *Neo*tuberosum programmes showed that clones of *S. andigena* were adapted and produced parents suitable for incorporation into modern potato. *Neo*tuberosum material has been employed in cultivar development. For example, both cultivars

Shelagh, highly resistant to late blight (The European Cultivated Potato Database 2009), and Rosa, resistant to early blight (Pelletier and Fry 1989), resulted from crosses between *S. tuberosum* and *neo*tuberosum (Bradshaw and Mackay 1994).

MAS technology is based on genetic markers related to genes expressing beneficial phenotypes (Mullins et al. 2006). This enables rapid and efficient selection for traits of interest, allows identification of the genetic components of the same trait, and potentially shortens the breeding process by several years. Despite its potential, there have been few instances of MAS use in potato breeding. This is largely because of the out-breeding, tetraploid genetics of cultivated potato, and the fact that many traits of interest to breeders are multigenic in nature and greatly affected by environment. MAS is in use for improvement of traits controlled by single dominant genes (Mullins et al. 2006). Molecular marker-based maps of potato have identified at least 19 major disease R-genes and numerous QTL for disease resistance (Gebhardt and Valkonen 2001), morphological, developmental, and quality traits that are single gene controlled (Mullins et al. 2006). For example, Gebhardt and Valkonen (2001) mentioned that a single class of genes that share conserved nucleotide-binding site (NBS) and leucine-rich repeat (LRR) motifs mediated inherited disease resistance and large-effect QTL for disease resistance. Based on map location, several NBS–LRR-type disease R-genes were cloned and characterized for MAS use. Caromel et al. (2005) identified two separate QTL for potato cyst nematode resistance in the wild potato *S. sparsipilum* and inserted them into the potato genotype *S. tuberosum* Casper H3. This work apparently resulted in highly resistant phenotypes but has not yet lead to new cultivar release.

Genetic engineering of potato is an area of active current research, much of which is in the pharmaceutical arena. This chapter is limited to articles related to improvement of agronomic, disease resistance, and nutritional characteristics (for more information see: Hilder and Boulter 1999; Uzogara 2000; Punja 2001; Romer et al. 2002). Most of these studies focused on resistance to different pest, virus, and fungal diseases (Davies 2002; Mullins et al. 2006). Some examples include resistance to black scurf (*Rhizoctonia solani*; Broglie et al. 1991), Colorado potato beetle (*Leptinotarsa decemlineata*) through expression of the *Bacillus thuringiensis* (Bt) toxin gene (Adang et al. 1993; Perlak et al. 1993), late blight (Cornelissen and Melchers 1993; Song et al. 2003; Van der Vossen et al. 2003, 2005; Osusky et al. 2004), potato leaf roll virus (PLRV), potato tuber moth (*Phthorimaea operculella*; Davidson et al. 2002), potato virus Y (PVY; Hassairi et al. 1998), and soft rot (*Erwinia carotovora*; During et al. 1993). Examples of transgenic potato for increased nutritional characteristics include increase in total amino acid composition especially lysine, tyrosine, and sulphur-containing amino acids (Chakraborty et al. 2000), selective increase in the essential amino acid methionine (Zeh et al. 2001), increased inulin (Hellwege et al. 2000), increased β-carotene and lutein (Ducreux et al. 2005), and decrease in the glycoalkaloid solanine (Lukaszewicz et al. 2004).

Commercial release of genetically modified potato cultivars was reported in North America by Monsanto (Davies 2002). For example, Monsanto's NewLeaf™ transgenic potatoes were released for commercial use in 1998 (Davies 2002; Kaniewski and Thomas 2004). NewLeaf™ was the trade name for transformed cultivars, such as Atlantic, Superior, and Russet Burbank containing the Bt cry3A and cry3C genes (encoding Bt endotoxin proteins) that conferred resistance to Colorado potato beetle. NewLeaf™ Plus was a trade name applied to cultivars that in addition to resistance to *L. decemlineata* and other insects also had resistance to PLRV. Genetically modified potatoes were increased within the certification industry in Canada but public opposition and concerns resulted in withdrawal of all of the NewLeaf™ varieties from the Canadian certification programme and from the Canadian market in 2000 (Kaniewski and Thomas 2004).

Monsanto NewLeaf(™) transgenic potato were withdrawn from markets around the world in the year 2000 due to public worries and pharmaceutical regulators' concerns about interference with antibiotics by initiation of multiple resistance phenomena (Hickman and Roberts 2013). New release of transgenic potato developed by Simplot Co. under the trade name 'Innate' is under study by the US Food and Drug Administration (FDA; Dykes and Melton 2013). Simplot researchers claim that "Innate" potato, unlike Monsanto 'NewLeaf', do not contain foreign genes, only specific genes that have been silenced for specific traits. Meanwhile, the German chemical company BASF is seeking approval from the European Commission for growing a modified starch potato 'Amflora' for starch industry use only (Hickman and Roberts 2013).

Tissue culture technologies played and will continue to play a key role in potato improvement because of the highly heterozygous nature of the tetraploid cultivated potatoes and associated tetrasomic inheritance; *S. tuberosum* has four homologous copies of each of 12 chromosomes ($2n=4x=48$; Wenzel 1994; Bhojwani and Razdan 1996; Bradshow 2007a, b). Somatic hybridization of plants occurs via protoplast fusion. Somatic hybridization overcomes sexual incompatibility and male or female sterility encountered in conventional sexual crossing (Oryczyk et al. 2003; Guo et al. 2004). Somatic hybridization was used for a long time in potato improvement programmes and somatic hybrids have been reported (Trabelsi et al. 2005). For example, F_1 hybrid seeds were used to convert fertile potato cultivars into cytoplasmic male sterile seed parents (Perl et al. 1990). Cytoplasmic male sterility is a maternally inherited phenotype, identified by inability of a plant to produce functional pollen (Perl et al. 1990) and is a mitochondria-controlled trait. Cytoplasmic male sterility was transferred from an alloplasmic male sterile into two male fertile potato cultivars that were previously used as seed parents in F_1 hybrid seed production.

Somatic hybridization and protoplast technologies were used for developing virus- and insect-resistant cultivars (Novy 2007). Hybrids were produced from the *S. tuberosum* clone PI 245939 (PVY-resistant) and a *S. tuberosum* dihaploid × *S. berthaultii* cross using somatic hybridization (Novy and Helgeson 1994; Novy 2007). These tri-species hybrids gave poor tuber yield but vigorous plant growth so Novy and Helgeson backcrossed the somatic hybrids with *S. tuberosum* to improve yield. They identified three somatic hybrids with significant resistance to PVY compared with 'Russet Burbank' and two clones that were resistant to PLRV. This work may lead to the production of important clones for release as new specific virus-resistant or virus-tolerant cultivars.

Protoplast extraction from leaf mesophyll tissues was used by Shepard et al. (1980) in the production of ~2500 protoclones from 'Russet Burbank'. They identified ~2.4% of protoclones with improved agronomic as well as early (*Alternaria solani* Sorauer) and late blight resistance traits. Thompson et al. (1986) selected potato protoclones with increased resistance to PVY and PLRV. Cassells et al. (1983) in their comparative study of explants for calliclones, potato clones produced through a callus phase indirectly or directly on explant tissue, including stem pieces (lateral buds and stem segments), recommended the use of stem nodal cuttings and the involvement of advanced calliclone lines into potato improvement programmes. The Cassells et al. group selected 'AC LR Russet Burbank' among a population of calliclones resistant to PLRV and released it in Canada in 2002 (CFIA 2013).

Tissue culture-based technology for potato improvement depends on the identification of useful genetic variation among plants using intensive screening that follows in vitro regeneration techniques (Nassar et al. 2011). These variant plants were first called 'somaclonal variants' by Larkin and Scowcroft (1981). Somaclonal variation may reflect a combination of processes that lead to differences among regenerants from culture; variation between cells present in the explant (endogenous variation), or variation resulting from the culture process (exogenously caused variation) or a combination of both processes. Endogenous and exogenous variation could come about through genetic, epigenetic, or a combination of genetic and epigenetic changes that occur in meristematic tissues in vitro.

8.3 Potato Somatic Embryogenesis Technology

Somatic embryogenesis (SE) is a tissue culture technology first reported by Waris (1957) in fineleaf waterdropwort (*Oenanthe aquatica*) and Reinert (1958) and Steward (1958) in carrot (*Daucus carota*) and has now been described in at least 200 Gymnosperm and Angiosperm species (Tisserat et al. 1979; Evans et al. 1981; Tulecke 1987). It describes a developmental process involving one or more somatic cells that results in a morphological structure, similar to a zygotic embryo derived from a fertilized egg, capable of developing into a complete plant (Neumann et al. 2009). SE was defined as the process in which a bipolar structure arises through a series of stages characteristic for zygotic embryo development (globular, heart, torpedo, and cotyledonary stages) without vascular connections to maternal tissue (Sharp et al. 1980; Ammirato 1987; Terzi and Loschiavo 1990). It describes the developmental process of inducing somatic cells to form totipotent embryogenic cells that grow into a complete plantlet with shoot and root poles (Suprasanna and Bapat 2005).

Sharp et al. (1980) and Evans et al. (1981) distinguished direct and indirect SE. Direct SE is initiated from pre-embryogenically determined cells while indirect SE starts from cells which require redifferentiation (sometimes described as dedifferentiation, or reversion to their meristematic role as parenchyma cells) before expressing embryogenic competence. Simply, the induction of embryogenesis involves the transformation of single parenchyma cells into densely cytoplasmic cells with embryogenic competence (Kohlenbach 1978). In indirect SE, callus formation precedes the formation of embryos. Cells capable of direct SE are physiologically similar to those in zygotic embryos. In those cells, the genes necessary for zygotic embryogenesis are active to varying degrees (Sharp et al. 1980; Carman 1990). The simplicity of SE induction is a 'memory' of pathways either before or after 'competent' cell redifferentiation (Carman 1990). Somatic embryo production tends to be more prolific using indirect than direct SE (Pretova and Dedicova 1992) and is related to a conducive environment, the amount of callus, and the proportion of competent cells within this callus.

The first reports of somatic embryos in potato tissues were by Lam (1975) and Bragd-Aas (1977). Potato was considered to be recalcitrant in the initiation of somatic embryos (Litz and Gary 1995) until studies by Pretova and Dedicova (1992), De Garcia and Martinez (1995), and many others. Pretova and Dedicova (1992) reported 'SE' in the potato 'Desiree' from immature *zygotic* embryo sections. These SE formed on epidermis of the hypocotyl or cotyledons of the zygotic embryos. In a parallel process, young seedling cells, particularly in the hypocotyl epidermis, were competent for embryogenic development (Sharp et al. 1980, 1982; William and Maheswaran 1986).

Embryoids were induced from superficial parenchyma cells on callus derived from tuber and leaf explants and not embedded into the callus (Lam 1975; JayaSree et al. 2001). The productivity of somatic embryos from nodal cuttings of 18 potato cultivars and 3 wild species (*S. capibaccatum, S. polyadenium,* and *S. trifidum*) was low and genotype dependent (Seabrook et al. 2001). The Seabrook group emphasized that embryogenic callus cannot be subcultured onto fresh medium for somatic embryo regeneration and fresh explants must be callused to produce SE. These findings contradict the observations of Nassar et al. (2011) where embryogenic callus was recultured onto fresh medium, with high cytokinin and gibberellins and very low auxin, for continuous production of SE. Although as callus growth continued, the relative proportion of SE tended to decrease and shoot numbers increased. Regeneration of SE was initiated at the surface of embryogenic callus cells, supporting observations of early workers. In contrast, shoots (calliclones) were initiated beneath the callus surface. Embroids were harvested at 2 cm in height and subcultured onto fresh medium (Fig. 8.1; Nassar et al. 2011).

Somatic embryos at the early-torpedo stage developed into plantlets within approximately 2–4 weeks (Seabrook and Douglas 2001) and this was also observed by Nassar et al. (2011). Harvest of tiny potato embryoids has been done at the late-torpedo and/or the cotyledonary stages and these grew into plantlets (Nickle and Yeung 1994; Seabrook and Douglas 2001). Harvest of complete embryoids (with shoot and root) occurred after 3–4 weeks of subculture (Nassar et al. 2011). These somatic embryos resembled the seedlings of true potato seeds (TPS) with a vine-like growth habit, thin stems, and small leaves (Fig. 8.1c shows callused tuber explants supporting a 'forest' of SE).

Fig. 8.1 Stepwise pictogram shows the use of somatic embryogenesis in a potato breeding programme. The annual cycle of in vitro production of somatic embryos and increase through micropropagation (single node cuttings) for the purpose of storage in vitro and field evaluation. Following harvest, the first selection for improved French fry processing quality was based on graded tuber weight and tuber type. Following 5 months storage, the next selection was based on % glucose, French fry colour, phytonutrient content, and disease resistance. Somaclones with superior features were transferred from in vitro storage and increased via micropropagation for retesting as clonal lines the following year

8.4 Types of Explants Used to Produce Somatic Embryos

Somatic embryos have been produced from various explants of field-grown potato plants, including tuber discs (Lam 1975; Bragdo-Aas 1977; Nassar et al. 2011), immature zygotic embryos (Pretova and Dedicova 1992), and microspores (androgenesis; Dunwell and Sunderland 1973). Explants from greenhouse-grown potato plants were used for the production of SE including single-node stem cuttings (Reynolds 1986), shoot meristem tips (Powell and Uhrig 1987), microspores (Sopory et al. 1978; Johansson 1986), and cut leaves (JayaSree 2001). Also, embryoids were produced from explants obtained from potato plants grown in vitro including shoot meristem tips (Fiegert et al. 2000), single-node stem cuttings (De Garcia and Martinez 1995; Seabrook et al. 2001; Vargas et al. 2005, 2008; Sharma et al. 2007, 2008a), internodal stem cuttings, leaves, microtubers, and roots (Seabrook and Douglass 2001; Sharma and Millam 2004), and microtubers (Nassar et al. 2011).

Relative productivity of somatic embryos of the abovementioned explants was different from one study to another depending on genotype and media components. Few SE were obtained, after long induction periods averaging 3 months, from tuber discs, shoot tips, microspores, immature zygotic embryos, or leaves (Lam 1975; Bragdo-Aas 1977; Sopory et al. 1978; Johansson 1986; Powell and Uhrig 1987; De Garcia and Martinez 1995; JayaSree 2001; Seabrook et al. 2001). The average number of somatic embryos produced on one stem explant varied based on genotype, from 2.2 ('Yukon Gold') to 44.5 (advanced clone; F83065; Seabrook and Douglas 2001). An embryogenic cell suspension system was efficient in the production of SE from callus of nodal cuttings; approximately 600 SE were produced on 1 g callus in 50 ml of media (Vargas et al. 2005). One single nodal cutting produced an average from three replicates of 65 SE when callus was induced with 20 μM of 2,4-dichlorophenoxyacetic acid (2,4-D) pulse treatment for 60 min after 8 weeks on embryo-induction stage medium (Sharma et al. 2007). The greatest number of SE (> 1000) were produced from field-grown tuber and microtuber explants of 'Russet Burbank' and callus transplanted onto fresh SE-formation medium (Nassar et al. 2011). This improvement led to multiple harvests of SE every ~ 3 weeks after multiple rounds (three to four times) of subculture of callused explants onto fresh SE-formation medium.

8.5 Growth Regulators and Induction of SE

The indirect production of SE involves sequential use of two media; the first is for callus induction and the second is for SE initiation and growth (Lam 1975; De Garcia and Martinez 1995; JayaSree et al. 2001; Seabrook and Douglass 2001; Sharma and Millam 2004; Sharma et al. 2007; Nassar et al. 2011). The duration of each stage depends on the growth regulator content and ratio. For example, high auxin and high temperature were favourable for cell proliferation while high cytokinin and alternating low temperature promoted SE initiation (Lam 1977b). Enhanced SE production efficiency from different tissue explants and media component alterations has not been sufficiently tested. Molecular and biochemical involvement of plant hormones and growth regulators in the development of somatic embryogenic tissues has not been described clearly for potato. In addition, the role of plant growth regulators in the induction of somatic cells and tissues at different stages of maturity into embryogenic cells or tissues is still vague and needs more study.

The Murashige and Skoog (MS) basal salt medium (Murashige and Skoog 1962) supplemented with various auxins and cytokinins was the most commonly used medium to induce calli and somatic embryos on potato tissues (Table 8.1). Auxins are considered the controlling factor for embryo development, as endogenous auxin levels are critical during induction and expression of SE potential (Cooke et al. 1993; Sharma et al. 2007). Addition of 2,4-D increases the exogenous level which impacts the endogenous auxin level and promotes cell division and development of a hormonal gradient required for embryogenesis (Michalczuk et al. 1992). Then embryogenesis is expressed depending on plant species, genotype, and physiological condition of donor plant and explant (Sharma et al. 2007).

8.5.1 Callus Induction Stage

During callus induction, auxins such as 2,4-D and others (naphthalene acetic acid, NAA; indole acetic acid, IAA; or low concentrations of cytokinins (6-benzylaminopurine, BAP)) control embryogenesis. These auxins may be required to induce callusing but must be removed during embryo initiation due to their suppressive effect; auxin withdrawal is one trigger for SE (Suprasanna and Bapat 2005). Many studies were conducted using various potato tissues as explants. For example, callus was induced on tuber discs incubated in MS basal medium for 1–2 days in the dark and then transferred to 16-h light (108 μmol $m^{-2}s^{-1}$) at 26°C (Lam 1975). NAA at 2 mgl^{-1} was used for cell proliferation and formation of friable callus 3 ½ months after explantation (Lam 1977a). Regeneration of plantlets from isolated single cells of potato tubers of 'Superior' and *S. demissum* was achieved through several culture steps (Lam 1977b). First, friable callus was initiated on tuber discs on MS medium with 2,4-D (3 ppm) at 27°C with 145.8 μmol $m^{-2}s^{-1}$. Then a cell suspension was prepared from that callus by agitating 1 g callus under dim light (2.7 μmol $m^{-2}s^{-1}$) at 27°C. The cell suspension was recultured every 2–3 weeks. Seven to ten days after subculture, single cells were isolated and callused under dim light. Calli were transferred onto SE induction medium at 21°C and 145.8 μmol $m^{-2}s^{-1}$ for SE production and plantlet regeneration (Lam 1977b).

Incubation of potato tissues on a medium with 2,4-D was necessary for embryogenic competence to occur; meaning that the auxin played a key role in the embryogenic determination programming of cells (Pretova and Dedicova 1992; Pretova and Obert 2006). Stem nodal segments of 'Superior' were callused on MS basal salt medium supplemented with yeast extract (0.5 mgl^{-1}) and 2,4-D (4 mgl^{-1}) for 37 days at 23 ± 1° C with 16-h photoperiod (50 μmol $m^{-2}s^{-1}$). Friable green callus was produced 12–30 days after initiation, becoming brown and compact after three rounds (total of about 90 days) of subculture (De Garcia and Martinez 1995). Leaf explants of 'Jyothi' incubated at 25 ± 2°C and 16 h light from cool fluorescent lamps (50 μmol $m^{-2}s^{-1}$) on MS medium supplemented with 2,4-D (0.9 μM) and

Table 8.1 List of various media and growth regulators (mgl^{-1}) used to induce callus on different potato explants

Basal salts	2,4-D	IAA	BAP	NAA	TDZ	K	Z	P	GA_3	YE	Reference
NN	–	–	–	8	–	0.5	–	–	–	–	Bragdo-Aas 1977
MS	–	0.4	0.4	8	–	–	–	–	–	–	Bragdo-Aas 1977
MS and NN	2–3	0.1	0.5	2	–	0.1	–	–	0.2	–	Lam 1977a, 1977b
K3	–	0.1	1	–	–	–	–	–	–	–	Johansson 1986
MS	10	–	–	–	–	3	–	–	–	–	Reynolds 1986
MS	1	–	–	–	–	–	1	–	–	–	Johansson 1988
MS and B5	2	–	0.05	–	–	–	–	–	–	–	Pretova and Dedicova 1992
MS	4	–	–	–	–	–	–	–	–	0.5	De Garcia and Martinez 1995
MS	–	–	0.25	3	–	–	–	–	–	–	Fiegert et al. 2000
MS	0.9[a]	–	10[a]	1.1[a]	–	–	–	–	–	–	JayaSree et al. 2001
MS	–	19[a]	0.15[a]	–	0.15[a]	–	–	–	–	–	Seabrook and Douglass 2001
MS	–	19[a]	0.15[a]	–	0.15[a]	–	–	–	–	–	Seabrook et al. 2001
MS	11.5[a]	–	–	–	–	–	–	1.65[a]	–	–	Bordallo et al. 2004
MS	4	–	–	–	–	–	–	–	–	–	Vargas et al. 2005, 2008
MS	5[a]	19[a]	0.15[a]	–	0.15[a]	–	–	–	–	–	Sharma et al. 2004, 2007
MS	5[a]	–	–	–	–	–	–	–	–	–	Sharma et al. 2008a, 2008b
MS	–	–	17–74[a]	–	–	–	–	–	–	–	Sharma et al. 2010
MS	2	19[a]	–	–	0.15[a]	–	–	–	–	–	Nassar et al. 2011
MS	5	–	0.5	–	–	0.5	0.5	–	–	–	Sabeti et al. 2013

[a] Concentrations indicated by "a" were expressed in μML^{-1}

MS Murashige and Skoog medium, *NN* Nitsch and Nitsch organics, *2,4-D* 2,4-dichlorophenoxy acetic acid, *IAA* indole acetic acid, *BAP* 6-benzylaminopurine, *NAA* naphthalene acetic acid, *TDZ* thidiazuron, *K* kinetin, *Z* zeatin, *P* picloram, GA_3 gibberellic acid, *YE* yeast extract

BAP (10 μM) initiated nodular embryogenic callus on the cut ends of leaf explants, while using NAA (1.1 μM) with BAP (10 μM) produced compact embryogenic calli (JayaSree et al. 2001). Nodular embryogenic callus was induced during 7–14 days using IAA (19 μM) and either BAP (0.15 μM) or thidiazuron (TDZ; 0.15 μM) on MS medium at 19°C with 100 μmol $m^{-2}s^{-1}$ from cool white fluorescent lights (Seabrook et al. 2001; Seabrook and Douglass 2001).

Friable callus was induced on single-node cuttings (1–1.5 cm) after 2 months on MS solid medium supplemented with 2,4-D at 25 ± 1°C and 50 μmol $m^{-2}s^{-1}$ (Vargas et al. 2005, 2008). Then embryogenic cell suspension cultures were established from friable callus onto MS liquid medium with yeast extract (1 gl^{-1}), citric acid (50 mgl^{-1}), ascorbic acid (50 mgl^{-1}), kinetin (0.5 mgl^{-1}), and 2,4-D (0.5 mgl^{-1}) at 25 ± 1°C after 4 weeks (Vargas et al. 2005, 2008). Callus was induced after 2 weeks of incubation on MS medium supplemented with 2,4-D (5 μM; Sharma and Millam 2004; Sharma et al. 2007). Sharma et al. (2007) tried to improve or accelerate the somatic embryogenic process by reducing the time required for explants to callus using different concentrations of 2,4-D (2,4-D pulses) applied at various intervals. They concluded that pulse treatment of cultured explants with 20 μM of 2,4-D solution for 1 h could shorten the induction phase of somatic embryos by 2 weeks. However, while they succeeded in inducing the SE in a shorter time, they mentioned that the 2-week preparatory period was required for competent cells to form the somatic embryos (Sharma et al. 2007).

8.5.2 Embryoid Initiation Stage

Following the callus stage, come the embryo initiation and developmental (transitional) stages. These involve the addition of cytokinin (BAP, zeatin), gibberellin (GA_3), and relatively low concentrations of auxin (IAA, NAA) to the medium (Table 8.2). Cytokinins and GA_3 have beneficial effects on embryo initiation and development (Brown et al. 1995) by promoting shoot growth and elongation (JayaSree et al. 2001). Embryoids originate from parenchyma cells on the surface of the callus. These develop into plantlets superficially and once these SE plantlets are harvested, a new group of SE began to grow, also superficially (JayaSree et al. 2001; Nassar et al. 2011).

BAP was described as a critical hormone for shoot induction from parenchyma cells of potato tubers because it caused the greatest explant disc expansion and chlorophyll retention (Lam 1975), promoted protein synthesis by interacting with plant ribosomes (Berridge and Ralph 1970), and promoted cambial activity (Loomis and Torrey 1964). The use of BAP (4 μM) was necessary for initiation of SE on microspore explants (Sopory et al. 1978). Shoots were initiated 7 weeks after tuber disc explants of potato 'Superior' were cultured on a MS medium supplemented with 0.4 ppm of BAP. More than three shoots on average were obtained after 4 months. However, shoots were abnormal and without any roots (Lam 1975). Normal plantlets (true SE) were formed from tuber discs incubated on MS medium with specific

Table 8.2 List of different media recipes and growth regulators (mgl^{-1}) used to induce embryo formation on various potato explants

Basal salts	2,4-D	IAA	BAP	NAA	K	Z	YE	D	GA_3	Reference
MS and NN	–	0.4	0.4	–	0.8	–	–	–	0.4	Lam 1975
NN	–	0.4	0.4–1.2	–	0.8	–	–	–	–	Bragdo-Aas 1977
MS	–	0.4	0.4–1.2	–	0.8	–	–	–	–	Bragdo-Aas 1977
MS and NN	–	0.05	0.5	0.1	0.2	0.5	–	–	0.2	Lam 1977a, 1977b
MS	–	6[a]	4[a]	–	–	–	–	–	–	Sopory et al. 1978
MS	–	1	1	–	–	0.5	–	–	10	Cassells et al. 1983
MS	–	–	0.5	–	–	–	–	–	–	Johansson 1986
MS	10	–	–	–	1	–	–	–	–	Reynolds 1986
K3	–	0.1	1	–	–	1	–	–	–	Johansson 1988
MS and Monnier	–	–	0.05	–	–	–	–	2	–	Pretova and Dedicova 1992
MS	–	–	1	–	–	–	–	–	0.1	De Garcia and Martinez 1995
MS	–	–	–	–	–	0.05	–	–	0.1	Fiegert et al. 2000
MS	–	–	10[a]	–	–	4.6[a]	–	–	14.4[a]	JayaSree et al. 2001
MS	–	0.05[a]	–	–	–	12[a]	–	–	0.55[a]	Seabrook and Douglass 2001
MS	–	0.05[a]	–	–	–	12[a]	–	–	0.55[a]	Seabrook et al. 2001
MS	0.5	–	–		0.5	1	1000	–	–	Vargas et al. 2005, 2008
MS	–	0.05[a]	–	–	–	12[a]	–	–	0.55[a]	Sharma et al. 2004, 2007
MS	–	–	–	–	–	–	–	–	–	Sharma et al. 2008a, b
MS	–	–	–	–	–	5–69[a]	–	–	2.9[a]	Sharma et al. 2010
MS	–	0.05[a]	–	–	–	12[a]	–	–	0.55[a]	Nassar et al. 2011

[a] Concentrations indicated by "a" were expressed as μML^{-1}

2,4-D 2,4-dichlorophenoxy acetic acid, *IAA* indole acetic acid, *BAP* 6-benzylaminopurine, *NAA* naphthalene acetic acid, *K* kinetin, *Z* zeatin, *YE* yeast extract, *D* dicamba, *GA_3* gibberellic acid, *MS* Murashige and Skoog, *NN* Nitsch and Nitsch auxin–kinetin ratio (Lam 1977a). Zeatin increased the chlorophyll in callus and was required for plantlet formation (Lam 1977a). Calli of single cells (10–15 mm) of tuber tissues of 'Superior' and *S. demissum* were transferred onto a basal medium with 0.2 ppm of NAA for shoot production, which was observed after 5–7 weeks (Lam 1977b).

SE induction and growth was enhanced by cold treatment (7 °C for 3 days) and addition of *N, N*- dimethylsuccinamic acid and L-cysteine–HCl to anther cultures of 'Maria', 'Stina', and 'Elin', respectively (Johansson 1986). After the callusing phase, the greatest number of SE was obtained by placing anthers on cubes of sold medium surrounded with growth regulator-free liquid medium for 1 week and afterwards on a medium containing 2,4-D (Johansson 1988). Shoot regeneration from SE or callus was accomplished on the K3 medium (Nagy and Maliga 1976) with 1 % sucrose, 1 mgl^{-1} of zeatin, 1 mgl^{-1} of BAP, and 0.1 mgl^{-1} of IAA (Johansson 1988). Addition of 2,4-D or zeatin after the first week of callus induction from anther explants on a growth regulator free medium stimulated and enhanced embryo production (Johansson 1988). Addition of dicamba or 2,4-D to either MS or Monnier basal medium (Monnier 1978) induced somatic embryos at greater frequencies than vitamin B_5 with BAP and yeast extract (Pretova and Dedicova 1992). Callused leaf explants were transferred onto MS medium containing zeatin (4.6–22.8 µM), BAP (10.0 µM), and/or GA_3 (14.4 µM; JayaSree et al. 2001).

Media that contained zeatin and BAP, required for plant shoot development, produced the greatest number of SE directly from meristematic centres produced on nodular callus tissue after 2–4 weeks. The highest frequency of SE development from the initial globular stage to fully developed cotyledonary stage occurred in the presence of zeatin (22.8 µM) and BAP (10 µM). Low concentration (4.6–9.1 µM) of zeatin (and no auxin) produced globular embryos while greater concentrations (13.7–22.8 µM) supported further development of embryos to the torpedo and cotyledonary stages (JayaSree et al. 2001). Callus was subcultured three times, once every 30 days, on MS medium with 2,4-D (2 mgl^{-1}) under continuous light of 50 µmol $m^{-2}s^{-1}$ then it differentiated, forming SE on MS medium supplemented with BAP (1 mgl^{-1}) or GA_3 (0.1 mgl^{-1}; De Garcia and Martinez 1995). After the callusing stage, somatic embryos of 18 tetraploid potato cultivars, diploid and monoploid germplasm, and three wild species were initiated and produced within 14–28 days on MS salts with zeatin (12 µM), IAA (50 nM), and GA_3 (550 nM; Seabrook et al. 2001; Seabrook and Douglass 2001).

8.6 Histology of Somatic Embryos

Somatic embryos resemble zygotic embryos; these pass through the same embryogenic transition stages: globular, heart, torpedo, and late torpedo cotyledonary. SE transition stages were monitored under the scanning electron microscope and some secondary globular embryoids were formed in the hypocotyl area of the torpedo stage SE (De Garcia and Martinez 1995; JayaSree et al. 2001; Seabrook and Douglass 2001; Vargas et al. 2005). Sharma and Millam (2004) examined the histological stages of somatic embryos through a stereo microscope and photographed with a Nikon digital camera. They noticed first the globular-stage embryos accompanied by protoderm development (visible at early- and late-globular stages), cotyledonary initials (heart-shaped stage), and torpedo-shaped embryos (late-heart-shaped) which were similar to potato zygotic embryos and the mature cotyledons had differentiating apical meristems (Sharma and Millam 2004).

A histological study conducted by Vargas et al. (2005) showed that the competent callus cells that became embryogenic cells were isodiametric, with thick cell walls, dense cytoplasm, prominent nuclei, and many starch grains compared with non-embryogenic cells. Accumulation of extracellular proteins of different sizes occurred for different stages of embryogenesis tested using sodium dodecyl sulphate polyacrylamide gel electrophoresis (SDS-PAGE; Vargas et al. 2005). At the first stage (cell clusters, free cells, and cell suspension), a 78-kDa protein was dominant. At the second stage (embryogenic cell determination), 7–14 kDa proteins were detected. In the third stage (developmental stage), a 32-kDa protein was secreted from the cell. In the final stage (torpedo and mature embryos), 20–50 kD proteins were reported (Vargas et al. 2005).

8.7 Molecular Aspects of SE

The biochemical and molecular events of plant SE were reviewed by Suprasanna and Bapat (2005). Conversion of somatic cells into embryogenic cells is characterized by RNA and DNA synthesis, change in cell pH, increase in oxygen uptake and kinase activity, migration of nuclei towards the cell wall, changes in cytoskeleton, conversion of adenosine triphosphate (ATP) to adenosine diphosphate (ADP), and inactivation of cytosolic and maturation–promotion factors (Suprasanna and Bapat 2005). Differential gene expression was proposed as a mechanism of embryogenic capability but not studied (Seabrook et al. 2001; Seabrook and Douglass 2001; Figueroa et al. 2002). Genes at various stages of plant SE were grouped into five classes with a complex system during SE stages (for more information see Kommine et al. 2005; Suprasanna and Bapat 2005). Gene classes are expressed: (1) during the entire plant growth, (2) in the embryo induction stage (embryo specific) but not expressed at late stages, (3) during early embryogenesis after the induction stage, (4) during the expansion phase of the cotyledon and plantlet maturation, and (5) throughout the entire embryo initiation and production stages.

Sharma et al. (2008a) described the only study of gene expression during the various SE stages in potato using DNA microarray with functional classification on gene ontology (GO). They induced SE on single nodal segments of in vitro grown plantlets following two contradictory induction conditions: embryo forming (absence of 2,4-D at the SE initiation stage) and embryo inhibitory (presence of 2,4-D at the SE initiation stage). Circadian gene expression profiles of both contrasting conditions revealed 358 transcripts that showed differential gene expression patterns during the induction phase (callusing; 14 days long), 480 transcripts during the embryo transition phases (SE development and growth; 21 days long), and 44 transcripts common to both phases. Functional classification on GO showed that most of the transcripts were involved in apoptosis, development, reproduction, stress, and signal transduction, which were associated with the SE process but no mention was made of the relationship to zygotic embryogenesis (Sharma et al. 2008a).

In detail, Sharma et al. (2008a) studied five samples at different stages of callus and SE transition developmental stages; excised explants (0 time), embryogenic callus (14 days), SE transition stages; (21 and 28 days), and growth of SE (35 days). They reported that during SE production, the mechanical injury of explant excision and growth hormone stress changed the normal plant cell development-signalling pathway (Sharma et al. 2008a). Complementary DNA (cDNA) and GO profiling identified 358 and 480 differentially expressed transcripts for induction (comparing 0- and 14-day samples) and embryo transition (embryo-forming condition; multiple comparison between results of samples collected at 0, 14, 21, 28, and 35 days) and 44 transcripts were common to both stages. About 27 genes were identified during the induction phase with about 16 others that might also contribute.

Reduction in the expression of *proliferating cell nuclear antigen* (*PCNA*) gene homologue indicated that apoptosis precedes SE (Sharma et al. 2008a). Polarity in explants/cells was critical for embryogenesis to proceed and auxin movement in tissues controlled it (Sharma et al. 2008a). Auxin influx and efflux facilitators, *AUX1* (auxin influx carrier protein1/polar auxin transport inhibitor-resistant protein1), *pinformed* (*PIN*), and *P-glycoprotein* (*PGP*) control auxin polarity (Parry et al. 2001; Paponov et al. 2005; Geisler and Murphy 2006). Sharma et al. (2008a) identified upregulation in transcripts of *EMB30/GNOM* (an endosome regulator of vesicle trafficking and involved in the control of *PIN* localization) at the end of the induction phase, formation of pre-embryogenic masses, which indicates that most events for proper embryo development occurred during the induction phase (Sharma et al. 2008a). Also, the auxin influx carrier protein (a homologue of *AUX1*) and an ATP-binding cassette (ABC) transporter family protein were upregulated by the end of the induction phase. The ABC transporter family protein could be a *PGP4* homologue and was assumed to be involved in auxin influx (Terasaka et al. 2005; Santelia et al. 2005; Sharma et al. 2008a). During embryo transition phases, transcripts of two ABC transporter family proteins (possible homologues of *PGP*) and a polysaccharide *ABC* transporter permease showed significantly differential regulation. This suggests that while *PIN*-mediated auxin channelling was operative during 2,4-D-fortified induction leading to the generation of pro-embryogenic masses, the *PGP*-mediated alterations in auxin distribution were functioning for re-establishing *PAT* in the absence of exogenous auxin, leading to the transition of pro-embryogenic masses into somatic embryos (Sharma et al. 2008a). At the end of the induction phase, polarity of SE was ensured by downregulation of S-adenosyl-L-methionine:carboxyl methyltransferase (*SAM*; Sharma et al. 2008a).

During induction of embryos, SE-specific tubulin- and actin-binding protein genes (villin and profilin) were upregulated. At the late stages of embryo development, genes encoding glycine- and proline-rich proteins were expressed. DNA topoisomerase-I and glutamine synthetase showed enhanced expression trends throughout the embryo transition stages especially at the torpedo stage. Calcium-mediated signal transduction is important during SE in potato. During the induction phase, calmodulin-like domain protein kinase (*CPK2*) homologue expression was increased. *CPK2* was associated with greater levels of the calmodulin (CaM)-binding family of proteins throughout embryo transition stages (21, 28, and 35 days). Also, upregulation of the transcripts of the calcium exchanger (*CAX1*; involved in CaM transport) occurred during the induction phase.

Somatic embryogenesis receptor kinase (*SERK*) gene family was critical to signal transduction during SE. *SERK* was a molecular marker for the identification of competent and non-competent cells for SE (Schmidt et al. 1997). *SERK* encodes for a protein having an N-terminal domain with five *LRRs* acting as a protein-binding region. The *SERK* protein has a proline-rich region between the extracellular *LRR* domain of *SERK* and the membrane-spanning region. *SERK* transcripts were upregulated during the callus phase (Sharma et al. 2008a) and did not change during the subsequent embryo transition stages. *SERK* was expressed from the induced embryogenic cell stage to the globular stage but not detected in non-embryogenic stages (Schmidt et al. 1997; Sharma et al. 2008a, b). The *StSERK* expression, during the progression of SE, was further confirmed using real-time reverse transcription polymerase chain reaction (RT-PCR; Sharma et al. 2008b).

Oxidative stress affects SE induction where it activates mitogen-activated protein kinase (*MAPK*) cascades in plants (Kovtun et al. 2000). Transcripts of *MAPK* were upregulated during the induction and the embryo-forming phases (Sharma et al. 2008a). During SE, *KNAT1* (homeobox protein knotted-1 like 1), a *KN-1* ortholog, showed significant differential gene expression during the embryogenesis stages. Functionally, *KNAT1* was expressed during the cell progression and diversification process (Sharma et al. 2008a). *KNAT1* expression corresponds with the conversion of heart-stage to torpedo-stage embryos and could potentially be used as a developmental marker to characterise this particular embryo transition phase. Another homeodomain transcription factor *KNAT7* was also significantly upregulated throughout the embryoid stages (Sharma et al. 2008a).

Arabinogalactan proteins (AGPs), extracellular-expressed proteins, and lipid transfer proteins (LTPs) were highly expressed at 1 and 3 weeks of the embryo induction phase. LTP1 and LTP2 transcripts were highly expressed after 2 weeks of the embryo transition phase. Late embryogenesis abundant (LEA) protein ortholog showed a progressive trend in expression. Its expression peaked by the third week of the embryo transition phase, which suggests that the potato SE were accumulating appropriate gene products for protection against a period of desiccation, as would be expected in zygotic embryogeny (Sharma et al. 2008a). By 28 and 35 days of SE development, *germins*/GLPs gene was induced in the embryos (Sharma et al. 2008a).

A series of events occur during potato embryogenesis that was interpreted based on cDNA array data. This started with upregulation of a *CCAAT*-box binding transcription factor (homolog of *leafy cotyledon 1* (*LEC1*)) during the callus phase. *CCAAT*-box and *LEC1* were downregulated after 1 week into the embryo induction phase and peaked after 2 and 3 weeks, corresponding with the heart-shaped embryo and maturation phases, respectively. *FUSCA* and *abscisic acid insensitive 3* (*ABI3*), transcripts of associated seed storage protein genes, were upregulated during the late stages of embryogenesis. These data indicate that the *CCAAT*-box-binding factors (a potential *LEC1* ortholog in potato), *ABI3,* and *FUSCA* proteins may have a significant role during SE growth. Moreover, a gene encoding a *germin*-like protein homologue (germin-like protein 6; *GLP6*), which was classified under nutrient reservoir activity significantly upregulated after 2 and 3 weeks (Sharma et al. 2008a).

It was apparent that genes with alternate functions apart from SE are also involved in various stages of SE. For example, transcripts of the floral homeotic protein gene *APETALA2* (*AP2*) were significantly upregulated through callus induction and up to 2 weeks into the embryo-induction phase (Sharma et al. 2008a). *AP2* plays active roles in the formation of the floral meristem, specifying floral organ identity, and in the regulation of the expression of floral homeotic genes in *Arabidopsis*. Additionally, patatin-like protein transcripts were upregulated during the induction phase. Moreover, transcripts of *defective-in-antherdehiscence1* were highly expressed at the end of the callus phase (Sharma et al. 2008a). This gene product initiates jasmonic acid biosynthesis and has a regulatory function in anther dehiscence, pollen maturation, and flower opening (Ishiguro et al. 2001).

Gibberellin (GA) biosynthesis pathway was induced during SE. For example, the transcript profile of a GA_3 biosynthesis gene homolog, *ent*-Kaurenoic acid oxidase, was induced throughout all embryo-forming stages and peaked at 35 days. The absence of GA_3 from the medium inhibited the development of somatic embryos (Sharma 2006). Transcripts of *gibberellin oxidase-like protein* were found to be upregulated throughout the embryo transition stages and peaked at 35 days. Therefore, addition of GA_3 during embryogenesis is critical for efficient SE in potato (Sharma et al. 2008a). Another gene, xyloglucan endo-transglycosylase (*XET*), was induced at all embryo-forming stages. *XET* catalyzes the cleavage of xyloglucan, which causes a loosening of the cell wall. *No apical meristem* (*NAM*) was expressed during the transition from globular to heart stage of SE. Enhanced *scarecrow* (*SCR*) expression (post-torpedo stage) possibly limits the capacity of somatic embryos to enter dormancy and enhances root formation (Sharma et al. 2008a).

8.8 Use of SE

SE has versatile use in plant biotechnology. Extensive use of SE occurs in genetic improvement (molecular transformation) technology and in direct or indirect adventitious propagation of plants. It is also described as the best model to study early regulatory and morphogenetic events during plant development (Zimmerman 1993; Suprasanna and Bapat 2005). SE are used for large-scale production of potato through synthetic seed (synseed) derived from various tissue explants, which may also include cryopreservation, followed by regeneration of plants (Bajaj 1995; Smith 1995; Neumann et al. 2009). The unicellular origin of somatic embryos is believed to rule out the creation of chimeras (Gioregetti 1995; Sharma and Millam 2004). However, it is well documented that genetic variation is a common feature of callus growth, and indirect SE production is therefore liable to reflect the variation inherent in the callus it is derived from. For this reason, SE is generally considered unsuitable for clonal increase and maintenance of potato cultivars. This somatic (somaclonal) variation may be negative or positive and can be utilized to generate somatic lines with variant genetic traits for disease resistance and better agronomic or nutritional characteristics (Haberlach et al. 1985; Pretova and Dedicova 1992; De Garcia and Martinez 1995; JayaSree et al. 2001; Nassar et al. 2011, 2014).

Synthetic seeds have long been considered a promising alternative to conventional potato tubers or cut seed tubers because this technology produces specific virus-free, assumed genetically identical (genetically similar) materials, that are relatively easy to handle and transport (Bordallo et al. 2004; Sharma et al. 2007). Somatic embryos were hardened for 2 weeks before being desiccated and coated for use as synseed (Pond and Cameron 2003). This induces quiescence in the embryos and provides more handling flexibility in large-scale production systems (Gary et al. 1991). Both TPS and synseed require a relatively long growing season and are not generally recommended for the short season temperate zone, although they have been promoted by CIP (International Potato Center 2013) for more economical production of potato in the tropics (Engels et al. 1993; Bradshaw 2007b; Khourgami et al. 2012; CIP 2013).

If SE are to be used for synseed purposes, bioreactors would be necessary for large-scale SE production followed by encapsulation and field planting (Fiegert et al. 2000; Nyende et al. 2002; Schafer-Menuhr et al. 2003). Little progress has been done to develop synthetic seeds for potato due to problems of somaclonal variation (non-identical somatic embryos that may vary from cultivar type), long periods required for induction and production of embryos, and difficulty synchronizing embryo maturity (Fiegert et al. 2000; Bordallo et al. 2004). SE have been used in the production of monohaploids and diploids through anther culture (androgenesis; the in vitro culture of whole anthers or free microspores; Wenzel 1994).

8.9 SE and Potato Improvement (Making Use of Somaclonal Variation)

SE has been applied to developmental studies, crop improvement, and genetic transformation of a wide range of plant species (JayaSree et al. 2001). The advantage of using SE in potato improvement is the occurrence of potentially useful somatic variation from direct or indirect regeneration routes. Somatic variation is relatively greater from indirect SE and shoot production via callused explants. Somaclonal variation among individual somatic embryos or shoots could result from structural changes, change in chromosome numbers, point mutations, and changes in the expression of a gene due to structural changes in the chromosome (heterochromatin and effects of position). Variation could also be due to activation of transposable elements, chromatin loss, DNA amplification, somatic crossing over, and somatic reduction and structural changes in the cytoplasmic organelle DNA (Kaeppler et al. 2000) or combinations of these phenomena. The use of certain plant growth regulators, particularly over long culture intervals, increases the incidence of somatic variation. For example, using 2,4-D instead of NAA increased the frequency of abnormal plants (Shepard 1981). Several factors could affect the incidence of somaclonal variation, including genotype, explant source, in vitro period, and cell/tissue cultivation conditions (Larkin and Scowcroft 1981; Bordallo et al. 2004).

Production of haploids using embryogenesis is a tool of plant breeding (Johansson 1986). Monohaploid plants of *S. tuberosum* were produced through embryogenesis in microspore cultures from dihaploid donor plants (Jacobsen and Sopory 1978). Of 14 studies of potato SE, only one evaluated somaclones in the greenhouse and noted 'off-types' (Seabrook and Douglas 2001). Somatic embryoids of potato were not tested in the field until the recent work of Nassar et al. (2008, 2011, 2014). A wide range of different sizes and structures of potato embryoids were noticed within callus (De Garcia and Martinez 1995; JayaSree et al. 2001) which suggests different origins of the embryos. In calli of 'Bintje' a mixture of cells with different chromosomal numbers (24–200) occurred (SreeRamulu et al. 1983). Cytological examination of embryoids and plantlet root tips revealed monohaploids and dihaploids (Sopory et al. 1978).

Differences in chromosome number and mitotic abnormalities throughout callus formation and embryo differentiation indicated genetic variation among regenerants (Vargas et al. 2005). This group observed accumulation of extracellular proteins of various molecular weights at different stages of embryogenesis. However, cytogenetic analysis of plants obtained from embryogenic-cell-suspension culture of 'Désirée' including callus tissue, embryogenic cells at the globular, heart, or torpedo stages, and root apices of 20 regenerated plants were homogeneous when tested using random amplified polymorphic DNA (RAPD) analysis (Vargas et al. 2008). In addition, very low level of amplified fragment length polymorphism (AFLP) marker profile variation was seen (3 out of 451 bands) among 15 somatic embryos of 'Désirée' (Sharma et al. 2007). The low level of molecular variation observed could possibly be because of epigenetic variation (DNA methylation changes) that occurs during the process of SE (Sharma et al. 2007).

Phenotypic variation and nuclear microsatellite polymorphism (nDNA) were studied in (di)haploids and tetraploid genotypes of *S. tuberosum* through androgenesis (Sharma et al. 2010). About 3–7%, increase in nDNA content occurred in anther-derived tetraploids. The (di)haploids and their anther donors showed unique genotypes at the microsatellite loci *POTM1-2, STM0015,* and *STM0019b* while the tetraploids had the same allelic profiles with their anther donor except at *STM0019a* which was characterized by the absence of a 186-bp allele (Sharma et al. 2010). A specific new allele was detected for the locus *STWAX-2* where the standard donor alleles were replaced by a 230-bp allele in both C-13 and D4 (di)haploids. The over-expression of microsatellite variation in D4 showed triallelic profiles at the microsatellite loci *POTM1-2* and *STM0015* attributed to its chimeric structure, which might have been formed through incomplete fusion of two different pro-embryos during the first steps of microspore division (Sharma et al. 2010).

There has been relatively little effort invested in improving potato using SE techniques (Lam 1977a; De Garcia and Martinez 1995) until Nassar et al. (2011, 2014). This group developed an in vitro model system to study the periclinal chimeral status of 'Russet Burbank' using SE. The putative periclinal chimeral structure NB 'Russet Burbank' (New Brunswick clone of 'Russet Burbank') was studied over two field seasons using populations of intraclones (somatic embryos derived from specific tuber source tissues; periderm, cortex, and pith) produced through SE (Nassar et al. 2008). The tubers of intraclones derived from all histogenic layers were russet, which indicated that 'Russet Burbank' was not organized in a periclinal chimera arrangement for an L_1 and/or L_2, mutation of 'Burbank'. This did not definitively prove that the original mutation from 'Burbank' was not of this type, but showed that the current arrangement was not a periclinal chimera.

These efforts marked the start of a McGill University effort to improve NB 'Russet Burbank' for French fry processing through field and processing evaluation of somatic embryo-derived plants (Fig. 8.1). Approx. 800 intraclones were regenerated from field-grown tubers or microtubers. Approx 2–9% of intraclonal plants had similar yield and tuber type to ex vitro plantlet or field-tuber-derived plant controls but superior processing features (lower % glucose and better fry colour after 5 months storage). The single cell origin of these somatic variants suggested that they are more likely to be stable (Nassar et al. 2011). This seems to be the case as 25 advanced lines were repeatedly tested over 4 years as ex vitro plantlets and found to be stable in yield and in reducing sugars/fry results following 5 months storage (Nassar et al. 2011). Twenty-five advanced somaclones that were selected out of 800 on the basis of yield (mean tuber weight), tuber type, low reducing sugar levels, and good French fry colour after late storage (≥ 5 months) were examined for phytonutrient characteristics. SE technology produced enough variation for selection of nutritionally improved clones with greater antioxidant capacity and greater concentration of phenolic compounds; chlorogenic acid, caffeic acid derivatives, ferulic acid derivatives, rutin, and ascorbic acid with 2, 0.6, 0.5, 0.5, 0.4 and 0.1% of improvement, respectively, out of 800 clones (Nassar et al. 2014).

Comparing the percentages of improved lines with what has been reported in the literature and taking into consideration the challenges of conventional potato breeding, somaclonal selection appeared to be a useful strategy for both processing and phytonutrient improvement in potato. Selection among advanced somatic lines for multiple traits (improved processing, better phytonutrient status) was attained. These results suggest that field-based somaclonal selection of NB 'Russet Burbank' has the potential to significantly improve this potato cultivar. Nassar et al. (2011, 2014) recommended that SE technology be used to improve advanced lines within potato breeding programmes prior to registration.

8.10 Conclusions

While SE has been described for potato since the 1960s, genetic, epigenetic, and biochemical characterization of the process is still in its infancy. Several new techniques will help elucidate SE variation, including DNA microarray, RNA transcriptomic, and/or metabolomic and proteomic approaches. Further work should be directed toward the complete identification of gene(s) that control the initiation and regeneration of potato somatic embryos, relation between growth regulators in media and SE, and specific proteins related to embryogenesis. The work of Sharma et al. (2008a) illustrated some of the underlying mechanisms of somatic embryo stages yet embryo stages, organ formation, and embryo maturation require further study. We recommend incorporation of SE technology into potato breeding programmes for the improvement of advanced lines.

References

Adang MJ, Brody MS, Cardineau G, Eagan N, Roush RT, Shewmaker CK, Jones A, Oakes JV, McBride KE (1993) The reconstruction and expression of a *Bacillus thuringiensis* cryIIIA gene in protoplasts and potato plants. Plant Mol Biol 21:1131–1145

Ammirato PV (1987) Organizational events during somatic embryogenesis. In: Green CE, Somers DA, Hackett WP, Biesboer DD (eds) Plant biology (vol 3). Plant tissue and cell culture. Alan R. Liss, Co., New York, pp 57–81

Bajaj YPS (1995) Cryopreservation of somatic embryos. In: Bajaj YPS (ed) Biotechnology in agriculture and forestry. Somatic embryogenesis and synthetic seed I (vol 30). Springer, Berlin, pp 221–229

Berridge MV, Ralph RK (1970) On the significance of cytokinin binding to plant ribosomes. In: Carr DJ (ed) Plant growth substances. pp 248–255

Beukema HP, Zaag DE (1990) Introduction to potato production. Pudoc Wageningen, Netherlands, pp 186–199

Bhojwani SS, Razdan MK (1996) Plant tissue culture: theory and practice, a revised edition (chapter 9). Elsevier, Amsterdam, p 235

Bordallo PN, Silva DH, Maria J, Cruz CD, Fontes EP (2004) Somaclonal variation on in vitro callus culture potato cultivars. Hortic Bras Brasília 22:300–304

Bradshaw JE, Mackay GR (1994) Breeding strategies for clonally propagated potatoes. In: Bradshaw JE, GR Mackay (eds) Potato genetics (chapter 21). CAB International, Oxon, pp 471–473

Bradshow JE (2007a) The canon of potato science: 4. Tetrasomic inheritance. Potato Res 50:219–222

Bradshow JE (2007b) Potato-breeding strategy. In: Vreugdenhil D (ed) Potato biology and biotechnology: advances and perspectives (chapter 8). Elsevier BV, Amsterdam

Bragd-Aas M (1977) Regeneration of plants from callus of potato tubers. Acta Hortic 78:133–137

Broglie KI, Chet I, Holliday M, Cressman R, Biddle P, Knowlton S, Mauvais CJ, Broglie R (1991) Transgenic plants with enhanced resistance to the fungal pathogen *Rhizoctonia solani.* Science 254:1194–1197

Brown CR (2008) Breeding for phytonutrient enhancement of potato. Am J Potato Res 85:298–307

Brown DCW, Finstad KI, Watson EM (1995) Somatic embryogenesis in herbaceous dicots. In: Thorpe TA (ed) In vitro embryogenesis in plants. Kluwer, Dordrecht, pp 345–415

Calleberg EK, Johansson LB (1993) The effect of starch and incubation temperature in anther culture of potato. Plant Cell Tissue Organ Cult 32:27–34

Camire ME, Kubow S, Donnelly DJ (2009) Potatoes and human health. Crit Rev Food Sci 49:823–840
Carman JG (1990) Embryogenic cells in plant tissue cultures: occurrence and behavior. In Vitro Cell Dev Biol 26:746–753
Caromel B, Mugniery D, Kerlan M-C, Andrzejewski S, Palloix A, Ellisseche D, Rousselle-Bourgeois F, Lefebvre V (2005) Resistance quantitative trait loci originating from *Solanum sparsipilum* act independently on the sex ratio of *Globodera pallida* and together for developing a necrotic reaction. Mol Plant-Microbe Interact 18:1186–1194
Carputo D, Barone A, Cardi T, Sebastiano A, Frusciante L, Peloquin SJ (1997) Endosperm balance number manipulation for direct in vivo germplasm introgression to potato from a sexual isolated relative (*Solanum commerconii* Dunn.). Proc Nat Acad Sci U S A 94:12013–12017
Carroll CP (1982) A mass-selection method for the acclimatization and improvement of edible diploid potatoes in the United Kingdom. J Agric Sci (Cambridge) 99:631–640
Carroll CP, De Maine MJ (1989) The agronomic value of tetraploid F1 hybrids between potatoes of group Tuberosum and group *Phureja/Stenotomum*. Potato Res 32:447–456
Cassells AC, Gotez EM, Austin S (1983) Phenotypic variation in plants produced from lateral buds, stem explants and single-cell-derived callus of potato. Potato Res 26:367–372
CFIA (Canadian Food Inspection Agency) (2013) Potato varieties. Russet Burbank. http://www.inspection.gc.ca/english/plaveg/potpom/var/russetburbank/russetbe.html. Accessed May 2013
Chakraborty S, Chakraborty N, Datta A (2000) Increased nutritive value of transgenic potato by expressing a nonallergenic seed albumin gene from *Amaranthus hypochondriacus*. Proc Nat Acad Sci U S A 97:3724–3729
CIP (International Potato Center) (2013) True potato seed. http://cipotato.org/publications/pdf/002591.pdf. Accessed Aug 2013
Cooke TJ, Racusen RH, Cohen JD (1993) The role of auxin in plant embryogenesis. Plant Cell 5:1494–1495
Cornelissen B, Melchers L (1993) Strategies for control of fungal diseases with transgenic plants. Plant Phys 101:709–712
Cotes JM, Nustez CE, Martinez R, Estrada N (2002) Analyzing genotype by environment interaction in potato using yield-stability index. Am J Potato Res 79:211–218
Darmo E, Peloquin SJ (1991) Use of 2x tuberosum haploid-wild species hybrids to improve yield and quality in 4x cultivated potato. Euphytica 53:1–9
Davidson MM, Jacobs JM, Reader JK, Butler RC, Frater CM, Markwick NP, Wratten SD, Conner AJ (2002) Development and evaluation of potatoes transgenic for a cry1Ac9 gene conferring resistance to potato tuber moth. J Am Soc Hortic Sci 127:590–596
Davies HV (2002) Commercial developments with transgenic potato. In: Valpuesta V (ed) Fruit and vegetable biotechnology (chapter 9). Woodhead Publishing Limited, Abington Hall, pp 222–244
De Garcia E, Martinez S (1995) Somatic embryogenesis in *Solanum tuberosum* L. cv. Desiree from stem nodal sections. J Plant Phys 145:526–530
De Maine MJ (1995) The effects of inbreeding value of potato di-haploids. Ann Appl Biol 127:151–156
Douches DS, Maas D, Jastrzebski K, Chase RW (1996) Assessment of potato breeding progress in the USA over the last century. Crop Sci 36:1544–1552
Ducreux LJM, Morris WL, Hedley PE, Shepherd T, Davies HV, Millam S, Taylor MA (2005) Metabolic engineering of high carotenoid potato tubers containing enhanced levels of β-carotene and lutein. J Exp Bot 56:81–89
Dunwell JM, Sunderland N (1973) Anther culture of *Solanum tuberosum* L. Euphytica 22:317–323
During K, Porsch P, Fladung M, Lorz H (1993) Transgenic potato plants resistant to the phytopathogenic bacterium *Erwinia carotova*. Plant J 3:587–598
Dykes A, Melton M (2013) GMO potatoes: coming soon to a McDonald's near you? Truststream Media. http://truthstreammedia.com/gmo-potatoes-coming-soon-to-a-mcdonalds-near-you-2/. Accessed Aug 2013

Engels C, Bedewy REl, Sattelmacher B (1993) Seed tuber production from true potato seed (TPS) in Egypt and the influence of environmental conditions in different growing periods. Potato Res 36:195–203

Evans D, Sharp W, Flick C 1981. Growth and behavior of cell cultures-embryogenesis and organogenesis. In: Thorpe TA (ed) Plant tissue culture. Academic, New York, pp 45–113

FAO (2013) FAO STAT. http://faostat3.fao.org/home/index.html#DOWNLOAD. Accessed May 2013

Fiegert A, Mix-Wagner G, Vorlop K (2000) Regeneration of *Solanum tuberosum* L., cv. Tomensa: induction of somatic embryogenesis in liquid culture for the production of "artificial seeds". Landbauforschung Volkenrode 50:199–202

Figueroa FQ, Mendez-Zeel M, Teyer FS, Rojas-Herrera R, Loyola-Vargas VM (2002) Differential gene expression in embryogenic and non-embryogenic cell clusters from cell suspension cultures of *Coffea arabica*. J Plant Phys 159:1267–1270

Gardner EJ, Snustad P (1986) Genética. Rio de Janeiro, Guanabara, p 497

Gary DJ, Purohit A, Triglano RN (1991) Somatic embryogenesis and development of synthetic seed technology. Crit Rev Plant Sci 10:33–61

Gebhardt C, Valkonen JP (2001) Organization of genes controlling disease resistance in potato. Ann Rev Phytopathol 39:79–102

Geisler M, Murphy A (2006) The ABC of auxin transport: the role of p-glycoproteins in plant development. FEBS Lett 580:1094–1102

Giorgetti L, Vergara MR, Evangelista M, Lo Schiavo F, Terzi M, Ronchi VN (1995) On the occurrence of somatic meiosis in embryogenic carrot cell cultures. Mol Gen Genet 246:657–662

Glendinning DR (1987) Neo-tuberosum. Scottish Crop Research Institute Annual Report. pp 77–78

Guo WW, Cai XD, Grosser JW (2004) Somatic cell cybrids and hybrids in plant improvement. In: Daniell H, Chase CD (eds) Molecular biology and biotechnology of plant organelles. pp 635–659

Haberlach GT, Cohen BA, Reichert NA, Baer MA, Towill LE, Helgeson JP (1985) Isolation, culture and regeneration of protoplasts from potato and several related *Solanum* species. Plant Sci 39: 67–74

Harris PM (ed) (1992) The potato crop: the scientific basis for improvement (2nd edn). Chapman and Hall, London, pp 13, 41–46

Hassairi AK, Masmoudi J, Albouy C, Robaglia MJ, Ellouz R (1998) Transformation of two potato cultivars Spuntia and Claustar (*Solanum tuberosum*) with lettuce mosaic virus coat protein gene and heterologous immunity to potato virus Y. Plant Sci 136:31–42

Hawkes JG (1990) The potato: evolution, biodiversity, and genetic resources. Belhaven, London

Hellwege EM, Czapla S, Jahnke A, Willmitzer L, Heyer AG (2000) Transgenic potato (*Solanum tuberosum*) tubers synthesize the full spectrum of inulin molecules naturally occurring in globe artichoke (*Cynara scolymus*) roots. Proc Nat Acad Sci U S A 97:8699–8704

Hermsen JGTH, Ramanna MS (1973) Double-bridge hybrids of *Solanum bulbocastanum* and cultivars of *Solanum tuberosum*. Euphytica 22:457–466

Hickman M, Roberts G (2013) Fury as EU approves GM potato. The independent. http://www.independent.co.uk/environment/green-living/fury-as-eu-approves-gm-potato-1915833.html. Accessed Aug 2013

Hilder VA, Boulter D (1999) Genetic engineering of crop plants for insect resistance—a critical review. Crop Prot 18:177–191

Ishiguro S, Kawai-Oda A, Ueda J, Nishida I, Okada K (2001) The defective in anther ehiscence1 gene encodes a novel phospholipase A1 catalyzing the initial step of jasmonic acid biosynthesis, which synchronizes pollen maturation, anther dehiscence, and flower opening in Arabidopsis. Plant Cell 13:2191–2209

IYP (International Year of Potato) (2008) Potato and biodiversity. www.potato2008.org. Accessed May 2013

Jacobsen E, Schouten HJ (2008) Cisgenesis, a new tool for traditional plant breeding, should be exempted from the regulation on genetically modified organisms in a step-by-step approach. Potato Res 51:75–88

Jacobsen E, Sopory SK (1978) The influence and possible recombination of genotypes on the production of microspore embryoids in anther cultures of *Solanum tuberosum* and dihaploid hybrids. Theor Appl Genet 52:119–123

Jansky SH, Yerk GL, Peloquin SJ (1990) The use of potato haploids to put 2x wild species germplasm into a usable form. Plant Breed 104:290–294

JayaSree T, Pavan U, Ramesh M, Rao AV, Reddy KJM, Sadanandam A (2001) Somatic embryogenesis from cultures of potato. Plant Cell Tissue Org Cult 64:13–17

Johansson L (1986) Improved methods for induction of embryogenesis in anther cultures of *Solanum tuberosum*. Potato Res 29:179–190

Johansson L (1988) Increased induction of embryogenesis and regeneration in anther cultures of *Solanum tuberosum* L. Potato Res 31:145–149

Johnston GR, Rowberry RG (1981) Yukon gold: a new yellow fleshed, medium early, high quality table and French fry cultivar. Am Potato J 58:241–244

Jones PW, Cassells AC (1995) Criteria for decision making in crop improvement programmes—Technical considerations. Euphytica 85:465–476

Kaeppler SM, Kaeppler HF, Rhee Y (2000) Epigenetic aspects of somaclonal variation in plants. Plant Mol Biol 43:179–188

Kaniewski W, Thomas P (2004) The potato story. AgBioForum 7:41–46

Khourgami A, Rafiee M, Jafari H, Bitrafan Z (2012) Effect of transplant preparation method on yield and agronomic traits of true potato seed (TPS) progenies in Sahneh region. World Acad Sci EngTechnol 67:1–3

Knight TA (1814) On raising new and early varieties of the potato (*Solanum tuberosum*). Belf Mon Mag 12:137–138

Kohlenbach HW (1978) Regulation of embryogenesis in vitro. In: Schutte HR, Gross D (eds) Regulation of developmental processes in plants. VEB Kongress-und Werbedruck, Oberlungwitz

Komamine A, Murata N, Nomura K (2005) Mechanisms of somatic embryogenesis in carrot suspension cultures—morphology, physiology, biochemistry, and molecular biology. In Vitro Cell Dev Biol-Plant 41:6–10

Kovtun Y, Chiu WL, Tena G, Sheen J (2000) Functional analysis of oxidative stress-activated mitogen-activated protein kinase cascade in plants. Proc Nat Acad Sci U S A 97:2940–2945

Lam S-L (1975) Shoot formation in potato tuber discs in tissue culture. Am Potato J 52:103–106

Lam S-L (1977a) Plantlet formation from potato tuber discs in vitro. Am Potato J 54:465–468

Lam S-L (1977b) Regeneration of plantlets from single cells in potatoes. Am Potato J 54:575–580

Larkin PJ, Scowcroft WR (1981) Somaclonal variation—a novel source of variability from cell cultures for plant improvement. Appl Genet 60:197–214

Lauer F, Miller JC, Anderson N Jr, Banttari E, Kallio A, Munson S, Orr P, Preston D, Smallwood DG, Sowokinos J, Titrud G, Wenkel R, Weirsma J, Wildung D (1988) Krantz: a russet cultivar for the irrigated sands. Am Potato J 65:387–391

Litz RE, Gray DJ (1995) Somatic embryogenesis for agricultural improvement. World J Microbiol Biotechnol 11:416–425

Loomis RS, Torrey JG (1964) Chemical control of vascular cambium initiation in isolated radish roots. Proc Nat Acad Sci U S A 52:3

Lucier G, Budge A, Plummer C, Spurgeon C (1990) US Potato Statistics, 1949–1989. USDA, Economic Research Service 829: 1–8. US Gov. Print Office, Washington, DC

Lukaszewicz M, Matysiak-Kata I, Skala J, Fecka I, Cisowski W, Szopa J (2004) Antioxidant capacity manipulation in transgenic potato tuber by changes in phenolic compounds content. J Agric Food Chem 52:1526–1533

Maris B (1989) Analysis of an incomplete diallel cross among three ssp. tuberosum varieties and seven long-day adapted ssp. andigenaclones of potato (*Solanum tuberosum* L.). Euphytica 41:163–182

Michalczuk L, Cooke TJ, Cohen JD (1992) Auxin levels at different stages of carrot somatic embryogenesis. Phytochemistry 31:1097–1103

Monnier M (1978) Culture of zygotic embryos. In: Thorpe TA (ed). Frontiers of plant tissue culture. Proceeding 4th International Congress of Plant tissue and cell culture. University of Calgary, Alberta, Canada, International Association for Plant Tissue Culture 277–286
Mullins E, Milbourne D, Petti C, Doyle-Prestwich BM, Meade C (2006) Potato in the age of biotechnology. Trends Plant Sci 11:254–260
Murashige T, Skoog F (1962) A revised medium for rapid growth and bioassays with tobacco tissue cultures. Physiol Plant 15:473–497
Nagy JJ, Maliga P (1976) Callus induction and plant regeneration from mesophyll protoplasts of *Nicotiana sylvestris*. Zeitschrift fiir Pflanzenphysiologie 78:453–455
Nassar AMK, Ortez E, Leclerc Y, Donnelly DJ (2008) Periclinal chimeral status of NB 'Russet Burbank' potato. Am J Potato Res 85:432–437
Nassar AMK, Abdulnour J, Leclerc Y, Li X-Q, Donnelly DJ (2011) Intraclonal selection for improved processing of NB 'Russet Burbank' potato. Am J Potato Res 88:387–397
Nassar AMK, Kubow S, Leclerc YN, Donnelly DJ (2014) Somatic mining for phytonutrient improvement of 'Russet Burbank' potato. Am J Potato Res 91:89–100
Neele AEF, Barten JHM, Louwes KM (1988) Effects of plot size and selection intensity on efficiency of selection in the first clonal generation of potato. Euphytica S:27–34
Neumann K-H, Kumar A, Imani J (2009) Plant propagation-meristem cultures, somatic embryogenesis. In: Neumann K-H, Kumar A, Imani J Plant cell and tissue culture—a tool in biotechnology, basics and applications, Springer-Verlag, Berlin pp 75–137
Nickle TC, Yeung EC (1994) Further evidence of a role for abscisic acid in conversion of somatic embryos of *Daucus carota*. In Vitro Cell Dev Biol Plant 30P:96–103
Novy RG (2007) Industry highlights: the use of the wild potato species, *Solanum etuberosum* in developing virus—and insect-resistant potato varieties. In: Acquaah G (ed) Principles of plant genetics and breeding. Blackwell Publishing, MA, pp 174–177
Novy RG, Helgeson JP (1994) Resistance to potato virus Y in somatic hybrids between *Solanum etuberosum* and *S. tuberosum* x *S. berthaultii* hybrid. Theor Appl Genet 89:783–786
Nyende AB, Schittenhelm S, Mix-Wagner G, Greef JM (2002) Synthetic potato seeds offer the potential to improve Kenyan seed system. Landbauforsch Volk 42:141–8
Ortiz R (1998) Potato breeding via ploidy manipulations. Plant Breed Rev 16:14–87
Oryczyk W, Przetakiewicz J, Nadolska-Orczyk A (2003) Somatic hybrids of *Solanum tuberosum*—application to genetics and breeding. Plant Cell Tissue Organ Cult 74:1–13
Osusky M, Osuska L, Hancock RE, Kay WW, Misra S (2004) Transgenic potatoes expressing a novel cationic peptide are resistant to late blight and pink rot. Transgenic Res 13:181–190
Paponov IA, Teale WD, Trebar M, Blilou K, Palme K (2005) The PIN auxin efflux facilitators: evolutionary and functional perspectives. Trends Plant Sci 10:170–177
Parry G, Marchant A, May S, Swarup R, Swarup K, James N (2001) Quick on the uptake: characterization of a family of plant auxin influx carriers. J Plant Growth Regul 20:217–225
Pelletier JR, Fry WE (1989) Characterization of resistance to early blight in three potato cultivars: incubation period, lesion expansion rate, and spore production. Phytopathology 79:511–517
Peloquin SJ, Gabert AC, Ortiz R (1996) Nature of "pollinator" effect in potato haploid production. Ann Bot 77:539–542
Perl A, Aviv D, Galun E (1990) Protoplast-fusion-derived CMS potato cybrids: potential seed parents for hybrid, true-potato-seeds. J Heredity 81:438–442
Perlak F, Stone TB, Muskopf YM, Petersen LJ, Parker GB, McPherson SA, Wyman J, Love S, Reed G, Biever D, Fischhoff DA (1993) Genetically improved potatoes: protection from damage by Colorado potato beetles. Plant Mol Biol 22:313–321
Plaisted RL (1987) Advances and limitations in the utilization of Neotuberosum in potato breeding. In: Jellis GJ, Richardson DE (eds) The production of new potato varieties. Cambridge University Press, Cambridge, pp 186–196
Ponds S, Cameron H (2003) Artificial seeds. In: Thomas B, Murphy DJ, Murray BG (eds) Encyclopedia of applied plant sciences, vol 3. Elseveir, pp 1379–1388
Powell W, Uhrig H (1987) Anther culture of *Solanum tuberosum* genotypes. Plant Cell Tissue Organ Cult 11:13–24

Pretova A, Dedicova B (1992) Somatic embryogenesis in *Solanum tuberosum* L. cv. Désirée from unripe zygotic embryos. J Plant Physiol 139:539–542

Pretova A, Obert B. (2006) Some aspects of embryo development in vitro. Proc. Vth IS on In Vitro Culture and Hort Breeding. Fári, M.G., I. Holb, and Gy. D. Bisztray (eds). Acta Horticulture 725: 83–88.

Punja ZK (2001) Genetic engineering of plants to enhance resistance to fungal pathogens: a review of progress and future prospects. Canadian J Plant Pathol 23:216–235

Reinert J (1958) Morphogenese und ihre knotrolle an gewebekulturen aus carotten. Naturwiss 45:344–345

Reynolds T (1986) Somatic embryogenesis and organogenesis from callus cultures of *Solanum carolinense*. Am J Botany 73:914–918

Rokka VM, Ishimaru CA, Lapitan NLV, Pehu E (1998) Production of androgenic dihaploid lines of the disomic tetraploid potato species *Solanum acaule* ssp. *acaule*. Plant Cell Rep 18:89–93

Romer S, Lubeck J, Kauder F, Steiger S, Adamat C, Sandmann G (2002) Genetic engineering of a zeaxanthin-rich potato by antisense inactivation and co-suppression of carotenoid epoxidation. Metabol Eng 4:263–272

Sabeti M, Zarghami R, Ebrahim ZM (2013) Effects of explants and growth regulators on callogenesis and somatic embryogenesis of Agria potato cultivar. Int J AgriSci 3:213–221

Santelia D, Vincenzetti V, Azzarello E, Bovet L, Fukao Y, Duchtig P (2005) MDR-like ABC transporter AtPGP4 is involved in auxin-mediated lateral root and root hair development. FEBS Lett 579:5399–5406

Schafer-Menuhr A, Mix-Wagner G, Vorlop KD (2003) Regeneration of plants from cell suspensions cultures and encapsulated cell suspension cultures of *Solanum tuberosum* L. Clarissa. Landbauforsch Volk 53:53–59

Schmidt EDL, Guzzo F, Toonen MAJ, de Vries SC (1997) A leucine-rich repeat containing receptor-like kinase marks somatic plant cells competent to form embryos. Development 124:2049–2062

Seabrook JEA, Douglass LK (2001) Somatic embryogenesis on various potato tissues from a range of genotypes and ploidy levels. Plant Cell Rep 20:175–182

Seabrook JEA, Douglass LK, Tai GCC (2001) Segregation for somatic embryogenesis on stem-internode explants from potato seedlings. Plant Cell Tissue Organ Cult 65:69–73

Sharma SK (2006) The development of an efficient somatic embryogenesis system for the production of synthetic seed in potato. PhD Thesis, University of Dundee, Dundee, United Kingdom

Sharma SK, Millam S (2004) Somatic embryogenesis in *Solanum tuberosum* L.: a histological examination of key developmental stages. Plant Cell Rep 23:115–119

Sharma SK, Bryan GJ, Winfield MO, Millam S (2007) Stability of potato (*Solanum tuberosum* L.) plants regenerated via somatic embryos, axillary bud proliferated shoots, microtubers and true potato seeds: a comparative phenotypic, cytogenetic and molecular assessment. Planta 226:1449–1458

Sharma SK, Millam S, Hedley PE, McNicol J, Bryan GJ (2008a) Molecular regulation of somatic embryogenesis in potato: an auxin led perspective. Plant Mol Biol 68:185–201

Sharma SK, Millam S, Hein I, Bryan G (2008b) Cloning and molecular characterisation of a potato SERK gene transcriptionally induced during initiation of somatic embryogenesis. Planta 228:319–330

Sharma S, Sarkar D, Pandey SK (2010) Phenotypic characterization and nuclear microsatellite analysis reveal genomic changes and rearrangements underlying androgenesis in tetraploid potatoes (*Solanum tuberosum* L.). Euphytica 171:313–326

Sharp WR, Sondahl MR, Caldas LS, Maraffa SB (1980) The physiology of in vitro asexual embryogenesis. Hortic Rev 2:268–310

Sharp WR, Evans DA, Sondahl MR (1982) Application of somatic embryogenesis to crop improvement. In Fujiwara A (ed) Plant tissue culture. Maruzen, Japan, pp 759–762

Shepard JF (1981) Protoplasts as sources of disease resistance in plants. Annl Rev Phytopathol 19:145–155

Shepard JF, Bidney D, Shahin E (1980) Potato protoplasts in crop improvement. Science 208:17–24

Simmonds NW (1969) Prospects of potato improvement. Scott Plant Breed Stn Annl Rep 48:18–38
Smith MAL (1995) Machine vision analysis of plant cells and somatic embryos. In: Bajaj YPS (ed) Biotechnology in agriculture and forestry (vol 30). Somatic embryogenesis and synthetic seed. Springer, Berlin, pp 87–101
Song J, Bradeen JM, Naess SK, Raasch JA, Wielgus SM, Haberlach GT, Liu J, Kuang H, Austin-Phillips S, Buell CR, Helgeson JP, Jiang J (2003) Gene RB cloned from *Solanum bulbocastanum* confers broad spectrum resistance to potato late blight. Proc Natl Acad Sci U S A 100:9128–9133
Sopory SK, Jacobsen E, Wenzel G (1978) Production of monohaploid embryoids and plantlets in cultured anthers of *Solanum tuberosum*. Plant Sci Lett 12:47–54
SreeRamulu K, Dijkhuis P, Roest S (1983) Phenotypic variation and ploidy level of plants regenerated from protoplasts of tetraploid potato *Solanum tuberosum* L. cv. Bintje. Theor Appl Genet 65:329–388
Steward FC (1958) Growth and development of cultivated cells. III. Interpretations of the growth from free cell to carrot plant. Am J Bot 45:709–713
Suprasanna P, Bapat VA (2005) Differential gene expression during somatic embryogenesis. Plant cell monogram 2. Springer, Berlin
Tai GCC (2005) Haploids in the improvement of Solanaceous species. In Palmer CE, WA Keller, KJ Kasha (eds) Biotechnology in agriculture and forestry (vol 56). Springer, Berlin, pp 173–190.
Tarn TR, Tai TCC (1983) Tuberosum X *Tuberosum* and *Tuberosum* X *Andigena* potato hybrids: comparisons of families and parents, and breeding strategies for Andigena potatoes in long-day temperate environments. Theor Appl Genet 66:87–91
Terasaka K, Blakeslee JJ, Titapiwatanakun B, Peer WA, Bandyopadhyay A, Makam SN (2005) PGP4, an ATP binding cassette P-glycoprotein, catalyzes auxin transport in *Arabidopsis thaliana* roots. Plant Cell 17:2922–2939
Terzi M, LoSchiavo F (1990) Somatic embryogenesis. In: Bhojwani SS (ed) Plant cell culture: applications and limitations. Elsevier, Amsterdam, pp 54–66
The European Cultivated Potato Database (2009) Cv. Shelagh. http://www.europotato.org/display_description.php?variety_name=Shelagh. Accessed May 2013
Thompson AJ, Gunn RE, Jellis GI, Boulton RE, Lacey CND (1986) The evaluation of potato somaclones. In Semal J (ed) Somaclonal variations and crop improvement. Martinus Nijhoff, Dordrecht, pp 236–243
Tisserat B, Esan EB, Murashige T (1979) Somatic embryogenesis in angiosperms. Hortic Rev 1:1–78
Trabelsi S, Gargouri-Bouzid R, Vedel F, Nato A, Lakhoua L, Drira N (2005) Somatic hybrids between potato *Solanum tuberosum* and wild species *Solanum vernei* exhibit a recombination in the plastome. Plant Cell Tissue Organ Cult 83:1–11
Tulecke W (1987) Somatic embryogenesis in woody perennials. In Bongs JM, Durzan DJ (eds) Cell and tissue culture in forestry. Martinus Nijhoff Publication, Dordrecht, pp 61–91
Uhrig H, Salamini F (1987) Dihaploid plant production from 4X-genotypes of potato by the use of efficient anther plants producing tetraploid strains (4X EAPP-clones)—proposal of a breeding-methodology. Zeitgeschichte Pflanzenzücht 98:228–235
Uzogara SG (2000) The impact of genetic modification of human foods in the 21st century: a review. Biotechnol Adv 18:179–206
Van der Vossen EAG, Sikkema A, Hekkert BL, Gros J, Stevens P, Muskens M, Wouters D, Pereira A, Stiekema W, Allefs S (2003) An ancient R gene from the wild potato species *Solanum bulbocastanum* confers broad-spectrum resistance to *Phytophthora infestans* in cultivated potato and tomato. Plant J 36:867–882
Van der Vossen EAG, Gros J, Sikkema A, Muskens M, Wouters D, Wolters P, Pereira A, Allefs S (2005) The *Rpi-blb2* gene from *Solanum bulbocastanum* is an Mi-1 homolog conferring broad-spectrum late blight resistance in potato. Plant J 44:208–222
Vargas ET, De Garcia E, Oropeza M (2005) Somatic embryogenesis in *Solanum tuberosum* from cell suspension cultures: histological analysis and extracellular protein patterns. J Plant Physiol 162:449–456

Vargas TE, Xena N, Vidal MC, Oropeza M, De Garcia E (2008) Genetic stability of *Solanum tuberosum* L. cv. Desiree plantlets obtained from embryogenic cell suspension cultures. Interciencia 33:213–218

Visser RGF (2009) Future prospects for potato genome technology. 7th World Potato Congress Proceedings. Christchurch, New Zealand. http://www.potatocongress.org/wpc2009.cfm. Accessed May 2013

Walker RI (1955) Cytological and embryological studies in Solanum, section Tuberarium. Bull Torrey Bot Club 82:87–101

Waris H (1957) Kukkakasvin kemiallisesti aikaansaatu muodonmuutos [A chemical-induced change in the morphogenesis of a flowering plant]. Suomalainen Tiedeakatemia. Esitelmät ja Pöytäkirjat 117–127

Wenzel G (1994) Tissue culture. In: Bradshaw E, GR Mackay (eds) Potato genetics. CAB International, Wallingford, pp 173–195

Williams EG, Maheswaran G (1986) Somatic embryogenesis factors influencing coordinated behaviour of cells as an embryogenic group. Ann Bot 57:443–462

Zeh M, Casazza AP, Kreft O, Roessner U, Bieberich K, Willmitzer L, Hoefgen R, Hesse H (2001) Antisense inhibition of threonine synthase leads to high methionine content intransgenic potato plants. Plant Physiol 127:792–802

Zimmerman JL (1993) Somatic embryogenesis: a model for early development in higher plants. Plant Cell 5:1411–1423

Chapter 9
Brassica Ogu-INRA Cytoplasmic Male Sterility: An Example of Successful Plant Somatic Fusion for Hybrid Seed Production

Georges Pelletier and Francoise Budar

9.1 Introduction

The concept of hybrid vigor has gradually emerged since the eighteenth century when J.G. Koelreuter observed individual plants resulting from interspecific crosses in various genera, such as *Nicotiana, Dianthus, Verbascum, Mirabilis,* and *Datura* (Zirckle 1952). These observations were confirmed a century later by Darwin (1876) in vegetables, and by Beal (1880) in maize. In 1914, Schull coined the term "heterosis" to define this physiological state of vigor of plants produced from crosses between parental lines that originate from sufficiently different genetic pools (Schull 1952). Relatively high heterosis has been reported as a general feature of F_1 hybrids in *Brassica* species: up to 80 % higher seed production compared to the mid-parent value in winter oilseed rape (*Brassica napus*; Lefort-Buson and Dattée 1982), 90 % in Indian mustard (*Brassica juncea*; Pradhan et al. 1993), and 60 % in summer turnip rape (*Brassica rapa*; Falk et al. 1994).

However, it is only relatively recently that breeders have been able to take advantage of heterosis for plant breeding. In the case of plants that reproduce by seed, obtaining large amounts of F_1 hybrid seed requires controlling cross-fertilization on a very large scale. The best way to do so is to grow large female stocks that will be pollinated by a male parent and harvest F_1 seed from these stocks. Female stocks can be obtained mechanically in monoecious species and sometimes by chemical means in hermaphrodite species. However, the most efficient system involves maternally determined male sterility. Jones and Clarke described the first male-sterility system used for the production of hybrid varieties in onion (*Allium cepa*; Jones and Clarke 1943). In the following decades, similar systems were developed in a wide range of species such as sorghum (*Sorghum bicolor*), sugar beet (*Beta vulgaris*), corn (*Zea mays*), sunflower (*Helianthus annuus*), rice (*Oryza sativa*), carrot (*Daucus carota*), etc.

G. Pelletier (✉) · F. Budar
Institut Jean-Pierre Bourgin, INRA-AgroParisTech UMR 1318, Versailles, France
e-mail: Georges.Pelletier@versailles.inra.fr

X.-Q. Li et al. (eds.), *Somatic Genome Manipulation*,
DOI 10.1007/978-1-4939-2389-2_9

Studies on heterosis in *Brassica* have stimulated research on male-sterility systems, and they have greatly benefited from a wide range of biotechnological approaches such as interspecific/intergeneric crosses, in vitro culture, somatic hybridization, and genetic engineering. Only a few cases of spontaneous cytoplasmic male sterility (CMS) have been described in *Brassica* species: in *B. rapa* (Chinese cabbage; Okhawa and Shiga 1981) and in *B. napus* (rapeseed) with the "Polima" (*pol*) system (Fu 1981) and the "Shaan 2A" (*nap*) system (Li 1986 communication in Chinese cited in Fu and Yang 1995).

The first CMS system described in an open-pollinated radish (*Raphanus sativus*) cultivar (Ogura 1968) has since been reported to be widespread in wild radish populations in Japan (Yamagishi and Terachi 1994a, 1996) and in *Raphanus raphanistrum* in Asia as well as in Europe (Yamagishi and Terachi 1997; Giancola et al. 2007). This CMS therefore predates the divergence between the two radish species.

9.2 Somatic Hybridization in Brassicaceae for the Manipulation of Sex

9.2.1 Transfer of the Ogura Radish Sterility-inducing Cytoplasm into Brassica

In 1974, Bannerot et al. introduced cytoplasmic male-sterile radish cytoplasm into *Brassica oleracea* (cabbage) using interspecific crosses and in vitro embryo culture (Bannerot et al. 1974; Yamagishi and Terachi 1994a, 1996), followed by several backcrosses with *B. oleracea* as the male parent until the *B. oleracea* genome (nine chromosomes) was stabilized in the Ogura radish cytoplasm (Fig. 9.1). Unfortunately, the resulting male-sterile plants with a complete *Brassica* genome in an Ogura radish cytoplasmic background showed chlorophyll deficiency, which was particularly expressed at low temperatures, and therefore rendered this combination unsuitable for commercial hybrid variety production (Fig. 9.2). Moreover, *Brassica* plants possessing this cytoplasm had underdeveloped nectaries, with reduced nectar production and malformed ovaries and pods (Fig. 9.3). Therefore, their flowers were not attractive to honeybees, the natural pollinators needed to ensure hybrid seed production. Female fertility, in terms of seed set, of these initial Ogura CMS *B. oleracea* plants was therefore reduced.

Furthermore, hybridization of these Ogura CMS *B. oleracea* plants with a *B. napus* male parent resulted, after cytoplasm transfer, in an Ogura CMS *B. napus* that showed the same range of defects (Pelletier et al. 1983).

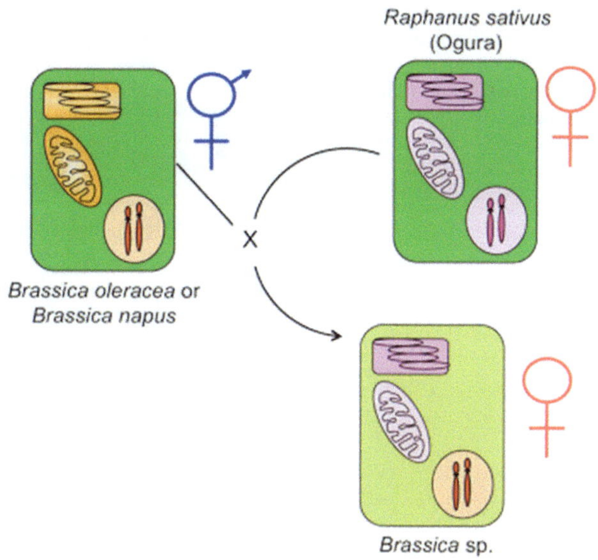

Fig. 9.1 An interspecific cross between an Ogura male-sterile radish with pollen of *Brassica* sp. followed by a series of backcrosses to *Brassica*, resulted in a male-sterile *Brassica* with Ogura radish plastids and mitochondria

Fig. 9.2 Chlorophyll deficiency (Y) observed in *Brassica* plants with (Ogura) radish cytoplasm, compared to normal plants (G)

9.2.2 *Somatic Hybridization for Eliminating Ogura Cytoplasm Defects*

The defects in chloroplast biogenesis and floral morphology in *Brassica* plants with the Ogura cytoplasm were assumed to result from negative interactions between the *Raphanus* cytoplasm and the *Brassica* nucleus. For eliminating them, protoplasts

Fig. 9.3 Malformed pods observed in *Brassica* plants with (Ogura) radish cytoplasm

from a normal *B. napus* line (Table 9.1) were fused with protoplasts from a CMS (Ogura radish cytoplasm) *B. napus,* and protoplasts from a normal *B. oleracea* line were fused with protoplasts from a CMS (Ogura radish cytoplasm) *B. oleracea* in two independent experiments (Pelletier et al. 1983, 1988).

In parallel, an atrazine-resistant *B. rapa* (previously known as *B. campestris*) biotype was discovered in Canada (Maltais and Bouchard 1978). This resistance is encoded by plastid DNA. Alloplasmic *B. napus* lines with *B. rapa* atrazine-resistant cytoplasm have been produced and have shown to be fully resistant to the herbicide (Beversdorf et al. 1980). Material carrying this plastid marker trait was instrumental in protoplast fusion experiments for selecting cybrids that had successfully received this particular *Brassica* type of chloroplasts.

9.2.3 *Experimental Procedure for Brassica Protoplast Isolation and Fusion*

To isolate protoplasts for cybrid production, seeds of parental lines were surface-sterilized and then germinated in vitro. Shoot tip cultures were generated and cultured on the same medium (medium A, see Table 9.2) under a 16-h photoperiod, 150 $\mu mol.m^{-2}.s^{-1}$ light intensity, and 24°C, and repeatedly subcultured at 3-4-week intervals. Protoplasts were then isolated from scarified leaf fragments from these axenic in vitro plants by overnight maceration in 0.2% cellulase R10 and 0.1%

Table 9.1 Recombination of cytoplasmic-associated traits in cybrids

Genotype	Cytoplasm	Nucleus	Parents of the fusion[a]	Associated traits
1	*B. napus*	*B. napus*	–	Male fertile, green leaves, normal flowers, atrazine susceptible
2	*B. oleracea*	*B. oleracea*	–	Male fertile, green leaves, normal flowers, atrazine susceptible
3	*B. rapa* atr^R	*B. napus*	–	Male fertile, green leaves, normal flowers, atrazine tolerant
4	*R. sativus*	*B. oleracea*	–	Male sterile, yellow leaves, abnormal flowers,[c] atrazine susceptible
5	Cybrid 58	*B. napus*	1+4[b]	Male sterile, green leaves, normal flowers, atrazine susceptible
6	Cybrid 17	*B. oleracea*	2+4	Male sterile, green leaves, normal flowers, atrazine susceptible
7	Cybrid 77	*B. napus*	3+4	Male sterile, green leaves, abnormal flowers, atrazine tolerant
8	Cybrid 13	*B. napus*	1+7	Unstable male sterile,[d] green leaves, normal flowers, atrazine susceptible
9	Cybrid 18	*B. napus*	1+7	Unstable male fertile,[d] green leaves, normal flowers, atrazine tolerant

R. sativus: Raphanus sativus, B. napus: Brassica napus, B. oleracea: Brassica oleracea, B. rapa atr^R: *Brassica rapa* atr^R

[a] When applicable

[b] Numbers refer to column 1

[c] Abnormal flowers means no nectar and or malformed ovaries (see Fig. 9.3)

[d] Cybrids 13 and 18 produced sexual progeny and somatic segregation of fully male-sterile and fully male-fertile plants, hence their phenotype was considered unstable

Table 9.2 The various media used in vitro for protoplast, tissue, and plant culture

Medium	Composition
A	MS medium, 10 g.l^{-1} glucose, 10 g.l^{-1} sucrose, 10 mg.l^{-1} gentamicin sulfate, 8 g.l^{-1} agar
B	Gamborg (B5) medium, 20 g.l^{-1} glucose, 70 g.l^{-1} mannitol, 1 mg.l^{-1} NAA, 1 mg.l^{-1} BA, O.25 mg.l^{-1} 2,4D, 10 mg.l^{-1} Tween 80
C	Half-strength B5 medium plus 200 mg.l^{-1} NH_4NO_3, 20 g.l^{-1} sucrose, 40 g.l^{-1} mannitol, 0.2 mg.l^{-1} NAA, 1 mg.l^{-1} BA
D	Half-strength B5 macro salts, MS microelements, 20 g.l^{-1} sucrose, 1 mg.l^{-1} 2,4D, 30 mg.l^{-1} adenine sulfate
E	MS medium, 10 g.l^{-1} sucrose, 20 g.l^{-1} mannitol, 1 mg.l^{-1} NAA, 1 mg.l^{-1} IPA, 0.02 mg.l^{-1} GA3, 8 g.l^{-1} agarose
F	MS medium, 10 g.l^{-1} sucrose, 0.1 mg.l^{-1} NAA, 0.5 mg.l^{-1} BA, 8 g.l^{-1} agarose
G	Half-strength MS medium, 10 g.l^{-1} glucose, 0.01 mg.l^{-1} NAA, 8 g.l^{-1} agar; pH adjusted to 5.8 before autoclaving

MS (Murashige and Skoog 1962), B5 (Gamborg et al. 1968), *NAA*: naphthalene acetic acid, *BA*: benzyl aminopurine, *2,4-D:* 2,4-dichlorophenoxyacetic acid, *IPA*: N^6 -(Δ^2 isopentenyl)-adenine, GA_3: gibberellic acid

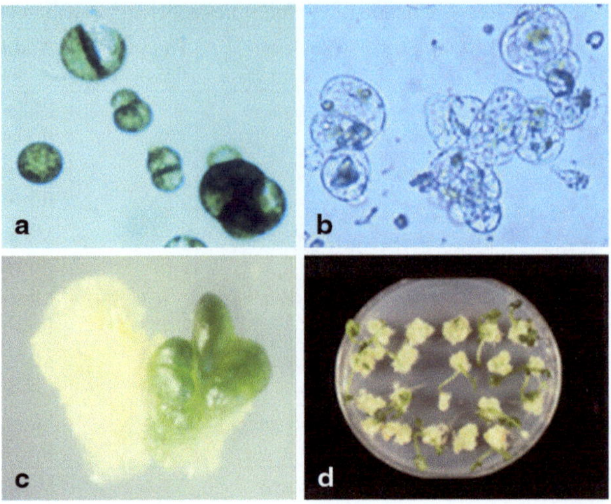

Fig. 9.4 **a** *Brassica* protoplast fusion, **b** cell colony formation, **c** derived callus with bud regeneration, and **d** bud growth into plantlets

pectolyase Y23 in culture medium B (Table 9.2). After washing in 2.5% KCl, 0.2% $CaCl_2.2H_2O$, protoplasts were fused (Fig. 9.4a) in drops of 25% polyethylene glycol 6000 in 3% $CaCl_2.2H_2O$ and subsequently further diluted directly in the petri dish with culture medium B to a cell density of $5 \times 10^4.ml^{-1}$.

A succession of different media was used (shown in Table 9.2) to stimulate the growth of protoplast-formed colonies and bud regeneration (Fig. 9.4b c, d). Protoplasts were grown in the dark for 3 days after fusion and then transferred to an illuminated growth chamber, then diluted on day 12 with an equal volume of medium C, and then with medium D on day 20. On day 30, colonies were plated onto medium E and emerging buds were placed onto medium F; after shoots developed, they were rooted on medium G.

No nucleus or cytoplasm inactivation is needed before protoplast fusion because the purpose was not to obtain a new combination of nuclei and cytoplasm but to allow genetic recombination between the two species' cytoplasms, as discovered previously in *Nicotiana tabacum* (Belliard et al. 1978, 1979).

9.2.4 Selection, Characterization, and Plastid Identification of Cybrids

Selection was carried out on regenerated plants by screening for new combinations of cytoplasmic traits that resulted from fusion events, as described in Table 9.1. More specifically, from the combination by protoplast fusion of *B. napus* or *B. oleracea* cytoplasms with Ogura *R. sativus* cytoplasm, cybrids were selected to have not only normal greening of leaves at low temperature but also well-formed male-sterile flowers. These cybrids were promising for developing male-sterile lines for F_1 hybrid seed production.

Raphanus and *Brassica* plastid DNAs are easily distinguished by their restriction patterns (Vedel et al. 1982; Yamagishi and Terachi 1997; Giancola et al. 2007) and the selected green, male-sterile plants contained only the *B. napus, B. rapa,* or *B. oleracea* chloroplast genomes (e.g., see cybrids 58, 17, 77 in Table 9.1). These results demonstrate that the chlorophyll deficiency in parental male-sterile *Brassica* lines was due to a detrimental interaction between *Raphanus* plastids and *Brassica* nuclei.

The plastid genomes of cybrids were identical to one parental genome in the case of *B. napus/B. rapa* cytoplasmic hybridization (e.g., cybrid 18 has *B. rapa* plastids and cybrid 13 has *B. napus* plastids; both were derived from the same fusion described in Table 9.1). Harboring one or the other parental plastid at random is a general feature of cybrid plants, indicating that the heteroplasmic state is very transitory. Nevertheless, in the cases where Ogura *Raphanus* plastids and *Brassica* plastids were combined by fusion in a *Brassica* nuclear background, the plastid type in recovered cybrids was always that of *Brassica,* indicating strong selection pressure against Ogura *Raphanus* plastids, which could result from the incompatible nucleo-plastid interaction observed in plant development (i.e., the dramatically reduced chlorophyll content).

The progeny of cybrids resulting from the combination of *B. napus* or *B. oleracea* cytoplasm with Ogura *R. sativus* cytoplasm shows complete stability of chloroplastic traits, i.e., normally green leaves and male sterility in subsequent sexual generations, indicating that the genetic determinants of the reverse traits (i.e., chlorophyll deficiency and male fertility) are lost after the cytoplasms are combined in protoplast fusion, and before the development of reproductive organs. Moreover, regarding flower morphology, abundant nectar production was recovered to some extent in cybrids 58 and 17 (and several others) but not in cybrid 77 (and several others), indicating that nectar production is not correlated with the plastid genome as is the case for leaf chlorophyll content.

Similar experiments on Ogura male sterility were performed by other laboratories in receipt of plant material obtained by Bannerot et al. in *B. napus* (Menczel et al. 1987; Jarl and Bornman 1988; Jarl et al. 1989;), *B. oleracea* (Kao et al. 1992; Walters et al. 1992; Walters and Earle 1993), and *B. juncea* (Kirti et al. 1995). When successful, these improved Ogura cytoplasms were introduced into other species such as *B. rapa* and *B. juncea* (Delourme et al. 1994) and vegetable *B. rapa* (Heath et al. 1994) via sexual crosses to produce male-sterile lines.

Protoplast fusions have also been used to transfer, or modify, other sterility-inducing cytoplasms into *Brassica,* such as the Kosena radish cytoplasm, subsequently shown to be a variant of the Ogura cytoplasm, in *B. napus* (Sakai and Imamura 1990, 1992), the "Anand" *B. rapa* cytoplasm in *B. oleracea* (Cardi and Earle 1997), the "polima" *B. napus* cytoplasm in *B. oleracea* (Yarrow et al. 1990), *Diplotaxis muralis* cytoplasm in *B. juncea* (Chatterjee et al. 1988), *Moricandia arvensis* cytoplasm in *B. juncea* (Prakash et al. 1998), and *Brassica tournefortii* cytoplasm in *B. napus* (Liu et al. 1995).

9.3 Molecular Analyses of Cybrid Mitochondrial Genomes

9.3.1 Mitochondrial Genomes of Cybrids

Restriction profiles of cybrid and parental mitochondrial DNAs have been compared (Chetrit et al. 1985). As observed in tobacco, each cybrid profile is unique and different from those of the parents (except cybrid 77, which is apparently identical to the Ogura radish mitochondrial profile). In these restriction profiles, there are some parental restriction fragments and some new ones, a signature of molecular exchanges and recombination between the two parental genomes.

The fact that cybrid mitochondrial genomes are recombined is valuable in two ways. First, the male-sterile determinant originating from the Ogura radish can be segregated from undesirable traits that likely result from a detrimental interaction between radish mitochondria and *Brassica* nuclei, such as low nectar production and abnormal flower development. Second, it opened up the possibility of associating a specific DNA sequence with the male-sterile trait, this sequence being present in sterile cybrids and absent or modified in fertile ones. The molecular characterization of genomic fragments originating from the radish mitochondrial genome that were linked to undesirable traits, and the identification of the male-sterility mitochondrial gene (see below) allowed the definition and patenting of the Ogu-INRA system in *Brassica* crops.

9.3.2 Identification of the Ogura CMS Gene

CMS-associated genes are most often unconserved mitochondrial open reading frames (ORFs) created by intragenomic recombination (Hanson 1991; Braun et al. 1992; Kubo et al. 2011). They have a chimeric structure in which parts of normal mitochondrial genes can be recognized (Hanson and Bentolila 2004). The Brassicaceae family may be the one that has benefitted the most from efforts to identify mitochondrial male-sterility genes, because at least five different CMS systems have been studied at the molecular level: "Ogura" (Bonhomme et al. 1991), *pol* (L'Homme and Brown 1993), *nap* (L'Homme et al. 1997), *tour* (Landgren et al. 1996), and *Mori* (Ashutosh et al. 2008).

The mitochondrial determinant for Ogura-derived male sterility was one of the first CMS genes identified. It was discovered by studying *Brassica* cybrids obtained via the protoplast fusion experiments listed in Table 9.1. The first set of fusion experiments gave *B. napus* and *B. oleracea* cybrid cytotypes, which were used in a second series of fusion experiments involving cybrids and parental lines ("back-fusion"; Pelletier et al. 1986). Cybrid 13, derived from such an experiment, provided the material for the molecular characterization of the sterility determinant. This cybrid was selected as male-sterile, but it repeatedly reverted to male fertility with constant minor variation in mitochondrial DNA restriction profiles. It was therefore possible to correlate a mitochondrial DNA region with the male-sterile trait and a DNA sequence (*orf138*) whose expression at the RNA level is strictly correlated with the sterile phenotype of *B. napus* and *B. oleracea* cybrids (Bonhomme et al. 1991, 1992). Expression of this mitochondrial gene was subsequently associated with the sterile phenotype of the original "Ogura" radish (Krishnasamy and Makaroff 1993). Interestingly, *orf125,* a gene very similar to *orf138,* has been identified in the mitochondrial genome of the Kosena radish and assumed to be responsible for male sterility in this cytoplasm (Iwabuchi et al. 1999). Furthermore, Kosena CMS is very likely a variant form of Ogura CMS, and *orf125* a sequence variant of *orf138* (Koizuka et al. 2000; Yamagishi and Terachi, 2001).

9.3.3 Contribution of New Recombined Genomic Arrangements to Our Understanding of Mitochondrial Gene Expression

orf138 is a chimeric gene. The end of its coding sequence and the 3′ flanking region are identical (70 nucleotides including the last 12 codons) to the 3′ untranslated region of *Arabidopsis thaliana orf557* (*ccb206*) that likely encodes a protein thought to be involved in cytochrome c biogenesis (Bellaoui et al. 1998). The *B. napus orf577* gene, homologous to the bacterial *ccl1* gene, is very similar to *A. thaliana orf557* and has in its 3′ untranslated region the same short and perfect homology to *orf138* (Menassa et al. 1997; Handa et al. 1996).

The *orf138* gene is co-transcribed with a gene called *orfB,* conserved among plant mitochondrial genomes, and which has been demonstrated to correspond to the subunit 8 of the ATP synthase complex (Gray et al. 1998; Sabar et al. 2003). The bi-cistronic structure of the locus is responsible for the stability of the CMS trait in the Ogura radish and in stable cybrids. Instability has been observed in cybrids harboring a second copy of the *orfB* gene, introduced through back-fusion to normal *B. napus* and derived from the *B. napus* mitochondrial genome, or in cybrids in which the *orf138* and *orfB* genes are no longer associated and the *orf138* gene is expressed as a monocistron, such as male-sterile plants derived from the unstable cybrid 18 (Bonhomme et al. 1991, 1992; Grelon et al. 1994; Bellaoui et al. 1998).

The unstable cybrid 13 not only helped identify the sterility gene but also shed light on the importance of posttranscriptional events such as degradation/stabilization of RNA in plant mitochondrial gene expression (Binder and Brennicke 2003). Indeed, in fertile revertants of cybrid 13 (13F plants), the *orf138* gene is present but found upstream of the *atpA* gene instead of the *orfB* gene, due to recombination that deleted the Ogura radish *orfB* gene. As mentioned above, these plants carry a *B. napus orfB* gene, unlinked to *orf138* that compensated this loss. In this new genomic organization, the *orf138* gene is no longer expressed (no detectable level of mRNA or protein), although it is transcribed as efficiently as in the sterile cybrid 13 plants because the transcript is rapidly degraded due to the lack of the stabilizing sequences found in the *orfB* 3′ UTR (Bonhomme et al. 1992; Bellaoui et al. 1997).

In contrast, male-sterile plants derived from cybrid 18, which carry a monocistronic *orf138* with very efficient stabilizing sequences in its 3′UTR, over-accumulate *orf138* mRNA and the ORF138 protein (Grelon et al. 1994; Bellaoui et al. 1997).

The *orf125* gene from Kosena radish is present in the same genomic organization as the *orf138* gene and is co-transcribed with *orfB* (Iwabuchi et al. 1999). Iwabuchi et al. (1999) also described a monocistronic *orf125* genomic organization in a sterile *B. napus* cybrid.

9.3.4 Deciphering the Mechanism of CMS in Brassica Cybrids

The ORF138 protein is bound to the inner mitochondrial membrane and forms oligomers (Grelon et al. 1994; Krishnasamy and Makaroff 1994; Duroc et al. 2005). It is present in all plant organs. In *B. napus* cybrids, its accumulation appears to follow the abundance of normal mitochondrial proteins in the different organs of the plant (Bellaoui et al. 1999). This result is not very surprising given that *orf138* expression is mediated by the promoter and posttranscriptional expression signals that drive the expression of the *orfB* gene, which encodes a subunit of the ATP synthase complex. However, as in many other CMS systems, it raises an intriguing question: Why is the phenotypic consequence of ORF138 accumulation restricted to pollen development?

In the tapetal cells of sterile anthers, mitochondria swell and lose their internal membrane structure at the tetrad stage (González-Melendi et al. 2008). This leads to premature degeneration of the tapetum and subsequent abortion of developing pollen. ORF138 forms an apparently homo-oligomeric complex in the inner membrane of mitochondria (Duroc et al. 2005, 2009). In contrast to what has been shown for G-CMS in beet (Ducos et al., 2001), sunflower CMS (Sabar et al. 2003), and WA-CMS in rice (Luo et al. 2013), ORF138 does not interfere with complexes of the electron transport chain (Duroc et al. 2009). Nevertheless, measurements of oxygen consumption on whole anthers and on mitochondria isolated from buds suggested that mitochondria are partially uncoupled in sterile reproductive organs, whereas no sign of decoupling is detected in vegetative tissues (Duroc et al. 2009).

9.4 The Restoration of Fertility in Ogura-related Systems

9.4.1 The Genetics of Restoration in Ogura and Ogu-INRA Plants

Restorer genes are generally dominant, except for the unique homozygous lethal restorer *rfl1* of maize, which was obtained by mutagenesis (Wen et al. 2003). Determinism is monogenic or oligogenic according to the male-sterility-inducing cytoplasm. In the case of the Ogura radish system, the genetics of restoration differs among crosses, suggesting that there are several redundant restoration loci in radish nuclear genotypes (Yamagishi and Terachi 1994b). In some cases, a single dominant restorer is sufficient to completely restore fertility of radish plants with the Ogura cytoplasm.

A restored *B. napus* has been obtained after crossing an alloplasmic rapeseed possessing the Ogura cytoplasm with a *Raphano-brassica* hybrid from an intergeneric cross, and subsequent chromosome doubling. The genetics of restoration in rapeseed carrying the unmodified Ogura cytoplasm is complex, involving several loci of which one is linked to a gene involved in petal color: white-flowered plants, a *Raphanus*-based character, were more fertile than yellow-flowered plants. Interestingly, the restoration of rapeseed cybrids with a corrected phenotype involves a unique locus, unlinked to petal color (Pelletier et al. 1986). The obtained restorer rapeseed genotypes show a unique radish introgression that carries the restoration locus, named *Rfo* (Delourme and Eber 1992; Delourme et al. 1998).

Two complementary genes were found to be necessary for restoring the Kosena radish CMS, whereas only one of them, named *Rfk1* or *Rfk,* is sufficient for the restoration of *B. napus* cybrids (Koizuka et al. 2000). This gene was subsequently found to be identical to *Rfo* (see below).

9.4.2 *Effect of Restorer Nuclear Background on the Expression of the Ogura CMS Mitochondrial Gene*

In the vast majority of CMS systems, fertility restoration acts via the inhibition of male-sterility gene expression, most often at the RNA level, and sometimes at the protein level (Budar and Pelletier 2001; Budar et al. 2003). The influence of a restoration gene on the expression of a mitochondrial chimeric gene is sometimes the only evidence of the implication of this mitochondrial gene in the sterile phenotype (Kubo et al. 2011).

Nuclear restorers have a dramatic effect on the accumulation of the ORF138 protein in buds and leaves in Ogura radish (Krishnasamy and Makaroff 1994). In Ogu-INRA *B. napus,* restoration is characterized by the decrease of ORF138 accumulation in anthers and its complete depletion from the tapetum (Bellaoui et al. 1999; Uyttewaal et al. 2008). In both species, the accumulation of *orf138* mRNA is not affected by the nuclear restorer gene indicating that the restorer acts either on mRNA translation efficiency or on ORF138 protein stability (Krishnasamy and Makaroff 1994; Bellaoui et al. 1999; Uyttewaal et al. 2008).

The *orf125* gene has an expression profile in Kosena CMS plants that is similar to that of the *orf138* gene in Ogura radish or Ogu-INRA Brassicas (Iwabuchi et al. 1999). In addition, as in the Ogura radish CMS or the Ogu-INRA *Brassica* cybrids, the restorer nuclear background induces a strong decrease in ORF125 protein accumulation, but does not affect the accumulation of *orf125* mRNA, either in Kosena radish or *B. napus* cybrids (Iwabuchi et al. 1999).

9.4.3 Identification of the Male-Fertility Restorer Gene

Rfo and *Rfk,* which are in fact the same gene, were independently identified by fine genetic mapping in radish (Brown et al. 2003; Desloire et al. 2003; Koizuka et al. 2003). This restorer gene encodes a pentatricopeptide repeat (PPR) protein which has been further characterized in a restored *B. napus* cybrid (Uyttewaal et al. 2008). The restoring protein interacts with *orf138* mRNA, and probably impairs its translation, thus preventing the deleterious effects of the ORF138 protein on tapetal cells.

9.5 The Use of the Ogu-INRA Hybridization System for Seed Production in Brassica Crops

Cabbage and rapeseed cybrids with the Ogura sterility determinant and corrected phenotypes have been evaluated in regard to their seed production and attractiveness to pollinators. These studies helped select two to three recombined cytoplasms in each crop, which were then provided to breeders in the early 1990s.

In cabbage, only the vegetative parts of the plant are of agricultural interest; therefore, there was no need to introduce the restorer gene in hybrids. The first cabbage F_1 cultivars using the Ogu-INRA system were registered in the official French seed catalogue in 1993. They display ~40% of heterosis, giving yields of around 200 tons per hectare (sauerkraut cabbage). In 1999, the French catalogue listed 65 F_1 cultivars of different cabbage types that have been obtained using the Ogu-INRA CMS system.

The emergence of restored rapeseed hybrids was impaired by undesirable traits linked to the introgression of the radish *Rfo* gene, such as high glucosinolates, poor female fertility, and poor genetic stability of the restorer rapeseed genotypes. Nevertheless, cultivars containing a mixture of female hybrid plants and fertile plants from one of the parental lines in a ~85:15% ratio (line–hybrid composite, LHC) led to the registration of the first rapeseed hybrid cultivar (Synergy) in 1994. However, this type of cultivar may induce losses in yield when temperatures are unfavorable during the pollination period, which limits pollinator visits.

Considering the opposition of European authorities to transgenic rapeseed and the very high cost of complying with regulations, it is not possible to use restored genotypes in which only the radish *Rfo* gene has been introduced via transformation in Europe, even though this would eliminate all the developmental defects. This is an example of regulation inconsistencies and society's general attitude towards genetically modified organisms because the same gene (together with many other unknown genes) is already present in restored hybrid rapeseed cultivars with introgressed radish genes for male-fertility restoration.

An intense breeding effort to produce restorer lines has led to an improved R113 restorer line, followed by R2000 (Primard-Brisset et al. 2005). The first restored hybrid cultivar was authorized in 1999 in France. In 2002, 14% of rapeseed crops were planted with restored hybrids versus 8 % with LHC. The improvement of restorers has completely supplanted LHC cultivars. The introduction of a dwarf mutation in one of the hybrid parents allowed the development of half-dwarf hybrid cultivars in the mid-2000s.

Plant material has been distributed without exclusivity to a large number of seed companies in the world and license fees represent only a small percentage of the commercial price of seed. This material is widely used in *Brassica* vegetables, but much more in rapeseed. According to our calculations, in 2009, 5.4×10^6 ha were planted with varieties containing the Ogu-INRA cytoplasm worldwide. Considering that the total rapeseed crop surface area of the main continents that use this system (Europe, North America, Australia) covers $\sim 15 \times 10^6$ ha, we estimate that the cytoplasms resulting from somatic hybridization are found in approximately one third of the rapeseed crops grown in these countries.

9.6 Conclusion

The main way of creating new CMS systems is to associate the nucleus of one species with the cytoplasm of a different origin, e.g., another species. This new association can be achieved either by crosses and backcrosses with the nuclear donor or by somatic cell genetics (cybrid production). The first method leads to several sources of CMS, which remain generally underexploited due to the multiple interactions between the cytoplasmic genes and the (foreign) nucleus with negative consequences on plant productivity. Cybrid production can result in improved CMS systems, due to plastid exchange and mitochondrial recombination.

More generally, in the absence of a method for transforming mitochondria in higher plants, the only possibility for manipulating the mitochondrial genome lies in cybrid production by protoplast fusion, which can be considered to be "cytoplasm breeding." This requires expertise in plant protoplast culture, which is becoming regrettably increasingly rare in today's plant biology laboratories.

Outside the family of Brassicaceae, examples of cybrid production for agricultural purposes can be found in tobacco (Horlow et al. 1993), rice (Akagi et al. 1995), or *Citrus* species (Bassene et al. 2008), illustrating how cellular and molecular biology can contribute significantly to plant breeding.

Acknowledgments We are very grateful to our colleagues at the INRA Rennes Center for their very impressive work on rapeseed breeding and for our long-running and very fruitful collaboration, particularly with Michel Renard and Régine Delourme. We acknowledge the excellent contributions of all colleagues and students involved in Versailles and Rennes INRA Centers who have worked for over 30 years on the Ogu-INRA CMS in Brassica. Our research on this CMS system has been partly funded by PROMOSOL, CETIOM, and Genoplante. We are also very grateful to Céline Duroc, joint head of the CETIOM technical center, for her help in retrieving data on the use of the Ogu-INRA system in agriculture.

References

Akagi H, Taguchi T, Fujimura T (1995) Stable inheritance and expression of the CMS traits introduced by asymmetric protoplast fusion. Theor Appl Genet 91:563–567. doi:10.1007/BF00223280

Ashutosh KP, Dinesh Kumar V, Sharma PC, Prakash S, Bhat SR (2008) A novel *orf108* co-transcribed with the atpA gene is associated with cytoplasmic male sterility in *Brassica juncea* carrying *Moricandia arvensis* cytoplasm. Plant Cell Physiol 49:284–289. doi:10.1093/pcp/pcm182

Bannerot H, Boulidard L, Cauderon Y, Tempé J (1974) Transfer of cytoplasmic male sterility from *Raphanus sativus* to *Brassica oleracea*. In: Proceedings of Eucarpia Meeting Cruciferae: vegetable crops section. Invergrownie: Scottish Hoticultural Research Intitute: 52–54

Bassene J, Berti L, Carcouet E, Dhuique-Mayer C, Fanciullino AL, Bouffin J, Ollitrault P, Froelicher Y (2008) Influence of mitochondria origin on fruit quality in a citrus cybrid. J Agric Food Chem 56:8635–8640

Beal WJ (1880) Indian corn. Report Michigan State Board of Agric ulture. 19:279–289.

Bellaoui M, Pelletier G, Budar F (1997) The steady-state level of mRNA from the Ogura cytoplasmic male sterility locus in *Brassica* cybrids is determined post-transcriptionally by its 3′ region. EMBO J 16:5057–5068. doi:10.1093/emboj/16.16.5057

Bellaoui M, Martin-Canadell A, Pelletier G, Budar F (1998) Low-copy-number molecules are produced by recombination, actively maintained and can be amplified in the mitochondrial genome of Brassicaceae: relationship to reversion of the male sterile phenotype in some cybrids. Mol Gen Genet 257:177–185

Bellaoui M, Grelon M, Pelletier G, Budar F (1999) The restorer Rfo gene acts post-translationally on the stability of the ORF138 Ogura CMS-associated protein in reproductive tissues of rapeseed cybrids. Plant Mol Biol 40:893–902

Belliard G, Pelletier G, Quetier F, Vedel F (1978) Morphological characteristics and chloroplast DNA distribution in different cytoplasmic parasexual hybrids of *Nicotiana tabacum*. Mol Gen Genet 165:231–237

Belliard G, Vedel F, Pelletier G (1979) Mitochondrial recombination in cytoplasmic hybrids of *Nicotiana tabacum* by protoplast fusion. Nature 281:401–403. doi:10.1038/281401a0

Beversdorf WD, Weiss-Lerman J, Erickson LR, Souza-Machado V (1980) Transfer of cytoplasmically inherited triazine resistance from birds-rape to cultivated oil seed rape (*B. campestris* and *B. napus*). Can J Genet Cytol 22:167–172

Binder S, Brennicke A (2003) Gene expression in plant mitochondria: transcriptional and post-transcriptional control. Philos Trans R Soc Lond, B, Biol Sci 358:181–8; discussion 188–9. doi:10.1098/rstb.2002.1179

Bonhomme S, Budar F, Férault M, Pelletier G (1991) A 2.5 kb NcoI fragment of Ogura radish mitochondrial DNA is correlated with cytoplasmic male-sterility in *Brassica* cybrids. Curr Genet 19:121–127. doi:10.1007/BF00326293

Bonhomme S, Budar F, Lancelin D, Small I, Defrance MC, Pelletier G (1992) Sequence and transcript analysis of the Nco2.5 Ogura-specific fragment correlated with cytoplasmic male sterility in *Brassica* cybrids. Mol Gen Genet 235:340–348

Braun CJ, Brown GG, Levings CSI (1992) Cytoplasmic male sterility. In: Hermann R (ed) Cell organelles. Plant Gene Research. Springer, New-York, pp 219–245

Brown GG, Formanová N, Jin H, Wargachuk R, Dendy C, Patil P, Laforest M, Zhang J, Cheung WY, Landry BS (2003) The radish Rfo restorer gene of Ogura cytoplasmic male sterility encodes a protein with multiple pentatricopeptide repeats. Plant J 35:262–272

Budar F, Pelletier G (2001) Male sterility in plants: occurrence, determinism, significance and use. C R Acad Sci III Sci Vie 324:543–550

Budar F, Touzet P, De Paepe R (2003) The nucleo-mitochondrial conflict in cytoplasmic male sterilities revisited. Genetica 117:3–16

Cardi T, Earle ED (1997) Production of new CMS *Brassica oleracea* by transfer of "Anand" cytoplasm from *B. rapa* through protoplast fusion. Theor Appl Genet 94:204–212

Chatterjee G, Sikdar SR, Das S, Sen SK (1988) Intergeneric somatic hybrid production through protoplast fusion between *Brassica juncea* and *Diplotaxis muralis* Theor Appl Genet 76:915–922

Chetrit P, Mathieu C, Vedel F, Pelletier G, Primard C (1985) Mitochondrial DNA polymorphism induced by protoplast fusion in Cruciferae. Theor Appl Genet 69:361–366

Darwin C (1876) The effects of cross and self fertilisation in the vegetable kingdom. John Murray, London

Delourme R, Eber F (1992) Linkage between an isozyme marker and a restorer gene in radish cytoplasmic male sterility of rapeseed (*Brassica napus* L.). Theor Appl Genet 85:222–228. doi: 10.1007/BF00222863

Delourme R, Eber F, Renard M (1994) Transfer of radish cytoplasmic male sterility from *Brassica napus* to *B. juncea* and *B. rapa*. Eucarpia Cruciferae Newsletter 16:79

Delourme R, Foisset N, Horvais R, Barret P, Champagne G, Cheung WY, Landry BS, Renard M (1998) Characterisation of the radish introgression carrying the Rfo restorer gene for the Ogu-INRA cytoplasmic male sterility in rapeseed (*Brassica napus* L.). Theor Appl Genet 97:129–134

Desloire S, Gherbi H, Laloui W, Marhadour S, Clouet V, Cattolico L, Falentin C, Giancola S, Renard M, Budar F, Small I, Caboche M, Delourme R, Bendahmane A (2003) Identification of the fertility restoration locus, Rfo, in radish, as a member of the pentatricopeptide-repeat protein family. EMBO Rep 4:588–594. doi:10.1038/sj.embor.embor848

Ducos E, Touzet P, Boutry M (2001) The male sterile G cytoplasm of wild beet displays modified mitochondrial respiratory complexes. Plant J 26:171–180

Duroc Y, Gaillard C, Hiard S, Defrance MC, Pelletier G, Budar F (2005) Biochemical and functional characterization of ORF138, a mitochondrial protein responsible for Ogura cytoplasmic male sterility in Brassiceae. Biochimie 87:1089–1100. doi:10.1016/j.biochi.2005.05.009

Duroc Y, Hiard S, Vrielynck N, Ragu S, Budar F (2009) The Ogura sterility-inducing protein forms a large complex without interfering with the oxidative phosphorylation components in rapeseed mitochondria. Plant Mol Biol 70:123–137. doi:10.1007/s11103-009-9461-6

Falk KC, Rakow G, Downey RK, Spurr DT (1994) Performance of inter-cultivar summer turnip rape hybrids in Saskatchewan. Can J Plant Sci 74:441–445

Fu D (1981) Production and research on rapeseed in the Peoples Republic in China. Eucarpia Cruciferae Newslett 6:6–7

Fu D, Yang GS (1995) Rapeseed heterosis breeding in China. In: 9th International Rapeseed Congress. Cambridge, pp 119–121

Gamborg OL, Miller RA, Ojima K (1968) Nutrient requirements of suspension cultures of soybean root cells. Exptl Cell Research 50:151–158

Giancola S, Rao Y, Chaillou S, Hiard S, Martin-Canadell A, Pelletier G, Budar F (2007) Cytoplasmic suppression of Ogura cytoplasmic male sterility in European natural populations of. *Raphanus raphanistrum* Theor Appl Genet 114:1333–1343. doi:10.1007/s00122-007-0520-6

González-Melendi P, Uyttewaal M, Morcillo CN, Hernández Mora JR, Fajardo S, Budar F, Lucas MM (2008) A light and electron microscopy analysis of the events leading to male sterility in Ogu-INRA CMS of rapeseed (*Brassica napus*). J Exp Bot 59:827–838. doi:10.1093/jxb/erm365

Gray MW, Lang BF, Cedergren R, Golding GB, Lemieux C, Sankoff D, Turmel M, Brossard N, Delage E, Littlejohn TG, Plante I, Rioux P, Saint-Louis D, Zhu Y, Burger G (1998) Genome structure and gene content in protist mitochondrial DNAs. Nucleic Acids Res 26:865–878

Grelon M, Budar F, Bonhomme S, Pelletier G (1994) Ogura cytoplasmic male-sterility (CMS)-associated orf138 is translated into a mitochondrial membrane polypeptide in male-sterile *Brassica* cybrids. Mol Gen Genet 243:540–547

Handa H, Bonnard G, Grienenberger JM (1996) The rapeseed mitochondrial gene encoding a homologue of the bacterial protein Ccl1 is divided into two independently transcribed reading frames. Mol Gen Genet 252:292–302

Hanson MR (1991) Plant mitochondrial mutations and male sterility. Annu Rev Genet 25:461–486. doi:10.1146/annurev.ge.25.120191.002333

Hanson MR, Bentolila S (2004) Interactions of mitochondrial and nuclear genes that affect male gametophyte development. Plant Cell 16 (Suppl):154–169. doi:10.1105/tpc.015966

Heath DW, Earle ED, Dickson MH (1994) Introgressing cold-tolerant Ogura cytoplasm from rapeseed into pak choi and Chinese cabbage. HortScience 29:202–203

Horlow C, Defrance MC, Pollien JM, Goujaud J, Delon R, Pelletier G (1993) Transfer of cytoplasmic male sterility by spontaneous androgenesis in tobacco (*Nicotiana tabacum* L.). Euphytica 66:45–53

Iwabuchi M, Koizuka N, Fujimoto H, Sakai T (1999) Identification and expression of the Kosena radish (*Raphanus sativus* cv. Kosena) homologue of the ogura radish CMS-associated gene, orf138. Plant Mol Biol 39:183–188

Jarl CI, Bornman CH (1988) Correction of chlorophyll-defective, male-sterile winter oilseed rape (*Brassica napus*) through organelle exchange: phenotypic evaluation of progeny. Hereditas 108:97–102

Jarl CI, Grinsven MQJM, Mark F (1989) Correction of chlorophyll-defective male-sterile winter oilseed rape (*Brassica napus*) through organelle exchange: molecular analysis of the cytoplasm of parental lines and corrected progeny. Theor Appl Genet 77:135–141. doi:10.1007/BF00292328

Jones HA, Clarke AE (1943) Inheritance of male sterility in the onion and the production of hybrid seed. Proc Amer Soc Hort Sci 43:189–194

Kao HM, Keller WA, Gleddie S, Brown GG (1992) Synthesis of *Brassica oleracea/Brassica napus* somatic hybrid plants with novel organeile DNA compositions. Theor Appl Genet 83:313–320

Kirti PB, Banga SS, Prakash S, Chopra VL (1995) Transfer of Ogu cytoplasmic male sterility to *Brassica juncea* and improvement of the male sterile line through somatic cell fusion. Theor Appl Genet 91:517–521

Koizuka N, Imai R, Iwabuchi M, Sakai T, Imamura J (2000) Genetic analysis of fertility restoration and accumulation of ORF125 mitochondrial protein in the Kosena radish (*Raphanus sativus* cv. Kosena) and a *Brassica napus* restorer line. Theor Appl Genet 100:949–955

Koizuka N, Imai R, Fujimoto H, Hayakawa T, Kimura Y, Kohno-Murase J, Sakai T, Kawasaki S, Imamura J (2003) Genetic characterization of a pentatricopeptide repeat protein gene, orf687, that restores fertility in the cytoplasmic male-sterile Kosena radish. Plant J 34:407–415

Krishnasamy S, Makaroff C (1993) Characterization of the radish mitochondrial orfB locus: possible relationship with male sterility in Ogura radish. Curr Genet 24:156–163

Krishnasamy S, Makaroff CA (1994) Organ-specific reduction in the abundance of a mitochondrial protein accompanies fertility restoration in cytoplasmic male-sterile radish. Plant Mol Biol 26:935–946

Kubo T, Kitazaki K, Matsunaga M, Itchoda N, Estiati A, Mikami T (2011) Male sterility-inducing mitochondrial genomes: how do they differ? Crit Rev Plant Sci 30:378–400

Landgren M, Zetterstrand M, Sundberg E, Glimelius K (1996) Alloplasmic male-sterile *Brassica* lines containing *B. tournefortii* mitochondria express an ORF 3′ of the atp6 gene and a 32 kDa protein. Plant Mol Biol 32:879–890

Lefort-Buson M, Dattée Y (1982) Genetic study of some agronomic characters in winter oilseed rape (*Brassica napus* L.) I Heterosis II genetic parameters. Agronomie 2:315–322

L'Homme Y, Brown GG (1993) Organizational differences between cytoplasmic male sterile and male fertile *Brassica* mitochondrial genomes are confined to a single transposed locus. Nucleic Acids Res 21:1903–1909

L'Homme Y, Stahl RJ, Li X-Q, Hameed A, Brown GG (1997) *Brassica* nap cytoplasmic male sterility is associated with expression of a mtDNA region containing a chimeric gene similar to the pol CMS-associated orf 224 gene. Curr Genet 31:321–335

Liu JH, Dixelius C, Eriksson I, Glimelius K (1995) *Brassica napus*(+) *B. tournefortii*, a somatic hybrid containing traits of agronomic importance for rapeseed breeding. Plant Sci 109:75–86

Luo D, Xu H, Liu Z, Guo J, Li H, Chen L, Fang C, Zhang Q, Bai M, Yao N, Wu H, Wu H, Ji C, Zheng H, Chen Y, Ye S, Li X, Zhao X, Li R, Liu YG (2013) A detrimental mitochondrial-nuclear interaction causes cytoplasmic male sterility in rice. Nat Genet 45:573–577. doi:10.1038/ng.2570

Maltais B, Bouchard CJ (1978) Une moutarde des oiseaux (*B. rapa*) résistante à l'atrazine. Phytoprotection 59:117–119

Menassa R, El-Rouby N, Brown GG (1997) An open reading frame for a protein involved in cytochrome C biogenesis is split into two parts in *Brassica* mitochondria. Curr Genet 31:70–79

Menczel L, Morgan A, Brown S, Maliga P (1987) Fusion-mediated combination of Ogura-type cytoplasmic male sterility with *Brassica napus* plastids using X-irradiated CMS protoplasts. Plant Cell Rep 6:98–101

Murashige T, Skoog F (1962) A revised medium for rapid growth and bioassays with tobacco tissue culture. Physiol Plant 15:473–497

Ogura H (1968) Studies on the new male sterility in Japanese radish, with special references to utilization of this sterility towards the practical raising of hybrid seeds. Mem Fac Agric Kagoshima Univ 6:39–78

Okhawa Y, Shiga T (1981) Possibility of hybrid seed production by use of cytoplasmic male sterility in *Brassica campestris*, Chinese cabbage. In: 1st Symp. Asian Veg. Res. Taiwan, pp 301–311

Pelletier G, Primard C, Vedel F, Chetrit P, Remy R, Rousselle P, Renard M (1983) Intergeneric cytoplasmic hybridization in Cruciferae by protoplast fusion. Mol Gen Genet 191:244–250

Pelletier G, Primard C, Vedel F, Chétrit P (1986) Genetic improvement of cytoplasmic traits through cytoplasmic hybridization in Crucifera. In: Horn W, Jensen CJ, Oldenbach W, Scheider O (eds) Genetic manipulation in plant breeding. Walter de Gruyter & Co, Berlin, pp 653–661

Pelletier G, Primard C, Férault M, Vedel F, Chétrit P, Renard M, Delourme R (1988) Use of protoplasts in plant breeding: cytoplasmic aspects. Plant Cell Tiss Organ Cult 12:173–180

Pradhan AK, Sodhi YS, Mukhopadhyay A, Pental D (1993) Heterosis breeding in Indian mustard (*Brassica juncea* L. Czern & Coss): analysis of component characters contributing to heterosis for yield. Euphytica 69:219–229. doi:10.1007/BF00022368

Prakash S, Kirti PB, Bhat SR, Gaikwad K, Kumar VD, Chopra VL (1998) A *Moricandia arvensis*-based cytoplasmic male sterility and fertility restoration system in. Brassica juncea Theor Appl Genet 97:488–492

Primard-Brisset C, Poupard JP, Horvais R, Eber F, Pelletier G, Renard M, Delourme R (2005) A new recombined double low restorer line for the Ogu-INRA cms in rapeseed (*Brassica napus* L.). Theor Appl Genet 111:736–746. doi:10.1007/s00122-005-2059-8

Sabar M, Gagliardi D, Balk J, Leaver CJ (2003) ORFB is a subunit of F1F(O)-ATP synthase: insight into the basis of cytoplasmic male sterility in sunflower. EMBO Rep 4:381–386. doi:10.1038/sj.embor.embor800

Sakai T, Imamura J (1990) Intergeneric transfer of cytoplasmic male sterility between *Raphanus sativus* (cms line) and *Brassica napus* through cytoplast-protoplast fusion. Theor Appl Genet 80:421–427

Sakai T, Imamura J (1992) Alteration of mitochondrial genomes containing atpA genes in the sexual progeny of cybrids between *Raphanus sativus* cms line and *Brassica napus* cv. Westar. Theor Appl Genet 84:923–929

Schull GH (1952) Beginnings of the heterosis concept. In: Gowen JW (ed) Heterosis. Iowa State College Press, Iowa, pp 14–48

Uyttewaal M, Arnal N, Quadrado M, Martin-Canadell A, Vrielynck N, Hiard S, Gherbi H, Bendahmane A, Budar F, Mireau H (2008) Characterization of *Raphanus sativus* Pentatricopeptide repeat proteins encoded by the fertility restorer locus for Ogura cytoplasmic male sterility. Plant Cell 20:3331–3345. doi:10.1105/tpc.107.057208

Vedel F, Mathieu C, Lebacq P, Ambard-Bretteville F, Rémy R, Pelletier G, Renard M, Rousselle P (1982) Comparative macromolecular analysis of the cytoplasms of normal and cytoplasmic male sterile. *Brassica napus* Theor Appl Genet 62:255–262. doi:10.1007/BF00276248

Walters TW, Earle ED (1993) Organellar segregation, rearrangement and recombination in protoplast fusion-derived *Brassica oleracea* calli. Theor Appl Genet 85:761–769. doi:10.1007/BF00225017

Walters TW, Mutschler MA, Earle ED (1992) Protoplast fusion-derived Ogura male sterile cauliflower with cold tolerance. Plant Cell Rep 10:624–628

Wen L, Ruesch KL, Ortega VM, Kamps TL, Gabay-Laughnan S, Chase CD (2003) A nuclear restorer-of-fertility mutation disrupts accumulation of mitochondrial ATP synthase subunit alpha in developing pollen of S male-sterile maize. Genetics 165:771–779

Yamagishi H, Terachi T (1994a) Molecular and biological studies on male -sterile cytoplasm in the Cruciferae. I. The origin and distribution of Ogura male-sterile cytoplasm in Japanese wild radishes (*Raphanus sativus* L.) revealed by PCR-aided assay of their mitochondrial DNAs. Theor Appl Genet 87:996–1000

Yamagishi H, Terachi T (1994b) Molecular and biological studies on male sterile cytoplasm in Cruciferae. II. The origin of Ogura male sterile cytoplasm inferred from the segregation pattern of male sterility in the F1 progeny of wild and cultivated radishes (*Raphanus sativus* L.). Euphytica 80:201–206. doi:10.1007/BF00039651

Yamagishi H, Terachi T (1996) Molecular and biological studies on male-sterile cytoplasm in the Cruciferae. III. Distribution of Ogura-type cytoplasm among Japanese wild radishes and Asian radish cultivars. Theor Appl Genet 93:325–332

Yamagishi H, Terachi T (1997) Molecular and biological studies on male-sterile cytoplasm in the Cruciferae. IV. Ogura-type cytoplasm found in the wild radish. *Raphanus raphanistrum* Plant Breed 116:323–329

Yamagishi H, Terachi T (2001) Intra-and inter-specific variations in the mitochondrial gene orf138 of Ogura-type male-sterile cytoplasm from *Raphanus sativus* and. *Raphanus raphanistrum* Theor Appl Genet 103:725–732

Yarrow SA, Burnett LA, Wildeman RP, Kemble RJ (1990) The transfer of "Polima"cytoplasmic male sterility from oilseed rape (*Brassica napus*) to broccoli (*B. oleracea*) by protoplast fusion. Plant Cell Rep 9:185–188

Zirckle C (1952) Early ideas on inbreeding and crossbreeding. In: Gowen JW (ed) Heterosis. Iowa State College Press, Iowa, pp 1–13

Chapter 10
Protoplast Technology in Genome Manipulation of Potato Through Somatic Cell Fusion

Veli-Matti Rokka

10.1 Introduction

There are more than 200 wild potato species named, yet only a limited number of them have been utilized in breeding for potato germplasm development. Many solanaceous species are diploids ($2n=2x=24$), and therefore cannot readily cross with tetraploid ($2n=4x=48$) cultivated potato (*Solanum tuberosum* L. ssp. *tuberosum*). Incompatibility is common, because many wild potato species do not hybridize with cultivated potato due to the differences in ploidy levels or in the endosperm balance numbers (EBNs; Johnston et al. 1980). The overwhelming breeding barriers caused by ploidy level differences and EBN variations could, however, be partly overcome by induction of haploid lines and through somatic hybridization (Rokka et al. 1995).

In breeding at the diploid level of potato, haploid lines induced from cultivated potato often show low fertility, thus, somatic hybridization is an alternative technique to combine genomes of the both haploid parental clones. Introduction of the analytic–synthetic breeding scheme by Wenzel et al. (1979) was the start of potato breeding using a combination of both anther culture and protoplast techniques to produce tetraploid potato hybrids to obtain maximized heterozygosity levels in the resulting breeding lines. Through cell fusion, all of the dominant characters of the haploid lines are theoretically expressed in the resulting somatic hybrids, which may subsequently have a capacity for development as a commercial variety.

Usually, protoplast fusion can be classified as symmetric and asymmetric. Symmetric fusion incorporates the whole genomes of both parents, whereas asymmetric fusion allows partial genome transfer from a donor to a recipient. In asymmetric fusion, a donor is usually subjected to irradiation or ultraviolet (UV) treatment for the subsequent somatic genome combination (Grosser and Gmitter 2011).

V.-M. Rokka (✉)
Natural Resources Institute Finland, Green Technology,
Myllytie 1, FI-31600 Jokioinen, Finland

X.-Q. Li et al. (eds.), *Somatic Genome Manipulation*,
DOI 10.1007/978-1-4939-2389-2_10

10.1.1 Technical Development in Protoplast Technology

The first successful protoplast fusions were induced using chemical compounds such as sodium nitrate (Power et al. 1970) and polyethylene glycol (PEG; Kao and Michayluk 1974). The first interspecific somatic hybrid plant between *Solanum* species, was the "pomato or topato," the artificially produced hybrid between tomato and potato (Melchers et al. 1978). These hybrids produced fibrous roots, no tubers, and only parthenocarpic fruits. The next applications were *S. chacoense* Bitt. + *S. tuberosum* (Butenko and Kuchko 1979) and *S. nigrum* L. + *S. tuberosum* (Binding et al. 1982) somatic hybrids. Since then, several different species have been hybridized through protoplast fusion with cultivated potato.

The first experiences in protoplast fusion between two haploid cultivated potato lines were not extremely promising (Wenzel et al. 1982), but the introduction of electrical cell fusion technique (Zimmermann and Scheurich 1981), optimizing of shoot regeneration protocols for potato, and development of more sophisticated molecular techniques for hybrid identification, enabled the production of higher numbers of somatic hybrids. Certain hybrid combinations are still difficult or even impossible to obtain, which may be caused by differences in the cytoplasmic compositions of the parental clones used in protoplast isolation (Lössl et al. 1999).

10.1.2 Somatic Genome Manipulation Applications to Transfer Valuable Breeding Traits to Potato

Related or distantly related genera of cultivated crops possess large reservoirs of genes having various desirable traits such as high-quality characters, disease resistance, and cytoplasmic male sterility (CMS; Liu et al. 2005). The wild *Solanum* species are an important source of resistance for various biotic and abiotic stresses (Helgeson and Haberlach 1999). Such germplasm can be integrated into potato breeding programs. Somatic hybridization can provide new opportunities for producing pre-breeding materials with increased genetic variability and transferring desirable agronomic traits into cultivated potatoes. For potato, a broad overview for utilization of somatic hybridization in genetics and plant breeding was reported by Orczyk et al. (2003).

Recently, somatic hybridization for potato genome improvement has been widely utilized. One of the most desirable traits for potato cultivation is to obtain resistance to late blight which is an economically important disease affecting potatoes caused by *Phytophthora infestans* (Mont) de Bary. For example, in breeding for resistance to late blight, the search for new sources of durable resistance from the nonhost wild *Solanum* species, such as non-tuberous *S. nigrum*, has been of interest (Zimnoch-Guzowska et al. 2003). The results obtained by Helgeson et al. (1998) indicated that effective resistance to the late blight of *S. bulbocastanum* Dun. can be transferred into potato breeding lines by somatic hybridization and that this resistance can then be further transmitted into potato breeding lines by sexual crossing.

Since then, *S. bulbocastanum* has generated an enormous level of interest and has been widely used in several applications to transfer resistance traits to cultivated potato (Naess et al. 2001; Iovene et al. 2007). In addition, to obtain late blight resistance from other genetic resources, interspecific somatic hybrids between cultivated potato and *S. pinnatisectum* Dun. (Sarkar et al. 2011) and *S. tarnii* Hawkes and Hjert. (Thieme et al. 2008) were produced. The hybrids of *S. pinnatisectum* + *S. tuberosum* showed high foliage resistance to late blight based on field assessment for two seasons (Sarkar et al. 2011). Similarly, based on another study, the somatic hybrids were also highly resistant to *P. infestans* (Polzerová et al. 2011).

A recent interesting example is *Solanum* × *michoacanum* (Bitter.) Rydb., which is a wild diploid potato species derived from a spontaneous cross of *S. bulbocastanum* and *S. pinnatisectum* with resistance to late blight. To introgress late blight resistance genes from *Solanum* × *michoacanum* into *S. tuberosum*, protoplast fusion was applied (Smyda et al. 2013). After two seasons of testing, the produced somatic hybrids were also resistant. The hybrids were reported to have adequate pollen stainability for use in crossing program and are a novel promising material for introgression resistance into the cultivated potato background (Smyda et al. 2013).

Somatic hybrid lines originating from fusion between *S. tuberosum* and *S. berthaultii* Hawkes displayed a better tolerance to salt stress compared to cultivated potato (Bidani et al. 2007). To study a source of resistance to bacterial wilt caused by *Ralstonia solanacearum* (Smith) (Yabuuchi et al.) *S. stenotomum* Juz. and Buk. was used (Fock et al. 2001). All the somatic hybrids tested showed a resistance level as high as that of the wild species (Fock et al. 2001). Similarly, *S. chacoense* was used to explore the molecular markers associated with bacterial wilt resistance, and to provide information for use of these hybrids in potato breeding programs (Chen et al. 2013).

10.2 Summary

One of the bottlenecks in using somatic hybrids is obtaining sufficient fertility levels in the produced primary hybrids and also in their progeny to create the desired gene introgression. The increased hybrid chromosome number and the incompatibility of genome recombination may cause disorders in both gamete and seed formation in the hybrids (Sarkar et al. 2011). In addition, the genome incompatibility level influences the genetic integration of the desired traits into the cultivated crops. One way to overcome these problems is to use successive backcrosses with cultivated lines and self-pollination to produce stable introgression lines (Xia 2009; Wang et al. 2013), or using repeated haploid induction and somatic hybridization (Rokka et al. 1995; 2005; Iovene et al. 2012). Thieme et al. (2010) produced progeny of hybrids by pollinating them with pollen from a cultivar, via selfing or cross-pollination. The results confirmed that protoplast electrofusion can be used to transfer many resistance characters from wild species into somatic hybrids. The obtained resistant somatic hybrids can now be used in pre-breeding studies, molecular characterization, and for increasing the genetic diversity available for potato breeding by marker-assisted combinatorial introgression into the potato gene pool (Thieme et al. 2010).

10.3 Procedure for Protoplast Fusion of Potato to Produce Symmetric Somatic Hybrids

10.3.1 Equipment and Tools

To prepare culture media, the following equipment, instruments, and tools are needed:

- Analytical balance with readability 0.1 mg
- Spatula set with different spoon sizes
- Weighing boats
- Manual pipettors with volume range from 10 μl to 5 ml
- Pipette tips with three volume ranges (sterile and nonsterile)
- Glass Erlenmeyer beakers with spout (different volume sizes)
- Glass measuring cylinders with different volume sizes
- Volumetric flasks with different volume sizes
- Glass decanter beakers with different volume sizes
- Schott laboratory glass bottles of different sizes
- Aluminum foil
- pH meter
- Magnetic stirrer
- Stirrer bars
- Magnet stir bar retriever
- Appropriate plant tissue culture vessels such as Microbox Combiness jars with lids (OS 140 Box + ODS Filter XXL) (Duchefa E 1674)
- Autoclave bags
- Autoclave
- Laminar flow cabinet
- Spray bottle for 70 % ethanol
- Syringes (10 ml) with luer lock connection (BD Discardit)
- Syringes (60 ml) with luer lock connection (BD Plastipak)
- Filter units for syringes (0.2 μm) (Schliecher and Schuell)
- Sterilized media bottles (100 ml)
- Sterilized media bottles (250 ml)
- Sterile plastic petri dishes, diameter 9 cm (Falcon)
- Parafilm sealing film
- Scissors
- Refrigerator (4 °C)
- Freezer (−18 °C)

10.3.1.1 To propagate the donor material for protoplast isolation, the following equipment, instruments, and tools are needed:

- Laminar flow cabinet
- Sterilizing lamp with burning heart
- Matches or gas lighter
- Stainless steel beaker (sterile)
- Spray bottle for 70 % ethanol
- Refrigerator
- Laboratory scalpel handles for blades (sterile)
- Scalpel blades no. 11
- Laboratory forceps, extra fine, length 150 mm (sterile)
- Laboratory forceps, extra fine, length 200 mm (sterile)
- Tool box for instruments (sterile)
- Parafilm sealing film
- Scissors
- Glass petri dishes, diameter 9 cm (sterile)
- Plant tissue culture vessels with media
- Growth chamber

10.3.1.2 In protoplast fusion, the following equipment, instruments, and tools are needed:

- Laboratory forceps, extra fine, length 200 mm (sterile) (BD 239, Aesculap)
- Tweezers, extra fine tip, length 110 mm (sterile) (BD 329, Aesculap)
- Laboratory scalpel handles for blades (sterile) (BB 73, Aesculap)
- Scalpel blades no. 10 (BB 511, Aesculap)
- Scalpel blades no. 11 (BB 510, Aesculap)
- Laminar flow cabinet
- Sterilizing lamp with burning heart
- Matches or gas lighter
- Stainless steel beaker (sterile)
- Spray bottle for 70 % ethanol
- Glass petri dishes, diameter 9 cm (sterile)
- Parafilm sealing film
- Scissors
- Sterile plastic petri dishes, diameter 3, 6, and 9 cm (Falcon)
- Growth chamber (Sanyo or Rumed)
- Sterilized ddH_2O in 500 ml media bottle
- Manual pipettors with volume range from 5 to 10 μl
- Pipette tips with three volume ranges (sterile)
- Pipette tips with cut ends for 5-ml manual pipettor (sterile)
- Glass low-form beakers with spout (250 ml), covered by aluminum foil (sterile)
- Syringes (10 ml) with luer lock connection (BD Discardit)

- Filter units for syringes (0.2 μm) (Schliecher and Schuell)
- Glass Pasteur pipettes (sterile)
- Aluminum foil
- Refrigerator
- Freezer (−18 °C)
- Shaker (Infors, HT TR-125)
- Inverted microscope (Olympus CK 2)
- Sterilized 5-cm-diameter plastic jars with nylon mesh cover
- Nylon mesh (48 μm)
- Glass centrifuge tubes, 15 ml (sterile)
- Rack for 15-ml centrifuge tubes
- Centrifuge swing out rotor (Sorvall TC)
- Fuchs Rosenthal haemocytometer with cover slips
- Microscope (stereo, dissecting binocular) with external light supply
- Electrofusion apparatus (Electro Cell fusion CFA 500, Krüss Gmbh)
- Electrofusion setup with chamber and cover slip
- Masking tape
- Microwave oven
- Plant growth chamber (Sanyo)
- Plastic store boxes (approximately 100 × 300 × 100)

10.3.2 Media, Solutions, and Chemicals

10.3.2.1 Preparation of Culture Medium for Donor Plants

MS 30 Medium, Combiness Jars, 1 l (Murashige and Skoog 1962)

- MS macroelements (10 ×) 100 ml
- MS microelements (100 ×) 10 ml
- Fe-ethylenediaminetetraacetic acid (EDTA) 5 ml
- MS vitamin stock (100 ×) 10 ml
- Casein hydrolysate 100 mg
- Sucrose 30 g
- α-Naphthalene acetic acid (NAA) 0.05 mg (50 μl from stock 1 mg/1 ml)
- Agar (bacteriological type A) 8 g

Add each ingredient, except agar, into an Erlenmeyer jar, add ddH_2O up to 500 ml, measure pH 5.6. Pour solution into a media bottle, add ddH_2O up to 1 l, and add agar. Stir properly, autoclave solution. Sterilize Combiness jars and lids separately in autoclave bags. Shake medium well, divide it into Combiness Microbox jars (200 ml per jar) in a laminar hood, add lid.

10.3.2.2 Preplasmolysis solution, 250 ml

0.53 M mannitol 24.138 g

Autoclave in media bottles of 100 ml in volume.

10.3.2.3 Enzyme Solution, 100 ml (Carlberg et al. 1983)

- K_3 macroelements (10 ×) 10.0 ml
- K_3 microelements (10 ×) 10.0 ml
- Fe-ethylenediaminetetraacetic acid (EDTA) 0.5 ml
- Mannitol 8.198 g
- Cellulysin (Calbiochem) 1 g
- Macerase (Calbiochem) 0.1 g

Mix the solution for 1–2 h in a magnetic stirrer to increase the dissolving and enzyme activity, and measure pH 5.6. Divide the final stock into 10 ml aliquots in 15-ml plastic centrifuge tubes. Store in a freezer at − 18 °C. Filter sterilize before use.

K_3 Macroelements, 10 × Strength, 250 ml (Menczel et al. 1981)

- KNO_3 6250 mg
- NH_4NO_3 625 mg
- $(NH_4)_2SO_4$ 335 mg
- $NaH_2PO_4 \times H_2O$ 375 mg
- $CaCl_2 \times 2H_2O$ 2250 mg
- $MgSO_4 \times 7H_2O$ 625 mg

Dissolve each component in the given order before adding the next chemical. Dissolve $CaCl_2 \times 2H_2O$ individually in a separate flask and then add to the mixture. Sterilize by autoclaving and store at 4 °C.

K_3 Microelements, 10 × Strength, 250 ml (Gamborg et al. 1968)

- $MnSO_4 \times H_2O$ 25 mg
- H_3BO_3 7.5 mg (7.5 ml from 1 mg/ml stock solution)
- $ZnSO_4 \times 7H_2O$ 5 mg (5 ml from 1 mg/ml stock solution)
- $Na_2MoO_4 \times 2H_2O$ 0.625 mg (625 μl from 1 mg/ml stock solution)
- $CuSO_4 \times 5H_2O$ 0.1 mg (100 μl from 1 mg/ml stock solution)
- $CoCl_2 \times 6H_2O$ 0.0625 mg (63 μl from 1 mg/ml stock solution)
- KI 1.875 mg (1.875 ml from 1 mg/ml stock solution)

Sterilize by autoclaving and store at 4 °C.

10.3.2.4 Protoplast Washing Solution, 250 ml

- Washing solution stock (10 ×) 25.0 ml
- $CaCl_2 \times 2H_2O$ 21.5 mg
- Mannitol 24.0 g

Measure pH 5.6. Autoclave in three separate media bottles of 100 ml in volume for use.

Washing Solution Stock, 10 × Strength, 250 ml (Frearson et al. 1973; Jones et al. 1989)

- KNO_3 475 mg
- $CaCl_2 \times 2H_2O$ 110 mg
- $MgSO_4 \times 7H_2O$ 93 mg
- KH_2PO_4 43 mg

Autoclave in three separate media bottles of 100 ml for use. Use the wash solution (10 ×) stock for the preparation of both protoplast washing solution and Percoll solution.

10.3.2.5 Percoll Solution (30 %, v/v), 100 ml

- Percoll dilution solution 70 ml
- Percoll (Pharmacia Fine Chemical AB) 30 ml

Combine Percoll dilution solution and Percoll. Measure pH 5.6. Filter sterilize in small amounts to sterile media bottles of 100 ml in volume.

Percoll Dilution Solution, 250 ml (Jones et al. 1989)

- Washing solution stock (10 %) 25 ml
- $CaCl_2 \times 2\ H_2O$ 215 mg
- Mannitol 33,75 g

Add ddH_2O up to 250 ml. Autoclave.

10.3.2.6 0.45 M Mannitol Solution, 250 ml

- Mannitol 20.495 g

Autoclave in media bottles of 100 ml in volume. Do not adjust pH.

10.3.2.7 Fusion Solution, 250 ml

- Mannitol 19.129 g

Autoclave in media bottles of 100 ml in volume. Do not adjust pH.

10.3.2.8 2 × V-KM Medium, 100 ml (Bokelmann and Roest 1983)

- V-KM macroelements (10 ×) 20.0 ml
- V-KM microelements (10 ×) 20.0 ml
- Fe-EDTA 1.0 ml
- V-KM sugar compounds (100 ×) 2.0 ml
- KM organic acids (100 ×) 2.0 ml
- V-KM vitamins (100 ×) 2.0 ml
- NAA (Sigma N 0640) 0.2 mg (=200 µl from 1 mg/ml stock)
- BA (Duchefa B 0904) 0.08 mg (=80 µl from 1 mg/ml stock)
- Coconut water (Sigma C 5915) 4.0 ml
- $CaCl_2 \times 2H_2O$ 147 mg
- Casein hydrolysate (Duchefa C 1301) 50 mg
- Glucose 16.4 g

Measure pH 5.8. Filter sterilize.

V-KM Macroelements, 10 × Strength, 250 ml

- KNO_3 3700 mg
- $MgSO_4 \times 7H_2O$ 2460 mg
- KH_2PO_4 170 mg

Autoclave and store at 4 °C.

V-KM Microelements, 10 × Strength, 250 ml

- H_3BO_3 7.5 mg (=7.5 ml from 1 mg/ml stock)
- $MnSO_4 \times 4H_2O$ 34 mg
- $ZnSO_4 \times 7H_2O$ 5 mg (5 ml from 1 mg/ml stock)
- $Na_2MoO_4 \times 2H_2O$ 0.625 mg (625 µl from 1 mg/ml stock)
- $CuSO_4 \times 5H_2O$ 0.0625 mg (63 µl from 1 mg/ml stock)
- $CoCl_2 \times 6H_2O$ 0.0625 mg (63 µl from 1 mg/ml stock)
- KI 1.875 mg (1.875 ml from1 mg/ml stock)

Autoclave and store at 4 °C.

V-KM Sugar Compounds, 100 × Strength, 100 ml

- Sucrose 2500 mg
- Mannitol 2500 mg
- Sorbitol 2500 mg
- Ribose 2500 mg
- Xylose 2500 mg
- Mannose 2500 mg
- Rhamnose 2500 mg
- Fructose 2500 mg
- Cellobiose 2500 mg

Divide the final stock into 1 ml aliquots into Eppendorf tubes, and store it in a freezer at −18 °C.

Kao and Michayluk (KM) Organic Acids, 100 × Strength, 100 ml

- Sodium pyruvate 200 mg
- Fumaric acid 400 mg
- Citric acid 400 mg
- Malic acid 400 mg

Divide the final stock into 1 ml aliquots into Eppendorf tubes, and store it in a freezer at −18 °C.

V-KM Vitamins, 100 × Strength, 100 ml

- D-Ca-panthotenate 10 mg
- Choline chloride 10 mg
- Ascorbic acid 20 mg
- Pyridoxine HCl 10 mg
- Thiamine HCl 10 mg
- *p*-Aminobenzoic acid 0.2 mg
- Folic acid 4 mg
- Biotin 0.1 mg
- Vitamin A 0.1 mg
- Vitamin D_3 0.1 mg
- Vitamin B_{12} 0.2 mg
- Nicotinamide 10 mg
- Myo-inositol 1000 mg

Divide the final stock into 1 ml aliquots into Eppendorf tubes, and store it in a freezer at −18°C.

10.3.2.9 Agarose Solution, 0.8 %, 100 ml

- Agaroosi, Type VII (Sigma A 6560) 800 mg

Add 100 ml of ddH_2O. Store in a media bottle. Autoclave. Before use, melt it in a microwave oven. And cool it down to 37 °C before use.

10.3.2.10 1 × V-KM Medium, 100 ml (Bokelmann and Roest 1983)

- V-KM macroelements (10 ×) 10.0 ml
- V-KM microelements (10 ×) 10.0 ml
- Fe-EDTA 0.5 ml
- V-KM sugar compounds (100 ×) 1.0 ml
- KM organic acids (100 ×) 1.0 ml
- V-KM vitamins (100 ×) 1.0 ml
- NAA (Sigma N 0640) 0.1 mg (= 100 µl from 1 mg/ml stock)
- 6-benzyl amino purine (BA) (Duchefa B 0904) 0.04 mg (=40 µl from 1 mg/ml stock)
- Coconut water (Sigma C 5915) 1 ml
- $CaCl_2 \times 2H_2O$ 73.5 mg
- Casein hydrolysate (Duchefa C 1301) 25 mg
- Glucose 8.2 g

Measure pH 5.8. Filter sterilize.

10.3.2.11 Cg Medium, 1 l (Creissen and Karp 1985)

- Cg macroelements (10 ×) 100 ml
- Cg microelements (10 ×) 100 ml
- Fe-EDTA 2.5 ml
- Cg organic compounds (100 ×) 10 ml
- Sucrose 2.5 g
- Mannitol 54.6 g
- MES (Sigma M 3023) 0.976 g
- NAA (Sigma N 0640) 0.1 mg (= 100 µl from 1 mg/ml stock)
- BA (Duchefa B 0904) 0.5 mg (= 500 µl from 1 mg/ml stock)
- Agar Noble (Difco) 7.0 g
- Glutamine 0.1 g

Measure pH 5.6. Autoclave. Add glutamine through filter sterilization to a cooled medium and then mix properly, then plate on 9-cm sterile petri dishes, 30 ml per plate.

Cg Macroelements, 10 × Strength, 1 l

- KNO_3 19.00 g
- NH_4Cl 1.07 g
- KH_2PO_4 1.70 g
- $CaCl_2 \times 2H_2O$ 4.40 g
- $MgSO_4 \times 7H_2O$ 3.70 g

Dissolve each component in the given order before adding the next chemical. Dissolve $CaCl_2 \times 2H_2O$ individually in a separate flask and then add to the mixture. Sterilize by autoclaving and store at 4 °C.

Cg Microelements, 10 × Strength, 1 l

- H_3BO_3 31 mg
- $MnCl_2 \times 4H_2O$ 99 mg
- $ZnSO_4 \times 7H_2O$ 46 mg
- KI 4.20 mg
- $Na_2MoO_4 \times 2H_2O$ 1.30 mg
- $CuSO_4 \times 5H_2O$ 0.13 mg
- $CoSO_4 \times 7H_2O$ 0.15 mg

Dissolve each component in the given order before adding the next chemical. Sterilize by autoclaving and store at 4 °C.

Cg Organic Compounds, 100 × Strength, 100 ml

- Thiamine HCl 5.0 mg
- Glycine 20.0 mg
- Nicotinic acid 50.0 mg
- Pyridoxine HCl 5.0 mg
- Folic acid 5.0 mg
- Biotin 0.5 mg
- Adenine hemisulfate 400 mg
- Myo-inositol 1000 mg

Store as 10 ml aliquots at −18 °C.

10.3.2.12 Medium D, 1 l (Shepard 1980)

- Medium D macroelements (10 ×) 100 ml
- Cg microelements (10 ×) 100 ml
- Fe-EDTA 2.5 ml

- Medium D organic compounds (100 ×)10 ml
- Mannitol 36.436 g
- Sucrose 2.5 g
- MES (Sigma M 3023) 0.98 g
- Adenine hemisulfate (Sigma A 2545) 80 mg
- Indole-3-acetic acid (IAA) (Duchefa I 0901) 0.1 mg (= 100 μl from 1 mg/ml stock)
- Agar Noble (Difco) 10 g
- Zeatine (Duchefa Z 1917) 1 mg (= 1 ml from 1 mg/ml stock)

Measure pH 5.6. Add zeatine through filter sterilization to a cooled medium and then mix properly. Plate on 9-cm sterile petri dishes, 30 ml per plate.

Medium D Macroelements, 10 × Strength, 1 l

- KNO_3 19.00 g
- NH_4Cl 2.675 g
- KH_2PO_4 1.70 g
- $CaCl_2 \times 2H_2O$ 4.40 g
- $MgSO_4 \times 7H_2O$ 3.70 g

Autoclave and store at 4 °C.

Medium D Organic Compounds, 100 × Strength, 100 ml

- Myo-inositol 1 g
- Thiamine HCl 5 mg
- Glycine 20 mg
- Nicotinic acid 50 mg
- Folic acid 5 mg
- Biotin 0.5 mg
- Caseinhydrolysate 1 g

Store as 10 ml aliquots at − 18 °C.

10.3.2.13 Medium SP, 1 l (Creissen and Karp 1985; Shepard 1980)

- Medium D macroelements (10 ×) 100 ml
- Cg microelements (10 ×) 100 ml
- Fe-EDTA 2.5 ml
- Medium D organic compounds (100 ×)10 ml
- Mannitol 36.436 g
- Sucrose 2.5 g

- MES (Sigma M 3023) 0.98 g
- Adeninehemisulfate (Sigma A2545) 80 mg
- BA (Duchefa B 0904) 0.25 mg (=250 µl from 1 mg/ml stock)
- Agar Noble (Difco) 10 g
- GA_3 (Duchefa G 0907) 1 mg (=1 ml from 1 mg/ml stock)

Measure pH 5.6. Add GA_3 through filter sterilization to a cooled medium and then mix properly. Plate on 9-cm sterile petri dishes, 30 ml per plate.

10.3.3 Methods

10.3.3.1 Donor Plant Material Culture

The potato genotypes selected for protoplast fusion are aseptically grown on MS30 medium (Murashige and Skoog 1962) containing 30 g/l of sucrose in the growth chambers for 4–6 weeks at 20 °C, light intensity of 25–50 µmol m^{-2} s^{-1} in a photoperiod of 16 h per day. The stock plants are propagated through single-node cuttings with one axillary bud using in vitro cultivation in Combiness Microboxes with green extra-large filter on the lid.

10.3.3.2 Protoplast Isolation

Material Pretreatment

Leaf material of 4–6-week-old plants is transferred using sterile forceps onto plastic 9-cm-diameter petri dish containing 10–20 ml of sterile ddH_2O. Do not allow leaves to dry in the laminar hood. One petri dish covered by leaves is enough for one protoplast isolation. From the remaining plants, new cultures can be done for future protoplast isolation. The leaf material is cut into small sections. After slicing, ddH_2O is removed using a 5-ml pipette with cut tip end. Replace by 10–20 ml of preplasmolysis solution. The dissected leaves are incubated in the preplasmolysis solution for 1 h.

Enzyme Treatment

Defrost the enzyme solution (10 ml) by incubating the tubes in cold water. After pretreatment of sliced leaves in the preplasmolysis solution, the liquid is removed by sterile Pasteur pipette, and the leaf segments are treated by 10–15 ml of enzyme solution sterilized by infiltration. The fusion parents are pretreated separately. The petri dishes with leaf samples are covered by aluminum foil and transferred to a shaker with 30 rpm for 16–18 h at 24 °C for protoplast isolation.

Protoplast Purification

On the following day, observe the protoplast yield as a result of enzyme digestion and the quality of protoplasts by inverted microscope. As a result of successful enzyme digestion, the protoplast suspensions of both fusion parents are filtered through 48 μm nylon sieve. The filtrate is centrifuged at $80 \times g$ for 5 min in four centrifuge tubes with a volume of 15 ml using swing-out rotor. After centrifugation, pour the enzyme solution off from the tubes. The remaining protoplast pellets in each four tube are each resuspended by a gentle shaking or using a pipette in 1.5 ml of protoplast washing solution to remove the pellets.

Add 2.5 ml of 30% (v/v) Percoll solution into eight centrifuge tubes (four tubes per genotype). Add 750 μl of resuspended protoplast solution as a layer using a Pasteur pipette on top of the Percoll solution. If the layer is relatively thin and the protoplast concentration is not extremely high, the viable protoplasts can be more effectively separated from dead protoplasts using centrifugation. The protoplasts are purified by centrifugation at $120–160 \times g$, 5 min. After centrifugation, the layer of viable protoplasts on the surface of the Percoll solution can be collected by a Pasteur pipette. The protoplasts of the same genotype are combined into one centrifuge tube. Do not mix the different genotypes at this stage. Add 10 ml of Protoplast washing solution, mix the sample gently, and centrifuge by $100 \times g$, 10 min. After centrifugation, the protoplast pellet is resuspended by 10 ml of 0.45 M mannitol solution and centrifuged by $80 \times g$, 5 min. Repeat the final wash by 0.45 M mannitol.

Protoplast Density

After the final wash, resuspend the pellet by 1 ml of 0.42 M mannitol (fusion solution). Take a small sample in a laminar and count the density by Fuchs–Rosenthal haemocytometer according to the instructions by manufacturer. Count the density using microscope. Dilute the protoplast samples with fusion solution to obtain 2.5×10^5 protoplasts per milliliter. Mix the protoplasts of the both partners by a transfer into a sterile beaker, in a ratio of 1:1.

10.3.3.3 Protoplast Fusion

Before protoplast fusion, surface sterilize the electrofusion chamber with its electrodes by 70% ethanol for 15–30 min. Transfer the electrofusion set using sterile forceps onto a petri dish in a laminar chamber and allow it to dry for another 30 min.

Transfer an inverted microscope into the laminar and surface-sterilize the desk. Transfer the lamellar electrofusion chamber to the microscope desk for observation. Connect the chamber to the elecrofusion apparatus. A volume of 500 μl of protoplast mixture in a density of 2.5×10^5 protoplasts per ml is transferred into the chamber. The protoplasts are aligned at 125 V cm^{-1} in an alternating current (AC) field for 70 s and fused using 1750–2500 V cm^{-1} direct current (DC) pulses (1 × 10 µs or 2 × 10 µs) using an electrofusion apparatus. Following fusion, the protoplasts (500 µl) are pipetted into 3.5-cm-diameter petri dish, and add 500 µl of 2× V-KM liquid culture medium. Immediately add 500 µl of 0.8% low-gelling temperature agarose solution from incubation chamber set 37 °C. Mix the solution gently and let agarose to solidify. Repeat the same procedure by another protoplast sample. Seal the plates with Parafilm.

10.3.3.4 Protoplast Culture

Allow the fused protoplasts to make a cell wall and their first cell divisions at 24 °C in the dark for 1 week. Observe the cultures by an inverted microscope after a week, and then resuspend the cultures with 3 ml per plate normal strength V-KM liquid medium using pipette tips with cut ends to create a homogenized solution. Let the culture grow for another 2 days. Then transfer the cultures to a dim light. Increase the light intensity slowly during the following 2 weeks to obtain the final light intensity of 25–50 µmol m^{-2} s^{-1}. After 3–5 weeks, the cultures will show microcolonies. If the density of microcolonies is high, add more V-KM solution during the first weeks of protoplast culture.

After a total of 4 weeks, the microcolonies are transferred with a 5-ml pipette tip onto Cg medium plated on 9-cm-diameter petri dishes. If the microcolony density is very high, divide the cultures onto several plates with Cg medium. Incubate the cultures at light intensity of 25–50 µmol m^{-2} s^{-1}, 16 h per day, 22 °C. The colonies will turn green within 1 month of culture, and they can be separately transferred using tweezers onto medium D. The calli are transferred to fresh medium D monthly for shoot regeneration. The first primordia will be developed after 3–8 months of culture and they are ready for transferring onto medium SP for shoot elongation. When the regenerated shoots are in length of 1–2 cm, they can be transferred to MS30 medium for maintenance.

The presented protocol in this chapter describes all the necessary working solutions in their appropriate volumes for practices of potato protoplast fusion. The procedure can be directly applied for production of sufficient numbers of symmetric intraspecific somatic hybrids between haploid potato lines derived from commercially grown cultivars (Rokka et al. 1996) or production of interspecific solanaceous species somatic hybrids (Rokka et al. 1994, 2005). The protocol has already been applied for species with a diverse genetic spectrum, and thereby it allows a wider utilization of the genetic resources comprising of the solanaceous species for genome manipulation of cultivated potato. In addition, somatic hybrids obtained through this procedure could be used as a source for production of novel phytochemicals (Väänänen et al. 2005), which may have high nutritious or pharmaceutical values in the future.

References

Bidani A, Nouri-Ellouz O, Lakhoua L, Sihachakr D, Cheniclet C, Mahjoub A, Drira N, Gargouri-Bouzid R (2007) Interspecific potato somatic hybrids between *Solanum berthaultii* and *Solanum tuberosum* L. showed recombinant plastome and improved tolerance to salinity. Plant Cell Tissue Org Cult 91:179–189

Binding H, Jain SM, Finger J, Mordhorst G, Nehl R, Gressel J (1982) Somatic hybridization of an atrazine resistant biotype of *Solanum nigrum* with *Solanum tuberosum*. Part 1. Clonal variation in morphology and in atrazine sensitivity. Theor Appl Genet 63:273–177

Bokelmann GS, Roest S (1983) Plant regeneration from protoplasts of potato (*Solanum tuberosum* cv. Binje). Z Pflanzenphysiol 109:259–265

Butenko RG, Kuchko AA (1979) Pysiological aspects of isolation, cultivation and hybridization of potato protoplasts. Fiziol Rastenii 26:1110–1119

Carlberg I, Glimelius K, Eriksson T (1983) Improved culture ability of potato protoplasts by use of activated charcoal. Plant Cell Rep 2:223–115

Chen L, Guo X, Xie C, He L, Cai X, Tian L, Song B, Liu J (2013) Nuclear and cytoplasmic genome components of *Solanum tuberosum* + *S. chacoense* somatic hybrids and three SSR alleles related to bacterial wilt resistance. Theor Appl Genet 126:1861–1872

Creissen GP, Karp A (1985) Karyotypic changes in potato plants regenerated from protoplasts. Plant Cell Tissue Org Cult 4:171–182

Fock I, Collonnier C, Luisetti J, Purwito A, Souvannavong V, Vedel F, Servaes A, Ambroise A, Kodja H, Ducreux G, Sihachakr D (2001) Use of *Solanum stenotomum* for introduction of resistance to bacterial wilt in somatic hybrids of potato. Plant Physiol Biochem 39:899–908

Frearson EM, Power JB, Cocking EC (1973) The isolation, culture and regeneration of Petunia leaf protoplasts. Dev Biol 33:130–137

Gamborg OL, Miller RA, Ojima K (1968) Nutrient requirements of suspension cultures of soybean root cells. Exp Cell Res 50:151–158

Grosser JW, Gmitter FG (2011) Protoplast fusion for production of tetraploids and triploids: applications for scion and rootstock breeding in citrus. Plant Cell Tissue Org Cult 104:343–357

Helgeson JP, Haberlach GT (1999) Somatic hybrids of *Solanum tuberosum* and related species. In: Altman A, Ziv M, Izhar S (eds) Plant biotechnology and in vitro biology in the 21st century, current plant science and biotechnology in agriculture. Springer, Netherlands, vol 36, pp 151–154

Helgeson JP, Pohlman JD, Austin S, Haberlach GT, Wielgus SM, Ronis D, Zambolim L, Tooley P, McGrath JM, James RV, Stevenson WR (1998) Somatic hybrids between *Solanum bulbocastanum* and potato: a new source of resistance to late blight. Theor Appl Genet 96:738–742

Iovene M, Savarese S, Cardi T, Frusciante L, Scotti N, Simon PW, Carputo D (2007) Nuclear and cytoplasmic genome composition of *Solanum bulbocastanum* (+) *S. tuberosum* somatic hybrids. Genome 50:443–450

Iovene M, Aversano, R, Savarese S, Caruso I, Di Matteo A, Cardi T, Frusciante L, Carputo D (2012) Interspecific somatic hybrids between *Solanum bulbocastanum* and *S. tuberosum* and their haploidization for potato breeding. Biol Plant 56:1–8

Johnston SA, den Nijs TPM, Peloquin SJ, Hanneman Jr RE (1980) The significance of genic balance to endosperm development in interspecific crosses. Theor Appl Genet 57:5–9

Jones H, Karp A, Jones MGK (1989) Isolation, culture, and regeneration of plants from potato protoplasts. Plant Cell Rep 8:307–311

Kao KN, Michayluk MR (1974) A method for high-frequency intergeneric fusion of plant protoplasts. Planta 115:355–367

Liu J, Xu X, Deng X (2005) Intergeneric somatic hybridization and its application to crop genetic improvement. Plant Cell Tissue Org Cult 82:19–44

Lössl A, Adler N, Horn R, Frei U, Wenzel G (1999) Chondriome-type characterization of potato: mt α, β, δ, σ, ε and novel plastid-mitochondrial configurations in somatic hybrids. Theor Appl Genet 98:1–10

Melchers G, Sacristán MD, Holder AA (1978) Somatic hybrid plants of potato and tomato regenerated from fused protoplasts. Carlsberg Res Commun 43:203–218

Menczel L, Nagy F, Kiss Z, Maliga P (1981) Streptomycin resistant and sensitive somatic hybrids of *N. tabacum* + *N. knightiana*: correlation and resistance to *N. tabacum* plastids. Theor Appl Genet 59:191–198

Murashige T, Skoog F (1962) A revised medium for rapid growth and bioassays with tobacco tissue culture. Physiol Plant 15:473–497

Naess S, Bradeen J, Wielgus S, Haberlach G, McGrath J, Helgeson J (2001) Analysis of the introgression of *Solanum bulbocastanum* DNA into potato breeding lines. Mol Genet Genom 265:694–704

Orczyk W, Przetakiewicz J, Nadolska-Orczyk A (2003) Somatic hybrids of *Solanum tuberosum*—application to genetics and breeding. Plant Cell Tissue Org Cult 74:1–13

Polzerová H, Patzak J, Greplová JM (2011) Early characterization of somatic hybrids from symmetric protoplast electrofusion of *Solanum pinnatisectum* Dun. and *Solanum tuberosum* L. Plant Cell Tissue Org Cult 104:163–170

Power JB, Cummins SE, Cocking EC (1970) Fusion of isolated protoplasts. Nature 255:1016–1018

Rokka V-M, Xu Y-S, Kankila J, Kuusela A, Pulli S, Pehu E (1994) Identification of somatic hybrids of dihaploid *Solanum tuberosum* lines and *S. brevidens* by species specific RAPD patterns and assessment of disease resistance of the hybrids. Euphytica 80:207–217

Rokka V-M, Valkonen JPT, Pehu E (1995) Production and characterization of haploids derived from somatic hybrids between *Solanum brevidens* and *S. tuberosum* through anther culture. Plant Sci 112:85–95

Rokka V-M, Xu Y-S, Tanhuanpää P, Pietilä L, Pehu E (1996) Electrofusion of protoplasts of anther-derived dihaploid lines of commercial potato cultivars. Agric Food Sci Finland 5:449–460

Rokka VM, Laurila J, Tauriainen A, Laakso I, Larkka J, Metzler M, Pietilä L (2005) Glycoalkaloid aglycone accumulations associated with infection by *Clavibacter michiganensis* ssp. *sepedonicus* in potato species *Solanum acaule* and *Solanum tuberosum* and their interspecific somatic hybrids. Plant Cell Rep 23:683–691

Sarkar D, Tiwari JK, Sharma S, Poonam, Sharma S, Gopal J, Singh BP, Luthra SK, Pandey SK, Pattanayak D (2011) Production and characterization of somatic hybrids between *Solanum tuberosum* L. and *S. pinnatisectum* Dun. Plant Cell Tissue Org Cult 107:427–440

Shepard JF (1980) Abscisid acid-enhanced shoot initiation in protoplast-derived calli of potato. Plant Sci Lett 18:327–333.

Smyda P, Jakuczun H, Dębski K, Śliwka J, Thieme R, Nachtigall M, Wasilewicz-Flis I, Zimnoch-Guzowska E (2013) Development of somatic hybrids *Solanum*×*michoacanum* Bitter.(Rydb.) (+) *S. tuberosum* L. and autofused 4× *S.* ×*michoacanum* plants as potential sources of late blight resistance for potato breeding. Plant Cell Rep 32:1231–1241

Thieme R, Rakosy-Tican E, Gavrilenko T, Antonova O, Schubert J, Nachtigall M, Heimbach U, Thieme T (2008) Novel somatic hybrids (*Solanum tuberosum* L. + *Solanum tarnii*) and their fertile BC_1 progenies express extreme resistance to potato virus Y and late blight. Theor Appl Genet 116:691–700

Thieme R, Rakosy-Tican E, Nachtigall M, Schubert J, Hammann T, Antonova O, Gavrilenko T, Heimbach U, Thieme T (2010) Characterization of the multiple resistance traits of somatic hybrids between *Solanum cardiophyllum* Lindl. and two commercial potato cultivars. Plant Cell Rep 29:1187–1201

Väänänen T, Ikonen T, Rokka V-M, Kuronen P, Serimaa R, Ollilainen V (2005) Influence of incorporated wild *Solanum* genomes on potato properties in terms of starch nanostructure and glycoalkaloid content. J Agric Food Chem 53:5313–5325

Wang J, Jiang J, Wang Y (2013) Protoplast fusion for crop improvement and breeding in China. Plant Cell Tissue Org Cult 112:131–142

Wenzel G, Schieder O, Przewozny T, Sopory S, Melchers G (1979). Comparison of single cell culture derived *Solanum tuberosum* L. plants and a model for their application in breeding programs. Theor Appl Genet 55:49–55

Wenzel G, Meyer C, Przewozny T, Uhrig H, Schieder O (1982) Incorporation of microspore and protoplast technique into potato breeding programs. In: Earle ED, Demarly Y (eds) Variability in plants regenerated from tissue culture. Praeger Publ, New York, pp 290–302

Xia G (2009) Progress of chromosome engineering mediated by asymmetric somatic hybridization. J Genet Genom 36:547–556

Zimmermann U, Scheurich P (1981) High frequency fusion of plant protoplasts by electric fields. Planta 151:26–32

Zimnoch-Guzowska E, Lebecka R, Kryszczuk A, Maciejewska U, Szczerbakowa A, Wielgat B (2003). Resistance to *Phytophthora infestans* in somatic hybrids of *Solanum nigrum* L. and diploid potato. Theor Appl Genet 107:43–48

Chapter 11
Strategic RNA Silencing for Plant Viral Resistance

Jin Ma, Changxiang Zhu, Fujiang Wen, Huimin Xu and Xiu-Qing Li

11.1 Introduction

Plant viruses are among the most important plant pathogens. Viral infections can cause a variety of disease symptoms in plants, such as stunting, mosaic patterns, yellowing, leaf rolling, ring spot, necrosis, wilting, and cause developmental abnormalities that can impact on crop quality and yield (Hull 2002). Until now, there have been no therapeutical measures available to counteract plant virus diseases in the field; the main control strategy is based on preventative measures. Sanitary cultural practices and insecticidal sprays can reduce infection by blocking viral transmission, but planting resistant varieties is the most economical and effective method of virus control. Conventional breeding for specific virus resistant varieties is a long-term process. The high rate of mutation of the viral genome can cause loss of resistance

X.-Q. Li (✉)
Potato Research Centre, Agriculture and Agri-Food Canada, 850 Lincoln Road,
P. O. Box 20180, E3B 4Z7, Fredericton, NB, Canada
e-mail: Xiu-Qing.Li@agr.gc.ca

C. Zhu (✉) · F. Wen
State Key Laboratory of Crop Biology, Shandong Key Laboratory of Crop Biology,
Shandong Agricultural University, 271018 Tai'an, People's Republic of China
e-mail: zhchx@sdau.edu.cn

J. Ma
Forestry and Environmental Management, University of New Brunswick, Fredericton,
NB E3B 6C2, Canada
e-mail: majin201082@gmail.com

J. Ma
Potato Research Center, Agriculture and Agri-Food Canada, P.O. Box 20280, Fredericton,
NB E3B 4Z7, Canada
e-mail: Xiu-Qing.Li@agr.gc.ca

H. Xu
Canadian Food Inspection Agency, Charlottetown Laboratory, 93 Mount Edward Road,
Charlottetown, PE C1A 5T1, Canada

X.-Q. Li et al. (eds.), *Somatic Genome Manipulation*,
DOI 10.1007/978-1-4939-2389-2_11

under field conditions (Lecoq et al. 2004). Genetic engineering for transgenic antiviral plants has become an effective approach for preventing plant virus disease.

In a pioneering study, Sanford and Johnston found that transgenic plants expressing virus genes are resistant to that virus and introduced the concept of pathogen-derived resistance (PDR) (Sanford and Johnston 1985). Using PDR, two strategies were rapidly developed to engineer virus-resistant plants: protein-mediated and RNA-mediated resistance (Prins et al. 2008). While the principles of protein-mediated resistance are still unclear, RNA-mediated virus resistance (RMVR) (Waterhouse et al. 1999) is related to RNA silencing and is the most powerful tool for engineering resistance to plant viruses.

RNA silencing in plants is equivalent to gene quelling in fungi (Cogoni et al. 1994) and RNA interference (RNAi) in animals (Fire et al. 1998) and is a defence mechanism against invading nucleic acid (Plasterk 2002). RNA silencing is a nucleotide sequence specific mRNA degradation process at the posttranscriptional level, and was therefore also named posttranscriptional gene silencing (PTGS) in plants (Eamens et al. 2008). The small RNAs (sRNAs), including small interfering RNA (siRNA) and micro RNA (miRNA), play major roles in plant RNA silencing. The generation of sRNA is from cleavage of double stranded RNA (dsRNA) or hairpin RNA (hpRNA) by Dicer-like ribonucleases. The major approaches in RNA silencing include sense/antisense virus gene(s), hpRNA, and artificial miRNA precursors. The effectiveness of RMVRs varies in the field because of physiological and environmental factors. Public apprehension regarding the perceived safety of transgenic crops limits the application of this technology (Fuchs and Gonsalves 2008).

This chapter summarizes the use of RNA silencing for plant virus resistance including discovery and mechanisms of RNA silencing and different methods to induce RNA silencing. The factors of effecting RNA silencing and the approaches to overcome the constraints are also discussed in this chapter.

11.2 Discovery and Mechanism of RNA Silencing

RNA silencing was first discovered in transgenic plants. In 1990, transgenic petunia (*Petunia hybrida*) plants overexpressing a chalcone synthase gene (CHS) had flowers with dramatic variation in pigmentation, including violet, violet–white sectors, or white compared with the control (violet) (Napoli et al. 1990). In some plant lines, both the transgenic and endogenous CHS genes were silenced to different degrees; a phenomenon referred to as 'co-suppression' (Napoli et al. 1990; Van Der Krol et al. 1990). In nuclear run-on experiments, it was found that the reduction in steady-state CHS mRNA was not the result of a transcriptional inactivation event (Van Blokland et al. 1994). These results suggest that CHS co-suppression is a typical PTGS. Similar observations using isolated nuclei were being made in transgenic plants with resistance to *Tobacco etch virus* (TEV) from expression of an untranslatable sequence coat protein (CP) mRNA of TEV (Lindbo and Dougherty 1992; Dougherty et al. 1994).

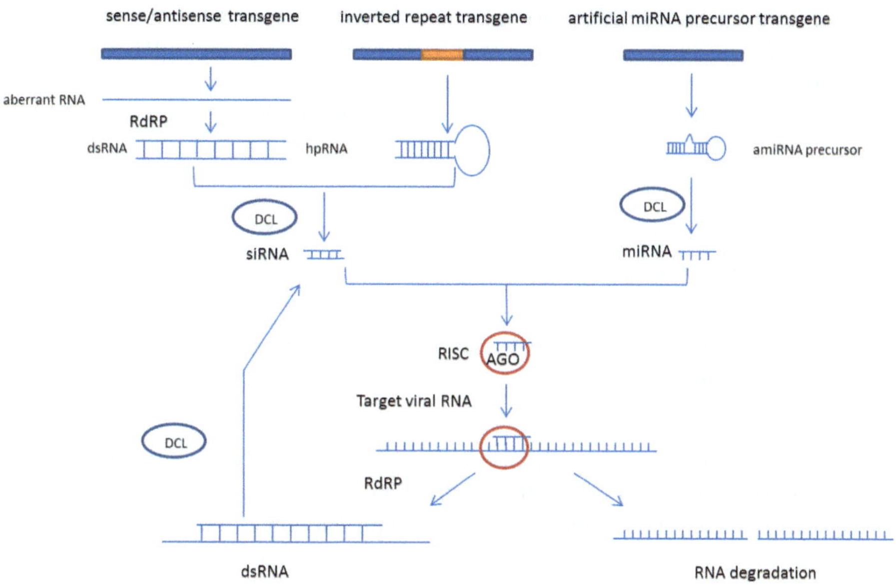

Fig. 11.1 RNA silencing mechanisms of sense/antisense, inverted repeats, and artificial miRNA (amiRNA) transgenes to confer plant viral resistance. The transcripts of sense or antisense transgene can be converted to double strand RNA (dsRNA) by plant-encoding RNA-dependent RNA polymerase (RDRP). The invert repeat transgene transcripts can form hairpin RNA (hpRNA), and amiRNA transgene generate microRNA (miRNA). The dsRNA and hpRNA can be processed by Dicer-like proteins (DCL) into 21–24 nt small interfering RNA (siRNA). The siRNA and miRNA are combined to members of the Argonaute (AGO) family to form RNA-induced silencing complex (RISC) that can degrade viral RNA. Alternatively, siRNA is used as primer for RDRP to transcribe target RNAs to generate additional dsRNAs and produce more siRNAs

The first studies expressed single strand RNA (transcripts of the transgenes) to trigger RNA silencing. However, it was soon found that dsRNA was more effective in inducing RNA silencing; as shown in nematodes (Fire et al. 1998), plants (Waterhouse et al. 1998), protozoa (Ngô et al. 1998), and insects (Kennerdell and Carthew 1998). Further studies showed that the siRNA generated from cleaving the dsRNA that triggered virus RNA degradation (Hamilton and Baulcombe 1999). Interestingly, the RNA silencing to a certain degree exists naturally in plants as an antiviral defence even without genetic transformation (Plasterk 2002). In transgenic plants expressing a viral sequence, some sRNA pre-exists before virus infection and can therefore block virus infection through sRNA-induced RNA silencing of viruses homologous to the sRNA. dsRNA or hpRNA is processed by Dicer-like (DCL) proteins into 21–24-nucleotide (nt) siRNA, a type of sRNA, which needs ATP participation (Nykänen et al. 2001). Then, sRNAs are combined with members of the Argonaute (AGO) family to form an RNA-induced silencing complex (RISC) (Fig. 11.1). In RISC, sRNAs guide direct RNA degradation, translational repression, or DNA methylation of homologous target genes (Wang et al. 2012). There are three basic RNA silencing pathways in plants: namely the siRNA pathway, the miRNA pathway, and the RNA-directed DNA methylation (RdDM) pathway. The RdDM pathway is used against DNA viruses, such as geminiviruses (Pooggin et al. 2003; Arago and Faria 2009). This chapter will focus on the siRNA and miRNA pathways which have been widely employed in plant antiviral genetic engineering, likely because the most devastating viruses in the major crops are RNA-type viruses.

11.3 Strategies of RNA Silencing in Plant Antiviral Genetic Engineering

11.3.1 Sense or Antisense Viral Sequence-Mediated Resistance

Transgenic plants expressing the *Tobacco mosaic virus* (TMV) CP gene showed resistance to TMV (Abel et al. 1986). This type of resistance, defined as protein-mediated resistance (PDR), led the way for the production of transgenic plants with resistance to virus using most types of viral genes, such as replicase (Golemboski et al. 1990), replication-associated protein (Hong and Stanley 1995), or movement protein (Lapidot et al. 1993; Malyshenko et al. 1993). Initially, it was believed that the viral proteins expressed from the transgenes conferred resistance in the plant, but subsequently it was found that expressed noncoding viral sense or antisense viral sequences also conferred resistance in plants (Waterhouse et al. 1998). Therefore, the resistance was derived not only from viral proteins but also from viral RNA itself. Similar studies have also shown that RMVR, which induced a form of PTGS, was the main mechanism in plant antiviral defence (Baulcombe 1996; Stam et al. 1997). In these early studies, the expressed viral proteins seemingly have no function in virus resistance. However, other studies of CP-mediated TMV resistance showed that the state of aggregation of CPs (Asurmendi et al. 2007) and the interference to accumulation of virus replication complexes by the mutated CP (Bendahmane et al. 2007) also affect plant virus resistance. Thus, despite the uncertain mechanisms of protein-mediated resistance, transgene resistance may be conferred due to the coexistence of both protein- and RNA-mediated resistance, although RNA-mediated resistance is likely the main mechanism.

In noncoding viral-sequence-mediated resistance, the length of viral sequence appeared to be important for triggering RMVR. For example, transgenic tobacco expressing >235 bp of the *Tomato spotted wilt virus* (TSWV) nucleocapsid (N) gene sequence showed TSWV resistance; whereas smaller N gene segments were ineffective (Pang et al. 1997). Expressing the 3′ end of potato Y virus-N strain (PVY^N) the CP gene in plants also indicated that the shortest effective length of the CP gene fragment which conferred resistance to tobacco could be somewhere between 202 bp and 417 bp from the 3′ end of the CP gene (Zhu et al. 2004). Furthermore, the expression level of transgenes could also affect the RMVR. The addition of flanking matrix attachment regions (MARs), gene expression enhancers, increased the efficiency and stability of RNA-mediated resistance to TSWV

(Levin et al. 2005) or PVY (Chi et al. 2005) in transgenic tobacco. In most studies, 5%–20% of the transgenic plants were resistant to the viruses, and the remaining plants were either susceptible or showed delayed symptom development. Further research is required to more efficiently trigger RNA silencing in RMVR.

11.3.2 Virus-Derived hpRNA-Mediated Resistance

RMVR is a process which is highly correlated with the accumulation of siRNAs. The instability of plant resistance using sense or antisense viral sequence transgene methods may be due to the low accumulation of siRNA. In transgenic plants containing sense or antisense viral sequences, plant hosts recognize and amplify the exogenous aberrant transgenic sequences, by plant-encoding RNA-dependent RNA polymerase (RDRP) into dsRNA, which serves as the substrate to trigger RNA silencing (Dalmay et al. 2000). Further exploiting the mechanism of RNA silencing proved that the inverted repeat transgene was transcribed into hpRNA, which then was further processed to produce siRNA that triggered RNA silencing (Fig. 11.1). The inverted repeat transgene approach is now widely used to achieve plant resistance to viruses.

The hpRNA consists of an inverted repeat of a fragment of the gene sequence (called 'stem') separated by a spacer (called 'loop'). Although the sequence of loop is not involved in RNA silencing, the source of loop sequence or stem–loop proportion have an effect on the resistance efficiency. In the PVY-resistance study, the percentage of PVY-resistant plants obtained by targeting the protease gene of PVY with different constructs was 58% for the hpRNA with an unrelated loop and 96% with a spliceable intron (Smith et al. 2000). Similar results occurred in *Cucumber mosaic virus* (CMV)-resistant transgenic plants (Chen et al. 2004). The stem–loop length ratio also affects plant virus resistance. When the proportion of stem–loop ratio was 4:1, 2:1, 1:1, about 60% of transgenic plants exhibited resistance to PVY^N. However, with the length of loop increased, the ratio of resistant plants lessened and when the proportion of stem–loop ratio was 1:8, only 9.52% of plants were resistant (Hirai et al. 2007; Li et al. 2008). Based on these results, a 4:1 stem–loop ratio is most appropriate to hpRNA formation, likely because it improves the stability of hpRNA, reduces the probability of ribozyme attack, and induces efficient gene silencing. For an effective antiviral transgenic construct, the 4:1–1:1 stem–loop ratios and a plant intron as the loop are recommended.

The sequence of double stranded stem plays a pivotal role in determining RNA silencing specificity; whereas, stem length affects the degree of resistance. For targeting the CMV RNA2 gene, the hpRNA stem using the full length of RNA2 and the hpRNA stem using a shorter fragment of RNA2 yielded 75 and 30% resistant plants, respectively (Chen et al. 2004). The PVY CP targeting hpRNA with 200 or 50 bp fragments as the stem obtained 82% or 69% resistant plants, respectively (Zhu et al. 2006; Li et al. 2007). The use of a long viral complementary DNA (cDNA) fragment as the hpRNA stem may offer a stronger gene silencing, but usually results in more off-target effect or virus recombination. The 50 bp as hpRNA stem is sufficient for obtaining resistance making it the ideal choice for the target sequence. Targeting different viral genes causes different degrees of resistance. In one study, eight different genes of PVY were targeted and the percentage of resistant plants obtained ranged from 33 % to 64 % (Chen et al. 2010). Not all invert repeat constructs targeting different genes of the *Rice stripe virus* (RSV) were equally effective in preventing RSV infection (Shimizu et al. 2011). It is essential to target the viral gene that plays an important role in viral proliferation. Furthermore, the targeting region of the virus has a more significant effect on the efficiency of RNA silencing. Targeting different regions of the same PVY CP gene can also produce different degrees of virus resistance. For example, targeting the middle and 3′ end was more effective than the 5′ end for RNA silencing (Li et al. 2007; Jiang et al. 2011b). Other studies also showed that siRNAs are preferentially generated from the 3′ end region of the transgene (Elbashir et al. 2001; Harborth et al. 2001; Holen et al. 2002). The analysis of secondary structures of target sequences illustrated a positive correlation between the local free energy of the secondary structure (sequences with less secondary structure usually have higher free energy) and the RNA silencing effectiveness (Jiang et al. 2011b). So, not only the structure of hpRNA itself but also the targeting regions affect plant resistance to viruses.

Plant viral resistance from hpRNA transgenes is effective but still far from complete resistance. Some transgenic plants show susceptibility or only delayed infection. Insertion at different loci can result in distinct cytoplasmic and nuclear RNAi activity. In a comparison between two insertion loci, a transgene-produced $hpRNA_{CMV}$ transcript was processed into substantial amounts of $siRNAs_{CMV}$ and the transgenic plants were virus resistant when the hpRNA gene was inserted into one locus, but the plants with $hpRNA_{CMV}$ at a second locus were susceptible to virus and failed to produce detectable $siRNAs_{CMV}$ even though the gene was transcribed (Dalakouras et al. 2011). The results also showed a correlation between the siRNA accumulation and hpRNA-mediated resistance. However, a recent report showed that accumulation of transgene-derived siRNAs is not sufficient for RNAi-mediated resistance by transforming Mexican lime plants (*Citrus aurantifolia*) with the 3′-terminal 549 nt of the *Citrus tristeza virus* (CTV) genome in intron-hairpin format. Analysis of intron-hairpin lines with single copy transgene insertion, showed that CTV resistance was correlated with low accumulation of the transgene-derived transcript rather than with high accumulation of transgene-derived siRNAs (LÓpez et al. 2010). A similar phenomenon occurred in RSV-resistant transgenic rice based on RNA silencing (Shimizu et al. 2011). These results suggested that only a fraction of the transgene-derived siRNAs are indeed functional for RNA silencing, with the other fraction being quickly degraded, like most virus-derived siRNAs in infected plants (Qu and Morris 2005).

Viruses counter the cellular antiviral mechanism by encoding diverse viral suppressors of RNA silencing (Díaz-Pendón and Ding 2008; Shi et al. 2008; Song et al. 2011). When these silencing suppressor genes were targeted by hpRNA such as in CTV (Fagoaga et al. 2006; Soler et al. 2012) and CMV (Chen et al. 2004), the plants became resistant to the viruses. However, these viral suppressor studies did not use multiple non-suppressor regions as control. In a study of hpRNA targeting on eight genes of PVY, including a silencing suppressor (HC-Pro), all induced tobacco plant resistance to PVY, but HC-Pro hpRNA induced the lowest percentage (33.4%) of resistant plants (Chen et al. 2010). However, since only one suppressor region was tested in that study, more suppressor genes should be tested statistically to confirm whether targeting other viral suppressor genes also induces lower resistance to viruses. To interpret the results of these previous studies, we hypothesize that virus silencing can be induced as long as the engineered RNA targets a critical region of the virus genome to interfere with genome replication although different proteins such as the silencing suppressors or CPs may also increase plant viral resistance through protein/enzyme level interaction(s). For confirmation, further RNA hpRNA research is required to statistically test this theory using several RNA silencing suppressors and several non-suppressor regions on the same construct background.

11.3.3 Artificial miRNA-Mediated Resistance

miRNAs generated from long miRNA precursors (pre-miRNA) can mediate RNA silencing similar to the effects of siRNA. Typically, pre-miRNAs are hairpin-shaped structures with bubbles on the stem because of sequence mismatches. This characteristic structure is processed by Dicer-like enzymes to produce the mature miRNA, a 21–24-nt single RNA that can be loaded into RISC. The miRNA–RISC mediates complementarity-dependent repression or degradation of target mRNA (Meister et al. 2004; Brodersen et al. 2008). It is possible to replace the sequence of the 21-nt mature miRNAs within the natural miRNA precursor, without disturbing its biogenesis and maturation, as long as the secondary structure of the pre-miRNA is maintained. The miRNA generated from this approach is called artificial microRNA or amiRNA. This has raised interest in creating amiRNA as a tool to silence genes of interest or virus. Compared with the long hpRNA- mediated resistance approach, amiRNAs have several advantages: First, a short amiRNA fragment may have fewer off-target effects. Second, the amiRNA uses the endogenous pre-miRNA gene as a template, which is more stable than non-endogenous ones to express in plant cells; Third, the high specificity in targeting is preferred for biosafety even through long hpRNA from virus sequence increases the probability of recombining with a second virus to achieve new functionalities.

The approach of amiRNA-mediated virus resistance was first achieved in *Arabidopsis thaliana* by targeting a viral silencing repressor gene sequence (Niu et al. 2006). The pre-miR159 was used to express two amiRNAs^{159} (amiR-p69^{159} or amiR-HC-Pro^{159}) for targeting the sequence of two viral silencing repressors, P69 of the *Turnip yellow mosaic virus* (TYMV) or HC-Pro of the turnip mosaic virus (TuMV). Transgenic plants expressing these two amiRNAs exhibited specific resistance to TYMV or TuMV, respectively. In *Nicotiana tabacum,* amiRNA (based on an *A. thaliana* miR171 precursor) was designed to target 2b of CMV and also conferred virus resistance. In this study, it was confirmed that expression of miRNA was more effective than expression of short hairpin sRNA in inhibiting virus function (Qu et al. 2007). Subsequently, the amiRNA approach was reported to confer high resistance against two different strains of CMV (Shandong and Q strains) (Duan et al. 2008). Recently, PVY or *Potato virus X* (PVX) resistance was obtained in *N. tabacum* expressing amiRNAs targeting HC-Pro of PVY or TGBp1/p25 of PVX (Ai et al. 2011).

A series of elements such as the different miRNA backbones, the targeting viral sequence of the 21-nt amiRNA, and the 21-nt amiRNA sequence itself affected amiRNA-mediated resistance efficiency. The miRNA precursor backbones that are highly expressed in organisms are the best for amiRNA function. By comparison of the three different amiRNA-mediated resistances, the different resistance levels of transgenic plants were positively correlated with the expression levels of amiRNAs (Ai et al. 2011). Concerning the study of targeting viral sequence, targeting 2a and 2b RNAs of CMV induced a higher antiviral ability than only targeting 2b (Qu et al. 2007; Zhang et al. 2011a). In addition, the amiRNA form had a different region on the 3′untranslated region (UTR) of CMV or the CP gene of PVY and both showed different resistance to CMV or PVY (Duan et al. 2008; Jiang et al. 2011a). The analysis suggested that the accessibility of the target site was a determinant of the efficacy of amiRNA (Duan et al. 2008) and the targeting sequences of loose structure (i.e., with no strong secondary structure) (Jiang et al. 2011a) were beneficial to the improvement of the virus resistance level. Two studies using PVX chimera with miRNA target sites (Simón-Mateo and García 2006) and amiRNA targeting TuMV (Lin et al. 2009) confirmed that only the 5′ region of miRNAs is critical for the initial target RNA binding in plants. Three different groups were defined according to the sensitivity of resistance breakdown to position mutations: critical (positions 3–6, 9, and 12), moderately critical (positions 2, 10, 11, 13, 15, and 18), and noncritical (the remaining), and the amiRNA-mediated specific resistance could be overcome by up to two mutations on the critical positions within the 21-nt sequence (Lin et al. 2009). So, to achieve highly efficient amiRNA-mediated resistance, appropriate miRNA backbones, less structural flanking region, and functionally conserved regions must all be considered.

11.4 Factors Affecting RNA Silencing-Mediated Resistance

Although RNA silencing as an antiviral strategy has been used in many crops (Dietzgen and Mitter 2006), high resistance in the greenhouse often broke down under field conditions. It is a common phenomenon for plants to be invaded by a complex of diverse pathogen sources in the field. In addition, due to the high similarity between targeting sequence and transgene, viruses that escape mutation may regenerate to counter resistance obtained by RNA silencing. Environmental conditions, such as low temperature, can decrease RNA silencing-based transgenic resistance to virus. Furthermore, public concerns about biosafety of transgenic plants severely limit the application of transgenic resistance strategy. These factors and solutions are discussed in the following section.

11.4.1 Multiple Viruses

In the field, plants are frequently subjected to multiple virus infections, resulting in intensified symptoms expression and virus accumulation, a phenomenon known as synergism. So, multiple virus resistance strategy based on RNA silencing has been developed to combat multiple virus infections. The hpRNA-mediated resistance approach involved constructing a single chimeric hpRNA comprised of 4 N gene segments (150 nt) of *Tospoviruses* from TSWV, *Groundnut ringspot virus* (GRSV), *Tomato chlorotic spot virus* (TCSV), and *Watermelon silver mottle virus* (WSMoV) that displayed a broad-spectrum resistance against tospoviruses (Bucher et al. 2006). Similarly, a single chimeric hpRNA fused with CP gene segments from PVY, TMV, and CMV also conferred resistance to these three viruses (Zhu et al. 2009). Short hairpin RNA (shRNA) that targeted highly similar regions for PNY^{N} and TEV-SD1 conferred resistance against both viruses (Zhang et al. 2013). Furthermore, co-expression of two amiRNA precursors on the same construct to generate a dimeric amiRNA could be used to produce two amiRNAs and provide resistance to two different viruses (Niu et al. 2006). Another study has also reported that transgenic *N. tabacum* expressing amiRNAs targeting HC-Pro and p25 from *A. thaliana* pre-miR159a showed highly specific resistance against PVY and PVX (Ai et al. 2011). Therefore, chimeric hpRNA, shRNA, and multimeric amiRNA strategies are all extremely useful to address the issue of multiple virus infections under field condition.

11.4.2 Virus Escape from Targeting

Viruses have a great capacity to generate biological variation by mutation, which causes viral sequence diversity among different isolates derived from the same virus. The sequence similarity between the transgene sequence and the targeting virus sequence is critical in RNA silencing-mediated resistance. Virus may escape from RNA silencing by mutation. Two engineering strategies can be adopted to prevent resistance loss due to virus mutation. One is to choose highly conserved virus sequences from a virus as the effective fragment target; the other is to construct the chimeric genes using parts of different genes of one virus with the objective to target different genes of the same virus simultaneously. For the first strategy, transgenic tobacco plants containing a hpRNA construct of the PVY replicase sequence were challenged with a variety of virus strains (Xu et al. 2009; Gaba et al. 2010). The transgenic tobacco line was resistant to PVY strains with about 90 % sequence homology to the introduced transgenes. The latter strategy has applications for animal viruses, especially on rapidly mutating animal viruses, such as *Human immunodeficiency virus* (HIV) (Song et al. 2003) or hepatitis B virus (HBV) (Bian et al. 2012). In plants, transgenic plants expressing a CP/SP chimeric gene of RSV were more resistant to two RSV isolates than the transgenic plant expressing a single CP or SP gene (Ma et al. 2011).

When the same TuMV was allowed to evolve by serial passages on virus partially resistant transgenic plants, the virus developed resistance to the amiRNA-induced silencing approximately seven times faster than viruses evolving in the wild plant. Virus-sensitive *A. thaliana,* and the resistance-breaking mutations were usually on the central positions of the target (e.g. positions 11 and 12) (Lafforgue et al. 2011). A subsequent ultradeep sequencing study of the TuMV-evolved lineages showed that some substantial nucleotide diversity was generated at every site of the amiRNA-target sequence along the evolution phase and the variant that eventually broke resistance was sampled among the many coexisting ones. The frequency of mutant alleles potentially breaking resistance was significantly greater in virus lineages replicating in the partially resistant transgenic lines than in wild-type *A. thaliana* (Martínez et al. 2012). Like hpRNA-mediated resistance, two strategies can also be used in amiRNA-mediated resistance. While the study of the latter one found the expression of multiple amiRNAs targeting conserved motifs in viral genomes has also been shown to overcome virus escape mutation for both negative- and positive-sense plant RNA viruses (Fahim et al. 2012; Kung et al. 2012). Recently, similar frequencies of virus mutations escaping from silencing were detected from viruses in transgenic plants expressing two amiRNAs complementary to independent targets (both are moderately conserved) and from transgenic plants expressing amiRNAs complementary to highly conserved RNA motifs of the viral genome (Lafforgue et al. 2013). Further research is required to compare transgenic plants expressing amiRNAs targeting one or two highly conserved viral genome sequences.

11.4.3 Temperature

Temperature can strongly affect plant–virus interactions. Under high temperatures, symptoms are frequently attenuated and virus titres in infected plants are decreased. In contrast, outbreaks of virus diseases are frequently associated with low temperatures. Meanwhile, low temperature can also affect siRNA-mediated resistance. A study of RNA silencing induced by viruses or transgenes demonstrated that siRNA accumulation could be inhibited at low temperatures (15 °C) and enhanced with rising temperatures (Szittya et al. 2003). However, inhibition of RNA silencing or decrease of siRNA concentration in low temperature has not always been observed. Transgene antisense-induced RNA silencing was not inhibited in potato plants at low temperature (Sós-Hegedus et al. 2005). Moreover, transgenic tobacco plants transformed separately with inverted repeat constructs targeting the sequences of TMV movement protein gene and the CMV replication protein gene, exhibited at both 15 and 24 °C similar high levels of resistance to TMV or CMV (Hu et al. 2011). On the other hand, the accumulation of miRNA is believed to be independent of changing temperatures and the transgenic lines expressing virus-derived amiRNA retain their resistance (Szittya et al. 2003; Niu et al. 2006).

11.4.4 Biosafety

Although transgenic plants mediated by RNA silencing exhibit efficient resistance to diverse viruses, potential safety issues are raised with the release of antiviral transgenic plants. Since transgenic papaya (*Carica papaya*) with resistance to the *Papaya ringspot virus* (PRSV) has proven to be a great success in the past few years, there have been many different crops protected against a whole range of viruses through the antiviral pathways of RNA silencing (Tepfer 2002; Fuchs and Gonsalves 2008). Perceived risks such as heteroencapsidation, recombination, gene flow, effect on nontarget organisms, and synergism may be avoided by the amiRNA strategy because only short sequences of virus were delivered into plant cells. However, risk assessment still needs to be conducted for the release of transgenic crops. In order to overcome this problem, a transient RNA silencing system was developed in plants by directly delivering hpRNA molecules into plant tissues without genetic transformation of the plants.

This nontransgenic strategy was first applied by directly delivering dsRNA to leaf cells either by mechanical inoculation or via an *Agrobacterium*-mediated transient expression assay (Tenllado and Díaz-Ruíz 2001). Meanwhile, another study found that a bacterial strain lacking the dsRNA-specific endonuclease RNase III could be cultured to produce high levels of specific dsRNAs, and those dsRNAs could effectively trigger strong and gene-specific gene silencing when fed to *Caenorhabditis elegans* (Timmons et al. 2001). Subsequently, this method was designed for bacterial systems to biosynthesize dsRNA in vivo, and crude extracts were inoculated into plants via spraying onto plant surfaces (Tenllado et al. 2003). More studies on nontransgenic approaches were conducted, including producing dsRNA using different dsRNA prokaryotic expression systems (Yin et al. 2009), resistant efficiency mediated by dsRNA derived from different viral function genes (Sun et al. 2010b), and different regions of the same gene (Sun et al. 2010a). The nontransgenic dsRNA approach, as confirmed in both a dicot plant (tobacco) and a monocot plant (maize), provided the possibility to protect a variety of crops from a diverse group of plant viruses (Gan et al. 2010). However, in contrast to the heritable resistance mediated by transgene RNA silencing, the transient approach did not confer long-term protection (Tenllado et al. 2003). Therefore, continuous spraying is required for the maintenance of protection.

11.5 Concluding Remarks and Future Directions

Current understanding of RNA silencing pathways (summarized in Fig. 11.1) have already provided new platforms for developing strategic molecular tools for plant antiviral defence. The hpRNA and amiRNA systems, developments based on our knowledge of two basic sRNA pathways in plants (siRNA and miRNA), have already proven to be effective tools for genetic engineering of RNA virus resistance. Based on the character of viral genomes, proper structural design of hpRNA or amiRNA plays a key role for obtaining highly efficient and inheritable resistance.

RNA silencing-mediated resistance to DNA viruses in plants has been rarely studied. In hpRNA- mediated resistance to geminiviruses, which have single-stranded circular DNA genomes, the resistance mechanism was related to RdDM and siRNA pathways of RNA silencing (Zhang et al. 2011b). However, more details about DNA virus resistance still needs to be uncovered to obtain high resistance efficiency to DNA viruses.

Transgenic plants with high resistance or immunity to RNA viruses or geminivirus have been obtained in several plants species, but most of the successful examples were obtained in greenhouses. Multiple field factors such as synergism, virus escape from targeting, and low temperature, may cause the loss of resistance in nature. Highly conserved sequences and chimeric constructs are both effective in countering multiple viruses or virus escape. The recent evolution experiment showed that viruses commonly mutate in central nucleotides of the amiRNA target. But similar research is not done on the siRNA target in plants. Finding the relevance between different siRNA target positions and resistance effectiveness is important for siRNA-mediated resistance to viruses.

Moreover, the antiviral strategy of spraying hpRNA onto plant surfaces avoids the public concern about transgenic crops and expands the application of RNA silencing mechanisms. Designing chimeric hpRNA to target multiple viruses is the next testing step. Many factors like spray titre and spray time also need to be tested in the field before application. It is important to determine how to overcome factors affecting RNA silencing-mediated resistance in the field and to improve the biosafety of antiviral spray activities.

References

Abel PP, Nelson RS, De B, Hoffmann N, Rogers SG, Fraley RT, Beachy RN (1986) Delay of disease development in transgenic plants that express the tobacco mosaic virus coat protein gene. Science 232:738–743

Ai T, Zhang L, Gao Z, Zhu CX, Guo X (2011) Highly efficient virus resistance mediated by artificial microRNAs that target the suppressor of PVX and PVY in plants. Plant Biol 13:304–316

Arago FJL, Faria JC (2009) First transgenic geminivirus-resistant plant in the field. Nat Biotechnol 27:1086–1088

Asurmendi S, Berg RH, Smith TJ, Bendahmane M, Beachy RN (2007) Aggregation of TMV CP plays a role in CP functions and in coat-protein-mediated resistance. Virology 366:98–106

Baulcombe DC (1996) Mechanisms of pathogen-derived resistance to viruses in transgenic plants. Plant Cell 8:1833–1844

Bendahmane M, Chen I, Asurmendi S, Bazzini AA, Szecsi J, Beachy RN (2007) Coat protein-mediated resistance to TMV infection of *Nicotiana tabacum* involves multiple modes of interference by coat protein. Virology 366:107–116

Bian Z, Xiao A, Cao M, Liu M, Liu S, Jiao Y, Yan W, Qi Z, Zheng Z (2012) Anti-HBV efficacy of combined siRNAs targeting viral gene and heat shock cognate 70. Virol J 9:275

Brodersen P, Sakvarelidze-Achard L, Bruun-Rasmussen M, Dunoyer P, Yamamoto YY, Sieburth L, Voinnet O (2008) Widespread translational inhibition by plant miRNAs and siRNAs. Science 320:1185–1190

Bucher E, Lohuis D, van Poppel PMJA, Geerts-Dimitriadou C, Goldbach R, Prins M (2006) Multiple virus resistance at a high frequency using a single transgene construct. J Gen Virol 87:3697–3701

Chen YK, Lohuis D, Goldbach R, Prins M (2004) High frequency induction of RNA-mediated resistance against *Cucumber mosaic virus* using inverted repeat constructs. Mol Breed 14:215–226

Chen X, Liu J, Xu L, Jiang F, Xie X, Zhu C, Wen F (2010) Inhibiting virus Infection by RNA Interference of the eight functional genes of the *Potato virus Y* genome. J Phytopathol 158:776–784

Chi S, Song Y, Zhu C, Zheng C, Liu X, Wen F (2005) Effect of matrix attachment regions on the RNA-mediated resistance induced by PVY^{N}-CP gene. Acta Phytopathol Sin 35:345–351

Cogoni C, Romano N, Macino G (1994) Suppression of gene expression by homologous transgenes. Antonie van Leeuwenhoek 65:205–209

Dalakouras A, Tzanopoulou M, Tsagris M, Wassenegger M, Kalantidis K (2011) Hairpin transcription does not necessarily lead to efficient triggering of the RNAi pathway. Transgenic Res 20:293–304

Dalmay T, Hamilton A, Rudd S, Angell S, Baulcombe DC (2000) An RNA-dependent RNA polymerase gene in arabidopsis is required for posttranscriptional gene silencing mediated by a transgene but not by a virus. Cell 101:543–553

Díaz-Pendón JA, Ding SW (2008) Direct and indirect roles of viral suppressors of RNA silencing in pathogenesis. Annu Rev Phytopathol 46:303–326

Dietzgen RG, Mitter N (2006) Transgenic gene silencing strategies for virus control. Australas Plant Pathol 35:605–618

Dougherty WG, Lindbo JA, Smith HA, Parks TD, Swaney S, Proebsting WM (1994) RNA-mediated virus resistance in transgenic plants: exploitation of a cellular pathway possibly involved in RNA degradation. Mol Plant-Microbe Interact 7:544–552

Duan CG, Wang CH, Fang RX, Guo HS (2008) Artificial microRNAs highly accessible to targets confer efficient virus resistance in plants. J Virol 82:11084–11095

Eamens A, Wang MB, Smith NA, Waterhouse PM (2008) RNA silencing in plants: yesterday, today, and tomorrow. Plant Physiol 147:456–468

Elbashir SM, Harborth J, Lendeckel W, Yalcin A, Weber K, Tuschl T (2001) Duplexes of 21-nucleotide RNAs mediate RNA interference in cultured mammalian cells. Nature 411:494–498

Fagoaga C, López C, De Mendoza AH, Moreno P, Navarro L, Flores R, Peña L (2006) Post-transcriptional gene silencing of the p23 silencing suppressor of *Citrus tristeza virus* confers resistance to the virus in transgenic Mexican lime. Plant Mol Biol 60:153–165

Fahim M, Millar AA, Wood CC, Larkin PJ (2012) Resistance to *Wheat streak mosaic virus* generated by expression of an artificial polycistronic microRNA in wheat. Plant Biotechnol J 10:150–163

Fire A, Xu S, Montgomery MK, Kostas SA, Driver SE, Mello CC (1998) Potent and specific genetic interference by double-stranded RNA in *Caenorhabditis elegans*. Nature 391:806–811

Fuchs M, Gonsalves D (2008) Safety of virus-resistant transgenic plants two decades after their introduction: lessons from realistic field risk assessment studies. Annu Rev Phytopathol 45:173–202

Gaba V, Rosner A, Maslenin L, Leibman D, Singer S, Kukurt E, Shiboleth YM, Gal-On A (2010) Hairpin-based virus resistance depends on the sequence similarity between challenge virus and discrete, highly accumulating siRNA species. Eur J Plant Pathol 128:153–164

Gan D, Zhang J, Jiang H, Jiang T, Zhu S, Cheng B (2010) Bacterially expressed dsRNA protects maize against SCMV infection. Plant Cell Rep 29:1261–1268

Golemboski DB, Lomonossoff GP, Zaitlin M (1990) Plants transformed with a *Tobacco mosaic virus* nonstructural gene sequence are resistant to the virus. Proc Natl Acad Sci U S A 87:6311–6315

Hamilton AJ, Baulcombe DC (1999) A species of small antisense RNA in posttranscriptional gene silencing in plants. Science 286:950–952.

Harborth J, Elbashir SM, Bechert K, Tuschl T, Weber K (2001) Identification of essential genes in cultured mammalian cells using small interfering RNAs. J Cell Sci 114:4557–4565

Hirai S, Oka SI, Adachi E, Kodama H (2007) The effects of spacer sequences on silencing efficiency of plant RNAi vectors. Plant Cell Rep 26:651–659

Holen T, Amarzguioui M, Wiiger MT, Babaie E, Prydz H (2002) Positional effects of short interfering RNAs targeting the human coagulation trigger tissue factor. Nucleic Acids Res 30:1757–1766

Hong Y, Stanley J (1995) Regulation of African cassava mosaic virus complementary-sense gene expression by N-terminal sequences of the replication-associated protein AC1. J Gen Virol 76:2415–2422

Hu Q, Niu Y, Zhang K, Liu Y, Zhou X (2011) Virus-derived transgenes expressing hairpin RNA give immunity to *Tobacco mosaic virus* and *Cucumber mosaic virus*. Virol J 8:41

Hull R (2002) Matthews' plant virology. Academic, San Diego

Jiang F, Song Y, Han Q, Zhu C, Wen F (2011a) The choice of target site is crucial in artificial miRNA-mediated virus resistance in transgenic Nicotiana tabacum. Physiol Mol Plant Pathol 76:2–8

Jiang F, Wu B, Zhang C, Song Y, An H, Zhu C, Wen F (2011b) Special origin of stem sequence influence the resistance of hairpin expressing plants against PVY. Biol Plant 55:528–535

Kennerdell JR, Carthew RW (1998) Use of dsRNA-mediated genetic interference to demonstrate that frizzled and frizzled 2 act in the wingless pathway. Cell 95:1017–1026

Kung YJ, Lin SS, Huang YL, Chen TC, Harish SS, Chua NH, Yeh SD (2012) Multiple artificial microRNAs targeting conserved motifs of the replicase gene confer robust transgenic resistance to negative-sense single-stranded RNA plant virus. Mol Plant Pathol 13:303–317

Lafforgue G, Martínez F, Sardanyés J, de la Iglesia F, Niu QW, Lin SS, Solé RV, Chua NH, Daròs JA, Elena SF (2011) Tempo and mode of plant RNA virus escape from RNA interference-mediated resistance. J Virol 85:9686–9695

Lafforgue G, Martínez F, Niu QW, Chua NH, Daròs J, Elena SF (2013) Improving the effectiveness of artificial microRNA (amiR)-mediated resistance against *Turnip mosaic virus* by combining two amiRs or by targeting highly conserved viral genomic regions. J Virol 87:8254–8256

Lapidot M, Gafny R, Ding B, Wolf S, Lucas WJ, Beachy RN (1993) A dysfunctional movement protein of tobacco mosaic virus that partially modifies the plasmodesmata and limits virus spread in transgenic plants. Plant J 4:959–970

Lecoq H, Moury B, Desbiez C, Palloix A, Pitrat M (2004) Durable virus resistance in plants through conventional approaches: a challenge. Virus Res 100:31–39

Levin JS, Thompson WF, Csinos AS, Stephenson MG, Weissinger AK (2005) Matrix attachment regions increase the efficiency and stability of RNA-mediated resistance to *Tomato spotted wilt virus* in transgenic tobacco. Transgenic Res 14:193–206

Li P, Song Y, Liu X, Zhu C, Wen F (2007) Study of virus resistance mediate d by inverted repeats derived from 5' and 3' ends of coat protein gene of *Potato virus Y*. Acta Phytopathol Sin 37:69–76.

Li Y, Song Y, Zhu C, Wen F (2008) Effect of stem-loop proportion of hpRNA on the RNA-mediated virus resistance. Acta Phytopathol Sin 38:468–477

Lin SS, Wu HW, Elena SF, Chen KC, Niu QW, Yeh SD, Chen CC, Chua NH (2009) Molecular evolution of a viral non-coding sequence under the selective pressure of amiRNA-mediated silencing. PLoS Pathog 5(2):e1000312

Lindbo JA, Dougherty WG (1992) Untranslatable transcripts of the tobacco etch virus coat protein gene sequence can interfere with tobacco etch virus replication in transgenic plants and protoplasts. Virology 189:725–733

LÓpez C, Cervera M, Fagoaga C, Moreno P, Navarro L, Flores R, Peña L (2010) Accumulation of transgene-derived siRNAs is not sufficient for RNAi-mediated protection against *Citrus tristeza virus* in transgenic Mexican lime. Mol Plant Pathol 11:33–41

Ma J, Song Y, Wu B, Jiang M, Li K, Zhu C, Wen F (2011) Production of transgenic rice new germplasm with strong resistance against two isolations of *Rice stripe virus* by RNA interference. Transgenic Res 20:1367–1377

Malyshenko SI, Kondakova A, Nazarova JV, Kaplan IB, Taliansky ME, Atabekov JG (1993) Reduction of *Tobacco mosaic virus* accumulation in transgenic plants producing non-functional viral transport proteins. J Gen Virol 74:1149–1156
Martínez F, Lafforgue G, Morelli MJ, González-Candelas F, Chua NH, Daròs JA, Elena SF (2012) Ultradeep sequencing analysis of population dynamics of virus escape mutants in RNAi-mediated resistant plants. Mol Biol Evol 29:3297–3307
Meister G, Landthaler M, Dorsett Y, Tuschl T (2004) Sequence-specific inhibition of microRNA- and siRNA-induced RNA silencing. RNA 10:544–550
Napoli C, Lemieux C, Jorgensen R (1990) Introduction of a chimeric chalcone synthase gene into petunia results in reversible co-suppression of homologous genes in trans. Plant Cell 2:279–289
Ngô H, Tschudi C, Gull K, Ullu E (1998) Double-stranded RNA induces mRNA degradation in trypanosoma brucei. Proc Natl Acad Sci USA 95:14687–14692
Niu QW, Lin SS, Reyes JL, Chen KC, Wu HW, Yeh SD, Chua NH (2006) Expression of artificial microRNAs in transgenic Arabidopsis thaliana confers virus resistance. Nat Biotechnol 24:1420–1428
Nykänen A, Haley B, Zamore PD (2001) ATP requirements and small interfering RNA structure in the RNA interference pathway. Cell 107:309–321
Pang SZ, Jan FJ, Gonsalves D (1997) Nontarget DNA sequences reduce the transgene length necessary for RNA-mediated tospovirus resistance in transgenic plants. Proc Natl Acad Sci U S A 94:8261–8266
Plasterk RHA (2002) RNA silencing: the genome's immune system. Science 296:1263–1265
Pooggin M, Shivaprasad PV, Veluthambi K, Hohn T (2003) RNAi targeting of DNA virus in plants. Nat Biotechnol 21:131–132
Prins M, Laimer M, Noris E, Schubert J, Wassenegger M, Tepfer M (2008) Strategies for antiviral resistance in transgenic plants. Mol Plant Pathol 9:73–83
Qu F, Morris TJ (2005) Suppressors of RNA silencing encoded by plant viruses and their role in viral infections. FEBS Lett 579:5958–5964
Qu J, Ye J, Fang R (2007) Artificial microRNA-mediated virus resistance in plants. J Virol 81:6690–6699
Sanford JC, Johnston SA (1985) The concept of parasite-derived resistance: deriving resistance genes from the parasite's own genome. J Theor Biol 113:395–405
Shi Y, Gu M, Fan Z, Hong Y (2008) RNA silencing suppressors: how viruses fight back. Future Virol 3:125–133
Shimizu T, Nakazono-Nagaoka E, Uehara-Ichiki T, Sasaya T, Omura T (2011) Targeting specific genes for RNA interference is crucial to the development of strong resistance to *Rice stripe virus*. Plant Biotechnol J 9:503–512
Simón-Mateo C, García JA (2006) MicroRNA-guided processing impairs Plum pox virus replication, but the virus readily evolves to escape this silencing mechanism. J Virol 80:2429–2436
Smith NA, Singh SP, Wang MB, Stoutjesdijk PA, Green AG, Waterhouse PM (2000) Total silencing by intronspliced hairpin RNAs. Nature 407:319–32.
Soler N, Plomer M, Fagoaga C, Moreno P, Navarro L, Flores R, Peña L (2012) Transformation of Mexican lime with an intron-hairpin construct expressing untranslatable versions of the genes coding for the three silencing suppressors of *Citrus tristeza virus* confers complete resistance to the virus. Plant Biotechnol J 10:597–608
Song E, Lee SK, Dykxhoorn DM, Novina C, Zhang D, Crawford K, Cerny J, Sharp PA, Lieberman J, Manjunath N, Shankar P (2003) Sustained small interfering RNA-mediated human immunodeficiency virus type 1 inhibition in primary macrophages. J Virol 77:7174–7181
Song L, Gao S, Jiang W, Chen S, Liu Y, Zhou L, Huang W (2011) Silencing suppressors: viral weapons for countering host cell defenses. Protein Cell 2:273–281
Sós-Hegedus A, Lovas Á, Kondrák M, Kovács G, Bánfalvi Z (2005) Active RNA silencing at low temperature indicates distinct pathways for antisense-mediated gene-silencing in potato. Plant Mol Biol 59:595–602
Stam M, Mol JNM, Kooter JM (1997) The silence of genes in transgenic plants. Ann Bot 79:3–12

Sun ZN, Song YZ, Yin GH, Zhu CX, Wen FJ (2010a) Hprnas derived from different regions of the nib gene have different abilities to protect tobacco from infection with *Potato virus Y*. J Phytopathol 158:566–568

Sun ZN, Yin GH, Song YZ, An HL, Zhu CX, Wen FJ (2010b) Bacterially expressed double-stranded RNAs against hot-spot sequences of tobacco mosaic virus or *Potato virus Y* genome have different ability to protect tobacco from viral infection. Appl Biochem Biotechnol 162:1901–1914

Szittya G, Silhavy D, Molnár A, Havelda Z, Lovas Á, Lakatos L, Bánfalvi Z, Burgyán J (2003) Low temperature inhibits RNA silencing-mediated defence by the control of siRNA generation. EMBO J 22:633–640

Tenllado F, Díaz-Ruíz JR (2001) Double-stranded RNA-mediated interference with plant virus infection. J Virol 75:12288–12297

Tenllado F, Martínez-García B, Vargas M, Díaz-Ruíz JR (2003) Crude extracts of bacterially expressed dsRNA can be used to protect plants against virus infections. BMC Biotechnol 3:3. doi:10.1186/1472-6750-3-3

Tepfer M (2002) Risk assessment of virus-resistant transgenic plants. Annu Rev Phytopathol 40:467–491

Timmons L, Court DL, Fire A (2001) Ingestion of bacterially expressed dsRNAs can produce specific and potent genetic interference in *Caenorhabditis elegans*. Gene 263:103–112

Van Blokland R Van der Geest N Mol JNM Kooter JM (1994) Transgene-mediated suppression of chalcone synthase expression in petunia hybrida results from an increase in RNA turnover. Plant J 6:861–877

Van Der Krol AR Mur LA Beld M Mol JNM Stuitje AR (1990) Flavonoid genes in petunia: addition of a limited number of gene copies may lead to a suppression of gene expression. Plant Cell 2:291–299

Wang MB, Masuta C, Smith NA, Shimura H (2012) RNA silencing and plant viral diseases. Mol Plant-Microbe Interact 25:1275–1285

Waterhouse PM, Graham MW, Wang MB (1998) Virus resistance and gene silencing in plants can be induced by simultaneous expression of sense and antisense RNA. Proc Natl Acad Sci U S A 95:13959–13964

Waterhouse PM, Smith NA, Wang MB (1999) Virus resistance and gene silencing: killing the messenger. Trends Plant Sci 4:452–457

Xu L, Song Y, Zhu J, Guo X, Zhu C, Wen F (2009) Conserved sequences of replicase gene-mediated resistance to potyvirus through RNA silencing. J Plant Biol 52:550–559

Yin G, Sun Z, Liu N, Zhang L, Song Y, Zhu C, Wen F (2009) Production of double-stranded RNA for interference with TMV infection utilizing a bacterial prokaryotic expression system. Appl Microbiol Biotechnol 84:323–333

Zhang X, Li H, Zhang J, Zhang C, Gong P, Ziaf K, Xiao F, Ye Z (2011a) Expression of artificial microRNAs in tomato confers efficient and stable virus resistance in a cell-autonomous manner. Transgenic Res 20:569–581

Zhang Z, Chen H, Huang X, Xia R, Zhao Q, Lai J, Teng K, Li Y, Liang L, Du Q, Zhou X, Guo H, Xie Q (2011b) BSCTV C2 attenuates the degradation of SAMDC1 to suppress DNA methylation-mediated gene silencing in Arabidopsis. Plant Cell 23:273–288

Zhang L, Xie X, Song Y, Jiang F, Zhu C, Wen F (2013) Viral resistance mediated by shRNA depends on the sequence similarity and mismatched sites between the target sequence and siRNA. Biol Plantrum 57:547–554

Zhu J, Zhu X, Wen F, Bai Q, Zhu C, Song Y (2004) Effect of cDNA fragments in different length derived from *Potato virus Y* coat protein gene on the induction of RNA-mediated virus resistance. Sci China Series C: Life Sci 47:382–388

Zhu X, Zhu C, Song Y, Wen F, Liu H, Li X (2006) Resistance to *Potato virus Y* mediated by the 3′ end segments of coat protein gene in the transgenic tobacco plants. Sci Agri Sin 39:1153–1158

Zhu CX, Song YZ, Yin GH, Wen FJ (2009) Induction of RNA-mediated multiple virus resistance to *Potato virus Y*, *Tobacco mosaic virus* and *Cucumber mosaic virus*. J Phytopathol 157:101–107

Chapter 12
Targeted Gene Mutation in Plants

Kunling Chen and Caixia Gao

12.1 Introduction

Targeted genome modification (TGM) has been widely used to investigate gene function and expand biotechnology applications in yeast (Scherer and Davis 1979), fruit fly (Bibikova et al. 2002), mice (Capecchi 2005), human cell lines (Urnov et al. 2005), plants, and many other organisms (Remy et al. 2010; Urnov et al. 2010). Reliable and efficient methods for obtaining site-specific modifications in plants are long sought-after goals for basic plant research and crop improvement (Pennisi 2010). Although first demonstrated in the late 1980s, TGM in plants was far from routine because of its extremely low efficiency (ranging from 10^{-3} to 10^{-6}; Paszkowski et al. 1988; Halfter et al. 1992; Tzfira and White 2005). The critical step in TGM is the introduction of DNA double-strand breaks (DSBs) at given genomic sites. The DSBs can be repaired by either nonhomologous end-joining (NHEJ) or homologous recombination (HR) pathways. NHEJ is often imprecise and can introduce mutations at target sites resulting in the loss of gene function. In contrast, HR uses a homologous DNA template which is provided by the investigator to invoke the cell's own repair machinery and can be employed to create site-specific sequence modifications or targeted insertions (Moynahan and Jasin 2010). Through NHEJ or HR, sequence-specific nucleases can be used to perform TGMs including mutations, insertions, replacements, and chromosome rearrangements (Fig. 12.1). Three powerful techniques, including zinc-finger nucleases (ZFNs), transcription activator-like nucleases (TALENs) and RNA-guided endonucleases (RGENs) genome engineering based on the type II prokaryotic clustered regularly inter-

C. Gao (✉)
State Key Laboratory of Plant Cell and Chromosome Engineering, Institute of Genetics and Developmental Biology, Chinese Academy of Sciences, 100101 Beijing, China
e-mail: cxgao@genetics.ac.cn

K. Chen
Institute of Genetics and Developmental Biology, Chinese Academy of Sciences, Beijing, China

X.-Q. Li et al. (eds.), *Somatic Genome Manipulation*,
DOI 10.1007/978-1-4939-2389-2_12

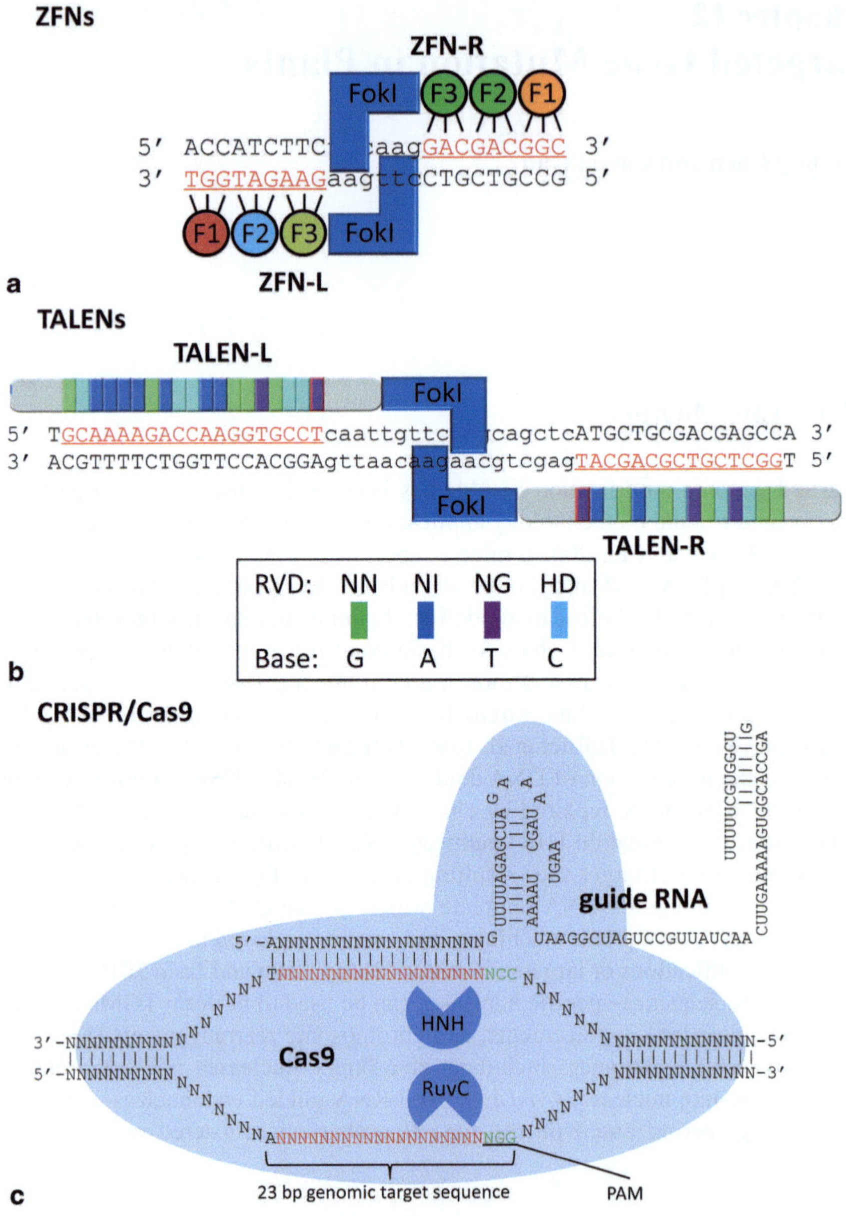

Fig. 12.1 The three customizable sequence-specific molecular scissors. **a** Zinc-finger nucleases (ZFNs). **b** Transcription activator-like effector nucleases (TALENs). **c** Clustered regularly interspaced short palindromic repeat (CRISPR)/CRISPR-associated protein 9 (Cas9). Both ZFNs and TALENs possess a FOKI nuclease domain for DNA cleavage, whereas the CRISPR/Cas system recruits a Cas nuclease for DNA cleavage. A zinc-finger module contacts three nucleotides of the target sequence on average, a TAL effector module binds one nucleotide, and an sgRNA interacts with one nucleotide of the DNA target site. TAL effector transcription activator-like effector, sgRNA single-guide RNA

spaced short palindromic repeat (CRISPR)/CRISPR-associated (Cas) system, have been developed for targeted DNA sequence modification of many organisms. For more than a decade, ZFNs have been widely used. These are engineered sequence-specific nucleases, i.e., chimeric proteins consisting of a customizable, zinc-finger based, DNA-binding domain (DBD) fused to the nonspecific cleavage domain of a FokI restriction endonuclease. ZFNs have been used to modify endogenous genes in a wide range of organisms, cell types, and plant species including *Arabidopsis* (Lloyd et al. 2005; de Pater et al. 2009; Osakabe et al. 2010; Zhang et al. 2010), tobacco (Cai et al. 2009; Townsend et al. 2009; Marton et al. 2010), maize (Shukla et al. 2009), petunia (Marton et al. 2010), and soybean (Curtin et al. 2011). Major constraints on ZFN application include the limited number of available target sites, more context-dependence effects between the repeat units, low targeting efficiency and specificity, and frequent off-target effects caused partly by nonspecific DNA binding (DeFrancesco 2012). TALENs have emerged as alternatives to ZFNs for gene targeting, and have been shown to have great potential for precise genome manipulation (Christian et al. 2010). Like ZFNs, TALENs consist of an engineered-specific transcription activator-like effector (TALE) DBD and FokI cleavage domain. The customizable TALE DBD, composed of several nearly identical tandem repeat arrays, can target any given sequence according to a simple repeat variable diresidue (RVD)-nucleotide recognition code (Bogdanove et al. 2010; Bogdanove and Voytas 2011). Within the past 3 years, TALEN-mediated genome modification has been widely adopted in yeast (Li et al. 2011), nematode (Wood 2011), fruit fly (Liu et al. 2012), rat (Tesson et al. 2011; Tong et al. 2012), human somatic and pluripotent cells (Hockemeyer et al. 2011; Miller et al. 2011), silkworm (Watanabe et al. 2012), livestock (Carlson et al. 2012), plants (Mahfouz et al. 2011; Li et al. 2012b; Shan et al. 2013; Zhang et al. 2013), *Xenopus* embryos (Lei 2012), zebra fish (Huang et al. 2011; Sander et al. 2011; Bedell 2012; Cade et al. 2012; Dahlem et al. 2012; Moore 2012), and many other organisms, and is recognized as a major breakthrough in the core technology of genome engineering (Varshney et al. 2011; Schierling et al. 2012). More recently, researchers from several groups have successfully employed RNA-guided genome engineering based on the type II prokaryotic CRISPR/Cas system (RGENs). In this system, only one customized guide RNA (gRNA) is required to target a specific sequence and the same Cas9 endonuclease to cleave the foreign DNA. This research on RNA-guided editing represents another breakthrough for genome engineering (Jinek et al. 2012; Cho et al. 2013; Cong et al. 2013; Hwang et al. 2013; Jiang et al. 2013; Mali et al. 2013). Genome editing with engineered nucleases was named the 2011 "Method of the Year" (Varshney et al. 2011), and 1 year later genome engineering using TALENs and CRISPR/Cas systems was crowned the 2012 "Breakthrough of the Year" by *Science* (Schierling et al. 2012). In this chapter, we discuss these different approaches for genome engineering used, in use, or for future use, for plant gene targeting.

12.2 Gene Targeting in Plants Using Zink-Finger Nucleases

ZFNs were the first custom-designed nucleases engineered to cut at specific DNA sequences in gene targeting in the late 1990s (Bibikova et al. 2003). ZFNs combine a DNA-recognition domain composed of three or four Cys_2His_2 zinc finger (ZF) modules with the nonspecific cleavage domain (N) of the FokI restriction endonuclease. Each ZF usually recognizes a 3-bp-long sequence, and binds DNA by inserting the a-helix into the major groove of the double helix. A ZF array (ZFA) normally consists of three or four fingers and can be engineered to bind diverse target DNA sequences within a genome of interest. The cleavage domain functions typically as a dimer, and the two ZFAs are designed to bind within a distance of 5–7 nucleotides from each other, producing an enzyme which is capable of targeting a unique DNA sequence, so that they can interact and induce the targeted DSB (Fig. 12.2a). Several zinc-finger engineering methods have been used to create novel ZFNs. Some groups have made ZFNs using "modular assembly," a method for engineering multi-finger arrays, which treats individual fingers as independent units. The success rate of modular assembly for making three- or four-finger assays has been reported to be low due to its failure to account for the context-dependent activity of zinc-finger domains in an array. Later, the oligomerized pool engineering (OPEN) method was efficiently used to make ZFNs in plants for targeted genome editing (Zhang et al. 2010). OPEN strategy uses a collection of zinc-finger pools, each consisting of a small number of different fingers designed to bind to a particular 3-bp unit, and an in vivo-based selection method. The labor and expertise required to screen combinatorial libraries have limited its broad adoption. An easier and effective alternative for making ZFNs, called context-dependent assembly (CoDA), uses archives of preselected two-finger units that function well together, obtained from OPEN (Sander et al. 2010). The arrays using the CoDA method can be rapidly assembled by simple cloning strategies, or DNA encoding the fingers can be synthesized commercially. Although CoDA accounts for context dependence between adjacent fingers, CoDA has some limitations compared to the other methods. For example, modular assembly is less effective than CoDA, but it can potentially be used to target sites that CoDA currently cannot target. With the current archive of CoDA units, a potential ZFN target site can be found approximately once in every 500 bp of random sequence (Sander et al. 2010). As more pools of finger variants for screening libraries in OPEN and two-finger units for assembly in ZF arrays in CoDA become available, CoDA will foster broader adoption of ZFN technology in genome-wide alterations. ZFN-mediated targeting has been successfully employed to modify many genes in plants (Table 12.1). However, the use of ZFNs is still constrained for most molecular laboratories due to the difficulty of production, high costs, and modest efficacy in many applications.

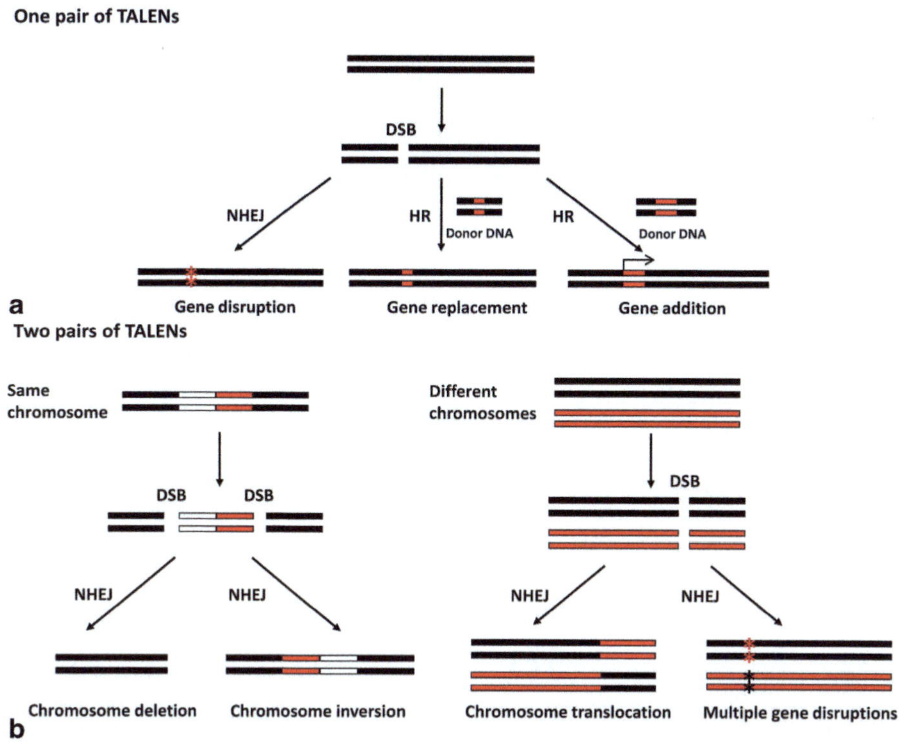

Fig. 12.2 Genome modification with TALENs. **a** Genome modification with one pair of TALENs. Double-strand breaks (DSBs) induced by one pair of TALENs will activate DNA repair mechanisms by either nonhomologous end joining (NHEJ) or homologous recombination (HR). Through NHEJ, DSBs lead to frame-shift mutations with small insertions or deletions (indels) and result in gene disruption (*left panel*); HR, stimulated by the homologous DNA template, leads to gene replacement (*middle panel*) or gene insertion (*right panel*). **b** Genome modification with two pairs of TALENs. The introduction of two pairs of TALENs targeted at two sites on the same chromosome can cause large deletions or inversions (*left panel*). The introduction of two nuclease-induced DSBs on different chromosomes can lead to translocations or multiple gene disruptions (*right panel*)

12.3 Gene Targeting in Plants Using Transcription Activator-Like Nucleases

TALENs are the most recent reagent of choice for efficiently modifying eukaryotic genomes in a targeted fashion (Baker 2012). Like ZFNs, TALENs are composed of an engineered array of DBDs fused with a nonspecific FokI nuclease domain (Christian et al. 2010). TALEs are major virulence factors secreted by the plant pathogenic bacterial genus *Xanthomonas,* which causes disease in plants such as rice and cotton (Boch and Bonas 2010; Bogdanove et al. 2010). TALEs are injected

Table 12.1 Examples of targeted gene modification using ZFNs or TALENs in plants

Plant species	Target gene	Modification type	ZFNs/TALENs	Mutations	Reference
Arabidopsis	Artificial sites	NHEJ	ZFN	Mutagenesis	(Lloyd et al. 2005)
Tobacco	*mGUS:NPTII*	NHEJ, HR	ZFN	Mutagenesis, Insertion	(Wright et al. 2005)
Tobacco	*mPAT*	HR	ZFN	Insertion	(Cai et al., 2009)
Tobacco	*CHR50*	HR	ZFN	Insertion	(Cai et al. 2009)
Arabidopsis	Artificial sites	NHEJ	ZFN	Mutagenesis	(de Pater et al. 2009)
Maize	*IPK1*	HR	ZFN	Insertion	(Shukla et al. 2009)
Tobacco	*SuRA, SuRB*	HR	ZFN	Insertion	(Townsend et al. 2009)
Arabidopsis	*mGUS*	NHEJ	ZFN	Mutagenesis	(Tovkach et al. 2009)
Tobacco	*mGUS*	NHEJ	ZFN	Mutagenesis	(Tovkach et al. 2009)
Tobacco	*mGUS*	NHEJ	ZFN	Mutagenesis	(Marton et al. 2010)
petunia	*mGUS*	NHEJ	ZFN	Mutagenesis	(Marton et al. 2010)
Arabidopsis	*ADH1*	NHEJ	ZFN	Mutagenesis	(Zhang et al. 2010)
Arabidopsis	*TT4*	NHEJ	ZFN	Mutagenesis	(Zhang et al. 2010)
Arabidopsis	*ABI4*	NHEJ	ZFN	Mutagenesis	(Osakabe et al. 2010)
Soybean	*DCL*	NHEJ	ZFN	Mutagenesis	(Curtin et al. 2011)
Soybean	*DCL4b*	NHEJ	ZFN	Mutagenesis	(Curtin et al. 2011)
Soybean	*RDR*	NHEJ	ZFN	Mutagenesis	(Curtin et al. 2011)
Soybean	*HEN*	NHEJ	ZFN	Mutagenesis	(Curtin et al. 2011)
Arabidopsis	*ADH1*	NHEJ	TALENs	Mutagenesis	(Cermak et al. 2011)
Tobacco	*EBE of Hax3*	NHEJ	TALENs	Mutagenesis	(Mahfouz et al. 2011)
Rice	*EBE of AvrXa7 and PthXo3*	NHEJ	TALENs	Mutagenesis	(Li et al. 2012b)
Tobacco	*SuRA, SuRB*	NHEJ, HR	TALENs	Mutagenesis, insertion and replacement	(Zhang et al. 2013)
Rice	*OsDEP1*	NHEJ	TALENs	Mutagenesis, large deletion, inversion	(Shan et al. 2013)

Table 12.1 (continued)

Plant species	Target gene	Modification type	ZFNs/ TALENs	Mutations	Reference
Rice	*OsBADH2*	NHEJ	TALENs	Mutagenesis	(Shan et al. 2013)
Rice	*OsCKX2*	NHEJ	TALENs	Mutagenesis	(Shan et al. 2013)
Rice	*OsSD1*	NHEJ	TALENs	Mutagenesis	(Shan et al. 2013)
Brachypodium	*BdABA1*	NHEJ	TALENs	Mutagenesis	(Shan et al. 2013)
Brachypodium	*BdCKX2*	NHEJ	TALENs	Mutagenesis	(Shan et al. 2013)
Brachypodium	*BdSMC6*	NHEJ	TALENs	Mutagenesis	(Shan et al. 2013)
Brachypodium	*BdSPL*	NHEJ	TALENs	Mutagenesis	(Shan et al. 2013)
Brachypodium	*BdSBP*	NHEJ	TALENs	Mutagenesis	(Shan et al. 2013)
Brachypodium	*BdCOI1*	NHEJ	TALENs	Mutagenesis	(Shan et al. 2013)
Brachypodium	*BdRHT*	NHEJ	TALENs	Mutagenesis	(Shan et al. 2013)
Brachypodium	*BdHTA1*	NHEJ	TALENs	Mutagenesis	(Shan et al. 2013)

TALENs transcription activator-like nucleases, *NHEJ* nonhomologous end joining, *ZFN* zinc-finger nucleases, *HR* homologous recombination

into host cells through the type III secretion system and interfere with cellular activities by activating the transcription of specific target genes (Bogdanove et al. 2010). They have specific structural features, including secretion and translocation signals in the N-terminal region, nuclear localization signals (NLS) and an acidic transcription–activation domain (AD) in the C-terminal region and a central DBD with 33~35 nearly identical long amino acid repeats, followed by the last module which contains only 20 amino acids (referred to as "half repeat"; Fig. 12.2a). The repeat variable diresidue (RVD) at positions 12 and 13 of each repeat dictates the specificity of repeat binding to a nucleotide in the DNA target (Fig. 12.2a). The two hypervariable residues in the RVD loops have different biochemical roles. The second amino acid of the RVD (position 13) mediates specific recognition of the sense-strand DNA base, whereas the first amino acid (position 12) does not directly contact the DNA but instead helps to stabilize the repeat structure (Deng et al. 2012; Mak et al. 2012). This RVD–nucleotide preference was first identified as the DNA sequence recognition code of TALEs in 2009 by two independent groups (Boch et al. 2009; Moscou and Bogdanove 2009). According to this code, the HD repeat specifies cytosine (C), NG specifies thymine (T), NI specifies adenine (A), NN specifies guanine (G) or adenine (A), N* specifies cytosine (C), IG specifies thymine (T), and NS specifies adenine (A), cytosine (C), guanine (G), or thymine (T) (Fig. 12.2a). Although many natural RVDs have been discovered, four of them, HD, NN, NI, and NG, account for 75 % of the total (Moscou and Bogdanove 2009). The DNA-binding specificity of a TAL effector is determined by its repeat number and the sequence of the RVD: The repeat number determines the length of the target sequence, while the RVD corresponds directly to the nucleotide in the target site. Moreover, recognition sites are always preceded by a T before the first repeat in the array, and this is the only critical rule for TALE targeting (Boch et al. 2009).

The long and highly repetitive nature of the DBDs of TALEs makes it a major challenge to construct engineered TALENs by ordinary polymerase chain reaction (PCR) and traditional cloning techniques. Although custom-engineered TALEs have become available commercially through Cellectis Bioresearch and Life Technologies, they are too expensive to adopt routinely like DNA sequencing. In the past 3 years, with more understanding of the features of TALENs, a variety of rapid TALEN assembly methods have been invented by a number of molecular genetics laboratories. Among them, the Golden Gate assembly method that uses ligation-based strategies is the most popular. Golden Gate assembly has the advantage of being simple, fast, and inexpensive, and has been widely used (Cermak et al. 2011; Li et al. 2011; Morbitzer et al. 2011; Sander et al. 2011; Weber et al. 2011). It employs type IIS restriction enzymes to create multiple sticky ends. Cloning is carried out by digestion and ligation in the same reaction mixture, and researchers can easily ligate up to ten TALE repeats in one reaction. For example, a custom TALEN with 15–31 repeats can be efficiently designed and made by two-step ligation using the Golden Gate assembly protocol invented by the Voytas laboratory (Cermak et al. 2011). The Golden Gate assembly method depends on sequencing to identify the correct clone, and it usually requires 3–5 days to accomplish assembly. Up to now, most TALENs targeting plant genes have been constructed by the Golden Gate method (summarized in Table 12.1). Researchers can obtain reagent kits for the above platforms from the nonprofit plasmid distribution service Addgene (http://www.addgene.org/talen/). Conveniently, sources of TALENs including target design and prediction, assembly methods, reference collections, protocols, and materials, and even newsgroups for discussion, are easily found on websites such as TALEN-NT (Doyle et al. 2012), idTALE (Li et al. 2012a), and EENdb (Xiao et al. 2013). Thus, very straightforward working methods for generating TALENs are being developed.

Although TALEN technology benefits from the discovery of the DNA recognition code which TALEs have adopted to regulate effector gene expression in plants, the achievements of genome modification in plants are lagging behind those in other organisms such as zebra fish (Huang et al. 2011; Sander et al. 2011; Cade et al. 2012; Moore 2012). Up to now, TALEN-mediated genome modification has been adopted in four plant species—*Arabidopsis* (Christian et al. 2010; Cermak et al. 2011; Li et al. 2012a), tobacco (Mahfouz et al. 2011; Li et al. 2012a; Zhang et al. 2013), rice (Li et al. 2012b; Shan et al. 2013), and *Brachypodium* (Shan et al. 2013). Most of the resulting mutations reported in plants until now are NHEJ-dependent genome mutations. The initial research was carried out by Christian et al. in Voytas' laboratory, 1 year after the recognition code was deciphered. These workers assembled pairs of TALENs targeting the *Arabidopsis* gene *ADH1* and proved that they were active in a yeast assay (Christian et al. 2010). One year later, Cermak et al. described a method and reagents for efficiently assembling TALEN constructs with custom repeat arrays. The same *Arabidopsis* gene *ADH1* was successfully TALEN targeted in *Arabidopsis* protoplasts using this method (Cermak et al. 2011). In the same year, Zhu's group generated a de novo Hax3 TALE-based hybrid nuclease (dHax3N) by fusing a modified dHax3 DBD sequence to a FokI C terminal cleavage domain (Mahfouz et al. 2011). When transiently expressed in tobacco leaves, the modified Hax3N created DSBs in its artificial target sequence in vivo. Furthermore, Li et al. (2012a) assembled dTALEN heterodimers binding to the coding sequence (CDS) of *AtPDS3* and demonstrated their cleavage activity in vitro. The high point of the application of TALEN technology to crop improvement so far has been the production of disease-resistant rice by high-efficiency TALE-based gene editing (Li et al. 2012b). Bacterial blight is a devastating disease responsible for large losses of rice yields, and the main cause of the virulence was the TALE secreted by *Xanthomonas oryzae* pv. *oryzae* (Xoo), which transcriptionally activates specific rice disease-susceptibility (*S*) genes. Li et al. (2012b) used TALENs to edit a specific *S* gene, the sucrose-efflux transporter gene (*Os11N3,* also named *OsSWEET14*), by disrupting the effector-binding element (EBE) in its promoter region without changing its expression. TALENs targeted at an essential sequence covering the natural EBEs of *AvrXa7* and *PthXo3* were designed and transformed into rice cells. Several independent mutant lines with disrupted EBE sequences were obtained, and the progeny of some of them displayed resistance specific to *AvrXa7* and *PthXo3,* and morphologically normal phenotypes.

Targeted gene insertions and replacements have been accomplished by the Voytas laboratory using highly efficient TALE scaffolds optimized by employing tobacco protoplasts and TALENs targeting the acetolactate synthase (*ALS*) gene (Zhang et al. 2013). The high frequencies of targeted mutations (30 %) and gene insertions (14 %) hold promise of permitting genome modification by protoplast transformation without the need for selection or enrichment regimes. In addition, when TALENs were co-transformed with donor DNA carrying a 6-bp variant of *surA* or *surB,* targeted gene replacements (4 %) in calli were achieved. The protoplast single-strand annealing (SSA) assay has been shown to provide a reliable means of assessing TALEN activity at endogenous chromosomal target sites. The Voytas's laboratory has provided plant biologists with an efficient plant genome engineering system using TALENs, including the Golden Gate assembly method (Cermak et al. 2011), optimal scaffolds of increased effectiveness (Zhang et al. 2013), and a nuclease activity validation method using protoplasts (Zhang et al. 2013). Their work thus offers hope of being able to perform plant genome engineering in general molecular biology laboratories.

Large-scale highly efficient TGM by TALENs in rice and *Brachypodium* has been achieved in our laboratory (Shan et al. 2013). We obtained targeted gene knockouts of four rice genes and eight *Brachypodium* genes, with mutation rates reaching > 30 % as measured both by restriction enzyme digestion assays and sequencing. Our reagents and cloning processes are seamlessly integrated into a published Golden Gate TALEN engineering platform, which has been used by more than 800 laboratories worldwide (Pennisi 2012). Instead of some laborious plant transformation procedure, a rapid, accurate, and reliable protoplast transient assay, based on PCR/restriction enzyme screening, was used to quickly screen active nuclease activities independently at each endogenous target site in rice and *Brachypodium*. The active TALENs were subsequently transformed into embryonic cells of rice or *Brachypodium* using the *Agrobacterium* transformation method. We identified a total of 127 mutant sequences in both rice and *Brachypodium,* and each was characterized as a small deletion, insertion, or nucleotide substitution. Most mutations were small deletions ranging from 1 to 20 bp, and all of the mutations occurred in the spacer region between the TALEN-binding sites. These high frequencies allowed us to isolate large genomic deletions (i.e., 1.3 kb) by simultaneous expression of two pairs of TALENs. This was the first report that demonstrated the production of genomic deletions of large, predetermined endogenous DNA segments in a targeted fashion in plants. In addition, preliminary experiments show that TALENs can induce insertion and deletion (indel) mutations that disrupt genes in *Arabidopsis,* petunia, cotton, maize, and wheat (unpublished data). It will not be surprising if, in the near future, TALEN technology is extended to potato, soybean, sorghum, oilseed rapes, and other plant species or crops of commercial value.

12.4 Gene Targeting Using RNA-Guided Endonucleases

Although genome-editing technologies such as ZFNs and TALENs have proved their abilities to enable genome modifications, there remains a need for new technologies that are robust, affordable, and easy to engineer. Recent advances in the study of the type II prokaryotic CRISPR adaptive immune system provide an alternative genome-editing strategy. CRISPR is a compound of bacterial and archaeal immunity. Three types (I, II, and III) of CRISPR systems with different features have been identified. The type II CRISPR systems, widespread in bacteria, use a single endonuclease, Cas9, which has attracted much attention due to its potential use in genomic engineering (Barrangou et al. 2007; Garneau et al. 2010; Deltcheva et al. 2011; Fig. 12.2b). This enzyme functions as part of a defense system against foreign DNA, in which it has two active sites that each cleave one strand of a double-stranded DNA molecule. The Cas9 endonuclease forms a complex with two RNAs called CRISPR RNA (crRNA) and trans-activating crRNA (tracrRNA), which form a targeted RGEN and guide Cas9 to recognize and cleave target DNA. The site-specific double-stranded DNA breaks are then repaired by either NHEJ or HR. The specificity of RGENs can be customized by replacing a short synthetic RNA molecule without changing the protein component. Several groups have now reported the application of RGENs as a simple and programmable tool to sequence-specific genome editing of bacteria, human cells, and whole zebra fish. Four of the groups showed that RNA-guided Cas9 can mediate genomic cleavage in human cells (Cho et al. 2013; Cong et al. 2013; Jinek et al. 2013; Mali et al. 2013). They reported that RNA-guided genome targeting in human cells is relatively simple to execute and the mutagenesis frequencies are similar to those obtained previously with ZFNs or TALENs. Moreover, Cho et al., Cong et al., and Mali et al. also demonstrated that the simultaneous introduction of multiple gRNAs into human cells can achieve multiplex gene editing of multiple target loci. Two other independent studies using this in vivo zebra fish system have been provided by Hwang at al. and Chang et al., who report the generation of somatic mutations in many loci by the Cas9/gRNA nuclease complex with efficiencies similar to that of ZFNs and TALENs (Chang et al. 2013; Hwang et al. 2013). Working in bacteria, Jiang et al. successfully use a dual-RNA:Cas9-directed system alone or together with an editing template to eliminate cleavage or introduce the desired mutations (Jiang et al. 2013).

In contrast to genome editing with ZFNs or TALENs, the target specificity of the RGENs is programmed with a small RNA, without any need for enzyme engineering. The design process is much simpler, as target site selection is determined solely by base complementarity to the guide DNA, and the Cas9 protein does not require reengineering for each new target site. Moreover, the high target specificity is determined by a 20-bp RNA–DNA interaction, which is encoded by a short sequence of approximately 100 bp. gRNA is therefore much easier and simpler to engineer than ZFNs or TALENs. Lastly, the short length of the gRNA sequence makes it easier to deliver into cells than the longer and highly repetitive ZFN/TALEN encoding vectors. The ease of the CRISPR/Cas system to modify genomic sequences greatly exceeds that of comparable ZFNs and TALENs, while offering similar or greater efficiencies. These reports described significant advances that hold real promise for application of CRISPR/Cas systems to other organisms, such as plants, for which plasmids can be introduced. It is expected that these CRISPR/Cas-directed genome-editing technologies will be broadly useful in the basic and applied plant sciences.

12.5 Conclusions and Perspectives

Gene-editing nucleases (i.e., ZFNs and TALENs) and the RGENs-mediated TGM are rapidly becoming powerful tools for genome engineering. In particular, in the past three years, there has been an explosion in the number and diversity of applications of these technologies (Joung and Sander 2012; Marx 2012; Mussolino and Cathomen 2012; Pennisi 2012). Different types of mutations, such as deletions, insertions, single-nucleotide substitution, and replacement, can be generated. Traditional transgene methods typically transform and integrate foreign genes into host cells at random. The biological effects may therefore be related to the integration sites rather than the transgenes themselves, so it is hard to define gene function unambiguously. Targeted gene insertion by ZFNs, TALENs, or RGENs is a good HR-dependent approach to integrate genes of interest into genomes at given sites and avoid random insertions. Using these targeted gene insertion technologies, researchers can select sites that are conducive to high levels of gene expression, and integrate different genes at the same locus under the control of identical regulatory elements to detect subtle differences in gene function. Furthermore, it could be used to insert multiple genes at the same chromosomal locus to stack genes. This makes it easier to combine all of the genes of interest at the same locus and then transfer them into other germplasm by crossing, to accelerate plant molecular breeding (Curtin et al. 2012; Tzfira et al. 2012). Targeted tagging is another example of targeted gene insertion. It specifically refers to inserting a tag or a reporter gene to label an endogenous gene. Zhang et al. used TALENs to generate an amyotrophic lateral sclerosis–green fluorescent protein (ALS–GFP) fusion protein expressed in tobacco protoplasts (Hockemeyer et al. 2011; Zhang et al. 2013).

In terms of targeted gene replacements, the precise modification is normally obtained by an HR-dependent DNA repair pathway, in which ZFNs, TALENs, or RGENs and donor DNA templates containing sequences homologous to those flanking the break site are delivered to cells simultaneously. In this way, DSBs induced by the TALENs can be immediately repaired using the donor DNA template and generate a specific change at a specific genomic location. Targeted gene replacements mediated by ZFNs, TALENs, or RGENs have been successfully used to modify predetermined sequences in yeast (Li et al. 2011), zebra fish (Bedell 2012), and rice (Zhang et al. 2013). Lastly, the development of large chromosomal rearrangement or simultaneous editing of several sites enables powerful applications across basic science, biotechnology, and medicine.

12.6 A Protocol for TALEN Construction

This protocol describes methods of TALEN Golden Gate assembly and combine vector construction (Cermak et al. 2011; Zhang et al. 2013).

12.6.1 Materials

12.6.1.1 Molecular Biology Reagents

1. TALEN plasmid library for assembly: Available from Addgene (http://www.addgene.org/TALeffector/goldengateV2/). In this protocol, a total of 57 plasmids are needed to be prepared, which include 40 module plasmids encoding RVDs (pHD1–10, pNN1–10, pNG1–10 and pNI1–10), four last repeat plasmid encoding half RVDs (pLR-HD, NN, NG and NI), ten array vectors (pFUS-A, pFUS-B1–9), two backbone vectors (pZHY500 and pZHY501), and one TALEN combine vector pZHY013 (Zhang et al. 2013).
2. T4 DNA ligase, restriction endonuclease BsaI and Esp3I (NEB), and other fast digested restriction enzymes including AflII, XbaI, BamHI, StuI, AatII, NheI, and BglII (Thermol).
3. Plasmid-Safe™ ATP-dependent DNase (Epicentre).
4. Axygen™ plasmid Miniprep kit and Axygen™ DNA gel extraction kit (Axygen).
5. Chemically competent cells of DH5α (TransGene).
6. Taq DNA polymerase (TransGene).
7. Other chemicals (Sigma).

Table 12.2 Primers used in this protocol

Name	Sequence	Usage
pCR8_F1	5′-ttgatgcctggcagttccct-3′	Colony PCR and sequencing primers for first cycle
pCR8_R1	5′-cgaaccgaacaggcttatgt-3′	
TAL_F1	5′-ttggcgtcggcaaacagtgg-3′	Colony PCR and sequencing primers for second cycle
TAL_R2	5′-ggcgacgaggtggtcgttgg-3′	

TAL transcription activator like, *PCR* polymerase chain reaction

12.6.1.2 Bacterial Media, Antibiotics and Additives

1. Antibiotic stock solutions (1000×): 100 mg/mL ampicillin, 50 mg/mL spectinomycin in double distilled water (ddH_2O) and 10 mg/mL tetracycline in 70% ethanol. Aliquot, store at −20 °C.
2. X-Gal and Isopropyl β-D-1-thiogalactopyranoside (IPTG) stocks: 20 mg/mL X-Gal in dimethyl sulfoxide (DMSO), 100 mM IPTG in ddH2O. Aliquot, store at −20 °C in dark.
3. Lysogeny broth (LB) liquid medium: Dissolve 10 g tryptone, 5 g yeast extract, 10 g NaCl in 1 L ddH2O. Autoclave to sterilize, allow media to cool to ~50 °C. Then add appreciated antibiotics and store at 4 °C.
4. LB solid medium: Same as LB Liquid medium, with addition of 15 g agar before autoclave. LB medium with IPTG/X-Gal: Spread 40 μL of 0.1 M IPTG and 40 μL 20 mg/mL X-gal on the surface of plate before use.
   a. LB/S plates, containing Spectinomycin (50 μg/mL)
   b. LB/AIX plates, containing Ampicillin (100 μg/mL), IPTG, and X-gal.
   c. LB/SIX plates, containing Spectinomycin (50 μg/mL), IPTG, and X-gal.

12.6.1.3 Primer Sequences

Primers pCR8_F1, pCR8_R1, TAL_F1, and TAL_R2 are used in this protocol (Table 12.2).

12.6.1.4 Equipment and Consumables

1. PCR thermocycler (BioRAD).
2. Environmentally controlled incubator (37 °C) for growing *Escherichia coli* (Eppendorf).
3. Environmentally controlled shaker (37 °C) for culturing *E. coli* (Eppendorf).
4. Nanodrop 2000 spectrophotometer for measuring plasmid concentration (Thermol).
5. Centrifuge for *E. coli* culture and plasmid purification (Eppendorf).

6. Heating water bath (42 °C; Changfeng).
7. Heating block (Eppendorf).
8. Standard equipment and regents for agrose gel electrophoresis (BioRAD).
9. A 0.2-mL thin-walled PCR tubes, 8-strip PCR tube, and 96 well PCR plate (Haimeng).

12.6.2 Methods

12.6.2.1 Preparation for Assembly

1. Set up plasmid library for TALEN constructs; each plasmid is diluted to 150 μg/μL before using.
2. Design considerations: There are numerous web-based software available for TALEN design and target finding. We use TAL effector nucleotide targeter (TALE-NT) 2.0 (https://tale-nt.cac.cornell.edu/), supplied by The Bogdanove laboratory, to design arrays of TALE to target our gene sequences.
3. Select candidate TALE-binding sites and design TALEN pairs, usually 2–3 sites per gene is enough to obtain expected mutation in one gene, and a specific restriction enzyme site in the spacer sequence is recommended for the TALEN activity assay.
4. Fill your TALEN RVD sequence (usually RVD number should vary from 15 to 20 and spacer length should be kept to 13–18 bp) in the Microsoft Excel spreadsheet of the Golden Gate TAL assembly form supplied by Addgene. Print the form before the experiment.

12.6.2.2 Day-1 of First Cycle Assembly

1. Choose your RVDs from the library according to your assembly form, pick plasmids for the first half of the RVDs 1–10 + destination vector pFUS_A; pick plasmids for the latter half of the RVDs 11 up to N–1 + destination vector pFUS_B#N-1. For example, if you prepare to assemble a TALEN of 18 RVDs, the first 10 RVDs should be ligated with pFUS_A, the left eight RVDs, except for the last half RVD, should be ligated with pFUS_B7.
2. Set up Golden Gate reaction separately in 0.2 mL PCR tubes:

   1 μL each module vector (RVDs)
   1 μL pFUS_A or pFUS_B (N-1)
   1 μL BsaI
   1 μL T4 DNA ligase
   2 μL 10X T4 DNA ligase buffer
   ? μL ddH_2O up to volume 20 μL
   20 μL reaction volume

3. Incubate at the following temperatures using a PCR thermal cycler:

   Step 1: 37 °C, 5 min
   Step 2: 16 °C,10 min (go back to step1 for nine more cycles)
   Step 3: 50 °C, 5 min
   Step 4: 80 °C, 5 min

4. Plasmid-Safe ATP-dependent DNase treatment to destroy all unligated double-stranded DNA (dsDNA) fragments: Once reaction above is finished, add 1 μL 10 mM ATP and 1 μL DNase into each tube and incubate at 37 °C for 1 h, then keep at 70 °C for 30 min to inactivate DNase.
5. Transform into competent cell with 10 μL reaction mixture, spread on LB/SIX plates and incubate in 37 °C incubator overnight.

12.6.2.3 Day 2 of First Cycle Assembly

1. Pick 1–3 white colonies from each plate and check by colony PCR using primers of PCR8_F1 and PCR8_R1.
2. Perform PCR with the following program:

   Step 1: 95 °C, 5 min
   Step 2: 95 °C, 30 s
   Step 3: 55 °C, 30 s
   Step 4: 72 °C, 2 min (go back to step 2 for 34 more cycles)
   Step 5: 72 °C, 5 min

3. Conduct agarose gel electrophoresis: A band of the expected size should be obtained, accompanied by smear and a "ladder" of bands starting at ~200 bp and every 100 bp up to ~500 bp.
4. Select right clones, start overnight culture on a 37 °C shaker.

12.6.2.4 Day 3 of Second Cycle Assembly

1. Miniprep plasmids and dilute to 150 ng/μL.
2. Use restriction enzymes AflII and XbaI to perform restriction digestion, right plasmid will send to sequencing with primer of pCR8_F1 or pCR8_R1.
3. Set up Golden Gate reaction for second cycle reaction:

   1 μL 150 ng of pFUS-A with first ten repeats cloned
   1 μL 150 ng of pFUSB (N-1) with 11-(N-1) repeats cloned
   1 μL 150 ng of respective pLR vector
   1 μL 150 ng of backbone vector (pZHY500 or pZHY501)
   1 μL Esp3I
   1 μL T4 Ligase

2 μL 10X T4 DNA ligase buffer
12 μL H_2O up to 20 μL
20 μL reaction volume

4. Incubate at the following temperatures using a PCR thermal cycler:

   Step 1: 37 °C, 5 min
   Step 2: 16 °C,15 min
   Step 3: 37 °C,15 min
   Step 4: 80 °C, 5 min

5. Transform into competent cell with 10 μL reaction mixture, spread on LB/AIX plates and incubate in 37 °C incubator overnight.

12.6.2.5 Day 4 of Second Cycle Assembly

1. Pick 1–3 white colonies from each plate and check by colony PCR using primers TAL_F1 andTAL_R2.
2. Perform PCR with the following program:

   Step 1: 95 °C, 5 min
   Step 2: 95 °C, 30 s
   Step 3: 55 °C, 30 s
   Step 4: 72 °C, 3 min (go back to step 2 for 35 cycles)
   Step 5: 72 °C, 5 min

3. Conduct agarose gel electrophoresis: a faint band around 2–3 KB, accompanied by a smear and a "ladder" of bands as in cycle 1.
4. Select right clones, start overnight culture on a 37 °C shaker.

12.6.2.6 Day 5 of Second-Cycle Assembly

Miniprep plasmids and perform restriction enzyme digestion with StuI and AatII, right clones will be sent for sequencing with primers of TAL_F1 or TAL_R2.

12.6.2.7 Clone TALEN Pairs into pZHY013 Sequentially

Clone left TALEN arrays into pZHY013 at XbaI/BamHI site at first, then clone right TALEN arrays into pZHY013 at NheI/BglII sites. These plasmids can be used as Gateway entry vectors to transferred TALEN pair into destination vectors with different usage.

References

Baker M (2012) Gene-editing nucleases. Nat Meth 9:23–26

Barrangou R, Fremaux C, Deveau H, Richards M, Boyaval P, Moineau S, Romero DA, Horvath P (2007) CRISPR provides acquired resistance against viruses in prokaryotes. Science 315:1709–1712

Bedell VM (2012) In vivo genome editing using a high-efficiency TALEN system. Nature 491:114–118

Bibikova M, Golic M, Golic KG, Carroll D (2002) Targeted chromosomal cleavage and mutagenesis in Drosophila using zinc-finger nucleases. Genetics 161:1169–1175

Bibikova M, Beumer K, Trautman JK, Carroll D (2003) Enhancing gene targeting with designed zinc finger nucleases. Science 300:764–764

Boch J, Bonas U (2010) Xanthomonas AvrBs3 family-type III effectors: discovery and function. Annu Rev Phytopathol 48:419–436

Boch J, Scholze H, Schornack S, Landgraf A, Hahn S, Kay S, Lahaye T, Nickstadt A, Bonas U (2009) Breaking the code of DNA binding specificity of TAL-type III effectors. Science 326:1509–1512

Bogdanove AJ, Voytas DF (2011) TAL effectors: customizable proteins for DNA targeting. Science 333:1843–1846

Bogdanove AJ, Schornack S, Lahaye T (2010) TAL effectors: finding plant genes for disease and defense. Curr Opin Plant Biol 13:394–401

Cade L, Reyon D, Hwang WY, Tsai SQ, Patel S, Khayter C, Joung JK, Sander JD, Peterson RT, Yeh JR (2012) Highly efficient generation of heritable zebrafish gene mutations using homo- and heterodimeric TALENs. Nucleic Acids Res 40:8001–8010

Cai CQ, Doyon Y, Ainley WM, Miller JC, DeKelver RC, Moehle EA, Rock JM, Lee YL, Garrison R, Schulenberg L, Blue R, Worden A, Baker L, Faraji F, Zhang L, Holmes MC, Rebar EJ, Collingwood TN, Rubin-Wilson B, Gregory PD, Urnov FD, Petolino JF (2009) Targeted transgene integration in plant cells using designed zinc finger nucleases. Plant Mol Biol 69:699–709

Capecchi MR (2005) Gene targeting in mice: functional analysis of the mammalian genome for the twenty-first century. Nat Rev Genet 6:507–512

Carlson DF, Tan W, Lillico SG, Stverakova D, Proudfoot C, Christian M, Voytas DF, Long CR, Whitelaw CB, Fahrenkrug SC (2012) Efficient TALEN-mediated gene knockout in livestock. Proc Natl Acad Sci U S A 109:17382–17387

Cermak T, Doyle EL, Christian M, Wang L, Zhang Y, Schmidt C, Baller JA, Somia NV, Bogdanove AJ, Voytas DF (2011) Efficient design and assembly of custom TALEN and other TAL effector-based constructs for DNA targeting. Nucleic Acids Res 39(12):e82

Chang N, Sun C, Gao L, Zhu D, Xu X, Zhu X, Xiong J-W, Xi JJ (2013) Genome editing with RNA-guided Cas9 nuclease in zebrafish embryos. Cell Res 23:465–472

Cho SW, Kim S, Kim JM, Kim J-S (2013) Targeted genome engineering in human cells with the Cas9 RNA-guided endonuclease. Nat Biotechnol 31:230–232

Christian M, Cermak T, Doyle EL, Schmidt C, Zhang F, Hummel A, Bogdanove AJ, Voytas DF (2010) Targeting DNA double-strand breaks with TAL effector nucleases. Genetics 186:757–761

Cong L, Ran FA, Cox D, Lin S, Barretto R, Habib N, Hsu PD, Wu X, Jiang W, Marraffini LA (2013) Multiplex genome engineering using CRISPR/Cas systems. Science 339:819–823

Curtin SJ, Zhang F, Sander JD, Haun WJ, Starker C, Baltes NJ, Reyon D, Dahlborg EJ, Goodwin MJ, Coffman AP, Dobbs D, Joung JK, Voytas DF, Stupar RM (2011) Targeted mutagenesis of duplicated genes in soybean with zinc-finger nucleases. Plant Physiol 156:466–473

Curtin SJ, Voytas DF, Stupar RM (2012) Genome engineering of crops with designer nucleases. Plant Genome J 5:42

Dahlem TJ, Hoshijima K, Jurynec MJ, Gunther D, Starker CG, Locke AS, Weis AM, Voytas DF, Grunwald DJ (2012) Simple methods for generating and detecting locus-specific mutations induced with TALENs in the zebrafish genome. PLoS Genet 8:e1002861

DeFrancesco L (2012) Erratum: move over ZFNs. Nat Biotechnol 30:112–112
Deltcheva E, Chylinski K, Sharma CM, Gonzales K, Chao Y, Pirzada ZA, Eckert MR, Vogel J, Charpentier E (2011) CRISPR RNA maturation by trans-encoded small RNA and host factor RNase III. Nature 471:602–607
Deng D, Yan C, Pan X, Mahfouz M, Wang J, Zhu JK, Shi Y, Yan N (2012) Structural basis for sequence-specific recognition of DNA by TAL effectors. Science 335:720–723
de Pater S, Neuteboom LW, Pinas JE, Hooykaas PJ, van der Zaal BJ (2009) ZFN-induced mutagenesis and gene-targeting in Arabidopsis through Agrobacterium-mediated floral dip transformation. Plant Biotechnol J 7:821–835
Doyle EL, Booher NJ, Standage DS, Voytas DF, Brendel VP, Vandyk JK, Bogdanove AJ (2012) TAL Effector-Nucleotide Targeter (TALE-NT) 2.0: tools for TAL effector design and target prediction. Nucleic Acids Res 40:W117–W122
Garneau JE, Dupuis M-È, Villion M, Romero DA, Barrangou R, Boyaval P, Fremaux C, Horvath P, Magadán AH, Moineau S (2010) The CRISPR/Cas bacterial immune system cleaves bacteriophage and plasmid DNA. Nature 468:67–71
Halfter U, Morris PC, Willmitzer L (1992) Gene targeting in Arabidopsis thaliana. Mol Gen Genet MGG 231:186–193
Hockemeyer D, Wang H, Kiani S, Lai CS, Gao Q, Cassady JP, Cost GJ, Zhang L, Santiago Y, Miller JC, Zeitler B, Cherone JM, Meng X, Hinkley SJ, Rebar EJ, Gregory PD, Urnov FD, Jaenisch R (2011) Genetic engineering of human pluripotent cells using TALE nucleases. Nat Biotechnol 29:731–734
Huang P, Xiao1. A, Zhou. M, Zhu. Z, Lin. S and Zhang B (2011) Heritable gene targeting in zebrafish using customized TALENs. Nature Biotech 29:699–700
Hwang WY, Fu Y, Reyon D, Maeder ML, Tsai SQ, Sander JD, Peterson RT, Yeh JJ, Joung JK (2013) Efficient genome editing in zebrafish using a CRISPR-Cas system. Nat Biotechnol 31:227–229
Jiang W, Bikard D, Cox D, Zhang F, Marraffini LA (2013) RNA-guided editing of bacterial genomes using CRISPR-Cas systems. Nat Biotechnol 31:233–239
Jinek M, Chylinski K, Fonfara I, Hauer M, Doudna JA, Charpentier E (2012) A programmable dual-RNA-guided DNA endonuclease in adaptive bacterial immunity. Science 337:816–821
Jinek M, East A, Cheng A, Lin S, Ma E, Doudna J (2013) RNA-programmed genome editing in human cells. eLife Sciences 2:e00471
Joung JK, Sander JD (2012) TALENs: a widely applicable technology for targeted genome editing. Nat Rev Mol Cell Biol 14:49–55
Lei Y (2012) Efficient targeted gene disruption in Xenopus embryos using engineered transcription activator-like effector nucleases (TALENs). Proc Natl Acad Sci U S A 109:17484–17489
Li T, Huang S, Zhao X, Wright DA, Carpenter S, Spalding MH, Weeks DP, Yang B (2011) Modularly assembled designer TAL effector nucleases for targeted gene knockout and gene replacement in eukaryotes. Nucleic Acids Res 39:6315–6325
Li L, Piatek MJ, Atef A, Piatek A, Wibowo A, Fang X, Sabir JS, Zhu JK, Mahfouz MM (2012a) Rapid and highly efficient construction of TALE-based transcriptional regulators and nucleases for genome modification. Plant Mol Biol 78:407–416
Li T, Liu B, Spalding MH, Weeks DP, Yang B (2012b) High-efficiency TALEN-based gene editing produces disease-resistant rice. Nature Biotech 30:390–392
Liu J, Li C, Yu Z, Huang P, Wu H, Wei C, Zhu N, Shen Y, Chen Y, Zhang B, Deng WM, Jiao R (2012) Efficient and specific modifications of the Drosophila genome by means of an easy TALEN strategy. J Genet Genomics 39:209–215
Lloyd A, Plaisier CL, Carroll D, Drews GN (2005) Targeted mutagenesis using zinc-finger nucleases in Arabidopsis. Proc Natl Acad Sci USA 102:2232–2237
Mahfouz MM, Li L, Shamimuzzaman M, Wibowo A, Fang X, Zhu JK (2011) De novo-engineered transcription activator-like effector (TALE) hybrid nuclease with novel DNA binding specificity creates double-strand breaks. Proc Natl Acad Sci U S A 108:2623–2628
Mak AN, Bradley P, Cernadas RA, Bogdanove AJ, Stoddard BL (2012) The crystal structure of TAL effector PthXo1 bound to its DNA target. Science 335:716–719

Mali P, Yang L, Esvelt KM, Aach J, Guell M, DiCarlo JE, Norville JE, Church GM (2013) RNA-guided human genome engineering via Cas9. Science 339:823–826
Marton I, Zuker A, Shklarman E, Zeevi V, Tovkach A, Roffe S, Ovadis M, Tzfira T, Vainstein A (2010) Nontransgenic genome modification in plant cells. Plant Physiol 154:1079–1087
Marx V (2012) Genome-editing tools storm ahead. Nat Methods 9:1055–1059
Miller JC, Tan S, Qiao G, Barlow KA, Wang J, Xia DF, Meng X, Paschon DE, Leung E, Hinkley SJ, Dulay GP, Hua KL, Ankoudinova I, Cost GJ, Urnov FD, Zhang HS, Holmes MC, Zhang L, Gregory PD, Rebar EJ (2011) A TALE nuclease architecture for efficient genome editing. Nat Biotechnol 29:143–148
Moore FE (2012) Improved somatic mutagenesis in zebrafish using transcription activator-like effector nucleases (TALENs). PLoS ONE 7:e37877
Morbitzer R, Elsaesser J, Hausner J, Lahaye T (2011) Assembly of custom TALE-type DNA binding domains by modular cloning. Nucleic Acids Res 39:5790–5799
Moscou MJ, Bogdanove AJ (2009) A simple cipher governs DNA recognition by TAL effectors. Science 326:1501
Moynahan ME, Jasin M (2010) Mitotic homologous recombination maintains genomic stability and suppresses tumorigenesis. Nat Rev Mol Cell Biol 11:196–207
Mussolino C, Cathomen T (2012) TALE nucleases: tailored genome engineering made easy. Curr Opin Biotechnol 23:644–650
Osakabe K, Osakabe Y, Toki S (2010) Site-directed mutagenesis in Arabidopsis using custom-designed zinc finger nucleases. Proc Natl Acad Sci U S A 107:12034–12039
Paszkowski J, Baur M, Bogucki A, Potrykus I (1988) Gene targeting in plants. EMBO J 7:4021
Pennisi E (2010) Sowing the seeds for the ideal crop. Science 327:802–803
Pennisi E (2012) The tale of the TALEs. Science 338:1408–1411
Remy S, Tesson L, Menoret S, Usal C, Scharenberg AM, Anegon I (2010) Zinc-finger nucleases: a powerful tool for genetic engineering of animals. Transgenic Res 19:363–371
Sander JD, Dahlborg EJ, Goodwin MJ, Cade L, Zhang F, Cifuentes D, Curtin SJ, Blackburn JS, Thibodeau-Beganny S, Qi Y (2010) Selection-free zinc-finger-nuclease engineering by context-dependent assembly (CoDA). Nat Methods 8:67–69
Sander JD, Cade L, Khayter C, Reyon D, Peterson RT, Joung JK, Yeh JR (2011) Targeted gene disruption in somatic zebrafish cells using engineered TALENs. Nat Biotechnol 29:697–698
Scherer S, Davis RW (1979) Replacement of chromosome segments with altered DNA sequences constructed in vitro. Proc Natl Acad Sci U S A 76:4951–4955
Schierling B, Dannemann N, Gabsalilow L, Wende W, Cathomen T, Pingoud A (2012) A novel zinc-finger nuclease platform with a sequence-specific cleavage module. Nucleic Acids Res 40:2623–2638
Shan Q, Wang Y, Chen K, Liang Z, Li J, Zhang Y, Zhang K, Liu J, Voytas DF, Zheng X, Zhang Y, Gao C (2013) Rapid and efficient gene modification in rice and Brachypodium using TALENs. Mol Plant 6(4):1365–1368
Shukla VK, Doyon Y, Miller JC, DeKelver RC, Moehle EA, Worden SE, Mitchell JC, Arnold NL, Gopalan S, Meng XD, Choi VM, Rock JM, Wu YY, Katibah GE, Zhifang G, McCaskill D, Simpson MA, Blakeslee B, Greenwalt SA, Butler HJ, Hinkley SJ, Zhang L, Rebar EJ, Gregory PD, Urnov FD (2009) Precise genome modification in the crop species *Zea mays* using zinc-finger nucleases. Nature 459:437–156
Tesson L, Usal C, Menoret S, Leung E, Niles BJ, Remy S, Santiago Y, Vincent AI, Meng X, Zhang L, Gregory PD, Anegon I, Cost GJ (2011) Knockout rats generated by embryo microinjection of TALENs. Nat Biotechnol 29:695–696
Tong C, Huang G, Ashton C, Wu H, Yan H, Ying QL (2012) Rapid and cost-effective gene targeting in rat embryonic stem cells by TALENs. J Genet Genomics 39:275–280
Tovkach A, Zeevi V, Tzfira T (2009) A toolbox and procedural notes for characterizing novel zinc finger nucleases for genome editing in plant cells. Plant J 57:747–757
Townsend JA, Wright DA, Winfrey RJ, Fu FL, Maeder ML, Joung JK, Voytas DF (2009) High-frequency modification of plant genes using engineered zinc-finger nucleases. Nature 459:442–161

Tzfira T, White C (2005) Towards targeted mutagenesis and gene replacement in plants. Trends Biotechnol 23:567–569

Tzfira T, Weinthal D, Marton I, Zeevi V, Zuker A, Vainstein A (2012) Genome modifications in plant cells by custom-made restriction enzymes. Plant Biotechnol J 10:373–389

Urnov FD, Miller JC, Lee YL, Beausejour CM, Rock JM, Augustus S, Jamieson AC, Porteus MH, Gregory PD, Holmes MC (2005) Highly efficient endogenous human gene correction using designed zinc-finger nucleases. Nature 435:646–651

Urnov FD, Rebar EJ, Holmes MC, Zhang HS, Gregory PD (2010) Genome editing with engineered zinc finger nucleases. Nat Rev Genet 11:636–646

Varshney RK, Bansal KC, Aggarwal PK, Datta SK, Craufurd PQ (2011) Agricultural biotechnology for crop improvement in a variable climate: hope or hype? Trends Plant Sci 16:363–371

Watanabe T, Ochiai H, Sakuma T, Horch HW, Hamaguchi N, Nakamura T, Bando T, Ohuchi H, Yamamoto T, Noji S, Mito T (2012) Non-transgenic genome modifications in a hemimetabolous insect using zinc-finger and TAL effector nucleases. Nat Commun 3:1017

Weber E, Gruetzner R, Werner S, Engler C, Marillonnet S (2011) Assembly of designer TAL effectors by Golden Gate cloning. PLoS One 6:e19722

Wood AJ (2011) Targeted genome editing across species using ZFNs and TALENs. Science 333:307

Wright DA, Townsend JA, Winfrey RJ, Jr., Irwin PA, Rajagopal J, Lonosky PM, Hall BD, Jondle MD, Voytas DF (2005) High-frequency homologous recombination in plants mediated by zinc-finger nucleases. Plant J 44:693–705

Xiao A, Wu Y, Yang Z, Hu Y, Wang W, Zhang Y, Kong L, Gao G, Zhu Z, Lin S, Zhang B (2013) EENdb: a database and knowledge base of ZFNs and TALENs for endonuclease engineering. Nucleic Acids Res 41:D415–D422

Zhang F, Maeder ML, Unger-Wallace E, Hoshaw JP, Reyon D, Christian M, Li XH, Pierick CJ, Dobbs D, Peterson T, Joung JK, Voytas DF (2010) High frequency targeted mutagenesis in Arabidopsis thaliana using zinc finger nucleases. Proc Natl Acad Sci U S A 107:12028–12033

Zhang Y, Zhang F, Li X, Baller JA, Qi Y, Starker CG, Bogdanove AJ, Voytas DF (2013) Transcription activator-like effector nucleases enable efficient plant genome engineering. Plant Physiol 161:20–27

Part III
General Experimental and Bioinformatic Technologies

Chapter 13
Mitochondrial Genetic Manipulation

Daria Mileshina, Adnan Khan Niazi, Frédérique Weber-Lotfi, José Gualberto and André Dietrich

13.1 Introduction

In the era of high-throughput genomics and new-generation sequencing, mitochondrial genetics might seem accessory, given the small number of genes that it accounts for. This is clearly not the case. The functional efficiency of mitochondria, and their cross talk with the other cellular compartments are essential for cell homeostasis and require appropriate contribution of the organelle genetic system. Mitochondrial DNA (mtDNA) organization and expression show strong differences among eukaryotic organisms, but are always complex and original. Genetic manipulation of mitochondria is thus appealing for a whole range of fundamental investigations. Moreover, mtDNA mutations in humans cause severe neurodegenerative diseases that are currently incurable and await the development of gene therapy strategies (e.g., Greaves et al. 2012). The number of pathogenic mutations identified already ranges over 300 (http://www.mitomap.org), referring to missense mutations in protein-coding genes, point mutations in transfer RNA (tRNA) or ribosomal RNA (rRNA) genes and duplications or deletions. In plants, mitochondrial genome rearrangements are the basis of cytoplasmic male sterility (CMS), a key genetic tool for the production of hybrids in breeding protocols (Frei et al. 2004; Chase 2007). Plant biotechnology based on organelle transgenesis also presents major advantages: (i) the product of the transgene is compartmentalized, opening the possibility to express proteins otherwise toxic for the cell and (ii) due to the maternal inheritance of mitochondria in most species, the transgene will not be spread through the pollen. Given

A. Dietrich (✉)
Institut de Biologie Moléculaire des Plantes du CNRS, 12 rue du Général Zimmer, 67084 Strasbourg, France
e-mail: andre.dietrich@ibmp-cnrs.unistra.fr

D. Mileshina · A. K. Niazi · F. Weber-Lotfi · J. Gualberto · A. Dietrich
Institut de Biologie Moléculaire des Plantes, CNRS and Université de Strasbourg, 12 rue du Général Zimmer, 67084 Strasbourg, France

X.-Q. Li et al. (eds.), *Somatic Genome Manipulation*,
DOI 10.1007/978-1-4939-2389-2_13

the importance of all these issues, the manipulation of mitochondrial genetics has been a long-standing goal but, up to now, regular transformation of these organelles has only been achieved in a couple of unicellular organisms through conventional biolistics (e.g., Bonnefoy et al. 2007; Zhou et al. 2010). The topic thus continuously challenged the inventiveness of the investigators, giving rise to a wide range of strategies progressing towards mitochondrial transformation or circumventing the problem through allotopic expression of proteins or RNAs. These strategies and their applications are the subject of the present chapter.

13.2 Transformation of the Mitochondrial Genome

13.2.1 Delivery of Nucleic Acids into Isolated Mitochondria

A number of approaches have been established to model and exploit mitochondrial transfection with isolated organelles from various organisms (Fig. 13.1). These studies provided important information on the mechanisms of DNA integration and expression, as well as on RNA editing and splicing.

13.2.1.1 Electroporation

One of the first described approaches to transfer DNA into isolated mitochondria was electroporation. Relatively large DNA molecules were thus introduced into mouse organelles (Collombet et al. 1997; Yoon and Koob 2003). Gene constructs were also successfully electroporated into isolated trypanosomatid mitochondria and expressed *in organello* (Estevez et al. 1999). In plants, Farré and Araya (2001) developed such a procedure for isolated wheat mitochondria and investigated the molecular and biochemical mechanisms occurring during transcription, splicing, and RNA editing. Using a cognate *cox2* chimeric gene construct, *cis*-recognition elements required for RNA editing in plant mitochondria were determined (Farré et al. 2001; Choury et al. 2004). It was also confirmed that editing and splicing of *cox2* transcripts are not linked in mitochondria (Farré and Araya 2002). Electroporation-mediated transfer of DNA into isolated mitochondria and *in organello* transcription were further used in additional plant species to investigate the sequences involved in transcription initiation and posttranscriptional modifications, or to map transcript ends (Staudinger and Kempken 2003; Choury et al. 2005; Staudinger et al. 2005; Bolle and Kempken 2006; Choury and Araya 2006; Bolle et al. 2007; Kempken et al. 2007). In vitro-transcribed RNA was also electroporated into isolated plant mitochondria. The introduced transcripts were properly spliced and edited, implying that RNA processing is not coupled with transcription (Hinrichsen et al. 2009). Methods related to this approach are detailed in Kempken et al. 2007.

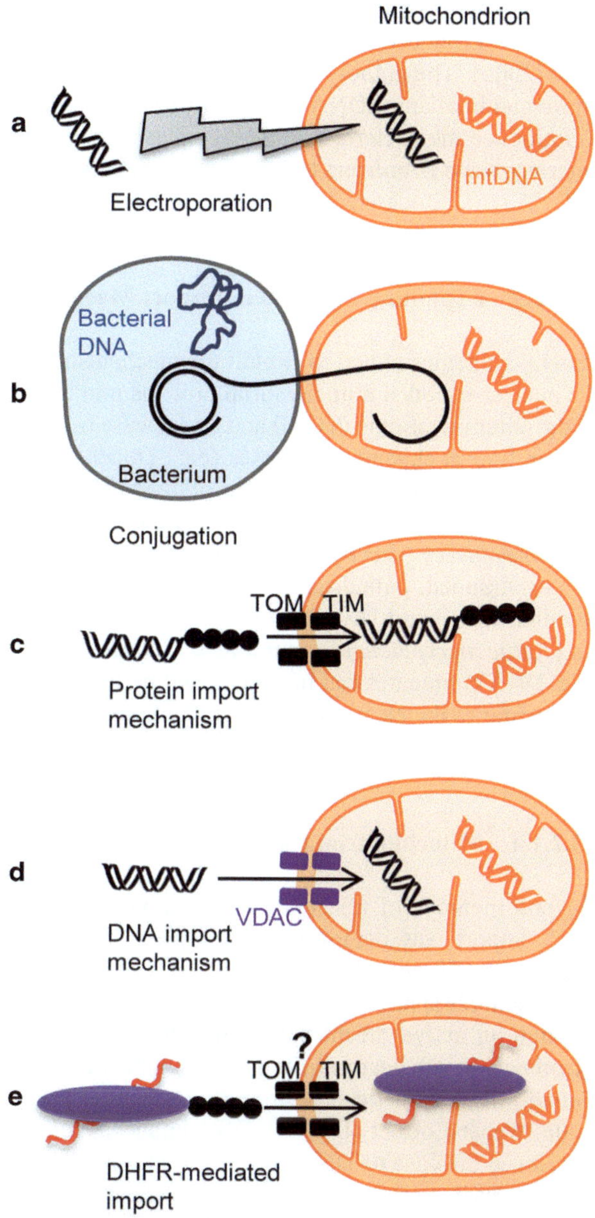

Fig. 13.1 Strategies to incorporate DNA into isolated mitochondria. DNA can be **a** introduced by electroporation, **b** amplified in bacteria as a mobilizable plasmid and transferred by conjugation, **c** fused to a mitochondrial-targeting sequence and translocated through the protein import pathway, **d** taken up thanks to the natural competence of mitochondria, or **e** driven by an RNA-binding protein carrying a mitochondrial-targeting sequence. *VDAC* voltage-dependent anion channel, *TOM* translocase of the outer membrane, *TIM* translocase of the inner membrane

13.2.1.2 Bacterial Conjugation

An original approach to introduce DNA into the mitochondria was described as bacteria-to-mitochondria conjugation (Yoon and Koob 2005). Conjugation is driven entirely by the donor cell and only involves its molecular machinery (Waters 1999). Thus, conjugative-competent *Escherichia coli* cells could transfer a DNA construct containing an origin of transfer (*oriT*) into the matrix of purified mouse mitochondria. The introduced construct was actively transcribed in the organelles (Yoon and Koob 2005). The authors also developed a method for cloning and replicating the entire mammalian mtDNA in *E. coli* (Yoon et al. 2009), with the longer-term aim to introduce the bacteria into the cytoplasm of mammalian cells as a recombinant enteroinvasive *E. coli* and to look for intracellular conjugative transfer of the cloned mtDNA into mitochondria (Yoon and Koob 2012).

13.2.1.3 Exploiting the Protein Import Machinery

Several strategies aimed to exploit the mechanisms that ensure the translocation of the nuclear-encoded mitochondrial proteins into the organelles, i.e., the translocase of the outer membrane (TOM) and translocase of the inner membrane (TIM) multiprotein import channel complexes (e.g., Dudek et al. 2012 for a recent review). A single-stranded or double-stranded oligonucleotide covalently linked to the C-terminus of a mitochondrial precursor protein could be introduced into isolated yeast (*Saccharomyces cerevisiae*) mitochondria (Vestweber and Schatz 1989). Similarly, double-stranded, palindromic DNA molecules of 17 or 322 bp attached to an N-terminal mitochondrial-targeting peptide were successfully transferred into isolated rat (*Rattus norvegicus*) liver mitochondria (Seibel et al. 1995). The artificial peptide–DNA conjugates can be recognized by the mitochondrial processing peptidases, leading to the cleavage of the targeting peptide and avoiding an interference of the latter with the functionality of the DNA in the organelles (Seibel et al. 1999).

13.2.1.4 Mitochondrial Natural Competence

Beyond methods of assisted mitochondrial transfection, a natural mechanism of DNA import was demonstrated for plant, mammalian, and yeast mitochondria (Koulintchenko et al. 2003, 2006; Weber-Lotfi et al. 2009). It was shown that isolated mitochondria have the property of taking up naked, linear, double-stranded DNA through an active mechanism. How the DNA crosses the mitochondrial double membrane is not clear yet. Translocation through the outer membrane likely occurs though the voltage-dependent anion channel (VDAC; Koulintchenko et al. 2003, 2006; Weber-Lotfi et al. 2009). DNA import appears to be a sequence-independent process in the case of DNA constructs up to a few kb (Koulintchenko et al. 2003). With increasing size (>6–7 kb), the import substrates are translocated in a sequence-dependent manner, at least for plant mitochondria (Ibrahim et al. 2011). Thus, the uptake of an 11.6 kb *Brassica napus* linear mitochondrial plasmid comprising its terminal-inverted repeats was much more efficient than that of the plasmid deprived of these terminal sequences. DNA imported through this natural mechanism can be transcribed *in organello* when carrying an appropriate mitochondrial promoter, and the transcripts can be accurately processed (Koulintchenko et al. 2003, 2006). The exogenous DNA can also be repaired upon uptake into plant or mammalian mitochondria (Boesch et al. 2009, 2010). Moreover, an imported construct, comprising a marker gene flanked by sequences homologous to the mtDNA, integrated into the resident genome in plant mitochondria through homologous recombination (Mileshina et al. 2011).

13.2.1.5 Carrier Protein-Driven RNA Import

Finally, a further strategy for the introduction of nucleic acids into mitochondria was described by Sieber et al. (2011) who used an RNA-binding protein as a vehicle to import RNAs up to 800 nt in size into isolated mitochondria. It was shown that the mammalian dihydrofolate reductase (DHFR), fused to a mitochondrial-targeting sequence (MTS), is able to bind RNAs in a nonspecific manner. Following binding, RNAs are carried into isolated plant, mammalian, or yeast mitochondria. Upon import, a tRNA precursor was correctly processed, leading to a mature tRNA. Similarly, after DHFR-driven uptake, the unedited version of the potato mitochondrial *atp9* messenger RNA (mRNA) was partially edited *in organello* (Sieber et al. 2011).

13.2.2 DNA Targeting into Mitochondria in Whole Cells: Cell Biology Strategies

The development of methodologies to transfect mitochondria in vivo has now progressed beyond unicellular organisms. No strategy is definitely established in the field, but in several cases success has been claimed (Fig. 13.2).

13.2.2.1 Biolistics

Based on the methods developed for nuclear transformation, early attempts to genetically transform mitochondria in vivo were through the "biolistics" (from "biological ballistics") approach, i.e., by bombardment of the cells with DNA-coated particles. *S. cerevisiae* was the first organism in which genetic transformation of the organelles was successfully achieved and used to generate defined alterations in the mtDNA (Fox et al. 1988; Johnston et al. 1988). DNA sequences delivered into yeast mitochondria are subsequently incorporated into the mtDNA by the homologous recombination mechanisms present in the organelles. Mitochondrial transformation in the unicellular green alga *Chlamydomonas reinhardtii* mediated by microprojectile bombardment was subsequently obtained (Randolph-Anderson et al. 1993), but the follow-up of these seminal experiments came only more recently (Remacle et al. 2006). Although foreign DNA has been successfully introduced into mitochondria in *S. cerevisiae* and *C. reinhardtii* by biolistics, this approach has failed so far in animals and higher plants. Various other strategies have thus been explored.

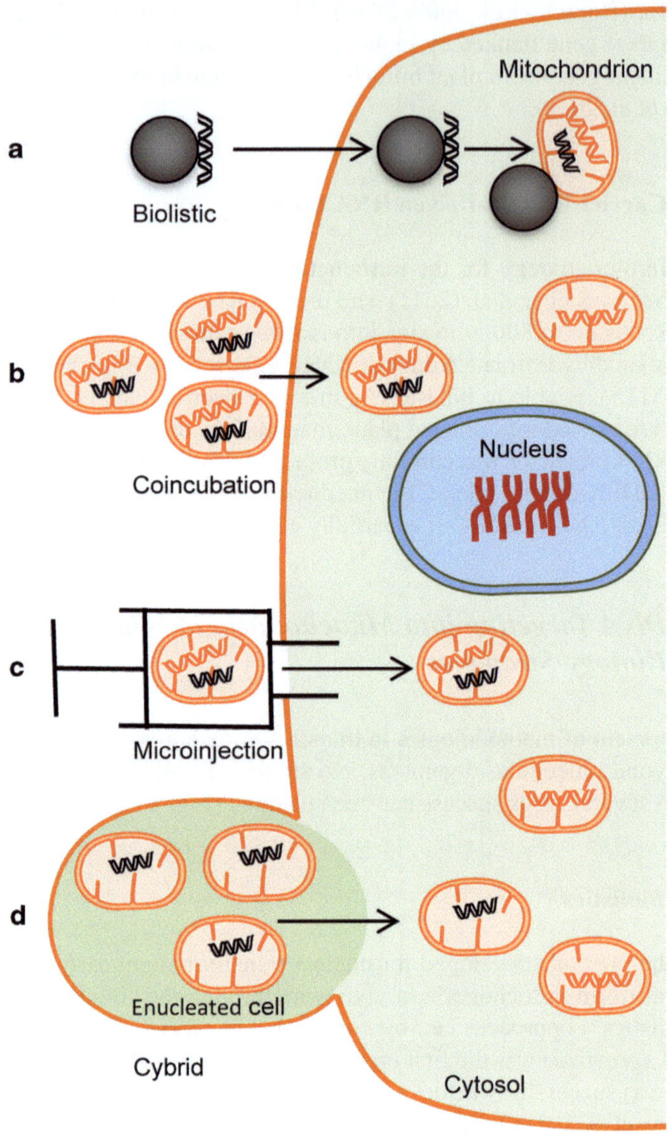

Fig. 13.2 Cell biology strategies to transfer DNA into mitochondria in whole cells. **a** Cells can be bombarded with DNA-coated tungsten or gold particles; isolated mitochondria carrying the DNA of interest can be introduced into recipient cells **b** by co-incubation or **c** by microinjection; **d** enucleated cells with mitochondria carrying the DNA of interest can be fused with recipient cells to form cybrids

13.2.2.2 Uptake of Intact Mitochondria

In early experiments, Clark and Shay (1982) co-incubated mammalian cells with purified mitochondria carrying a mitochondrial mutation able to provide chloramphenicol resistance. Presumably incorporated by endocytosis, the exogenous organelles transferred the resistance to the cells (Clark and Shay 1982; Ber et al. 1984). This concept too was only recently reactivated when Katrangi et al. (2007) showed that isolated mitochondria can be engulfed by mammalian cells during co-incubation. Uptake of isolated murine organelles restored respiration in human A549 lung carcinoma cells devoid of mtDNA (Katrangi et al. 2007).

13.2.2.3 Microinjection of Intact Mitochondria

It was further shown that isolated mitochondria can be introduced into recipient cells by microinjection. After initial assays by King and Attardi (1988, 1989), mitochondria isolated from *Mus spretus* liver samples were microinjected into fertilized ova obtained from superovulated *Mus musculus domesticus* females. The presence of the foreign mitochondria was readily detected after in vitro culture to the blastocyst stage (Pinkert et al. 1997; Pinkert and Trounce 2002). However, the level of foreign mitochondria considerably decreased during later developmental stages.

13.2.2.4 Cybrid Generation

Another cell biology-based method to transfer mtDNA into cells was pioneered by King and Attardi (1989) and put forward by Kagawa and Hayashi (1997) as a possible strategy for gene therapy. The original approach relies on the fusion of mammalian cells carrying DNA-deprived mitochondria (*rho*^{0} cells) with enucleated cells (i.e., cytoplasts), to generate mtDNA-repopulated so-called cybrids. It was subsequently extended to cytoplasts from patients with mutant mtDNA (King et al. 1992). The correction of a mitochondrial deficiency was demonstrated in a cell model by transfer of “healthy” mitochondria through cytoplast fusion in vitro (Kagawa et al. 2001). A similar approach allowed to generate mouse embryos carrying two mtDNA genotypes. Lines of heteroplasmic mice were subsequently derived from such embryos, and segregation of the mtDNA could be analyzed (Jenuth et al. 1996). A recent study (Yang and Koob 2012) combined microinjection and cytoplast fusion. Isolated mitochondria were first microinjected into rodent oocytes or into single-cell embryos. These are more convenient due to their large size. A “mitocytoplast” fraction was then pulled from the microinjected cells and fused to recipient cells, resulting in delivery of the organelles of interest and restoration of mitochondrial function (Yang and Koob 2012).

13.2.3 DNA Targeting into Mitochondria in Whole Cells: Specialized Nanocarriers

In a context where biolistics failed in most organisms, and cell biology approaches seemed too complex to apply to gene therapy, a variety of alternative strategies emerged that were aimed at developing vehicles able to carry DNA into the cells and towards mitochondria (Fig. 13.3).

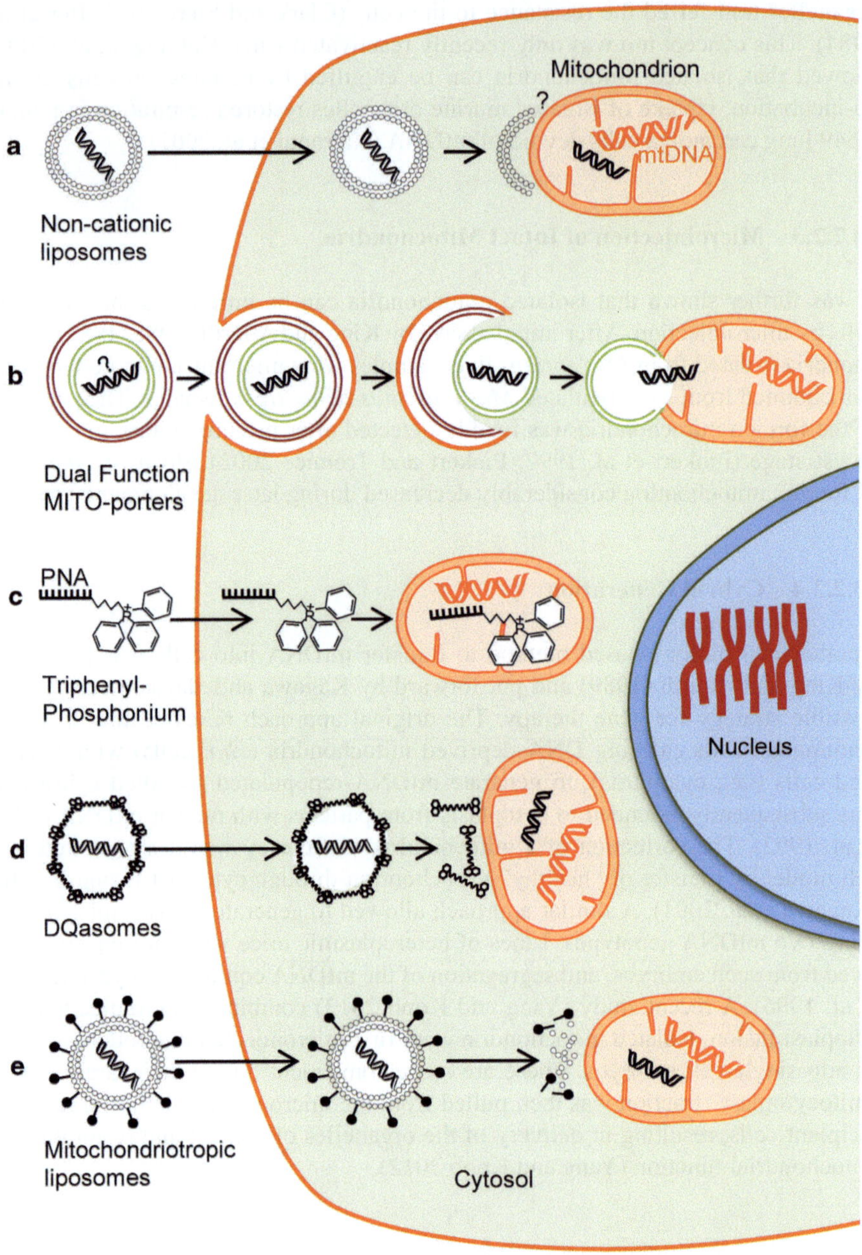

Fig. 13.3 Nanocarrier-based strategies to transfer DNA into mitochondria in whole cells. DNA can be **a** bound to noncationic liposomes, **b** incorporated into MITO-Porter vesicles that will fuse successively with the endosomal membranes and the mitochondrial membranes, **c** conjugated with a phenyl-phosphonium derivative, **d** complexed with dequalinium vesicles, or **e** loaded on mitochondriotropic liposomes. Whether DNA can be delivered through MITO-Porters remains to be established

13.2.3.1 Noncationic Liposomes

Cudd et al. showed that conventional noncationic liposomes seem to associate with liver mitochondria (Cudd et al. 1984; Cudd and Nicolau 1985, 1986). When, for example, liposomes containing uranyl acetate were intravenously injected into mice, they were not only present in cytoplasmic vacuoles but also associated with mitochondria and possibly delivered their aqueous content into the organelles (Cudd and Nicolau 1986). Also, Inoki et al. (2000) could show co-localization of proteoliposomes containing a mitochondrial membrane fraction with endogenous mitochondria in fertilized mouse eggs.

13.2.3.2 MITO-Porters

Subsequent modifications of proteoliposomes led to the design of carriers called MITO-Porters, liposome-based vesicles composed of fusogenic lipids and carrying octaarginine (R8) moieties at the surface (Yamada et al. 2008). A dual-function (DF) MITO-Porter version was recently described that possesses mitochondrion-fusogenic inner and endosome-fusogenic outer envelopes and has the ability to pass through endosomal and mitochondrial membranes via stepwise membrane fusion (Yamada and Harashima 2012a). MITO-Porters reportedly delivered the nucleic acid stain propidium iodide (Yasuzaki et al. 2010) and proteins such as green fluorescent protein (GFP; Yamada et al. 2008) and DNase I (Yamada et al. 2011; Yamada and Harashima 2012a) into the mitochondrial matrix. DNA delivery is expected to be possible, but has not been established yet. In an ultimate design, MITO-Porters were combined with a mitochondrial-targeting peptide, which enhanced specificity (Yamada and Harashima 2012b).

13.2.3.3 Amphiphile Carriers

Electron transport along the respiratory chain of the mitochondrial inner membrane creates a transmembrane, negative inside potential (Mitchell 1961). Consequently, positively charged molecules are attracted by mitochondria, but cannot enter the organelle matrix because the inner mitochondrial membrane is impermeable to polar molecules. However, certain amphiphile compounds are able to cross both mitochondrial membranes and accumulate in the mitochondrial matrix in response to the negative membrane potential. This led to the idea of using such amphiphilic compounds as carriers for the delivery of DNA into mitochondria. Among the first described mitochondriotropic cationic amphiphiles were phosphonium salts, such as methyltriphenylphosphonium, which were shown by Liberman et al. (1969) to be rapidly taken up by mitochondria in living cells. In an attempt to inhibit the replication of mtDNA carrying a pathogenic mutation, a lipophilic phosphonium cation was conjugated to an 11-mer peptide nucleic acid (PNA). PNAs are synthetic DNA analogues with the deoxyribose phosphate backbone replaced by aminoethyl glycine units connected to conventional bases by a methyl carbonyl linker (Hyrup and Nielsen 1996). They bind selectively and with high affinity to complementary DNA (cDNA) or RNA sequences. The phosphonium–PNA conjugate was imported into both isolated mitochondria and mitochondria in human cells (Muratovska et al. 2001). However, although the conjugate could efficiently interfere with DNA replication in vitro, in cells it was unable to significantly influence the copy number of mutated mtDNA.

13.2.3.4 DQAsomes and Mitochondriotropic Liposomes

Different mitochondriotropics were described to be able to drive bound DNA into mammalian cells (Horobin et al. 2007). These include dequalinium, a dicationic compound resembling "bola" form electrolytes, as it is an asymmetrical molecule with two charge centers separated by a relatively large distance. Dequalinium forms spherical aggregates termed "DQAsomes", with diameters varying between 70 and 700 nm. Such vesicles effectively bind DNA, forming stable DQAsome/DNA complexes (Weissig et al. 1998). It was shown that DQAsome/DNA complexes do not release any DNA in the presence of liposomes resembling the cytosolic membranes. At the same time, up to 75% of the DNA is displaced from its DQAsomal carrier upon contact of the complexes with liposomes resembling both mitochondrial membranes (Weissig et al. 2000) or intact mouse liver mitochondria (Weissig et al. 2001). Finally, DQAsome/DNA complexes proved able to enter mammalian cells by endocytosis and to escape from endosomes without losing their DNA load. They subsequently transport the DNA to the mitochondria, where at least part of it is released from the carrier. The released DNA remains strongly associated with the organelles (D'Souza et al. 2003), raising the hypothesis that it can be taken up through the import mechanism characterized with isolated mitochondria (2.1.4). In agreement with such a view, further studies brought evidence for DQAsome-mediated delivery of a DNA construct containing a recoded GFP gene into the mitochondria in mammalian cell lines (Lyrawati et al. 2011). Nevertheless, the efficiency was low, and dequalinium showed strong toxicity for mammalian cells (Lyrawati et al. 2011). As an alternative, mitochondriotropic liposomes composed of phospholipids, DNA-binding lipids, and surface-exposed amphiphilic cations were developed (Boddapati et al. 2005; Weissig et al. 2006; D'Souza et al. 2007). Like DQAsomes, such liposomes are able to enter mammalian cells and carry DNA towards mitochondria, where it remains associated. Improved formulations included a mitochondrial fraction or mitochondrial lipids (Wagle et al. 2011). Mitochondriotropic liposomes proved less toxic, especially when containing a mitochondrial fraction (Wagle et al. 2011), but molecular data confirming full incorporation of the DNA into the organelles are lacking.

13.2.4 DNA Targeting into Mitochondria in Whole Cells: Carriers Exploiting the Protein or RNA Import Pathways

13.2.4.1 Further Conjugates

Carrier systems based on the mitochondrial protein import pathway have also been assayed in vivo (Fig. 13.4). A three-component assembly was designed in which a mitochondrial-targeting peptide was conjugated to a fragment of PNA, the latter carrying a cDNA oligonucleotide of interest (Flierl et al. 2003). Such complexes were reported to be imported into isolated mammalian mitochondria and into the mitochondria of permeabilized cultured mammalian cells. In a similar design, the mitochondrial-targeting peptide was conjugated to a polyethylenimine moiety that carried the DNA (Lee et al. 2007; Fig. 13.4). The resulting complex appeared to co-localize with mitochondria in living cells.

13.2.4.2 Protofection

Protofection was originally mentioned in a review (Khan and Bennett 2004). To carry DNA into mammalian mitochondria in vivo, the proposed strategy challenged the protein import pathway and was based on the combination of a protein transduction domain (PTD; Suzuki 2012) with an MTS, both fused to the TFAM mitochondrial transcription factor (Iyer et al. 2009; Thomas et al. 2011), yielding a carrier called mitochondrial-targeting domain (MTD)–TFAM. In such a system (Fig. 13.4), the PTD ensures transduction into the cells, the MTS promotes mitochondrial targeting and import, whereas TFAM carries the DNA. With this strategy, the mitochondria-deficient phenotype of Parkinson's disease cybrids was reportedly rescued upon co-localization of MTD–TFAM carrying human mtDNA with mitochondria (Keeney et al. 2009). Also, partial restoration of organelle function in human cells carrying mitochondrial mutations was achieved by targeting healthy donor mtDNA loaded on MTD–TFAM (Iyer et al. 2012a). However, the MTD–TFAM protein alone promoted a similar boosting of mitochondrial functions in cell cultures and mice (Iyer et al. 2009; Keeney et al. 2009; Thomas et al. 2011). Conversely, with the aim of developing stem cell-based model systems, the MTD–TFAM carrier was used to provide mitochondria with mutated mtDNA copies in human neural progenitor cells (Iyer et al. 2012b). Like a number of other approaches, this strategy suffers from a lack of direct molecular evidence confirming the delivery of the DNA to the inside of the organelles.

13.2.4.3 Viral Carrier

Still building on the protein import pathway, but with a completely different approach, Yu et al. (2012a) developed a viral carrier system by redirecting virions of

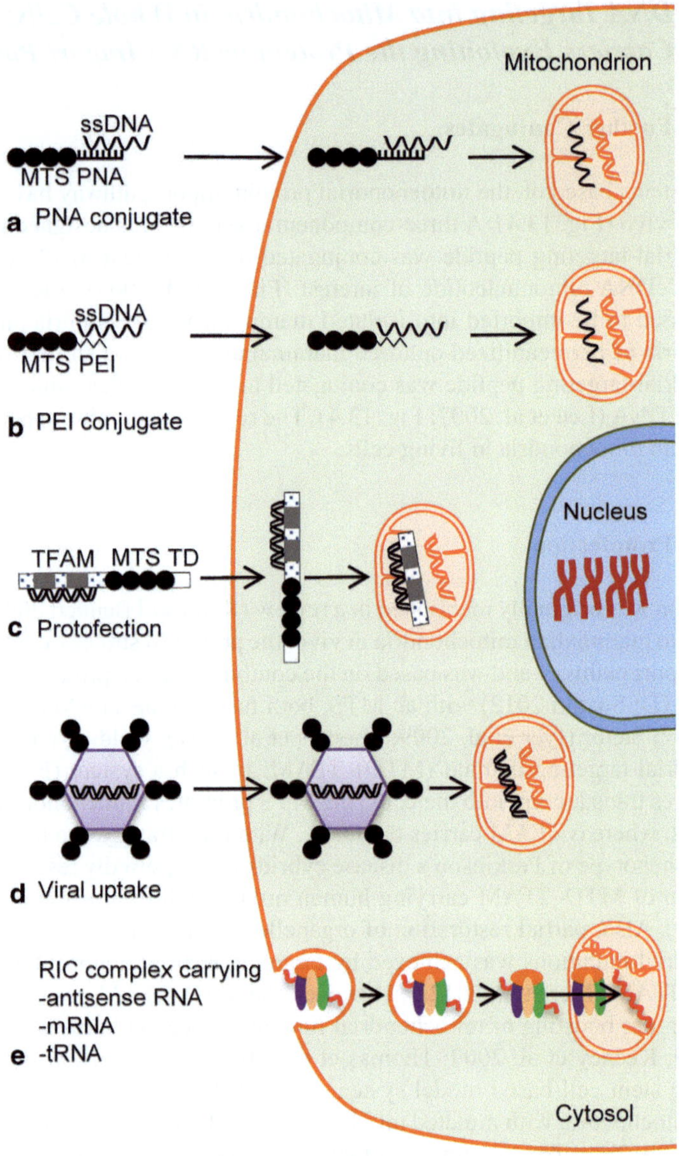

Fig. 13.4 Strategies to transfer DNA into mitochondria in whole cells, based on the protein or RNA import pathway. DNA can be **a** annealed to a PNA moiety conjugated to a mitochondrial-targeting sequence (*MTS*), **b** complexed with polyethylenimine (*PEI*) conjugated to a mitochondrial-targeting sequence, **c** loaded on the TFAM factor fused to a mitochondrial-targeting sequence and a transduction domain (*TD*), **d** incorporated into adeno-associated virus particles carrying mitochondrial-targeting sequences, or **e** loaded into the *L. tropica* RNA import complex (*RIC*). *TFAM* transcription factor A, mitochondrial, *PNA* peptide nucleic acid, *tRNA* transfer RNA, *mRNA* messenger RNA, *ssDNA* single-stranded DNA

the adeno-associated virus (AAV) to mitochondria (Fig. 13.4). The AAV is a single-stranded DNA parvovirus. Its genome encodes nonstructural replication (REP) proteins and the viral protein (VP)1, VP2, and VP3 capsid proteins. The VP2 protein can tolerate large extra peptide sequences at the N-terminus, so that an MTS could be added, generating an organelle-targeted AAV. The wild-type *ND4* cDNA under the control of the mitochondrial heavy-strand promoter was subsequently linked to an AAV genome backbone and packaged into the organelle-targeted capsid. Delivery of such virions into cybrid cell lines carrying an homoplasmic-mutated *ND4* mitochondrial gene was reported to result in organelle localization, wild-type ND4 protein synthesis, and rescue of defective respiration (Yu et al. 2012a). Similarly, the strategy appeared to rescue visual dysfunction caused by mutant *ND4* in mouse model systems and the adeno-associated virus cassette was proposed as a general platform for mitochondrial gene targeting (Yu et al. 2012a). In these assays, mitochondria were not purified and analyzed, but a number of direct and indirect observations supported that delivery of the exogenous DNA to the inside of the organelles had taken place. In reciprocal experiments aiming to develop an animal model system for Leber hereditairy optic neuropathy (LHON), mitochondrial targeting of a mutant *ND4* gene through the adeno-associated virus strategy induced loss of visual function and optic atrophy in mice (Yu et al. 2012b).

13.2.4.4 RNA Import Complex

Mitochondrial tRNA import was implied in the 1970s through studies in the ciliated protozoan *Tetrahymena pyriformis* (Chiu et al. 1974; Suyama 1986). It was later confirmed that in many organisms, mitochondria do not encode all tRNAs needed for protein synthesis. The missing tRNAs are imported from the cytosol (reviewed for instance in Schneider and Maréchal-Drouard 2000; Salinas et al. 2008; Schneider 2011). The number and identity of the imported tRNAs, as well as the uptake mechanisms, are variable, depending on the organism. In *Trypanosoma* and *Leishmania,* the mtDNA does not code for any tRNA, and these are all imported from the cytosol (Simpson et al. 1989; Hancock and Hajduk 1990). Studies in *Leishmania tropica* have put forward a particular mechanism for mitochondrial tRNA import, independent from the import of proteins. It would involve a specific complex in the inner membrane, the RNA import complex or "RIC," that would include several respiratory components (Bhattacharyya et al. 2003; Chatterjee et al. 2006; Mukherjee et al. 2007). Mahata et al. (2006) incorporated isolated RIC into human cells by caveolin-1-dependent endocytosis. It was implied that the complex incorporates into the inner membrane of mitochondria and enables the import of cytosolic tRNAs. The latter included $tRNA^{Lys}$, which suggested that the strategy would be suitable to rescue the myoclonic epilepsy with ragged-red fibers (MERRF) syndrome due to a mutation in the mitochondrial $tRNA^{Lys}$ gene. Further studies described organelle import of short antisense RNAs (Mukherjee et al. 2008) or long polycistronic-coding RNAs (Jash and Adhya 2011; Mahato et al. 2011) loaded onto RIC (or a derived sub-complex) and incorporated into human cells (Fig. 13.4), resulting in decreased respiration or mitochondrial functional rescue, respectively. Finally, application of the strategy to distribute RNAs into the mitochondria of different tissues in an animal model system was claimed (Jash et al. 2012). However, it should be noted that these results as a whole are highly controversial in the field (e.g. Schneider 2011). An editorial expression of concern (Schekman 2010) has been published about one of the manuscripts (Goswami et al. 2006). The RIC complex seems oddly specific to *L. tropica,* as no other laboratory working in the field could gather any evidence for its existence, even in species as close as *Leishmania tarentolae*. Some of the putative RIC components were shown to be dispensable for mitochondrial tRNA import (Paris et al. 2009; Cristodero et al. 2010). Also, that exogenous RIC would reach the organelles, cross the outer membrane as a huge complex, and functionally incorporate into the inner membrane raises many questions and would need robust molecular evidence.

13.2.5 Maintenance of Exogenous DNA in Mitochondria

After transfection, the exogenous DNA should be maintained in the organelles, either by integration into the mitochondrial genome or through independent replication. Homologous recombination is established in plant mitochondria and marker gene integration has been obtained with isolated organelles (Mileshina et al. 2011). The challenge is more open for mammalian mitochondria, as for these the mechanism of mtDNA replication is still discussed (see references in Ruhanen et al. 2011) and mtDNA recombination remains controversial (Elson and Lightowlers 2006; Fan et al. 2012). Approaches aiming at mitochondrial transformation have thus to be adapted to the size, structure, and genetics of the mtDNA, which vary widely according to the species.

13.2.5.1 Integration into the Main Mitochondrial Genome

In *C. reinhardtii,* the mitochondrial genome is a small linear DNA molecule 15.8 kb in length that encodes only eight proteins, two rRNAs, and three tRNAs. Telomer-like sequences at the ends are believed to be important for replication (Vahrenholz et al. 1993). Several respiratory-deficient mutants were described (Remacle et al. 2001), where one of the telomeres and some gene coding sequences were lost (*cob* gene, mutants *dum1–4, 11, 14,* and *16,* or *nad4* and *nad5* genes, mutant *dum24*). These mutants could be successfully transformed with linear or circular constructs that reconstituted the complete genome by homologous recombination (Yamasaki et al. 2005; Remacle et al. 2006; Fig. 13.5). These experiments also showed that, in *C. reinhardtii* mitochondria, DNA exchange can occur within homologous regions as short as 60 bp (Remacle et al. 2006). In yeast, transformation of mitochondria via recombination was also possible, by exploiting the occurrence of *rho*$^-$ mutations that correspond to the deletion of large portions of the mtDNA (Dujon 1981;

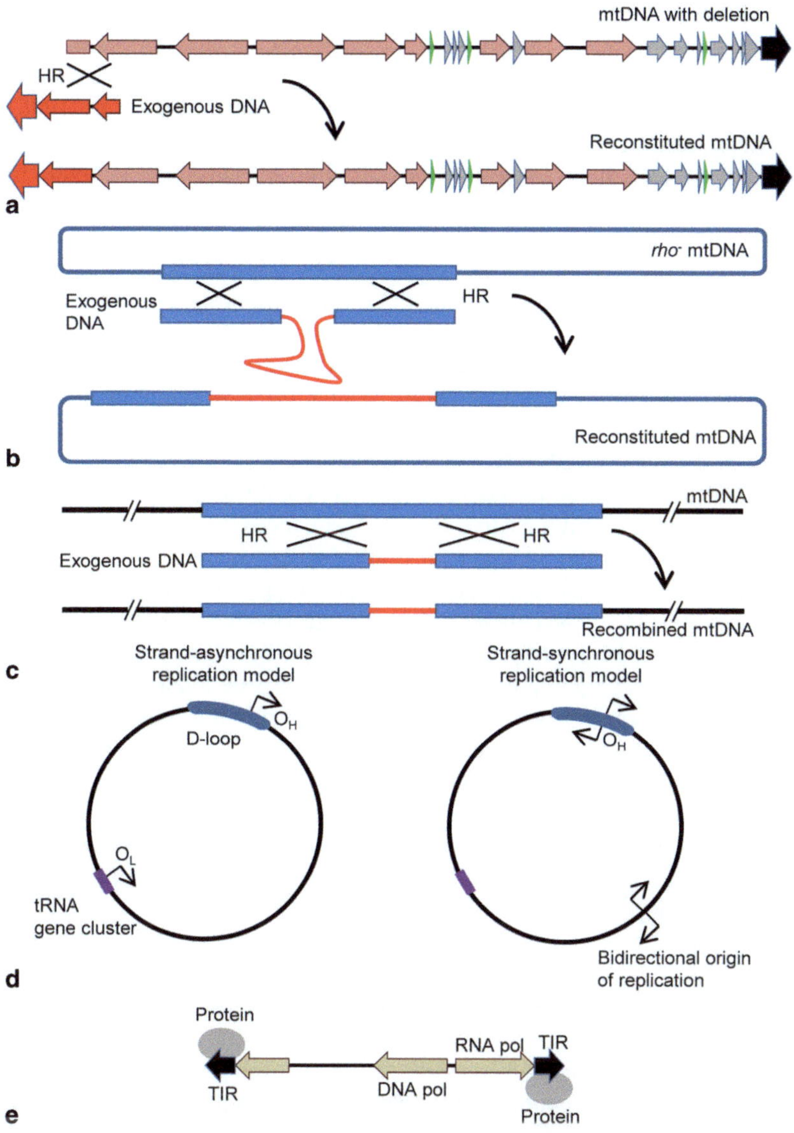

Fig. 13.5 Strategies to maintain exogenous DNA in mitochondria. In **a** *C. reinhardtii* and **b** *S. cerevisiae,* mutant mtDNA with a deletion was complemented by homologous recombination with exogenous DNA bringing the missing information; **c** a marker gene was integrated into the mtDNA following homologous recombination with exogenous DNA imported into isolated plant mitochondria; **d** exogenous DNA might be maintainable by independent replication in mammalian mitochondria if containing the appropriate origins, but two replication models are conflicting: Strand-asynchronous replication from distant heavy-strand (O_H) and light-strand (O_L) origins and strand-synchronous replication from dispersed bidirectional origins; **e** independently replicating natural mitochondrial linear plasmids might serve as a basis to generate customized replicons for plant organelles. *TIR* terminal-inverted repeat, *mtDNA* Mitochondrial DNA, *tRNA* transfer RNA, *DNA/RNA pol* DNA/RNA polymerase, *HR* homologous recombination

Fig. 13.5). In *rho*⁻ strains, the mtDNA replicates independently from protein synthesis and shows no clear requirement for a specific replication origin sequence. Due to this property, mitochondria of yeast can be transformed with bacterial plasmid DNAs that subsequently propagate as *rho*⁻ molecules (Fox et al. 1988). The obtained *rho*⁻ strains are then used to mate with a *rho*⁺ strain to produce a strain of interest. In this case, homologous regions of 50 bp are sufficient to obtain recombinant mitochondrial genomes. In other examples, linear donor DNA with as little as 260 bp of homologous sequence flanking each side of a deletion mutation could efficiently recombine with *rho*⁺ recipient mtDNA to yield the desired new strain (Bonnefoy and Fox 2000; Bonnefoy et al. 2001). In higher plants, up to now, there is no functional protocol for stable mitochondrial transformation. The genomes are large and complex. They contain many large and small repeated sequences that can lead to alternative mtDNA configurations by homologous recombination. Possible approaches will rely on this very active recombination system, responsible for the plasticity of the plant mtDNA (Kühn and Gualberto 2012). Several genes regulate plant mtDNA recombination. Under nonselective conditions, they can repress undesired recombination involving small repeated sequences that lead to aberrant mt genomes (Zaegel et al. 2006; Shedge et al. 2007; Miller-Messmer et al. 2012). On the other hand, several of these factors, and additional ones, are required for recombination-mediated repair of DNA breaks induced by genotoxic stress (Parent et al. 2011; Janicka et al. 2012; Miller-Messmer et al. 2012). It was shown that a marker gene flanked by sequences homologous to the mtDNA can be integrated into the main mitochondrial genome *in organello,* upon introduction of the construct into isolated higher-plant mitochondria (Mileshina et al. 2011; Fig. 13.5). For recombination, flanking sequences of at least 500 bp were necessary. It is possible that an increase in integration efficiency can be achieved in mutants deficient in recombination surveillance genes, or in plants treated with organellar-specific genotoxins. However, for in vivo transformation, an efficient selection protocol of the transformants still needs to be developed.

13.2.5.2 Transformation with Replicative-Competent Constructs

In *C. reinhardtii,* one of the successful transformation strategies was to replace the whole mitochondrial genome (Randolph-Anderson et al. 1993; Yamasaki et al. 2005). In these examples, the mutant mtDNA of the recipient strain could be replaced by partially purified DNA of *C. reinhardtii* or *Chlamydomonas smithii,* restoring respiratory competence. A similar strategy might be the method of choice for transformation of animal mitochondria, because of the apparent absence of recombination activity that could mediate integration by recombination (Elson and Lightowlers 2006). The mammalian mtDNA is a small, circular molecule and its size varies from 13 to 19 kb according to the species. These genomes are very compact and contain a single noncoding sequence, the "displacement loop" (D-loop). How they replicate is still controversial (Pohjoismaki et al. 2010; Wanrooij et al. 2012). A strand-asynchronous model of replication (Lee and Clayton 1997, 1998) considers that the D-loop region contains the replication origin of the heavy strand (Shadel and Clayton 1997), while the origin of replication of the light strand would be located in the cluster of tRNA genes (Clayton 2003; Falkenberg et al. 2007). However, a synchronous replication model proposes that synthesis of both strands proceeds from multiple, bidirectional replication forks (Holt et al. 2000; Yang et al. 2002; Bowmaker et al. 2003; Yasukawa et al. 2005). Research on these mechanisms is constantly progressing, and it should become possible to build replicons containing the whole mitochondrial genome, or just the required replication origins, so as to transform mammalian mitochondria and eventually complement mutant mtDNA copies (Fig. 13.5). In plants too, an alternative strategy to transform mitochondria would be to create replicons independent from the main mtDNA. These replicons could be based on linear plasmids naturally existing in plant mitochondria and replicating independently (Handa 2008). Linear mitochondrial plasmids are double-stranded molecules with defined ends that usually contain terminal-inverted repeats (TIRs), with proteins covalently attached to their 5' termini (Fig. 13.5). The size of the TIRs varies from 170 bp (maize 2.3 kb plasmid) to 407 bp (sugar beet 10.4 kb plasmid). All linear mitochondrial plasmid DNAs contain one or more open reading frames that are transcribed and translated, including polymerases that might be required for their replication.

13.2.6 Selection of Mitochondrial Transformants

Genetic transformation of organelle genomes presents special requirements with respect to selectable markers, as multiple genome copies must be converted to the transformed genotype for efficient expression of the marker. Organisms in which the mitochondrial function is expendable allow for the selection of transformants based on the replacement of mutant organelle genes with wild-type copies that restore organelle function (Boynton et al. 1988; Johnston et al. 1988; Randolph-Anderson et al. 1993). Alternative strategies are required when organelle genome mutations are unavailable due to lethality, and targeted disruption of genes essential for organelle function is not accessible. Such strategies have been developed for the recovery of plastid genome transformants in land plants (Svab et al. 1990; Svab and Maliga 1993) and *C. reinhardtii* (Newman et al. 1990; Goldschmidt-Clermont 1991; Kindle et al. 1991).

13.2.6.1 Use of rho^0 and rho– Cells in Yeast

Organelle transformation frequencies are not very high in yeast but the availability of efficient selection markers facilitates the isolation of transformants (Bonnefoy and Fox 2007; Fig. 13.6). *S. cerevisiae* cells are able to survive if they lack part of (*rho*$^-$) or the entire (*rho*^{0}) mitochondrial genome. Moreover, *rho*$^-$ mtDNA can replicate independently of protein synthesis and a specific origin of replication

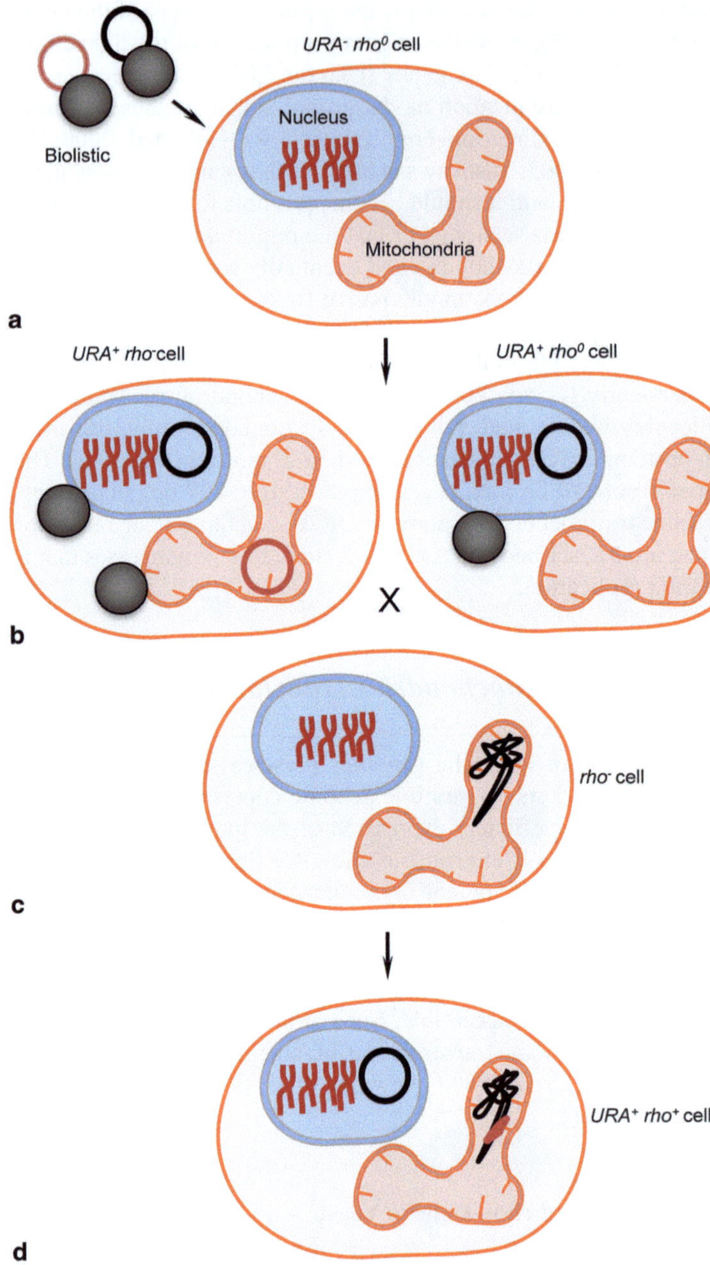

Fig. 13.6 Strategy for the selection of mitochondrial transformants in *S. cerevisiae* (I). **a** *Rho*^{0} cells are transformed by biolistics with a plasmid carrying a nuclear marker and a plasmid including a sequence destined to complement a mitochondrial mutation; a mixture of *rho*$^-$ and *rho*^{0} transformed cells is **b** selected through the nuclear marker and **c** crossed with a *rho*$^-$ mutant strain, **d** generating *rho*$^+$ mitochondrial transformants by complementation of the mutation with the exogenous sequence. *URA* uracil nutritional selection marker

is not required. Such mutants show no respiration and are not able to grow on a nonfermentable carbon source (Dujon 1981). These properties make *rho*$^{-}$ or *rho*^{0} mitochondrial mutants good instruments for the selection of yeast organelle transformants. In particular, *rho*^{0} cells can serve as recipients for co-transformation, allowing a two-step selection strategy. For this, the cells undergo a bombardment with a mix of two plasmids, the first one carrying a selectable nuclear marker and the second a portion of mitochondrial DNA of interest (Fox et al. 1988; Bonnefoy and Fox 2007). Primary selection is performed for nuclear transformants. For example, strains deficient in orotidine-5'-phosphate decarboxylase (URA3) were used as recipient cells. This enzyme catalyzes the sixth enzymatic step in de novo biosynthesis of pyrimidines, converting orotidine 5'-monophosphate (OMP) into uridine monophosphate. Thus, a URA– mutant is unable to grow on a medium-lacking uracil. In a second step, selection for mitochondrial transformants is performed among the URA3^{+} clones. For this, the latter are crossed with a strain carrying a mutation in the organelle gene of interest, resulting in respiring diploids. For example, a *rho*^{0} strain was transformed with a pBR322-*oxi1* plasmid and the resulting *rho*$^{-}$ strain was used to mate with an *oxi1* mutant (Fox et al. 1988). In other cases, selection of nuclear transformants can be followed by direct identification of mitochondrial transformants through DNA analysis (Bonnefoy and Fox 2007).

13.2.6.2 Barnase/Barstar Selection

Mireau et al. (2003) proposed a completely different strategy for the selection of mitochondrial transformants, based on the generation of an artificial mitochondrial defect by targeting the cytotoxic protein Barnase, an RNase, into the organelles (Fig. 13.7). They showed that Barnase targeted to mitochondria prevented growth of yeast cells in glycerol–ethanol media and induced irreversible, but not lethal, *rho*$^{-}$ petite deletion mutations in the mtDNA. Concomitant expression, in mitochondria, of recoded Barstar, a protein that inhibits Barnase (Hartley 1988), gave double transformants able to grow in glycerol–ethanol media. Thus, mitochondrial transformants having integrated the synthetic recoded Barstar gene into their mtDNA can be selected through resistance to mitochondrially targeted Barnase (Mireau et al. 2003).

13.2.6.3 Rescue of Deleterious Mutations in *Chlamydomonas reinhardtii*

C. reinhardtii is a model system for mitochondrial gene function in plant cells. Its small 15.8 kb linear mitochondrial genome is well known (Vahrenholz et al. 1993). The main strategy to select *C. reinhardtii* mitochondrial transformants is based on the restoration of organelle function in respiratory-deficient mutants (Fig. 13.8). These can be divided into two groups, according to differences in their growth under heterotrophic conditions: (i) mutants that still grow in the dark using acetate as a reduced carbon source and (ii) mutants that have lost their ability to grow in

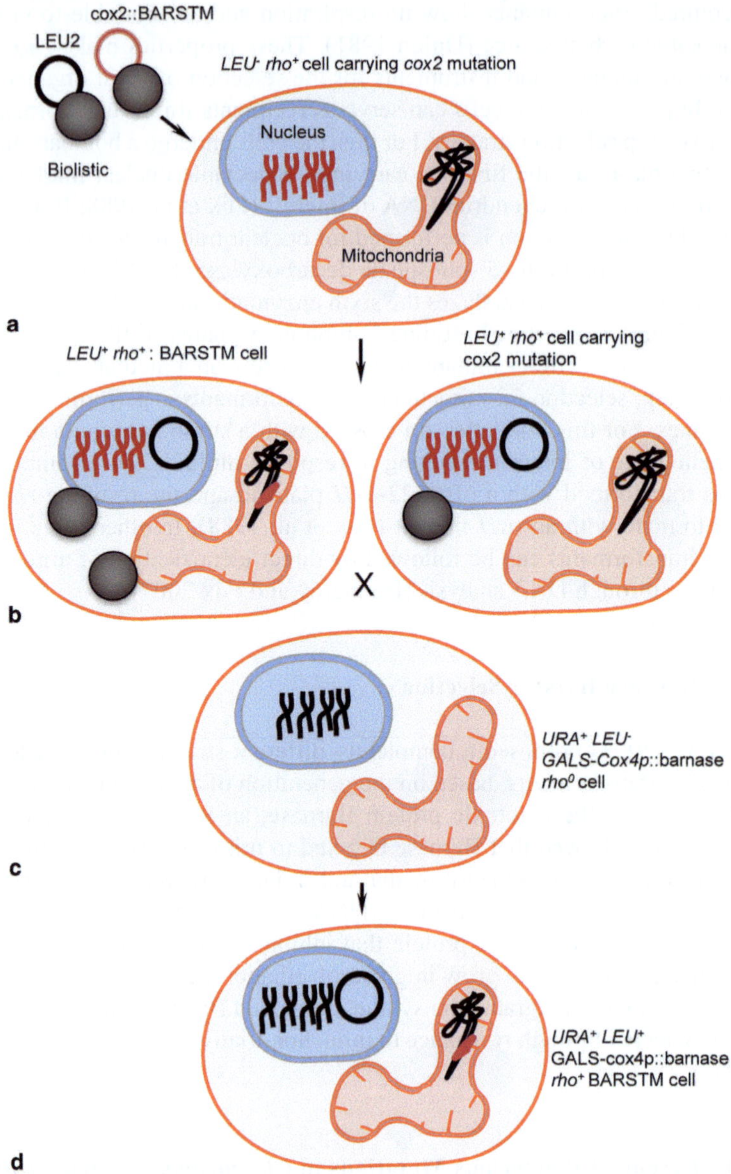

Fig. 13.7 Strategy for the selection of mitochondrial transformants in *S. cerevisiae* (II). **a** *Rho*$^+$ cells are transformed by biolistics with a plasmid carrying a nuclear marker and a plasmid-encoding barstar recoded for mitochondrial expression; a mixture of cells carrying or not the mitochondrial transformation is **b** selected through the nuclear marker and **c** crossed with a *rho*^{0} strain expressing the barnase nuclease fused to a mitochondrial-targeting sequence, **d** generating mitochondrial transformants able to resist to mitochondrially targeted barnase thanks to the expression of barstar, the barnase inhibitor, in the organelles; **e** in the example presented, the initial cells also carry an mtDNA *cox2* mutation that is complemented by the exogenous DNA upon mitochondrial transformation. *mtDNA* mitochondrial DNA

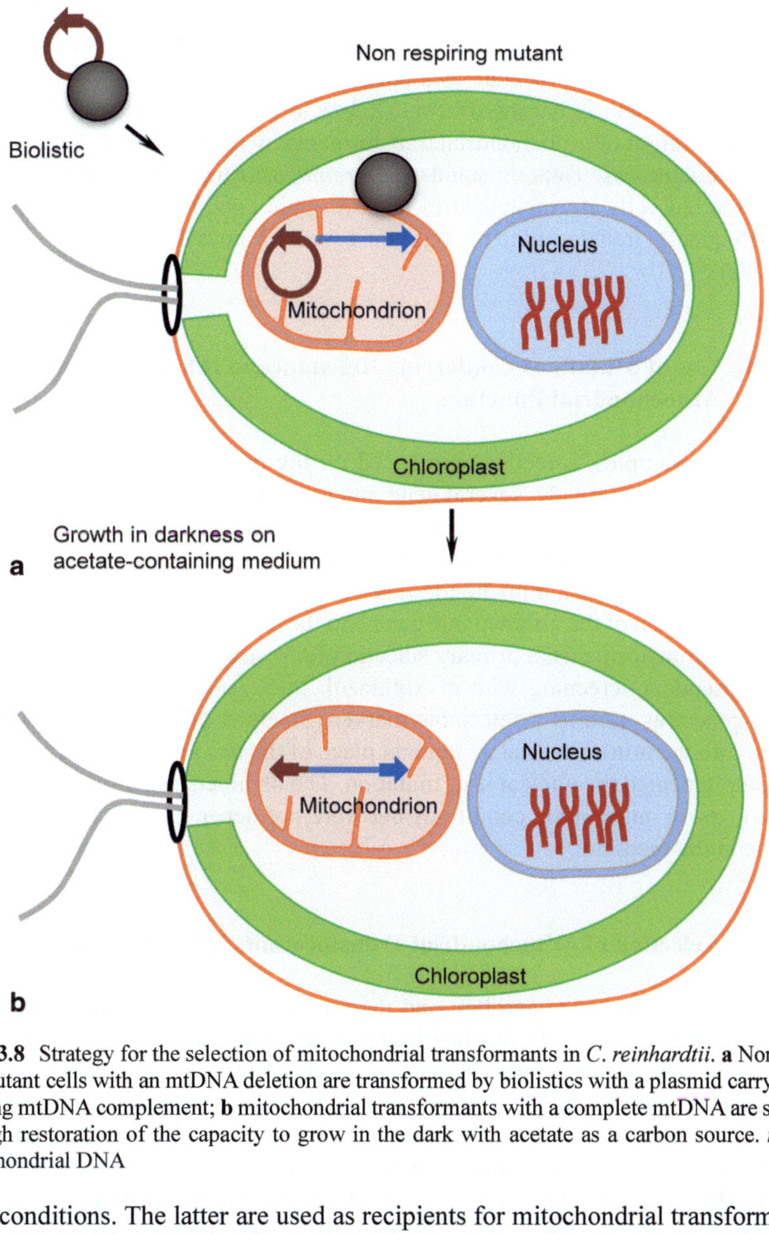

Fig. 13.8 Strategy for the selection of mitochondrial transformants in *C. reinhardtii*. **a** Nonrespiring mutant cells with an mtDNA deletion are transformed by biolistics with a plasmid carrying the missing mtDNA complement; **b** mitochondrial transformants with a complete mtDNA are selected through restoration of the capacity to grow in the dark with acetate as a carbon source. *mtDNA* mitochondrial DNA

such conditions. The latter are used as recipients for mitochondrial transformation. Several mutations altering the mitochondrial genes encoding apocytochrome b (*cob* gene) or subunit 1 of cytochrome *c* oxidase (*cox1* gene) have been characterized (Matagne et al. 1989; Dorthu et al. 1992; Randolph-Anderson et al. 1993; Colin et al. 1995). Mutant cells are homoplasmic for the mutations, and do not retain any mtDNA copies carrying the corresponding functional gene. The mutants lack the cytochrome pathway of the mitochondrial electron transport chain, but their respiratory activity is partially maintained via the alternative oxidase. Phenotypically, the mutants have lost their capacity to grow under heterotrophic conditions (darkness and addition of acetate as a reduced carbon source), whereas their photoautotrophic growth is barely affected (Remacle et al. 1998). These characteristics allow selection of mitochondrial transformants by restoration of the capacity to grow in the darkness. Thus, in standard *C. reinhardtii* transformation procedures, the bombarded cells are incubated in heterotrophic conditions during 6–8 weeks and subsequently transferred to the light for a week (Yamasaki et al. 2005; Remacle et al. 2006).

13.2.6.4 Use of Mutations Conferring Resistance to Inhibitors of Mitochondrial Function

In previous examples, selection was based on the restoration of respiration in a mutant strain. Alternatively, several gene mutations result in properties that can be used for selection of mitochondrial transformants. In *C. reinhardtii,* a point mutation in the *cob* gene that confers resistance to mucidine and myxothiazol (Bennoun et al. 1991) was used as a mitochondrial selection marker (Remacle et al. 2006). In that case, insertion of the mutated *cob* gene into the *dum11* strain allowed a two-step selection of transformants: a primary selection for restoration of growth in the dark and a subsequent screening with myxothiazol. In *S. cerevisiae,* a recoded *ARG8* nuclear gene was used as a selectable marker (Steele et al. 1996). This gene was inserted into the mitochondrial genome in place of the *cox3* gene, where its expression complemented the nuclear *arg*$^{-}$ mutation. The obtained transformants were not able to grow in autotrophic conditions but were not dependent on the presence of arginine in the medium.

13.2.6.5 Selection of Mitochondrial Transformants in Higher Plants

Higher plants are obligate aerobes, and there is currently no established selection strategy for the recovery of mitochondrial transformants. A potential mitochondrial selection marker was proposed by Ortega et al. (2000). The *cob* gene in organelle genomes encodes highly conserved amino acids (glycine-43 and phenylalanine-135). Mutation of at least one of these amino acids confers resistance to complex III inhibitors such as antimycine A or myxothiazol in mammals, fungi and algae (di Rago and Colson 1988; Howell and Gilbert 1988; di Rago et al. 1989; Bennoun et al. 1991). The authors proposed that corresponding mutations in the tobacco *cob* gene could confer resistance to such inhibitors and could be used in the selection of tobacco cells carrying a genetically transformed mitochondrial genome (Ortega et al. 2000). However, so far there has been no subsequent follow-up of this work in the literature. It can be noted that isopentenyl transferase, an enzyme in cytokinin biosynthesis, has been proposed in a patent as another potential marker for selection of mitochondrial transformants (see Sect. 4.7). Based on studies with isolated plant mitochondria, inserting a marker gene into the mtDNA through homologous recombination should be feasible in plant cells (Mileshina et al. 2011). The Barnase–Barstar strategy (2.6.2) was also thought to be promising for selection of mitochondrial transformants in plant species (Mireau et al. 2003). Controlled expression of a nuclear gene for barnase was achieved (Baroux et al. 2001a; Baroux et al. 2001b), but this system has not yet been further developed.

13.2.6.6 Selection of Mitochondrial Transformants in Mammalian Cells

Selectable markers for mitochondrial transformation in mammalian systems would be helpful too, at least for fundamental studies. As mentioned above (2.2.4), it is also possible to generate mammalian *rho*^{0} cells, i.e., cells with mitochondria totally lacking DNA (King and Attardi 1989). Restoration of organelle function in such mtDNA-deprived cells could be a marker for the repopulation of mitochondria with transfected exogenous DNA. On the other hand, mutations resulting in chloramphenicol resistance were identified in the human mitochondrial *16S rRNA* gene, the most efficient being a T to C transition at position 1321 (Blanc et al. 1981). An accordingly mutated *16S rRNA* gene could be included into the constructs to be transfected into mitochondria. Resistance to chloramphenicol would indicate expression of the mutated *16S rRNA* and proper integration into active mitochondrial ribosomes, providing a full functional test of the strategy. Finally, recoded drug-resistance genes encoding neomycin or hygromycin phosphotransferase have recently been put forward as potential selectable markers for mammalian mitochondrial transformation (Yoon and Koob 2008, 2011). When nuclearly expressed from transgenes with a standard sequence and organelle targeted via a MTS, these enzymes proved able to confer resistance to high antibiotic levels (Yoon and Koob 2008). Conversely, a neomycin phosphotransferase gene recoded for mitochondrial translation produced no background drug resistance upon nuclear transformation (Yoon and Koob 2011).

13.3 Allotopic Expression and Mitochondrial Targeting

13.3.1 Allotopic Expression of Proteins

In the absence of established methods to insert information into the mitochondrial genome, different approaches have been developed on the basis of nuclear transformation and mitochondrial targeting of the products expressed from the transgenes. Mitochondrial genomes encode only a small part of the hundreds of organellar proteins. The majority of the latter are synthesized in the cytosol from nuclear mRNAs and imported into the organelles. As mentioned above, the import into the matrix is driven by the N-terminal MTS (Mossmann et al. 2012). It is therefore possible to manipulate mitochondrial heteroplasmy and mutated mtDNA copy number through organelle targeting of enzymes encoded by nuclear transgenes (Fig. 13.9).

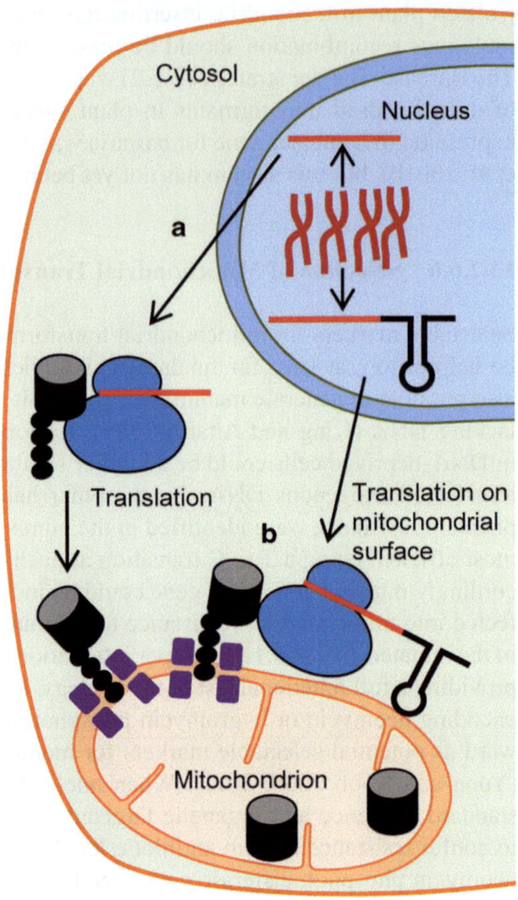

Fig. 13.9 Allotopic expression and mitochondrial targeting of proteins. Protein genes normally carried by the mtDNA (encoding oxidative phosphorylation chain subunits, CMS polypeptides) and genes encoding heterologous proteins, restriction endonucleases, or specific zinc-finger proteins were expressed in the nucleus after addition of a sequence coding for a mitochondrial-targeting peptide; the mRNAs were either **a** translated on free cytosolic polysomes or **b** specifically targeted to the mitochondrial surface, leading to membrane-bound polysomes; translocation into the organelles was through the regular protein import pathway. *mRNA* messenger RNA, *mtDNA* mitochondrial DNA, *CMS* cytoplasmic male sterility

13.3.1.1 Expression of Restriction Enzymes

Some mtDNA point mutations create a unique restriction site; hence, the idea to decrease the level of mutated genome copies by targeting a restriction endonuclease into the organelles. The feasibility of such an approach was tested by Srivastava and Moraes (2001). They expressed the *Pst*I restriction enzyme with an MTS in a rodent cell line containing one mtDNA haplotype with two sites for *Pst*I and another haplotype having none. Expression of mitochondrially targeted *Pst*I caused a significant shift in mtDNA heteroplasmy in favor of the haplotype without *Pst*I sites. The approach was further assessed with the *Sma*I enzyme that recognizes the pathogenic 8993T>G mutation in the human *MT-ATP6* gene. Expression and mitochondrial targeting of this restriction endonuclease in cybrids carrying the mutation resulted in specific elimination of the mutant mtDNA and restoration of both normal intracellular adenosine triphosphate (ATP) levels and mitochondrial membrane potential (Tanaka et al. 2002). More mtDNA mutations were subsequently addressed through mitochondrial targeting of appropriate restriction enzymes in mammalian cell lines and animal models (Bayona-Bafaluy et al. 2005; Alexeyev et al. 2008; Bacman et al. 2010;). In each case, an efficient shift in the mtDNA heteroplasmy was obtained. Bacman et al. (2007) also promoted a "differential multiple cleavage site" model in which a mitochondrially targeted restriction enzyme recognized a different number of cleavage sites in two subpopulations of mtDNA.

13.3.1.2 Expression of Zinc-Finger Proteins

When the addressed mutation does not create a restriction site, it might be possible to design specific zinc-finger proteins (ZFPs). Such proteins can bind any predetermined DNA sequence through appropriate zinc-finger domains (Papworth et al. 2006). To develop a system for monitoring site-specific alterations of the mtDNA, a ZFP recognizing the T8993T>G mutation was fused with the DNA-modifying activity of the hDNMT3a methylase (Minczuk et al. 2006). Import into mitochondria in human cells was ensured by an MTS combined with a nuclear export signal. Specific binding of the chimeric methylase to the mutated site resulted in selective methylation of cytosines adjacent to the mutation, bringing a proof of principle for targeted mtDNA alteration through the zinc-finger technology. In further experiments, zinc-finger peptides were linked to the catalytic cleavage domain of the *FokI* restriction endonuclease, yielding a site-specific nuclease to selectively deplete mutated mtDNA copies from heteroplasmic cells (Minczuk et al. 2008). More prospects of the ZFP strategy for mtDNA manipulation were put forward by Minczuk et al. (2010).

13.3.2 Allotopic Expression of RNAs

13.3.2.1 Import of tRNAs

As mentioned above (2.4.4), in many organisms, mitochondria import a variable number of tRNAs from the cytosol (Schneider 2011). These tRNAs are shared between the cytosol and mitochondria and are involved in protein synthesis in both compartments. Only few cases have been reported in mammals, including mitochondrial import of a cytosolic $tRNA^{Lys}$ in marsupials (Dörner et al. 2001), and of $tRNAs^{Gln}$ in rats and humans (Rubio et al. 2008). In *S. cerevisiae,* cytosolic $tRNA^{Lys}$(CUU; Tarassov and Entelis 1992) and $tRNAs^{Gln}$ (Rinehart et al. 2005) were also found in mitochondria. Strategies based on tRNA import were therefore developed to manipulate the mitochondrial genetic system (Fig. 13.10), especially using the extensive knowledge gained on the mechanism of $tRNA^{Lys}$(CUU) uptake into yeast mitochondria (Entelis et al. 2006; Brandina et al. 2007). Mitochondria isolated from HepG2 human cells proved capable to import the yeast $tRNA^{Lys}$(CUU) and mutated forms thereof (Kolesnikova et al. 2000). Nuclear expression of yeast $tRNA^{Lys}$ derivatives in transformed HepG2 cells subsequently established the in

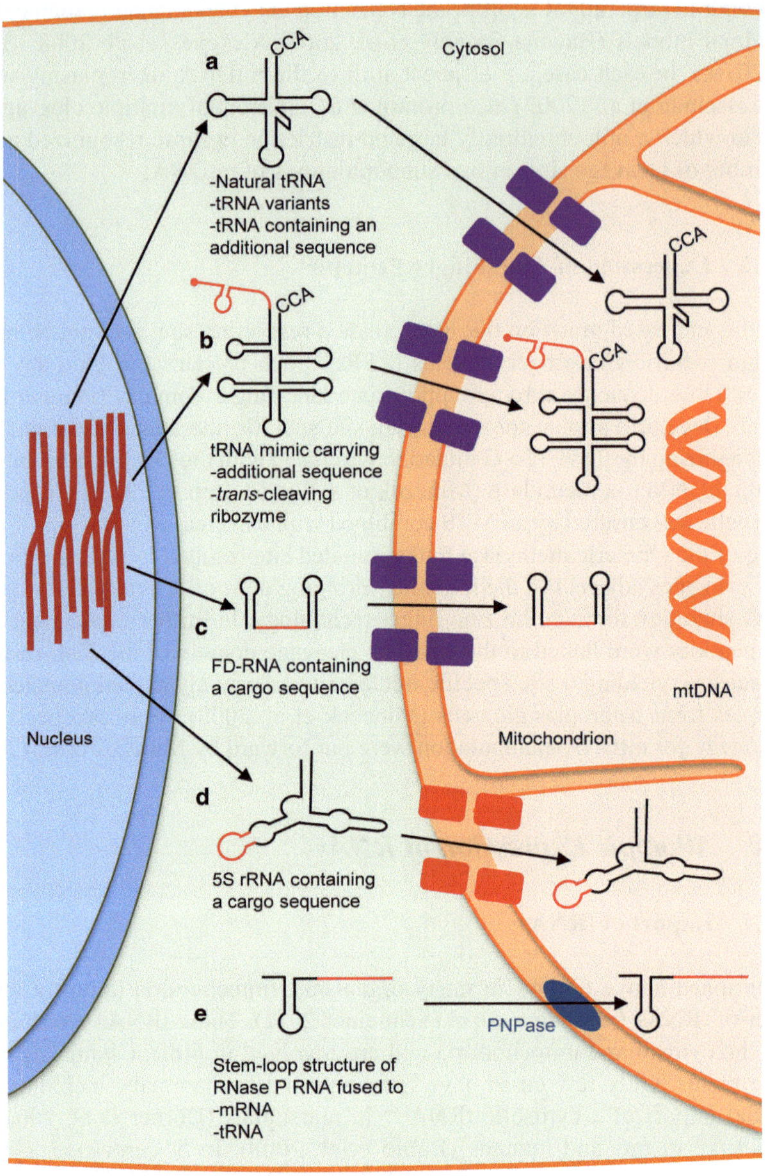

Fig. 13.10 Allotopic expression and mitochondrial targeting of customized RNAs. **a** Genes encoding tRNA variants or tRNAs carrying an additional sequence, **b** cargo sequences or catalytic RNAs combined with a tRNA mimic, and **c** oligoribonucleotides flanked with tRNA-derived hairpins (FD-RNAs) were nuclearly expressed and translocated into mitochondria through the natural tRNA import pathway; **d** oligoribonucleotides were also embedded into the 5S rRNA and translocated into the organelles through the corresponding import pathway; **e** alternatively, tRNAs or mRNAs were combined with a stem-loop structure derived from the RNase P RNA and translocated into mitochondria through the PNPase-mediated import pathway. *tRNA* transfer *RNA, rRNA* ribosomal *RNA, mRNA* messenger *RNA*

vivo organelle import of these tRNAs and their participation in mitochondrial protein synthesis (Kolesnikova et al. 2004). Further derivatives for which the aminoacylation identity had been switched from lysine to leucine were in turn expressed from nuclear transgenes in human cells and recovered in mitochondria (Karicheva et al. 2011). All these data opened up therapeutic prospects for the complementation of mutated mitochondrial tRNA genes. Notably, short extra sequences embedded in a tRNA could also be transported into the organelles in transgenic *L. tarentolae* cell lines (Sbicego et al. 1998) or in plants (Small et al. 1992; Fig. 13. 10).

13.3.2.2 Cargo RNA Import Driven by a tRNA Mimic

In plants, between one third and one half of the mitochondrial tRNAs are imported from the cytosol (Maréchal-Drouard et al. 1990; Kumar et al. 1996), hence the idea of using a mimic of a naturally imported tRNA as a shuttle to drive cargo sequences into the organelles. Such a strategy was developed with the tRNA-like structure (TLS) that forms the 3'-end of the *Turnip yellow mosaic virus* (TYMV) genomic RNA (Matsuda and Dreher 2004). The TYMV TLS consists of the last 82 nucleotides of the genomic RNA and its functionality is enhanced by an upstream pseudoknot. It is a valine acceptor, and thus it mimics a tRNA that is imported into mitochondria in all plant species studied so far. The shuttle used comprised the TLS together with the upstream pseudoknot and was named "PKTLS" (Val et al. 2011). Upon nuclear expression of an appropriate gene construct, the PKTLS tRNA mimic proved capable of driving a 5'-end trailor sequence into mitochondria in plant cells, a process mediated by the natural tRNA uptake pathway (Val et al. 2011; Fig. 13.10). PKTLS-mediated organelle import of a functional sequence in the form of a *trans*-cleaving hammerhead ribozyme directed against a mitochondrial mRNA was subsequently obtained in plant-cell suspensions and in whole plants. The strategy was efficient and caused a strong knockdown of the target RNA. Furthermore, other, nontarget mitochondrial RNAs were also affected, implying the existence of organelle RNA coordination mechanisms (Val et al. 2011). The approach thus appears suitable for organelle-related functional and regulatory studies in plants. It is potentially applicable to other organisms in which mitochondrial tRNA import occurs.

13.3.2.3 Import of Recombinant 5S rRNA

Yoshionari et al. (1994) detected the nuclear-encoded 5S rRNA in bovine mitochondria. Subsequent studies confirmed this observation and established that mammalian organelles naturally take up a significant portion of the 5S rRNA (Magalhaes et al. 1998; Entelis et al. 2001). The import determinants of the RNA were characterized by importing into human mitochondria, both in vitro and in vivo, 5S rRNA variants obtained by site-directed mutagenesis (Smirnov et al. 2008). Two regions are required for import: The first is located in domain α, the second is associated with domain γ. Domain β is not essential for uptake, which raises the idea that it might accommodate additional sequences and make the 5S rRNA a mitochondrial import shuttle for short RNAs in mammalian cells (Smirnov et al. 2008). The hypothesis was successfully tested by expressing in human cells a 5S rRNA variant in which an unrelated sequence of 13 nucleotides replaced loop C and part of helix III. The mutant RNA was imported into mitochondria, with an efficiency even higher than that of the wild-type rRNA (Smirnov et al. 2008; Fig. 13.10).

13.4 Applications

Some of the approaches described above have already been applied in strategies that aimed at mastering of male sterility in plants or gene therapy of mitochondrial diseases in humans.

13.4.1 Import of CMS-Associated Proteins

In plants, allotopic expression (Fig. 13.9) was used to generate male sterile lines through nuclear transformation with mitochondrial *orfs* encoding CMS-associated proteins, the addition of an MTS providing organelle targeting of the polypeptides (Table 13.1). Contrasting results were reported. Male sterile plants were obtained upon nuclear expression of the CMS-associated bean *orf239* (He et al. 1996), beet *orf129* (Yamamoto et al. 2008), chili pepper *orf456* (Kim et al. 2007), mustard *orf220* (Yang et al. 2010), rice *orf79* (Wang et al. 2006) or wall-rocket *orf108* (Kumar et al. 2012). Conversely, similar attempts with the proteins associated with maize CMS-T, petunia CMS or radish and *Brassica* Ogura CMS failed to yield sterile plants (von Allmen et al. 1991; Chaumont et al. 1995; Wintz et al. 1995; Duroc et al. 2006). Also puzzling, in some cases, expression of the CMS *orf* resulted in aberrant pollen function and a semi-sterile or male-sterile phenotype regardless of the presence or not of an MTS (He et al. 1996; Kumar et al. 2012). It seems, for instance, that the bean ORF239 protein can associate with the cell wall and therefore may cause pollen disruption even if it is not targeted to mitochondria. In maize CMS-T, sterility is correlated with the presence of the mitochondrial *T-urf13* gene (Dewey et al. 1986). When this gene, fused to an MTS-encoding sequence, was expressed in tobacco from a nuclear transgene, the URF13 protein actually accumulated in mitochondria, but there was no correlation with a male sterile phenotype (Chaumont et al. 1995). The lack of male sterility following nuclear expression of several CMS genes (von Allmen et al. 1991; Chaumont et al. 1995; Wintz et al. 1995; Duroc et al. 2006;) is not clearly understood. Inappropriate timing and/or spatial control of transgene expression is often put forward as an explanation. Also, too little of the protein may accumulate in cases in which a quantitative threshold must be exceeded for activity (Bellaoui et al. 1998). Alternatively, inappropriate

Table 13.1 Nuclear expression and mitochondrial targeting of CMS-associated proteins

Protein	References	Acceptor species	Main results reported
T-urf13 CMS-T *Zea mays*	von Allmen et al. 1991	*Nicotiana tabacum*	Fertile plants
T-urf13 CMS-T *Zea mays*	Chaumont et al. 1995	*Nicotiana tabacum*	No correlation of protein expression with CMS
urfS *Petunia parodii*	Wintz et al. 1995	*Petunia hybrida* *Nicotiana tabacum*	Protein located in soluble and membrane fractions. Fertility of plants was not affected
orf239 *Phaseolus vulgaris*	He et al. 1996	*Nicotiana tabacum*	Semi-sterile or male sterile plants
orf138-GFP *Brassica napus*	Duroc et al. 2006	*Arabidopsis thaliana*	Fertile plants
orf79 *Oryza sativa*	Wang et al. 2006	*Oryza sativa*	Transgenic plants exhibited a semi-male-sterility phenotype wherein ~50 % or more of the pollen grains were aborted
orf456 *Capsicum annuum*	Kim et al. 2007	*Arabidopsis thaliana*	Male-sterile plants. Defects on the exine layer and vacuolated pollen phenotypes were observed
orf129 *Beta vulgaris*	Yamamoto et al. 2008	*Nicotiana tabacum*	1 of 4 fully male-sterile plant; 1 of 4 fertile plant; 2 of 4 semi-fertile plants
orf220 *Brassica juncea*	Yang et al. 2010	*Brassica juncea*	Male-sterile plants
orf108 *Brassica juncea*	Kumar et al. 2012	*Arabidopsis thaliana*	50 % pollen sterility

CMS cytoplasmic male sterility, *GFP* green fluorescent protein

folding or submitochondrial location of the CMS protein may be responsible. Finally, mitochondrial targeting may be inefficient when the CMS polypeptide is highly hydrophobic. In a reciprocal strategy, CMS mutants with mtDNA variation were generated from protoplast culture in *Nicotiana sylvestris* (Li et al. 1988), and male fertility and complex I functionality were restored through targeting the NAD7 subunit into mitochondria in the mutants (Pineau et al. 2005).

13.4.2 Import of Oxidative Phosphorylation Complex Subunits

A series of approaches have been designed to rescue mtDNA mutations in protein genes through nuclear expression and mitochondrial import of the corresponding polypeptides (Fig. 13.9, Table 13.2). First attempts of this strategy were developed in yeast. The mitochondrial gene for ATP synthase subunit 8 (*ATP8*) was appropriately recoded, fused to an MTS-encoding sequence, and nuclearly expressed in mutant yeast cells lacking endogenous *ATP8*. The resulting ATP8 protein was imported and functionally assembled into the ATP synthase complex, which complemented the respiratory growth defect (Nagley et al. 1988). Experiments aiming to target

Table 13.2 Nuclear expression and mitochondrial targeting of oxidative phosphorylation chain subunits or substituting proteins

Protein	References	Acceptor cells/ organisms	Main results reported
ATP8 *Saccharomyces cerevisiae*	Nagley et al. 1988	*Saccharomyces cerevisiae*	Protein was functionally assembled into the ATPase complex
ATP8 Human	Oca-Cossio et al. 2003	HeLa COS-7	Protein was properly located in mitochondria
ATP6 Human	Manfredi et al. 2002	Cybrids carrying a 8993T>G transversion	Protein was functionally assembled into the ATPase complex. Rescue of a deficiency in ATP synthesis
ATP6 *Chlamydomonas reinhardtii*	Ojaimi et al. 2002	Human embryonic kidney 293T cells, monkey COS7 kidney cells, human cybrids carrying a 8993T>G tranversion	Partial rescue of a deficiency in ATP synthesis
ATP6 Human *Chlamydomonas reinhardtii*	Bokori-Brown and Holt 2006	Human cybrids carrying a 8993T>G tranversion	No integration of the imported protein into mature functional ATPase. No rescue of the mutated MT-ATP6
ATP6 Human HeLa	Kaltimbacher et al. 2006	HeLa	Protein was localized to mitochondria in vivo
ATP6 Human	Bonnet et al. 2007	Human fibroblasts harboring a 8993T>G mutation	Protein was functional in ATPase complex. Rescue of ATPase activity
Cyt b Human	Oca-Cossio et al. 2003	HeLa COS-7	High-level expression: fiber-like structures and loss of mitochondrial membrane potential, protein aggregates with the cytoskeleton. Low-level expression: protein localized on the surface of mitochondria
ND4 Human	Guy et al. 2002	Cybrids carrying a 11778G>A mutation	Rescue of complex I-dependent respiration
ND4 Human	Oca-Cossio et al. 2003	HeLa COS-7	Protein did not co-localize with Mito-Tracker staining and caused loss of mitochondrial membrane potential
ND4 Human	Bonnet et al. 2008	Fibroblasts from a LHON patient carrying a 11778G>A mutation	Complete and long-term restoration of mitochondrial function
ND6 Human	Perales-Clemente et al. 2011	*Mus musculus* cells with a homoplasmic mutation in the *Nd6* gene	Protein was not imported into the mitochondria. False rescue, selection of remaining wild-type mtDNA

Table 13.2 (continued)

Protein	References	Acceptor cells/ organisms	Main results reported
Mutated ND4 11778G>A Human	Qi et al. 2007	*Rattus norvegicus* retinal ganglion cell-like cells and DBA/1J *Mus musculus*	Optic nerve degeneration
Mutated ND4 11778G>A Human	Ellouze et al. 2008	Male Wistar and Long Evans *Rattus norvegicus*	Significant decrease of retinal ganglion cells associated with impaired visual function
Mutated ATP6 8993T>G *Mus musculus*	Dunn and Pinkert 2012	C57BL/6 *Mus musculus*	Protein was co-localized with mitochondria. Triggered neuromuscular and motor deficiencies. Mitochondrial functions tested similar to wild-type
Ndi1 *Saccharomyces cerevisiae*	Yagi et al. 2006a, b	Ten complex I-deficient mammalian cell lines	Protein was correctly targeted to the matrix side of the inner mitochondrial membrane. Restoration of NADH oxidase activity to the complex I-deficient cells
Ndi1 *Saccharomyces cerevisiae*	Seo et al. 2006	Male C57BL/6 *Mus musculus*	Rescue of complex I deficiency
AOX *Ciona intestinalis*	Dassa et al. 2009	Human HEK 293-derived complex III deficient cells	Compensation of the growth defect and oxidant-sensitivity of the cells

ATP adenosine triphosphate, *LHON* Leber hereditary optic neuropathy, *MT-ATP6* mitochondrial ATP6 gene

into mitochondria nuclearly expressed ATP6 subunit were subsequently developed in human cells. Successful expression of a human (Manfredi et al. 2002) or *C. reinhardtii* (Ojaimi et al. 2002) recoded *ATP6* transgene and mitochondrial targeting of its product were reported to partially suppress a respiratory-deficient phenotype caused by a pathogenic mutation in the mtDNA. However, similar experiments in human cells with both the human and the *C. reinhardtii ATP6* gene failed to achieve integration of the imported protein into functional ATP synthase and rescue of the mutated *MT-ATP6* gene (Bokori-Brown and Holt 2006). Conflicting conclusions were also drawn upon allotopic expression of complex I subunit 4 (ND4). In one study (Guy et al. 2002), rescue of mitochondrial deficiency due to an ND4 mutation causing Leber Hereditary Optic Neuropathy (LHON) was observed, whereas in other experiments (Oca-Cossio et al. 2003) mitochondrial import of the allotopically expressed ND4 protein could not be obtained. A similar study involving the ND6 polypeptide (subunit 6 of complex I) in mouse cells also highlighted the absence of organelle uptake and complex integration of the protein and concluded that apparently complemented cells were actually revertants for the *nd6* mutation

(Perales-Clemente et al. 2011). To develop an animal model for mitochondrial diseases, mutant and wild-type ATP6 were allotopically expressed and organelle targeted in mice (Dunn and Pinkert 2012). Expression of mutated ATP6 triggered neuromuscular and motor deficiencies. A comparable system was set up in mouse with mutated human ND4 (Qi et al. 2007). Altogether, these assays showed the limits and constraints of the strategy (Oca-Cossio et al. 2003; Perales-Clemente et al. 2011). In a twist to this approach, assays were developed to optimize organelle targeting of such highly hydrophobic mitochondrial proteins that are otherwise encoded by the mtDNA. It is known that many nuclear mRNAs encoding mitochondrial proteins localize to the organelle surface in yeast, mammals, and plants, so that translation and import can be coupled (Marc et al. 2002; Sylvestre et al. 2003; Michaud et al. 2010). Localization involves signals carried by the 3'-UTR (Sylvestre et al. 2003). Kaltimbacher et al. (2006) reported that targeting the nuclear transgene-derived *ATP6* mRNA with appropriate *cis*-acting signals to the outer mitochondrial membrane in human cells indeed facilitated organelle translocation of the gene product. The strategy was thus applied to fibroblasts harboring a mutation in the mitochondrial *ATP6* gene, which led to long-lasting rescue of mitochondrial dysfunction (Bonnet et al. 2007). Similar results were obtained upon mitochondrial surface targeting of allotopically expressed *ND1* (subunit 1 of complex I) or *ND4* transcripts in fibroblasts carrying mtDNA mutations in the corresponding genes (Bonnet et al. 2007, 2008). That such an optimized allotopic expression approach might become an effective treatment for mitochondrial deficiency resulting from mtDNA mutation was finally supported through a rat LHON model (Ellouze et al. 2008).

13.4.3 Import of Substituting Proteins

Instead of compensating mutations in mtDNA-encoded subunits through uptake of wild-type polypeptides, so-called xenotropic expression strategies aimed to bypass the deficient functions with proteins originating from other organisms (Table 13.2). Yeast mitochondria do not have complex I but use three NADH dehydrogenases located at both sides of the inner membrane (Bakker et al. 2001). Among these, the single-subunit NADH dehydrogenase Ndi1 transfers electrons from mitochondrial NADH to the quinone pool and thus has the potential to bypass complex I. This enzyme was targeted into mitochondria and indeed functionally restored NADH oxidase activity in complex I-deficient mammalian cells (Yagi et al. 2006a, b). Rescue of complex I deficiency was further validated in a mouse model of Parkinson's disease (Seo et al. 2006), which might open therapeutic prospects. On the other hand, many species, but not mammals, encode an alternative oxidase (AOX) that introduces a branch point into the respiratory electron transport chain and bypasses complexes III and IV (McDonald et al. 2009; Van Aken et al. 2009). Following nuclear expression and mitochondrial targeting, the *Ciona intestinalis* AOX was reported to compensate for the growth defect and oxidant sensitivity of complex III-deficient human cells (Dassa et al. 2009).

13.4.4 Import of tRNAs

Upon validation of allotopic expression and mitochondrial import of tRNA variants in human cells (3.2.1), the strategy was applied to complement mutations in mitochondrial tRNA genes through import of corrective, functional tRNAs (Fig. 13.10). The approach was first conducted with human cybrid cells and patient-derived fibroblasts carrying a MERRF syndrome-associated mutation in the mitochondrial $tRNA^{Lys}$ gene. A partial restoration of mitochondrial functions was obtained upon expression and organelle uptake of yeast $tRNA^{Lys}$ derivatives, demonstrating the potential of the strategy for the treatment of mtDNA diseases (Kolesnikova et al. 2004). In subsequent assays, the aminoacylation specificity of the recombinant tRNAs was switched from lysine to leucine, so as to address pathogenic mutations in mitochondrial $tRNA^{Leu}$. This in turn improved mitochondrial functions in human cybrids impaired by a $tRNA^{Leu}$ mutation associated with the mitochondrial encephalomyopathy, lactic acidosis, and stroke-like episodes (MELAS) syndrome (Karicheva et al. 2011). The therapeutic potential of customized tRNA targeting was thereby much broadened, with the extension to tRNAs with altered specificities.

13.4.5 Import of Antigenomic Oligoribonucleotides

Following successful organelle translocation of a 5S rRNA variant carrying a 13 nucleotide exogenous sequence in human cells (3.2.3), the approach (Fig. 13.10) was used to manipulate mitochondrial heteroplasmy in human cybrids bearing a large mtDNA deletion associated with the Kearns Sayre syndrome (KSS). For that, antigenomic sequences targeting the junction between the deletion boundaries, and thus expected to inhibit replication of the deleted mtDNA copies, were included into the β domain of the 5S rRNA. The recombinant 5S rRNAs expressed from nuclear transgenes in KSS cybrids were imported into mitochondria and allowed a stable shift of mtDNA heteroplasmy in favor of wild-type copies, with an improvement of mitochondrial translation (Comte et al. 2012). In addition, analyzing the import of yeast tRNA derivatives into human mitochondria led to the characterization of short synthetic sequences comprising two tRNA-derived hairpins joined by a short-linker domain. Such RNAs, named FD-RNAs, showed a high efficiency of mitochondrial import (Kolesnikova et al. 2010). The central linker can be replaced by a short sequence of interest without hindering import, providing a further mitochondrial shuttle system (Fig. 13.10). On that basis, FD-RNAs in which the linker was replaced by antigenomic sequences against the above-described mtDNA deletion were constructed and introduced into the KSS cybrids by transient transfection. In the cells, the synthetic RNAs were taken up by the organelles and triggered a specific replication pause of the deleted KSS mtDNA copies, in turn providing a shift of the heteroplasmy towards wild-type mtDNA (Comte et al. 2012). Although the effect of these approaches was only partial, it might be sufficient to decrease and keep the level of mutated mtDNA copies below the threshold that causes the symptoms of a mitochondrial disease (Smirnov et al. 2008; Comte et al. 2012).

13.4.6 PNPase-Facilitated RNA Targeting

Besides tRNAs and 5S rRNA, mammalian mitochondria were reported to import the RNA component of RNase mitochondrial RNA peptidase (MRP; Chang and Clayton, 1987, 1989) and that of RNase P (Doersen et al. 1985). The role of the latter is currently unclear, as it was shown in the meanwhile that human mitochondrial RNase P activity is carried by proteins only (Holzmann et al. 2008). Nevertheless, mitochondrial import of these RNAs was further reported and seems to involve polynucleotide phosphorylase (PNPase; Wang et al. 2010). PNPase, a 3'–5' exoribonuclease localized in the mitochondrial intermembrane space, would thus have another role in regulating the import of selected nucleus-encoded RNAs, including the RNase P/RNase MRP RNAs and the 5S rRNA (Wang et al. 2010, 2012a). It was proposed that PNPase recognizes a stem-loop structure in the RNAs that it helps to import. This stem-loop structure would serve as a targeting signal for mitochondrial translocation. The model was drawn from extensive in vitro and in vivo assays with both yeast and mammalian organelles (Wang et al. 2010). The data pointed towards the possibility to use the characterized stem-loop structure as one more shuttle system to carry nuclear-encoded RNAs into mitochondria (Fig. 13.10). The RNase P stem loop indeed enabled import of tRNA precursors into isolated mammalian organelles. The precursors were processed to mature tRNAs and partially rescued the translation defect of isolated MERRF ($tRNA^{Lys}$ gene mutation) or MELAS ($tRNA^{Leu}$ gene mutation) mutant mitochondria (Wang et al. 2012b). However, in vivo functional rescue in MERRF and MELAS cybrids needed further specific sequences. Organelle import and protein synthesis recovery were obtained only when, in addition to the RNase P stem loop, the tRNA precursors were expressed with (i) an extension favoring processing-free nuclear escape and (ii) an mRNA 3'-UTR-derived localization signal that mediates targeting to the mitochondrial surface (4.2). It was reported that the RNase P stem loop also enables organelle import and processing of nuclear transgene-encoded mRNAs in mammalian cells, as assayed with a 683 nucleotide-long *COX2* mRNA (Wang et al. 2012b). This approach would thus have the potential to target defects in all mtDNA disorders.

13.4.7 Further Mitochondrial Transfection Strategies in the Pipeline

Reflecting the relevance of the issue for applied purposes, mitochondrial transformation is further claimed by a number of patents or patent submissions that essentially have no counterpart in the open literature. Patent application WO2009150441 (Moller and Xu 2009) claims biolistics-mediated mitochondrial transformation in plants, using cytokinin or auxin biosynthesis markers for selection, in combination or not with antibiotic resistance. Based on homologous recombination with the organelle DNA, the methodology includes a molecular strategy to eliminate the marker after selection. In patent WO/2010/061187, Malcuit and Sorokin (2010) claim mitochondrial transport of nuclear transgene-encoded RNA promoted by an organelle-targeted RNA-binding protein, still in plants. As further steps, retrotranscription of the transported RNA by an organelle-targeted reverse transcriptase would be followed by homologous recombination and integration into the mitochondrial genome. In another patent application (WO/2008/148223), Eudes and Chugh (2008) describe mitochondrial transfection driven by cell-penetrating peptides (CPPs). CPPs are cationic short peptides that form nanocomplexes with the DNA and translocate across plant cell membranes. Based on extensive analysis of known mitochondrial-targeting peptides, a novel class of CPPs that specifically target the organelles was defined. A reporter construct complexed with such specific CPPs was reportedly delivered into mitochondria in *Triticale* protoplasts or microspores and was expressed in the organelles (Eudes F., communication at the Plant and Animal Genome XX Conference, San Diego, CA, USA, 2012).

13.5 Conclusion

As clearly outlined in this chapter, developing mitochondrial genetic manipulation has mobilized much imagination and experimental investment over the years. Investigations have proceeded as far as mtDNA replacement in germ cells, with the idea of preventing diseases (Tachibana et al. 2012). A variety of strategies, already successful or still promising, have been put forward, so that the field actually looks rather prolific. The counterpart is that a number of these approaches so far remained confined to one laboratory, and some are openly controversial. Mitochondrial medicine is on the way (e.g., Cwerman-Thibault et al. 2011), but the current challenge is to validate independently confirmed and generally accepted biotechnological tools (Lightowlers 2011). The stake is worth the effort.

Acknowledgments Collaborations with the laboratories of Y. Konstantinov (SIFIBR, Irkutsk, Russia), R.N. Lightowlers (University of Newcastle, UK), G.G.M. D'Souza (MCPHS, Boston, USA) and J. Barciszewski (IBCH, Poznan, Poland) are kindly acknowledged. Our projects are funded by the French Centre National de la Recherche Scientifique (CNRS, UPR2357), the Université de Strasbourg (UdS), the Agence Nationale de la Recherche (ANR-06-MRAR-037-02, ANR-09-BLAN-0240-01) and the Ministère de la Recherche et de l'Enseignement Supérieur (Investissements d'Avenir/Laboratoire d'Excellence MitoCross).

References

Alexeyev MF, Venediktova N, Pastukh V, Shokolenko I, Bonilla G, Wilson GL (2008) Selective elimination of mutant mitochondrial genomes as therapeutic strategy for the treatment of NARP and MILS syndromes. Gene Ther 15:516–523

Bacman SR, Williams SL, Hernandez D, Moraes CT (2007) Modulating mtDNA heteroplasmy by mitochondria-targeted restriction endonucleases in a 'differential multiple cleavage-site' model. Gene Ther 14:1309–1318

Bacman SR, Williams SL, Garcia S, Moraes CT (2010) Organ-specific shifts in mtDNA heteroplasmy following systemic delivery of a mitochondria-targeted restriction endonuclease. Gene Ther 17:713–720

Bakker BM, Overkamp KM, van Maris AJ, Kotter P, Luttik MA, van Dijken JP, Pronk JT (2001) Stoichiometry and compartmentation of NADH metabolism in *Saccharomyces cerevisiae*. FEMS Microbiol Rev 25:15–37

Baroux C, Blanvillain R, Gallois P (2001a) Paternally inherited transgenes are down-regulated but retain low activity during early embryogenesis in *Arabidopsis*. FEBS Lett 509:11–16

Baroux C, Blanvillain R, Moore IR, Gallois P (2001b) Transactivation of BARNASE under the AtLTP1 promoter affects the basal pole of the embryo and shoot development of the adult plant in *Arabidopsis*. Plant J 28:503–515

Bayona-Bafaluy MP, Blits B, Battersby BJ, Shoubridge EA, Moraes CT (2005) Rapid directional shift of mitochondrial DNA heteroplasmy in animal tissues by a mitochondrially targeted restriction endonuclease. Proc Natl Acad Sci U S A 102:14392–14397

Bellaoui M, Martin-Canadell A, Pelletier G, Budar F (1998) Low-copy-number molecules are produced by recombination, actively maintained and can be amplified in the mitochondrial genome of Brassicaceae: relationship to reversion of the male sterile phenotype in some cybrids. Mol Gen Genet 257:177–185

Bennoun P, Delosme M, Kuck U (1991) Mitochondrial genetics of *Chlamydomonas reinhardtii*: resistance mutations marking the cytochrome b gene. Genetics 127:335–343

Ber R, Stauver MG, Shay JW (1984) Use of isolated mitochondria to transfer chloramphenicol resistance in hamster cells. Isr J Med Sci 20:244–248

Bhattacharyya SN, Chatterjee S, Goswami S, Tripathi G, Dey SN, Adhya S (2003) "Ping-pong" interactions between mitochondrial tRNA import receptors within a multiprotein complex. Mol Cell Biol 23:5217–5224

Blanc H, Adams CW, Wallace DC (1981) Different nucleotide changes in the large rRNA gene of the mitochondrial DNA confer chloramphenicol resistance on two human cell lines. Nucleic Acids Res 9:5785–5795

Boddapati SV, Tongcharoensirikul P, Hanson RN, D'Souza GG, Torchilin VP, Weissig V (2005) Mitochondriotropic liposomes. J Liposome Res 15:49–58

Boesch P, Ibrahim N, Paulus F, Cosset A, Tarasenko V, Dietrich A (2009) Plant mitochondria possess a short-patch base excision DNA repair pathway. Nucleic Acids Res 37:5690–5700

Boesch P, Ibrahim N, Dietrich A, Lightowlers RN (2010) Membrane association of mitochondrial DNA facilitates base excision repair in mammalian mitochondria. Nucleic Acids Res 38:1478–1488

Bokori-Brown M, Holt IJ (2006) Expression of algal nuclear ATP synthase subunit 6 in human cells results in protein targeting to mitochondria but no assembly into ATP synthase. Rejuvenation Res 9:455–469

Bolle N, Kempken F (2006) Mono- and dicotyledonous plant-specific RNA editing sites are correctly edited in both *in organello* systems. FEBS Lett 580:4443–4448

Bolle N, Hinrichsen I, Kempken F (2007) Plastid mRNAs are neither spliced nor edited in maize and cauliflower mitochondrial *in organello* systems. RNA 13:2061–2065

Bonnefoy N, Fox TD (2000) *In vivo* analysis of mutated initiation codons in the mitochondrial COX2 gene of *Saccharomyces cerevisiae* fused to the reporter gene ARG8m reveals lack of downstream reinitiation. Mol Gen Genet 262:1036–1046

Bonnefoy N, Fox TD (2007) Directed alteration of *Saccharomyces cerevisiae* mitochondrial DNA by biolistic transformation and homologous recombination. Methods Mol Biol 372:153–166

Bonnefoy N, Bsat N, Fox TD (2001) Mitochondrial translation of *Saccharomyces cerevisiae* COX2 mRNA is controlled by the nucleotide sequence specifying the pre-Cox2p leader peptide. Mol Cell Biol 21:2359–2372

Bonnefoy N, Remacle C, Fox TD (2007) Genetic transformation of *Saccharomyces cerevisiae* and *Chlamydomonas reinhardtii* mitochondria. Methods Cell Biol 80:525–548

Bonnet C, Kaltimbacher V, Ellouze S, Augustin S, Benit P, Forster V, Rustin P, Sahel JA, Corral-Debrinski M (2007) Allotopic mRNA localization to the mitochondrial surface rescues respiratory chain defects in fibroblasts harboring mitochondrial DNA mutations affecting complex I or v subunits. Rejuvenation Res 10:127–144

Bonnet C, Augustin S, Ellouze S, Benit P, Bouaita A, Rustin P, Sahel JA, Corral-Debrinski M (2008) The optimized allotopic expression of ND1 or ND4 genes restores respiratory chain complex I activity in fibroblasts harboring mutations in these genes. Biochim Biophys Acta 1783:1707–1717

Bowmaker M, Yang MY, Yasukawa T, Reyes A, Jacobs HT, Huberman JA, Holt IJ (2003) Mammalian mitochondrial DNA replicates bidirectionally from an initiation zone. J Biol Chem 278:50961–50969

Boynton JE, Gillham NW, Harris EH, Hosler JP, Johnson AM, Jones AR, Randolph-Anderson BL, Robertson D, Klein TM, Shark KB, Sanford JC (1988) Chloroplast transformation in *Chlamydomonas* with high velocity microprojectiles. Science 240:1534–1538

Brandina I, Smirnov A, Kolesnikova O, Entelis N, Krasheninnikov IA, Martin RP, Tarassov I (2007) tRNA import into yeast mitochondria is regulated by the ubiquitin-proteasome system. FEBS Lett 581:4248–4254

Chang DD, Clayton DA (1987) A mammalian mitochondrial RNA processing activity contains nucleus-encoded RNA. Science 235:1178–1184

Chang DD, Clayton DA (1989) Mouse RNAase MRP RNA is encoded by a nuclear gene and contains a decamer sequence complementary to a conserved region of mitochondrial RNA substrate. Cell 56:131–139

Chase CD (2007) Cytoplasmic male sterility: a window to the world of plant mitochondrial-nuclear interactions. Trends Genet 23:81–90

Chatterjee S, Home P, Mukherjee S, Mahata B, Goswami S, Dhar G, Adhya S (2006) An RNA-binding respiratory component mediates import of type II tRNAs into *Leishmania* mitochondria. J Biol Chem 281:25270–25277

Chaumont F, Bernier B, Buxant R, Williams ME, Levings CS 3rd, Boutry M (1995) Targeting the maize T-urf13 product into tobacco mitochondria confers methomyl sensitivity to mitochondrial respiration. Proc Natl Acad Sci U S A 92:1167–1171

Chiu N, Chiu AO, Suyama Y (1974) Three isoaccepting forms of leucyl transfer RNA in mitochondria. J Mol Biol 82:441–457

Choury D, Araya A (2006) RNA editing site recognition in heterologous plant mitochondria. Curr Genet 50:405–416

Choury D, Farre JC, Jordana X, Araya A (2004) Different patterns in the recognition of editing sites in plant mitochondria. Nucleic Acids Res 32:6397–6406

Choury D, Farre JC, Jordana X, Araya A (2005) Gene expression studies in isolated mitochondria: *Solanum tuberosum* rps10 is recognized by cognate potato but not by the transcription, splicing and editing machinery of wheat mitochondria. Nucleic Acids Res 33:7058–7065

Clark MA, Shay JW (1982) Mitochondrial transformation of mammalian cells. Nature 295:605–607

Clayton DA (2003) Mitochondrial DNA replication: what we know. IUBMB Life 55:213–217

Colin M, Dorthu MP, Duby F, Remacle C, Dinant M, Wolwertz MR, Duyckaerts C, Sluse F, Matagne RF (1995) Mutations affecting the mitochondrial genes encoding the cytochrome oxidase subunit I and apocytochrome b of *Chlamydomonas reinhardtii*. Mol Gen Genet 249:179–184

Collombet JM, Wheeler VC, Vogel F, Coutelle C (1997) Introduction of plasmid DNA into isolated mitochondria by electroporation. A novel approach toward gene correction for mitochondrial disorders. J Biol Chem 272:5342–5347

Comte C, Tonin Y, Heckel-Mager AM, Boucheham A, Smirnov A, Aure K, Lombes A, Martin RP, Entelis N, Tarassov I (2012) Mitochondrial targeting of recombinant RNAs modulates the level of a heteroplasmic mutation in human mitochondrial DNA associated with Kearns Sayre Syndrome. Nucleic Acids Res. doi:10.1093/nar/gks1965

Cristodero M, Seebeck T, Schneider A (2010) Mitochondrial translation is essential in bloodstream forms of *Trypanosoma brucei*. Mol Microbiol 78:757–769

Cudd A, Nicolau C (1985) Intracellular fate of liposome-encapsulated DNA in mouse liver. Analysis using electron microscope autoradiography and subcellular fractionation. Biochim Biophys Acta 845:477–491

Cudd A, Nicolau C (1986) Interaction of intravenously injected liposomes with mouse liver mitochondria. A fluorescence and electron microscopy study. Biochim Biophys Acta 860:201–214

Cudd A, Labbe H, Gervais M, Nicolau C (1984) Liposomes injected intravenously into mice associate with liver mitochondria. Biochim Biophys Acta 774:169–180

Cwerman-Thibault H, Sahel JA, Corral-Debrinski M (2011) Mitochondrial medicine: to a new era of gene therapy for mitochondrial DNA mutations. J Inherit Metab Dis 34:327–344

D'Souza GG, Rammohan R, Cheng SM, Torchilin VP, Weissig V (2003) DQAsome-mediated delivery of plasmid DNA toward mitochondria in living cells. J Control Release 92:189–197

D'Souza GG, Boddapati SV, Weissig V (2007) Gene therapy of the other genome: the challenges of treating mitochondrial DNA defects. Pharm Res 24:228–238

Dassa EP, Dufour E, Goncalves S, Paupe V, Hakkaart GA, Jacobs HT, Rustin P (2009) Expression of the alternative oxidase complements cytochrome c oxidase deficiency in human cells. EMBO Mol Med 1:30–36

Dewey RE, Levings CS 3rd, Timothy DH (1986) Novel recombinations in the maize mitochondrial genome produce a unique transcriptional unit in the Texas male-sterile cytoplasm. Cell 44:439–449

di Rago JP, Colson AM (1988) Molecular basis for resistance to antimycin and diuron, Q-cycle inhibitors acting at the Qi site in the mitochondrial ubiquinol-cytochrome c reductase in *Saccharomyces cerevisiae*. J Biol Chem 263:12564–12570

di Rago JP, Coppee JY, Colson AM (1989) Molecular basis for resistance to myxothiazol, mucidin (strobilurin A), and stigmatellin. Cytochrome b inhibitors acting at the center o of the mitochondrial ubiquinol-cytochrome c reductase in *Saccharomyces cerevisiae*. J Biol Chem 264:14543–14548

Doersen CJ, Guerrier-Takada C, Altman S, Attardi G (1985) Characterization of an RNase P activity from HeLa cell mitochondria. Comparison with the cytosol RNase P activity. J Biol Chem 260:5942–5949

Dörner M, Altmann M, Paabo S, Morl M (2001) Evidence for import of a lysyl-tRNA into marsupial mitochondria. Mol Biol Cell 12:2688–2698

Dorthu MP, Remy S, Michel-Wolwertz MR, Colleaux L, Breyer D, Beckers MC, Englebert S, Duyckaerts C, Sluse FE, Matagne RF (1992) Biochemical, genetic and molecular characterization of new respiratory-deficient mutants in *Chlamydomonas reinhardtii*. Plant Mol Biol 18:759–772

Dudek J, Rehling P, van der Laan M (2012) Mitochondrial protein import: common principles and physiological networks. Biochim Biophys Acta. doi:10.1016/j.bbamcr.2012.1005.1028

Dujon B (1981) Mitochondrial genetics and functions. In: Strathern JN, Jones EW, Broach JR (eds) The molecular biology of the yeast *Saccharomyces*: life cycle and inheritance. Cold Spring Harbor Laboratory, Cold Spring Harbor, pp 505–635

Dunn DA, Pinkert CA (2012) Nuclear expression of a mitochondrial DNA gene: mitochondrial targeting of allotopically expressed mutant ATP6 in transgenic mice. J Biomed Biotechnol 2012:541245

Duroc Y, Gaillard C, Hiard S, Tinchant C, Berthomé R, Pelletier R, Budar F (2006) Nuclear expression of a cytoplasmic male sterility gene modifies mitochondrial morphology in yeast and plant cells. Plant Sci 170:755–767

Ellouze S, Augustin S, Bouaita A, Bonnet C, Simonutti M, Forster V, Picaud S, Sahel JA, Corral-Debrinski M (2008) Optimized allotopic expression of the human mitochondrial ND4 prevents blindness in a rat model of mitochondrial dysfunction. Am J Hum Genet 83:373–387

Elson JL, Lightowlers RN (2006) Mitochondrial DNA clonality in the dock: can surveillance swing the case? Trends Genet 22:603–607

Entelis NS, Kolesnikova OA, Dogan S, Martin RP, Tarassov IA (2001) 5S rRNA and tRNA import into human mitochondria. Comparison of *in vitro* requirements. J Biol Chem 276:45642–45653

Entelis N, Brandina I, Kamenski P, Krasheninnikov IA, Martin RP, Tarassov I (2006) A glycolytic enzyme, enolase, is recruited as a cofactor of tRNA targeting toward mitochondria in *Saccharomyces cerevisiae*. Genes Dev 20:1609–1620

Estevez AM, Thiemann OH, Alfonzo JD, Simpson L (1999) T7 RNA polymerase-driven transcription in mitochondria of *Leishmania tarentolae* and *Trypanosoma brucei*. Mol Biochem Parasitol 103:251–259

Eudes F, Chugh A (2008) Nanocarrier-based plant transfection and transduction. Patent, Publication number WO/2008/148223, Application number PCT/CA2008/001112, Country CA

Falkenberg M, Larsson NG, Gustafsson CM (2007) DNA replication and transcription in mammalian mitochondria. Annu Rev Biochem 76:679–699

Fan W, Lin CS, Potluri P, Procaccio V, Wallace DC (2012) mtDNA lineage analysis of mouse L-cell lines reveals the accumulation of multiple mtDNA mutants and intermolecular recombination. Genes Dev 26:384–394

Farré JC, Araya A (2001) Gene expression in isolated plant mitochondria: high fidelity of transcription, splicing and editing of a transgene product in electroporated organelles. Nucleic Acids Res 29:2484–2491

Farré JC, Araya A (2002) RNA splicing in higher plant mitochondria: determination of functional elements in group II intron from a chimeric cox II gene in electroporated wheat mitochondria. Plant J 29:203–213

Farré JC, Leon G, Jordana X, Araya A (2001) *cis*-Recognition elements in plant mitochondrion RNA editing. Mol Cell Biol 21:6731–6737

Flierl A, Jackson C, Cottrell B, Murdock D, Seibel P, Wallace DC (2003) Targeted delivery of DNA to the mitochondrial compartment *via* import sequence-conjugated peptide nucleic acid. Mol Ther 7:550–557

Fox TD, Sanford JC, McMullin TW (1988) Plasmids can stably transform yeast mitochondria lacking endogenous mtDNA. Proc Natl Acad Sci USA 85:7288–7292

Frei U, Peiretti EG, Wenzel G (2004) Significance of cytoplasmic DNA in plant breeding. In: Janick J (ed) Plant breeding reviews. Wiley, Hoboken, pp 175–210

Goldschmidt-Clermont M (1991) Transgenic expression of aminoglycoside adenine transferase in the chloroplast: a selectable marker of site-directed transformation of *Chlamydomonas*. Nucleic Acids Res 19:4083–4089

Goswami S, Dhar G, Mukherjee S, Mahata B, Chatterjee S, Home P, Adhya S (2006) A bifunctional tRNA import receptor from *Leishmania* mitochondria. Proc Natl Acad Sci U S A 103:8354–8359

Greaves LC, Reeve AK, Taylor RW, Turnbull DM (2012) Mitochondrial DNA and Disease. J Pathol 226:274–286

Guy J, Qi X, Pallotti F, Schon EA, Manfredi G, Carelli V, Martinuzzi A, Hauswirth WW, Lewin AS (2002) Rescue of a mitochondrial deficiency causing Leber Hereditary Optic Neuropathy. Ann Neurol 52:534–542

Hancock K, Hajduk SL (1990) The mitochondrial tRNAs of *Trypanosoma brucei* are nuclear-encoded. J Biol Chem 265:19208–19215

Handa H (2008) Linear plasmids in plant mitochondria: peaceful coexistences or malicious invasions? Mitochondrion 8:15–25

Hartley RW (1988) Barnase and barstar. Expression of its cloned inhibitor permits expression of a cloned ribonuclease. J Mol Biol 202:913–915

He S, Abad AR, Gelvin SB, Mackenzie SA (1996) A cytoplasmic male sterility-associated mitochondrial protein causes pollen disruption in transgenic tobacco. Proc Natl Acad Sci U S A 93:11763–11768

Hinrichsen I, Bolle N, Paun L, Kempken F (2009) RNA processing in plant mitochondria is independent of transcription. Plant Mol Biol 70:663–668

Holt IJ, Lorimer HE, Jacobs HT (2000) Coupled leading- and lagging-strand synthesis of mammalian mitochondrial DNA. Cell 100:515–524

Holzmann J, Frank P, Loffler E, Bennett KL, Gerner C, Rossmanith W (2008) RNase P without RNA: identification and functional reconstitution of the human mitochondrial tRNA processing enzyme. Cell 135:462–474

Horobin RW, Trapp S, Weissig V (2007) Mitochondriotropics: a review of their mode of action, and their applications for drug and DNA delivery to mammalian mitochondria. J Control Release 121:125–136

Howell N, Gilbert K (1988) Mutational analysis of the mouse mitochondrial cytochrome b gene. J Mol Biol 203:607–618

Hyrup B, Nielsen PE (1996) Peptide nucleic acids (PNA): synthesis, properties and potential applications. Bioorg Med Chem 4:5–23

Ibrahim N, Handa H, Cosset A, Koulintchenko M, Konstantinov Y, Lightowlers RN, Dietrich A, Weber-Lotfi F (2011) DNA delivery to mitochondria: sequence specificity and energy enhancement. Pharm Res 28:2871–2882

Inoki Y, Hakamata Y, Hamamoto T, Kinouchi T, Yamazaki S, Kagawa Y, Endo H (2000) Proteoliposomes colocalized with endogenous mitochondria in mouse fertilized egg. Biochem Biophys Res Commun 278:183–191

Iyer S, Thomas RR, Portell FR, Dunham LD, Quigley CK, Bennett JP, Jr (2009) Recombinant mitochondrial transcription factor A with N-terminal mitochondrial transduction domain increases respiration and mitochondrial gene expression. Mitochondrion 9:196–203

Iyer S, Bergquist K, Young K, Gnaiger E, Rao RR, Bennett JP, Jr (2012a) Mitochondrial gene therapy improves respiration, biogenesis, and transcription in G11778A Leber's hereditary optic neuropathy and T8993G Leigh's syndrome cells. Hum Gene Ther 23:647–657

Iyer S, Xiao E, Alsayegh K, Eroshenko N, Riggs MJ, Bennett JP, Jr. and Rao RR (2012b) Mitochondrial gene replacement in human pluripotent stem cell-derived neural progenitors. Gene Ther 19:469–475

Janicka S, Kuhn K, Le Ret M, Bonnard G, Imbault P, Augustyniak H, Gualberto JM (2012) A RAD52-like single-stranded DNA binding protein affects mitochondrial DNA repair by recombination. Plant J. doi:10.1111/j.1365-1313X.2012.05097.x

Jash S, Adhya S (2011) Suppression of reactive oxygen species in cells with multiple mitochondrial DNA deletions by exogenous protein-coding RNAs. Mitochondrion 11:607–614

Jash S, Chowdhury T, Adhya S (2012) Modulation of mitochondrial respiratory capacity by carrier-mediated transfer of RNA *in vivo*. Mitochondrion 12:262–270

Jenuth JP, Peterson AC, Fu K, Shoubridge EA (1996) Random genetic drift in the female germline explains the rapid segregation of mammalian mitochondrial DNA. Nat Genet 14:146–151

Johnston SA, Anziano PQ, Shark K, Sanford JC, Butow RA (1988) Mitochondrial transformation in yeast by bombardment with microprojectiles. Science 240:1538–1541

Kagawa Y, Hayashi JI (1997) Gene therapy of mitochondrial diseases using human cytoplasts. Gene Ther 4:6–10

Kagawa Y, Inoki Y, Endo H (2001) Gene therapy by mitochondrial transfer. Adv Drug Deliv Rev 49:107–119

Kaltimbacher V, Bonnet C, Lecoeuvre G, Forster V, Sahel JA, Corral-Debrinski M (2006) mRNA localization to the mitochondrial surface allows the efficient translocation inside the organelle of a nuclear recoded ATP6 protein. RNA 12:1408–1417

Karicheva OZ, Kolesnikova OA, Schirtz T, Vysokikh MY, Mager-Heckel AM, Lombes A, Boucheham A, Krasheninnikov IA, Martin RP, Entelis N, Tarassov I (2011) Correction of the consequences of mitochondrial 3243A >G mutation in the MT-TL1 gene causing the MELAS syndrome by tRNA import into mitochondria. Nucleic Acids Res 39:8173–8186

Katrangi E, D'Souza G, Boddapati SV, Kulawiec M, Singh KK, Bigger B, Weissig V (2007) Xenogenic transfer of isolated murine mitochondria into human rho0 cells can improve respiratory function. Rejuvenation Res 10:561–570

Keeney PM, Quigley CK, Dunham LD, Papageorge CM, Iyer S, Thomas RR, Schwarz KM, Trimmer PA, Khan SM, Portell FR, Bergquist KE, Bennett JP, Jr (2009) Mitochondrial gene therapy augments mitochondrial physiology in a Parkinson's disease cell model. Hum Gene Ther 20:897–907

Kempken F, Bolle N, Forner J, Binder S (2007) Transcript end mapping and analysis of RNA editing in plant mitochondria. Methods Mol Biol 372:177–192

Khan SM, Bennett JP, Jr (2004) Development of mitochondrial gene replacement therapy. J Bioenerg Biomembr 36:387–393

Kim DH, Kang JG, Kim BD (2007) Isolation and characterization of the cytoplasmic male sterility-associated *orf456* gene of chili pepper (*Capsicum annuum* L.). Plant Mol Biol 63:519–532

Kindle KL, Richards KL, Stern DB (1991) Engineering the chloroplast genome: techniques and capabilities for chloroplast transformation in *Chlamydomonas reinhardtii*. Proc Natl Acad Sci U S A 88:1721–1725

King MP, Attardi G (1988) Injection of mitochondria into human cells leads to a rapid replacement of the endogenous mitochondrial DNA. Cell 52:811–819

King MP, Attardi G (1989) Human cells lacking mtDNA: repopulation with exogenous mitochondria by complementation. Science 246:500–503

King MP, Koga Y, Davidson M, Schon EA (1992) Defects in mitochondrial protein synthesis and respiratory chain activity segregate with the tRNA(Leu(UUR)) mutation associated with mitochondrial myopathy, encephalopathy, lactic acidosis, and stroke-like episodes. Mol Cell Biol 12:480–490

Kolesnikova OA, Entelis NS, Mireau H, Fox TD, Martin RP, Tarassov IA (2000) Suppression of mutations in mitochondrial DNA by tRNAs imported from the cytoplasm. Science 289:1931–1933

Kolesnikova OA, Entelis NS, Jacquin-Becker C, Goltzene F, Chrzanowska-Lightowlers ZM, Lightowlers RN, Martin RP, Tarassov I (2004) Nuclear DNA-encoded tRNAs targeted into mitochondria can rescue a mitochondrial DNA mutation associated with the MERRF syndrome in cultured human cells. Hum Mol Genet 13:2519–2534

Kolesnikova O, Kazakova H, Comte C, Steinberg S, Kamenski P, Martin RP, Tarassov I, Entelis N (2010) Selection of RNA aptamers imported into yeast and human mitochondria. RNA 16:926–941

Koulintchenko M, Konstantinov Y, Dietrich A (2003) Plant mitochondria actively import DNA *via* the permeability transition pore complex. EMBO J 22:1245–1254

Koulintchenko M, Temperley RJ, Mason PA, Dietrich A, Lightowlers RN (2006) Natural competence of mammalian mitochondria allows the molecular investigation of mitochondrial gene expression. Hum Mol Genet 15:143–154

Kühn K, Gualberto JM (2012) Recombination in the stability, repair and evolution of the mitochondrial genome. Adv Bot Res 63:215–252

Kumar R, Maréchal-Drouard L, Akama K, Small I (1996) Striking differences in mitochondrial tRNA import between different plant species. Mol Gen Genet 252:404–411

Kumar P, Vasupalli N, Srinivasan R, Bhat SR (2012) An evolutionarily conserved mitochondrial *orf108* is associated with cytoplasmic male sterility in different alloplasmic lines of *Brassica juncea* and induces male sterility in transgenic *Arabidopsis thaliana*. J Exp Bot 63:2921–2932

Lee DY, Clayton DA (1997) RNase mitochondrial RNA processing correctly cleaves a novel R loop at the mitochondrial DNA leading-strand origin of replication. Genes Dev 11:582–592

Lee DY, Clayton DA (1998) Initiation of mitochondrial DNA replication by transcription and R-loop processing. J Biol Chem 273:30614–30621

Lee M, Choi JS, Choi MJ, Pak YK, Rhee BD, Ko KS (2007) DNA delivery to the mitochondria sites using mitochondrial leader peptide conjugated polyethylenimine. J Drug Target 15:115–122

Li XQ, Chetrit P, Mathieu C, Vedel F, De Paepe R, Remy R, Ambard-Bretteville F (1988) Regeneration of cytoplasmic male sterile protoclones of *Nicotiana sylvestris* with mitochondrial variations. Curr Genet 13:261–266

Liberman EA, Topaly VP, Tsofina LM, Jasaitis AA, Skulachev VP (1969) Mechanism of coupling of oxidative phosphorylation and the membrane potential of mitochondria. Nature 222:1076–1078

Lightowlers RN (2011) Mitochondrial transformation: time for concerted action. EMBO Rep 12:480–481

Lyrawati D, Trounson A, Cram D (2011) Expression of GFP in the mitochondrial compartment using DQAsome-mediated delivery of an artificial mini-mitochondrial genome. Pharm Res 28:2848–2862

Magalhaes PJ, Andreu AL, Schon EA (1998) Evidence for the presence of 5S rRNA in mammalian mitochondria. Mol Biol Cell 9:2375–2382

Mahata B, Mukherjee S, Mishra S, Bandyopadhyay A, Adhya S (2006) Functional delivery of a cytosolic tRNA into mutant mitochondria of human cells. Science 314:471–474

Mahato B, Jash S, Adhya S (2011) RNA-mediated restoration of mitochondrial function in cells harboring a Kearns Sayre Syndrome mutation. Mitochondrion 11:564–574

Malcuit I, Sorokin A (2010) Method for the transformation of plant cell mitochondria. Patent, Publication number WO/2010/061187, Application number PCT/GB2009/002755, Country GB

Manfredi G, Fu J, Ojaimi J, Sadlock JE, Kwong JQ, Guy J, Schon EA (2002) Rescue of a deficiency in ATP synthesis by transfer of MTATP6, a mitochondrial DNA-encoded gene, to the nucleus. Nat Genet 30:394–399

Marc P, Margeot A, Devaux F, Blugeon C, Corral-Debrinski M, Jacq C (2002) Genome-wide analysis of mRNAs targeted to yeast mitochondria. EMBO Rep 3:159–164

Maréchal-Drouard L, Guillemaut P, Cosset A, Arbogast M, Weber F, Weil JH, Dietrich A (1990) Transfer RNAs of potato (*Solanum tuberosum*) mitochondria have different genetic origins. Nucleic Acids Res 18:3689–3696

Matagne RF, Michel-Wolwertz MR, Munaut C, Duyckaerts C, Sluse F (1989) Induction and characterization of mitochondrial DNA mutants in *Chlamydomonas reinhardtii*. J Cell Biol 108:1221–1226

Matsuda D, Dreher TW (2004) The tRNA-like structure of Turnip yellow mosaic virus RNA is a 3'-translational enhancer. Virology 321:36–46

McDonald AE, Vanlerberghe GC, Staples JF (2009) Alternative oxidase in animals: unique characteristics and taxonomic distribution. J Exp Biol 212:2627–2634

Michaud M, Marechal-Drouard L, Duchene AM (2010) RNA trafficking in plant cells: targeting of cytosolic mRNAs to the mitochondrial surface. Plant Mol Biol 73:697–704

Mileshina D, Koulintchenko M, Konstantinov Y, Dietrich A (2011) Transfection of plant mitochondria and *in organello* gene integration. Nucleic Acids Res 39:e115

Miller-Messmer M, Kuhn K, Bichara M, Le Ret M, Imbault P, Gualberto JM (2012) RecA-dependent DNA repair results in increased heteroplasmy of the *Arabidopsis* mitochondrial genome. Plant Physiol 159:211–226

Minczuk M, Papworth MA, Kolasinska P, Murphy MP, Klug A (2006) Sequence-specific modification of mitochondrial DNA using a chimeric zinc finger methylase. Proc Natl Acad Sci U S A 103:19689–19694

Minczuk M, Papworth MA, Miller JC, Murphy MP, Klug A (2008) Development of a single-chain, quasi-dimeric zinc-finger nuclease for the selective degradation of mutated human mitochondrial DNA. Nucleic Acids Res 36:3926–3938

Minczuk M, Kolasinska-Zwierz P, Murphy MP, Papworth MA (2010) Construction and testing of engineered zinc-finger proteins for sequence-specific modification of mtDNA. Nat Protoc 5:342–356

Mireau H, Arnal N, Fox TD (2003) Expression of Barstar as a selectable marker in yeast mitochondria. Mol Genet Genomics 270:1–8

Mitchell P (1961) Coupling of phosphorylation to electron and hydrogen transfer by a chemiosmotic type of mechanism. Nature 191:144–148

Moller SG, Xu XM (2009) Mitochondrial transformation. Patent, Publication number WO/2009/150441, Application number PCT/GB2009/001510, Country NO

Mossmann D, Meisinger C, Vogtle FN (2012) Processing of mitochondrial presequences. Biochim Biophys Acta 1819:1098–1106

Mukherjee S, Basu S, Home P, Dhar G, Adhya S (2007) Necessary and sufficient factors for the import of transfer RNA into the kinetoplast mitochondrion. EMBO Rep 8:589–595

Mukherjee S, Mahata B, Mahato B, Adhya S (2008) Targeted mRNA degradation by complex-mediated delivery of antisense RNAs to intracellular human mitochondria. Hum Mol Genet 17:1292–1298

Muratovska A, Lightowlers RN, Taylor RW, Turnbull DM, Smith RA, Wilce JA, Martin SW, Murphy MP (2001) Targeting peptide nucleic acid (PNA) oligomers to mitochondria within cells by conjugation to lipophilic cations: implications for mitochondrial DNA replication, expression and disease. Nucleic Acids Res 29:1852–1863

Nagley P, Farrell LB, Gearing DP, Nero D, Meltzer S, Devenish RJ (1988) Assembly of functional proton-translocating ATPase complex in yeast mitochondria with cytoplasmically synthesized subunit 8, a polypeptide normally encoded within the organelle. Proc Natl Acad Sci U S A 85:2091–2095

Newman SM, Boynton JE, Gillham NW, Randolph-Anderson BL, Johnson AM, Harris EH (1990) Transformation of chloroplast ribosomal RNA genes in *Chlamydomonas*: molecular and genetic characterization of integration events. Genetics 126:875–888

Oca-Cossio J, Kenyon L, Hao H, Moraes CT (2003) Limitations of allotopic expression of mitochondrial genes in mammalian cells. Genetics 165:707–720

Ojaimi J, Pan J, Santra S, Snell WJ, Schon EA (2002) An algal nucleus-encoded subunit of mitochondrial ATP synthase rescues a defect in the analogous human mitochondrial-encoded subunit. Mol Biol Cell 13:3836–3844

Ortega VM, Bohner JG, Chase CD (2000) The tobacco apocytochrome b gene predicts sensitivity to the respiratory inhibitors antimycin A and myxothiazol. Curr Genet 37:315–321

Papworth M, Kolasinska P, Minczuk M (2006) Designer zinc-finger proteins and their applications. Gene 366:27–38

Parent JS, Lepage E, Brisson N (2011) Divergent roles for the two PolI-like organelle DNA polymerases of *Arabidopsis*. Plant Physiol 156:254–262

Paris Z, Rubio MA, Lukes J, Alfonzo JD (2009) Mitochondrial tRNA import in *Trypanosoma brucei* is independent of thiolation and the Rieske protein. RNA 15:1398–1406

Perales-Clemente E, Fernandez-Silva P, Acin-Perez R, Perez-Martos A, Enriquez JA (2011) Allotopic expression of mitochondrial-encoded genes in mammals: achieved goal, undemonstrated mechanism or impossible task? Nucleic Acids Res 39:225–234

Pineau B, Mathieu C, Gérard-Hirne C, De Paepe R, Chétrit P (2005) Targeting the NAD7 subunit to mitochondria restores a functional complex I and a wild type phenotype in the *Nicotiana sylvestris* CMS II mutant lacking *nad7*. J Biol Chem 280:25994–26001

Pinkert CA, Trounce IA (2002) Production of transmitochondrial mice. Methods 26:348–357

Pinkert CA, Irwin MH, Johnson LW, Moffatt RJ (1997) Mitochondria transfer into mouse ova by microinjection. Transgenic Res 6:379–383

Pohjoismaki JL, Holmes JB, Wood SR, Yang MY, Yasukawa T, Reyes A, Bailey LJ, Cluett TJ, Goffart S, Willcox S, Rigby RE, Jackson AP, Spelbrink JN, Griffith JD, Crouch RJ, Jacobs HT, Holt IJ (2010) Mammalian mitochondrial DNA replication intermediates are essentially duplex but contain extensive tracts of RNA/DNA hybrid. J Mol Biol 397:1144–1155

Qi X, Sun L, Lewin AS, Hauswirth WW, Guy J (2007) The mutant human ND4 subunit of complex I induces optic neuropathy in the mouse. Invest Ophthalmol Vis Sci 48:1–10

Randolph-Anderson BL, Boynton JE, Gillham NW, Harris EH, Johnson AM, Dorthu MP, Matagne RF (1993) Further characterization of the respiratory deficient dum-1 mutation of *Chlamydomonas reinhardtii* and its use as a recipient for mitochondrial transformation. Mol Gen Genet 236:235–244

Remacle C, Colin M, Matagne RF (1998) Suppression of a^{+} 1 T mutation by a nearby substitution in the mitochondrial cox1 gene of *Chlamydomonas reinhardtii*: a new type of frameshift suppression in an organelle genome. Mol Gen Genet 259:294–298

Remacle C, Duby F, Cardol P, Matagne RF (2001) Mutations inactivating mitochondrial genes in *Chlamydomonas reinhardtii*. Biochem Soc Trans 29:442–446

Remacle C, Cardol P, Coosemans N, Gaisne M, Bonnefoy N (2006) High-efficiency biolistic transformation of *Chlamydomonas* mitochondria can be used to insert mutations in complex I genes. Proc Natl Acad Sci U S A 103:4771–4776

Rinehart J, Krett B, Rubio MA, Alfonzo JD, Soll D (2005) *Saccharomyces cerevisiae* imports the cytosolic pathway for Gln-tRNA synthesis into the mitochondrion. Genes Dev 19:583–592

Rubio MA, Rinehart JJ, Krett B, Duvezin-Caubet S, Reichert AS, Soll D, Alfonzo JD (2008) Mammalian mitochondria have the innate ability to import tRNAs by a mechanism distinct from protein import. Proc Natl Acad Sci U S A 105:9186–9191

Ruhanen H, Ushakov K, Yasukawa T (2011) Involvement of DNA ligase III and ribonuclease H1 in mitochondrial DNA replication in cultured human cells. Biochim Biophys Acta 1813:2000–2007

Salinas T, Duchene AM, Marechal-Drouard L (2008) Recent advances in tRNA mitochondrial import. Trends Biochem Sci 33:320–329

Sbicego S, Nabholz CE, Hauser R, Blum B, Schneider A (1998) *In vivo* import of unspliced tRNATyr containing synthetic introns of variable length into mitochondria of *Leishmania tarentolae*. Nucleic Acids Res 26:5251–5255

Schekman R (2010) Editorial expression of concern: a bifunctional tRNA import receptor from *Leishmania* mitochondria. Proc Natl Acad Sci U S A 107:9476

Schneider A (2011) Mitochondrial tRNA import and its consequences for mitochondrial translation. Annu Rev Biochem 80:1033–1053

Schneider A, Maréchal-Drouard L (2000) Mitochondrial tRNA import: are there distinct mechanisms? Trends Cell Biol 10:509–513

Seibel P, Trappe J, Villani G, Klopstock T, Papa S, Reichmann H (1995) Transfection of mitochondria: strategy towards a gene therapy of mitochondrial DNA diseases. Nucleic Acids Res 23:10–17

Seibel M, Bachmann C, Schmiedel J, Wilken N, Wilde F, Reichmann H, Isaya G, Seibel P, Pfeiler D (1999) Processing of artificial peptide-DNA-conjugates by the mitochondrial intermediate peptidase (MIP). Biol Chem 380:961–967

Seo BB, Nakamaru-Ogiso E, Flotte TR, Matsuno-Yagi A, Yagi T (2006) *In vivo* complementation of complex I by the yeast Ndi1 enzyme. Possible application for treatment of Parkinson disease. J Biol Chem 281:14250–14255

Shadel GS, Clayton DA (1997) Mitochondrial DNA maintenance in vertebrates. Annu Rev Biochem 66:409–435

Shedge V, Arrieta-Montiel M, Christensen AC, Mackenzie SA (2007) Plant mitochondrial recombination surveillance requires unusual *RecA* and *MutS* homologs. Plant Cell 19:1251–1264

Sieber F, Placido A, El Farouk-AmeqraneS, Duchene AM, Marechal-Drouard L (2011) A protein shuttle system to target RNA into mitochondria. Nucleic Acids Res 39:e96

Simpson AM, Suyama Y, Dewes H, Campbell DA, Simpson L (1989) Kinetoplastid mitochondria contain functional tRNAs which are encoded in nuclear DNA and also contain small minicircle and maxicircle transcripts of unknown function. Nucleic Acids Res 17:5427–5445

Small I, Maréchal-Drouard L, Masson J, Pelletier G, Cosset A, Weil JH, Dietrich A (1992) *In vivo* import of a normal or mutagenized heterologous transfer RNA into the mitochondria of transgenic plants: towards novel ways of influencing mitochondrial gene expression? EMBO J 11:1291–1296

Smirnov A, Tarassov I, Mager-Heckel AM, Letzelter M, Martin RP, Krasheninnikov IA, Entelis N (2008) Two distinct structural elements of 5S rRNA are needed for its import into human mitochondria. RNA 14:749–759

Srivastava S, Moraes CT (2001) Manipulating mitochondrial DNA heteroplasmy by a mitochondrially targeted restriction endonuclease. Hum Mol Genet 10:3093–3099

Staudinger M, Kempken F (2003) Electroporation of isolated higher-plant mitochondria: transcripts of an introduced *cox2* gene, but not an *atp6* gene, are edited *in organello*. Mol Genet Genomics 269:553–561

Staudinger M, Bolle N, Kempken F (2005) Mitochondrial electroporation and *in organello* RNA editing of chimeric *atp6* transcripts. Mol Genet Genomics 273:130–136

Steele DF, Butler CA, Fox TD (1996) Expression of a recoded nuclear gene inserted into yeast mitochondrial DNA is limited by mRNA-specific translational activation. Proc Natl Acad Sci U S A 93:5253–5257

Suyama Y (1986) Two dimensional polyacrylamide gel electrophoresis analysis of *Tetrahymena* mitochondrial tRNA. Curr Genet 10:411–420

Suzuki Y (2012) Exploring transduction mechanisms of protein transduction domains (PTDs) in living cells utilizing single-quantum dot tracking (SQT) technology. Sensors (Basel) 12:549–572

Svab Z, Maliga P (1993) High-frequency plastid transformation in tobacco by selection for a chimeric aadA gene. Proc Natl Acad Sci U S A 90:913–917

Svab Z, Hajdukiewicz P, Maliga P (1990) Stable transformation of plastids in higher plants. Proc Natl Acad Sci U S A 87:8526–8530

Sylvestre J, Margeot A, Jacq C, Dujardin G, Corral-Debrinski M (2003) The role of the 3' untranslated region in mRNA sorting to the vicinity of mitochondria is conserved from yeast to human cells. Mol Biol Cell 14:3848–3856

Tachibana M, Amato P, Sparman M, Woodward J, Sanchis DM, Ma H, Gutierrez NM, Tippner-Hedges R, Kang E, Lee HS, Ramsey C, Masterson K, Battaglia D, Lee D, Wu D, Jensen J, Patton P, Gokhale S, Stouffer R, Mitalipov S (2012) Towards germline gene therapy of inherited mitochondrial diseases. Nature. doi:10.1038/nature11647

Tanaka M, Borgeld HJ, Zhang J, Muramatsu S, Gong JS, Yoneda M, Maruyama W, Naoi M, Ibi T, Sahashi K, Shamoto M, Fuku N, Kurata M, Yamada Y, Nishizawa K, Akao Y, Ohishi N, Miyabayashi S, Umemoto H, Muramatsu T, Furukawa K, Kikuchi A, Nakano I, Ozawa K, Yagi K (2002) Gene therapy for mitochondrial disease by delivering restriction endonuclease *Sma*I into mitochondria. J Biomed Sci 9:534–541

Tarassov IA, Entelis NS (1992) Mitochondrially-imported cytoplasmic tRNA(Lys)(CUU) of *Saccharomyces cerevisiae*: *in vivo* and *in vitro* targetting systems. Nucleic Acids Res 20:1277–1281

Thomas RR, Khan SM, Portell FR, Smigrodzki RM, Bennett JP Jr (2011) Recombinant human mitochondrial transcription factor A stimulates mitochondrial biogenesis and ATP synthesis, improves motor function after MPTP, reduces oxidative stress and increases survival after endotoxin. Mitochondrion 11:108–118

Vahrenholz C, Riemen G, Pratje E, Dujon B, Michaelis G (1993) Mitochondrial DNA of *Chlamydomonas reinhardtii*: the structure of the ends of the linear 15.8-kb genome suggests mechanisms for DNA replication. Curr Genet 24:241–247

Val R, Wyszko E, Valentin C, Szymanski M, Cosset A, Alioua M, Dreher TW, Barciszewski J, Dietrich A (2011) Organelle trafficking of chimeric ribozymes and genetic manipulation of mitochondria. Nucleic Acids Res 39:9262–9274

Van Aken O, Giraud E, Clifton R, Whelan J (2009) Alternative oxidase: a target and regulator of stress responses. Physiol Plant 137:354–361

Vestweber D, Schatz G (1989) DNA-protein conjugates can enter mitochondria *via* the protein import pathway. Nature 338:170–172

von Allmen JM, Rottmann WH, Gengenbach BG, Harvey AJ, Lonsdale DM (1991) Transfer of methomyl and HmT-toxin sensitivity from T-cytoplasm maize to tobacco. Mol Gen Genet 229:405–412

Wagle MA, Martinville LE, D'Souza GG (2011) The utility of an isolated mitochondrial fraction in the preparation of liposomes for the specific delivery of bioactives to mitochondria in live mammalian cells. Pharm Res 28:2790–2796

Wang Z, Zou Y, Li X, Zhang Q, Chen L, Wu H, Su D, Chen Y, Guo J, Luo D, Long Y, Zhong Y, Liu YG (2006) Cytoplasmic male sterility of rice with boro II cytoplasm is caused by a cytotoxic peptide and is restored by two related PPR motif genes via distinct modes of mRNA silencing. Plant Cell 18:676–687

Wang G, Chen HW, Oktay Y, Zhang J, Allen EL, Smith GM, Fan KC, Hong JS, French SW, McCaffery JM, Lightowlers RN, Morse HC, 3rd, Koehler CM, Teitell MA (2010) PNPASE regulates RNA import into mitochondria. Cell 142:456–467

Wang G, Shimada E, Koehler CM, Teitell MA (2012a) PNPASE and RNA trafficking into mitochondria. Biochim Biophys Acta 1819:998–1007

Wang G, Shimada E, Zhang J, Hong JS, Smith GM, Teitell MA, Koehler CM (2012b) Correcting human mitochondrial mutations with targeted RNA import. Proc Natl Acad Sci U S A 109:4840–4845

Wanrooij S, Miralles Fuste J, Stewart JB, Wanrooij PH, Samuelsson T, Larsson NG, Gustafsson CM, Falkenberg M (2012) *In vivo* mutagenesis reveals that OriL is essential for mitochondrial DNA replication. EMBO Rep. doi:10.1038/embor.2012.1161

Waters VL (1999) Conjugative transfer in the dissemination of beta-lactam and aminoglycoside resistance. Front Biosci 4:D433–D456

Weber-Lotfi F, Ibrahim N, Boesch P, Cosset A, Konstantinov Y, Lightowlers RN, Dietrich A (2009) Developing a genetic approach to investigate the mechanism of mitochondrial competence for DNA import. Biochim Biophys Acta 1787:320–327

Weissig V, Lasch J, Erdos G, Meyer HW, Rowe TC, Hughes J (1998) DQAsomes: a novel potential drug and gene delivery system made from Dequalinium. Pharm Res 15:334–337

Weissig V, Lizano C, Torchilin VP (2000) Selective DNA release from DQAsome/DNA complexes at mitochondria-like membranes. Drug Deliv 7:1–5

Weissig V, D'Souza GG, Torchilin VP (2001) DQAsome/DNA complexes release DNA upon contact with isolated mouse liver mitochondria. J Control Release 75:401–408

Weissig V, Boddapati SV, Cheng SM, D'Souza GG (2006) Liposomes and liposome-like vesicles for drug and DNA delivery to mitochondria. J Liposome Res 16:249–264

Wintz H, Chen HC, Sutton CA, Conley CA, Cobb A, Ruth D, Hanson MR (1995) Expression of the CMS-associated urfS sequence in transgenic petunia and tobacco. Plant Mol Biol 28:83–92

Yagi T, Seo BB, Nakamaru-Ogiso E, Marella M, Barber-Singh J, Yamashita T, Kao MC, Matsuno-Yagi A (2006a) Can a single subunit yeast NADH dehydrogenase (Ndi1) remedy diseases caused by respiratory complex I defects? Rejuvenation Res 9:191–197

Yagi T, Seo BB, Nakvamaru-Ogiso E, Marella M, Barber-Singh J, Yamashita T, Matsuno-Yagi A (2006b) Possibility of transkingdom gene therapy for complex I diseases. Biochim Biophys Acta 1757:708–714

Yamada Y, Harashima H (2012a) Delivery of bioactive molecules to the mitochondrial genome using a membrane-fusing, liposome-based carrier, DF-MITO-Porter. Biomaterials 33:1589–1595

Yamada Y, Harashima H (2012b) Enhancement in selective mitochondrial association by direct modification of a mitochondrial targeting signal peptide on a liposomal based nanocarrier. Mitochondrion. doi:10.1016/j.mito.2012.1009.1001

Yamada Y, Akita H, Kamiya H, Kogure K, Yamamoto T, Shinohara Y, Yamashita K, Kobayashi H, Kikuchi H, Harashima H (2008) MITO-Porter: a liposome-based carrier system for delivery of macromolecules into mitochondria *via* membrane fusion. Biochim Biophys Acta 1778:423–432

Yamada Y, Furukawa R, Yasuzaki Y, Harashima H (2011) Dual function MITO-Porter, a nano carrier integrating both efficient cytoplasmic delivery and mitochondrial macromolecule delivery. Mol Ther 19:1449–1456

Yamamoto MP, Shinada H, Onodera Y, Komaki C, Mikami T, Kubo T (2008) A male sterility-associated mitochondrial protein in wild beets causes pollen disruption in transgenic plants. Plant J 54:1027–1036

Yamasaki T, Kurokawa S, Watanabe KI, Ikuta K, Ohama T (2005) Shared molecular characteristics of successfully transformed mitochondrial genomes in *Chlamydomonas reinhardtii*. Plant Mol Biol 58:515–527

Yang YW, Koob MD (2012) Transferring isolated mitochondria into tissue culture cells. Nucleic Acids Res. doi:10.1093/nar/gks1639

Yang MY, Bowmaker M, Reyes A, Vergani L, Angeli P, Gringeri E, Jacobs HT, Holt IJ (2002) Biased incorporation of ribonucleotides on the mitochondrial L-strand accounts for apparent strand-asymmetric DNA replication. Cell 111:495–505

Yang J, Liu X, Yang X, Zhang M (2010) Mitochondrially-targeted expression of a cytoplasmic male sterility-associated *orf220* gene causes male sterility in *Brassica juncea*. BMC Plant Biol 10:231

Yasukawa T, Yang MY, Jacobs HT, Holt IJ (2005) A bidirectional origin of replication maps to the major noncoding region of human mitochondrial DNA. Mol Cell 18:651–662

Yasuzaki Y, Yamada Y, Harashima H (2010) Mitochondrial matrix delivery using MITO-Porter, a liposome-based carrier that specifies fusion with mitochondrial membranes. Biochem Biophys Res Commun 397:181–186

Yoon YG, Koob MD (2003) Efficient cloning and engineering of entire mitochondrial genomes in *Escherichia coli* and transfer into transcriptionally active mitochondria. Nucleic Acids Res 31:1407–1415

Yoon YG, Koob MD (2005) Transformation of isolated mammalian mitochondria by bacterial conjugation. Nucleic Acids Res 33:e139

Yoon YG, Koob MD (2008) Selection by drug resistance proteins located in the mitochondria of mammalian cells. Mitochondrion 8:345–351

Yoon YG, Koob MD (2011) Toward genetic transformation of mitochondria in mammalian cells using a recoded drug-resistant selection marker. J Genet Genomics 38:173–179

Yoon YG, Koob MD (2012) Nonreplicating intracellular bacterial vector for conjugative DNA transfer into mitochondria. Pharm Res 29:1040–1045

Yoon YG, Yang YW, Koob MD (2009) PCR-based cloning of the complete mouse mitochondrial genome and stable engineering in *Escherichia coli*. Biotechnol Lett 31:1671–1676

Yoshionari S, Koike T, Yokogawa T, Nishikawa K, Ueda T, Miura K, Watanabe K (1994) Existence of nuclear-encoded 5S-rRNA in bovine mitochondria. FEBS Lett 338:137–142

Yu H, Koilkonda RD, Chou TH, Porciatti V, Ozdemir SS, Chiodo V, Boye SL, Boye SE, Hauswirth WW, Lewin AS, Guy J (2012a) Gene delivery to mitochondria by targeting modified adenoassociated virus suppresses Leber's hereditary optic neuropathy in a mouse model. Proc Natl Acad Sci U S A 109:E1238–E1247

Yu H, Ozdemir SS, Koilkonda RD, Chou TH, Porciatti V, Chiodo V, Boye SL, Hauswirth WW, Lewin AS, Guy J (2012b) Mutant NADH dehydrogenase subunit 4 gene delivery to mitochondria by targeting sequence-modified adeno-associated virus induces visual loss and optic atrophy in mice. Mol Vis 18:1668–1683

Zaegel V, Guermann B, Le Ret M, Andres C, Meyer D, Erhardt M, Canaday J, Gualberto JM, Imbault P (2006) The plant-specific ssDNA binding protein OSB1 is involved in the stoichiometric transmission of mitochondrial DNA in Arabidopsis. Plant Cell 18:3548–3563

Zhou J, Liu L, Chen J (2010) Mitochondrial DNA heteroplasmy in *Candida glabrata* after mitochondrial transformation. Eukaryot Cell 9:806–814

Chapter 14
Laboratory Methods for Investigating Nuclear and Cytoplasmic Genomes and Transcriptome

Xiu-Qing Li

14.1 Introduction

A plant somatic cell usually has DNA in several organelles (or compartments), including the nucleus, plastids (such as chloroplasts and amyloplasts), and mitochondria. Bacteria are propagated purely through asexual divisions (fission) and therefore can be considered as somatic cells in a broader sense, particularly in the study of somatic genome evolution. A bacterium has often one or more plasmids in addition to its chromosome. Plants sometimes also have plasmids in their organelles (Chase and Pring 1985). Separate extraction of DNA, as well as RNA, from these sources is often needed in genome manipulation, genome sequencing, DNA characterization, and gene expression analysis.

Although the microarray and second-generation sequencing approaches and related bioinformatic analysis are very powerful in genome characterization and gene expression profiling (Li 2011; Li 2015), other experimental approaches are often needed to validate the results because each technique has its strengths and weaknesses. For example, some mitochondrial and chloroplast genes are homologous; it is difficult to know whether an Illumina read of the identical region is actually from chloroplasts or mitochondria. Some organelle genes, and occasionally, portions of the chloroplast and mitochondrial genomes, can also have similar sequences in the nuclear genome (Brennicke et al. 1993; Qu et al. 2008; Rousseau-Gueutin et al. 2011; Xiang and Li 2015). Consequently, bioinformatics alone is nearly impossible to know whether the sequence reads in the second-generation sequencing are from the mitochondrial, chloroplast, or nuclear genome. In these cases, to correctly interpret the results, other techniques in addition to DNA sequencing are needed, such

X.-Q. Li (✉)
Potato Research Centre, Agriculture and Agri-Food Canada,
850 Lincoln Road, P.O. Box 20280, Fredericton, NB E3B 4Z7, Canada
e-mail: Xiu-Qing.Li@agr.gc.ca; lixiuqing2008@gmail.com

X.-Q. Li et al. (eds.), *Somatic Genome Manipulation*,
DOI 10.1007/978-1-4939-2389-2_14

as isolation of organelle-enriched DNA prior to sequencing or molecular hybridization. Therefore, this chapter describes several laboratory techniques, including both extraction and analysis of DNA from separate organelles.

Variation in gene expression can be at the level of both expression activity and processing patterns. An example is mitochondrial RNA (mtRNA) in maize (Dewey et al. 1986; Li et al. 1996) and brassicas (Singh et al. 1996; L'Homme et al. 1997; Li et al. 1998). In the analysis of cytoplasmic male sterility (CMS) and nuclear–cytoplasmic interaction, it is often necessary to extract RNA from samples enriched with mitochondria. Total DNA and RNA extraction is often required in the study of DNA ratios and interactions between nuclear and organellar genomes. Therefore, this chapter also describes protocols for preparation of total RNA, mtRNA, and for electrophoretic fingerprinting of these RNAs.

Sometimes microscopic techniques can generate information useful for assisting the interpretation on the organelle DNA/RNA sequence data. For example, when mitochondrial gene expression is low, a microscopic observation is often needed to verify when the mitochondrial density in the cell is also low. For plastid research, it may also be useful to analyze starch granule size and shape. Some quick and general studies such as microscopic observations are often helpful for experimental design and data interpretation of DNA and RNA analysis. Therefore, this chapter describes three microscopic techniques first, followed by the description of four DNA prep protocols and three RNA prep and analysis protocols.

In addition to the techniques described in this chapter, the readers are referred to Chap. 13 for mitochondrial genetic manipulation (Mileshina et al. 2015), Chap. 8 for potato somatic embryogenesis (Nassar et al. 2015), and Chap. 9 (Pelletier and Budar, 2015) and Chap. 10 (Rokka 2015) for protoplast fusion, and bioinformatic analysis (Li 2015).

14.2 In Situ Hybridization for Studying Tissue-Specific Gene Expression

14.2.1 Introduction to In Situ Hybridization

In situ hybridization using a gene-specific probe can effectively illustrate the gene expression patterns at tissue-specific, cell-specific, and development-specific levels. For example, the technique played a critical role in determining the expression patterns of genes involved in flower development in *Arabidopsis* (Liu et al., 2011) and other plants such as citrus (Zhang et al. 2011b) using either labeling with a 35S isotope or digoxigenin RNA labeling kit (Roche http://www.roche.com). The overall protocol is very long with in situ hybridization protocols available, with either radioactive probes (Meyerowitz 1987; Traas 2008; Wang et al. 2008) or nonradioactive probes (Ferrandiz and Sessions, 2008; Roche 2008). Regardless of the probe labeling approach, the tissue fixation, sectioning, and hybridization common stages are all critically important to the success of the in situ hybridization. Therefore, the in situ hybridization section in this chapter focuses on discussion of these technical aspects according to the author's firsthand experience.

14.2.2 Various Technical Points for Improving In Situ Hybridization Resolution

The in situ hybridization procedure consists of the following major steps: tissue sampling, fixation, dehydration, embedding, slide coating, tissue sectioning, probe preparation, pre-hybridization treatment, hybridization with a probe, post-hybridization washing, detection of hybridization signals, and photomicroscopy.

Plant sample preparation and tissue sectioning for in situ hybridization consist of the following major steps: tissue fixation, tissue dehydration, tissue clearing, wax infiltration into tissues, tissue embedding in wax, microscope slide coating with poly-L-lysine, and tissue sectioning on a microtome (tissue section thickness: 7–10 μm). Tissue fixation is usually in formalin–acetic acid–alcohol (FAA) solution. A 200 ml of FAA contains 20 ml of 37 % formaldehyde, 10 ml glacial acetic acid, 100 ml of 95 % ethanol, and 70 ml of dH_2O.

In FAA solution, the formaldehyde and alcohol have a shrinking effect, while acetic acid has a swelling effect, on cytoplasm. Formaldehyde (formalin) has strong fixative effect on the tissues, but is likely toxic. The ethanol in FAA may also vary in different studies. In general histological sectioning such as for verifying whether a plantlet regeneration in tissue culture is via somatic embryogenesis, the present author usually uses ethanol and acetic acid at 3:1 v:v (Li and Demarly 1996). It was reported that 90 % ethanol (with formalin 5 %, acetic acid 5.0 %, v:v:v) can also be used in fixing plant tissues for histological analysis (Karamian and Ranjbar 2008). However, for in situ hybridization analysis of cell-specific gene expression, it is a 50 % ethanol in the FAA solution that has been confirmed to give high-quality signals (Li et al. 1996). According to Langdale (2001), an FAA of 50 % ethanol, 10 % formaldehyde, and 5 % acetic acid works well in tissue fixation for in situ hybridization but 75 % ethanol with 25 % acetic acid gives no signal in in situ hybridization.

At the step of tissue fixation in FAA, plant tissue/organs must be small enough to allow the FAA solution to penetrate. The fixed tissue must then be completely dehydrated in a series of increasingly more concentrated ethanol solutions. The series of ethanol solutions include 50, 60, 70, 85, 95, and 100 %. The tissues are moved through the dehydration series at 10–30-min intervals, depending on tissue succulence. Woody material takes longer (30 min). The 100 % ethanol solution (absolute ethanol) should be replaced two to three times to ensure the complete removal of water from the tissue. The next step involves an organic solvent (xylene) which serves as means to transition the tissue from alcohol into paraffin. Following 100 % alcohol is 50:50 alcohol to xylene, and then several changes of 100 % xylene, again at 10–30-min intervals. Vials are then transferred to a warming oven (60 °C) for the steps involving tissue infiltration with liquid paraffin (such as Fisher Paraplast Plus, Cat. 23 021400, Fisher Scientific Co., www.fishersci.ca). The first step is 50:50

xylene to paraffin followed by at least three changes of 100 % paraffin at 30–45-min intervals to remove xylene residue from the tissues.

At the step of pouring the wax tissues into plastic boatlike embedding molds (Fisher Scientific Cat. No. 1841) to solidify, the samples should be oriented vertically or horizontally for sectioning. After removing the cooled wax blocks from the Peel-A-Way molds, the wax blocks are transferred in their plastic frame and fitted into a microtome. The microtome is adjusted to cut each section 7–10 μm thick.

Prior to sectioning, the slides should be coated with certain sticky material to ensure that plant tissue sections do not drop off during the hybridization and washing. The slides can be coated with poly-L-lysine (Sigma P-1399; in 10 mM Tris pH 8.0) by immerse slides into 100 mg/ml for 10 min, and then air-dry the slide in a dust-free area or container. After sectioning, the wax tissue ribbons are positioned onto microscope slides coated with poly-L-lysine, and then "baked" to the slide by setting the slide warmer to a temperature between 45 and 50 °C.

At the hybridization step after deparafinize, it is important to avoid bubbles between the slide and coverslip. This is because the tissues in the bubble areas will be less hybridized and can create artifacts of lower signal density. To avoid bubbles, use a sufficient volume of hybridization solution, such as 150 μl per slide, slowly distribute the solution on the slide, and slowly lower the coverslip to the slide.

Photomicroscopy must also be carefully done. An overall examination of the hybridization quality of the entire specimen should be conducted first. If the signal density varies, the slide should not be used in a comparative study due to risk of artifacts. If a round area on the slide has clearly lower signal intensity, it could be caused by a bubble. Gene expression comparisons should be restricted to specimens that pass the quality control check and the results should be verified from tissues either on different slides or at least in different regions of the same slide. An example of cell-specific expression in plants is shown in Fig. 14.1.

14.3 A Method for Studying Mitochondrial Density in Roots Using Rhodamine 123 and Confocal Microscopy

14.3.1 Introduction to Rhodamine and Confocal Microscopy

The cationic fluorescent dye rhodamine 123 specifically labels the negatively charged mitochondrial inner membrane. Loss of the negative potential across the membrane will result in loss of the dye and the fluorescence intensity, and therefore, rhodamine 123 has been used to monitor mitochondria in living cells (Chazotte 2011) such as in studying mitochondria within microspore tetrads of plants under a confocal microscope (Gambier and Mulcahy 1994) and used in flow cytometric

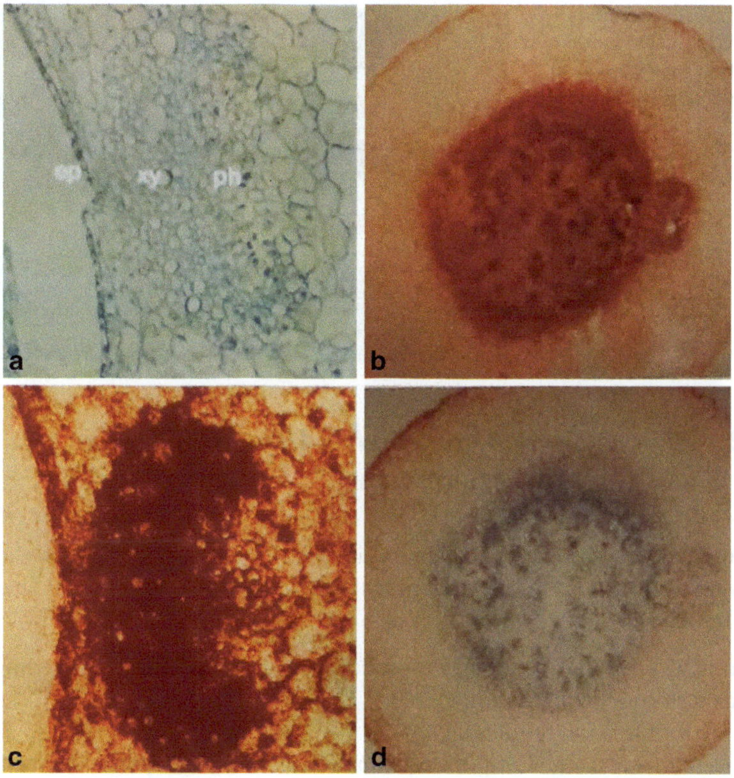

Fig. 14.1 In situ hybridization-detected cell-specific expression of mitochondrial genes in maize seedling tissues (reorganized from Li et al. (1996)). **a** Toluidine blue staining and *bright field* image of a coleoptile vascular bundle, **b** mitochondrial 18S rRNA (*rrn18*) gene expression in a coleoptile vascular bundle, **c** mitochondrial *RNAb* gene expression in a maize seedling stem, and **d** mitochondrial 26S rRNA (*rrn26*) gene in a maize seedling stem. **b**, **c**, and **d**: *bright field* and *dark field* co-exposure image. Note that *rrn18* is highly expressed in the vascular bundles; *RNAb* and *rrn26* have nearly opposite patterns from each other in tissue-specific expression

analysis of mitochondria (Petit 1992). Recently, it was reported that rhodamine B or 1-(Rhodamine B)-4-(2′-chloroacetyl)-piperazine amide" (RB-CAP) is more photostable than rhodamine 123 and still exhibits stringent selectivity in covalent labeling of mitochondria in mammalian living cells (Wu et al. 2012). However, RB-CAP has not yet been widely tested and used in mitochondrial histochemical analysis. Rhodamine-123-labeled mitochondria were green (emission 525 nm) under laser excitation at 488 nm, whereas chloroplasts under this wavelength are not green.

Confocal microscopy can increase optical resolution and contrast by using point illumination, and three-dimensional structures can be reconstructed from the images obtained. Therefore, the combination of confocal microscopy and rhodamine 123 staining is an effective approach to study mitochondrial distribution and function in both animal cells (Hall et al. 2013) and plant cells (Gambier and Mulcahy 1994; Li et al. 1996).

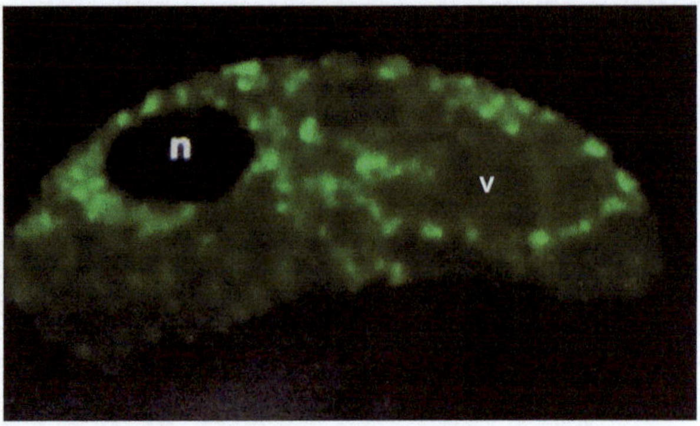

Fig. 14.2 A confocal microscope-reconstructed image of a maize root cap cell stained with rhodamine 123 (Li et al. 1996)

14.3.2 Illustration of Mitochondria Using Rhodamine and Confocal Microscopy

We examined tissue-specific mitochondrial distribution and density in rhodamine 123-stained maize root tips under confocal microscopy (Li et al. 1996). The method is as follows:

1) Germinate seedlings (e.g., maize seed germination for 3 days in vermiculite in the dark at room temperature).
2) Cut the root tips longitudinally.
3) Immerse in a layer of 15 μg/ml aqueous solution of rhodamine 123 for 5 h.
4) Observe under a confocal microscope (Leica Laser technik, Heidelberg, Germany) using an argon 488 nm laser for dye excitation. A photograph from a cell stained with rhodamine 123 and observed under a confocal microscope is shown in Fig. 14.2.

14.4 Using Polarized Microscope in Observing Starch Granules Without Staining

14.4.1 Introduction to Starch Granules

For plants, amyloplasts are one of the most important organelles, in terms of primary metabolism in plant biology and in terms of economic importance in the food supply for humans and animals. Starch granules form inside of amyloplasts and therefore are often characterized when studying chloroplast gene expression and metabolism. Plants store photosynthetic products in the form of starch grains in the amyloplasts, located in leaves and in storage organs (e.g., stem tubers such as potato, storage roots such as sweet potato, and seeds). There are several methods for studying starch granule size.

14.4.2 Method of Observation of Starch Granules (Using Potato Tuber as Example)

Since starch granules in cells are of different sizes, the smaller ones can be lost or washed away if the potato juice containing starch granules is placed directly between the slide and coverslip. Therefore, a hemocytometer or a small homemade chamber is needed on the slide. The following tape-hole method works well. Holes of 6 mm in diameter were made on a glassy tape (such as gel sealing tape, Cat# 11032018, Biometra, Goettingen) with a paper punch. The tape was then stuck onto a microscope slide. The holes in the tape were then used as round chambers for minimizing the shifting of small granules by the coverslip and for holding potato juices during observation under the microscope.

- Take a potato tuber and cut into pieces just before sampling.
- Use a garlic press to squeeze potato juice into a small container such as a microfuge tube.
- Add fourfold volume of distilled water.
- Mix well by tapping on the tube and immediately pipet 11 µl to the tape chamber on the slide, mix immediately by shaking the slide or using the pipet and put on the coverslip. No staining is required.
- Observe the starch granules under a light microscope equipped with a polarizing filter. Ideally, AxioVision Rel 4.7 software can be used (Carl Zeiss).
- Take pictures using a 200× magnification (under a 20× objective and a 10× eyepiece) if the study purpose is to measure the average length of the starch granules in the tuber. Each image acquired under 200× magnification usually has approximately 100–150 granules (Fig. 14.3). Three or more pictures per sample should be used to average starch granule size measurements in that juice sample.
- Measure the length of each starch granule using the software AxioVision if the picture was taken using the software. It is also possible to measure the relative length of the starch granules using a ruler on a print, but this is much more labor intensive.

Starch granules have considerable size variation, ranging from a few micrometers to approximately 80 µm. Measure all starch granules larger than 8 µm and calculate the average length.

The microscopic method assisted with the AxioVision software for estimating the average length of potato starch samples has very high reproducibility; the correlation coefficient between two experiments was $r = 0.92$ in our previous studies (Li et al. 2011; Zhang et al. 2011a).

For the sampling stage, potato juice/starch granules can be directly taken by scraping juice from the tuber strip with a razor blade and adding the juice directly to 7 µl water in the tape hole on the microscopic slide, and then immediately mixing well with the blade on the chamber. This razor-blade-scraping method is simpler but less reproducible and tends to collect larger granules than the squeezed juice method described above and in Li et al. (2011). The starch granules can be also

Fig. 14.3 Starch granules under a polarized microscope (using 200× magnification) from the potato cultivar Kennebec (Li, unpublished)

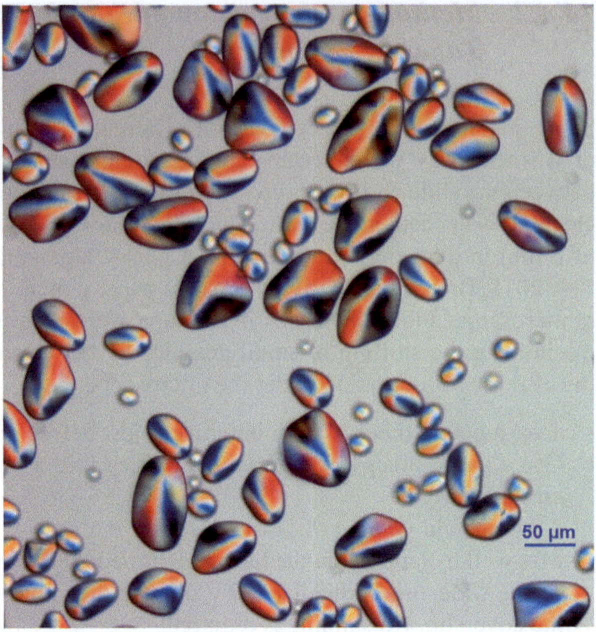

observed under 500X magnification using a 50× objective and a 10× eyepiece for fine detail of the starch granule surface. Since starch granules precipitate very rapidly and samples can vary in granule size within seconds, it is important to use exactly the same protocol for all the starch samples in order to get comparable results. Some practice with the sampling method is required before conducting the formal comparative experimental study.

14.5 Differential Preparation of Agrobacterium Ti Plasmid and Binary Plasmid

14.5.1 Introduction to Plasmid Preps from the Binary Vector System of Agrobacterium tumefaciens

Wild *Agrobacterium tumefaciens* strains usually cause crown gall disease in dicot plants (Hooykaas 1989). However, *A. tumefaciens* harboring “disarmed” tumor-inducing (Ti) plasmids (lacking plant oncogenes) can deliver foreign genes from cointegrated or binary vectors into dicot cells without causing crown gall disease. *A. tumefaciens* is widely used in plant genetic transformation, because some of the systems, particular for the agropine-type *A. tumefaciens,* can transform not only dicot plants but also monocot plants such as maize (Gould et al. 1991; Ritchie et al. 1993) and rice (Raineri et al. 1990; Li et al. 1992; Liu et al. 1992; Toki 1997; Hiei and Komari 2008). One of the most widely used agropine strains is EHA105, which was developed from EHA101 by removing the kanamycin resistance gene (Li et al. 1992; Hood et al., 1993).

The widely used system in *A. tumefaciens*-mediated transformation is a binary vector system, in which the agrobacterial cell has a large, modified Ti plasmid called a "helper Ti plasmid" and a much smaller plasmid usually called a "binary vector." The Ti plasmid is usually very large, ranging from, for example, 194 kb circular DNA (GI: 8572673; AF242881.1) for an *A. tumefaciens* octopine-type Ti plasmid to 245 kb circular DNA for an agropine-type Ti plasmid (DQ058764.1 GI:71849443). The helper Ti plasmid usually carries vir genes required for bacterial virulence (to infect plants). The binary vector is much smaller, is able to replicate in multiple hosts both *Escherichia coli* and *A. tumefaciens*; therefore, it can be used to insert selection marker genes responsible for resistance to antibiotics to ensure the presence of the plasmid in the cell on selection media. Since the binary vector contains the recombination borders of the transfer DNA (T-DNA) and is relatively small to allow efficient manipulation during DNA cloning, the binary vector is used to clone genes for transfer to the plant genome.

After the cloning manipulation of the binary vector in *E. coli*, the plasmid is usually transferred into an *A. tumefaciens* strain by freeze/thaw transformation, electroporation, or triparental mating (Wise et al. 2006). The binary vector transfer stage is to combine two or more plasmids into the same cell, and selection is only based on antibiotic resistance. Therefore, after the triparental mating, it is preferred to isolate the plasmid DNA and conduct restriction analysis to ensure the plasmids are still correct with no deletion, addition, or variation before using the strain for genetic transformation of plants. A protocol is required to rapidly prepare DNA from many colonies and check fingerprints using simple but high-resolution gel electrophoresis. Since the binary vector usually has greater copy number than the Ti plasmid in the cell, fingerprinting of the Ti plasmid is needed to minimize binary vector contamination.

Various difficulties are associated with plasmid DNA extraction from *A. tumefaciens*. Both binary and Ti plasmids, particularly the Ti plasmid, are maintained only at low copy number. The Ti plasmid is very large, of a size similar to plant mitochondrial DNA. Due to these difficulties, Ti plasmid extraction typically involves large (1 l) culture volumes using an alkaline lysis protocol (Currier and Nester 1976) followed by isopycnic gradient purification and dialysis (Sciaky et al. 1978). The DNA prepared is usually still a mixture of both plasmids, and Southern hybridization is usually required to analyze the Ti plasmid (Slusarenko 1990). Ti plasmid analysis is a laborious procedure that involves using radioactive isotopes. To solve the above problems and to meet the need for checking both the binary vector and the Ti plasmid, we developed improved rapid protocols (Li et al., 1995), derived from alkaline lysis methods (Currier and Nester 1976; Bimboim and Doly 1979), that yield plasmid DNA and produce restriction fingerprints with relative high resolution, as described below.

14.5.2 A Protocol for Rapid Micropreps of the Mixture of Both Ti and Binary Plasmids from *A. tumefaciens*

The following protocol (Li et al. 1995) can effectively prep both the Ti plasmid and the binary vector and can differentiate the two to a certain degree by playing the culture temperature according to the purpose of the plasmid preps.

- Grow *A. tumefaciens* bacteria in 0.8 ml in a 1.5 ml microcentrifuge tube (for microprep) or 5.0 ml in a 15 ml culture tube (for miniprep) yeast extract peptone (YEP) medium (Chilton et al. 1974; for 1 l medium: 10 g yeast extract, 10 g Bacto peptone, 5 g NaCl, pH 7.0; Chilton et al. 1974) on a rotary shaker at 200 rpm at 28 °C (or use 37 °C if elimination of Ti plasmid is required) for 16–24 h or until it reaches the OD_{600} = 1.0–1.5. At 37 °C, the Ti plasmid gets largely eliminated if there is no antibiotic selection to maintain the plasmid. The medium contains rifampicin at 20 μg/ml (for the chromosome) and an appropriate antibiotic for maintaining the binary vector such as kanamycin 100 μg/ml for maintaining the binary vector p35S-GUS-INT.
- Pellet the bacteria using a microcentrifuge at maximum speed for 1 min and resuspend the cell pellet by vortexing in 0.1 ml (for microprep) or 1 ml (for miniprep) buffer I (50 mM glucose, 25 mM Tris-HCl, pH 8.0, 10 mM ethylenediaminetetraacetic acid, EDTA).
- Lyse the bacteria with two volumes of a solution containing 0.2 M NaOH and 1 % sodium dodecyl sulfate (SDS), mix well, and set on ice for 15 min.
- Neutralize the mixture by adding 1.5 volumes of 5 M potassium acetate solution prechilled to −20 °C.
- Mix by gentle inversion and set on ice for 30 min.
- Remove cell debris by centrifuging at 10,000 g for 2 min (microprep) or for 5 min (miniprep).
- Collect the supernatant and centrifuge again at 10,000 g for 2 min (microprep) or for 5 min (miniprep).
- Treat the supernatants with equal volume to the collected supernatant using phenol-chloroform-isoamyl alcohol (25:24:1, saturated with 10 mM Tris, pH 8.0, 1 mM EDTA, Sigma-P2069) by inverting tubes for 4 min.
- Centrifuge for 5 min at 10,000 g to collect the supernatant (do not disturb the interface where there can be denatured proteins).
- Wash twice with equal volume of chloroform, centrifuge, and collect the supernatants.
- Precipitate DNA with an equal volume of −20 °C-prechilled isopropanol and maintain at −20°C for at least 1 h or overnight.
- Wash the pelleted DNA with 70 % ethanol, carefully remove the 70 % ethanol residues, and leave at room temperature on the bench to dry for 1 h. DNA yield can be significantly lower if the DNA pellet is not dried sufficiently.
- Resuspend the DNA in 10 to 100 μl buffer Tris/EDTA (TE; 10 mM Tris-HCl pH 8.0 and 1 mM EDTA). The yield of DNA obtained from the microprep is sufficient for one or two restriction analyses for Ti plasmid and for two to three restrictions for the binary vector analysis, whereas the miniprep yields enough DNA for up to ten restriction analyses.

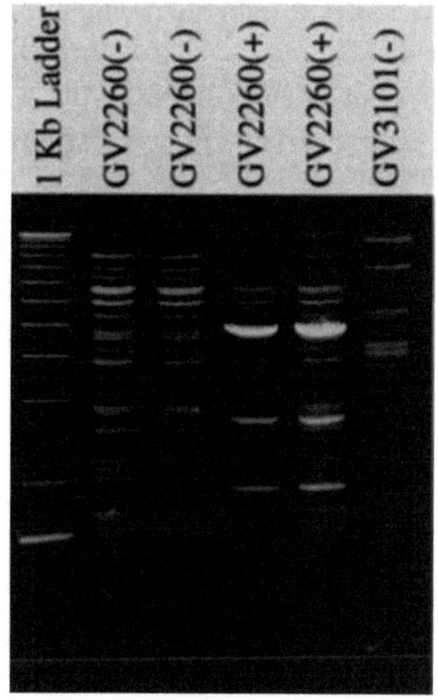

Fig. 14.4 Restriction analysis (*Pst*I) of the Ti plasmids of *A. tumefaciens* strains GV2260 and GV3101 (Li et al. 1995)

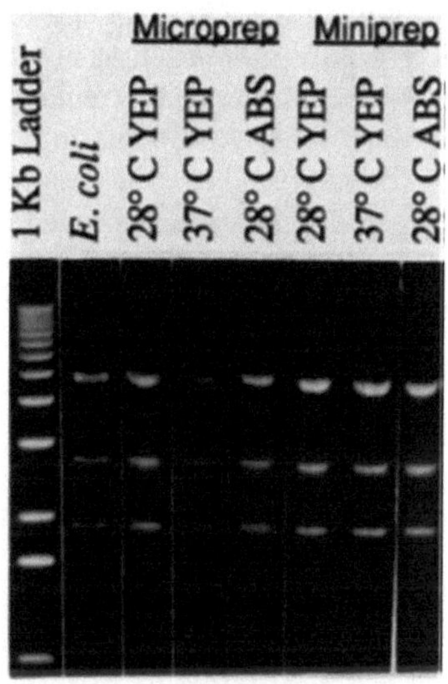

Fig. 14.5 Restriction analysis (*Pst*I) of the binary vector p35S-GUS-INT in *A. tumefaciens* strains GV2260 and GV3101, showing the effect of culture temperature on the elimination of Ti plasmid contamination (Li et al., 1995). An ethidium bromide stained 0.8 % agarose gel (20 × 25 × 0.65 cm) containing DNA prepared from *A. tumefaciens* GV2260/pGV2260/p35S-GUS-INT after restriction with *Pst* I. Each lane represents the DNA isolated from the equivalent of 0.8 ml of culture grown in the presence of kanamycin at 100 µg/ml. The BRL I-kb DNA ladder was used as a size marker in lane 1. Electrophoresis was performed for 18 h at 100 V in IX TBE buffer. TBE Tris/borate/EDTA

During the development of the protocol, restriction endonuclease digestions were performed for 1–2 h with five units of a restriction enzyme such as *Pst* I (Fig. 14.4 for Ti plasmid or Fig. 14.5 for binary vector-enriched DNA). Electrophoresis on an agarose minigel (8 × 6 × 0.5 cm) with 0.5 µg RNase A for 1 h at 105 V in 1X Tris/borate/EDTA (TBE) buffer (5X stock solution in 1 L of H_2O: 54 g of Tris base, 27.5 g of boric acid, 20 ml of 0.5 M pH 8.0 EDTA).

Technical notes: (1) The use of 37 °C to largely remove the Ti plasmid can increase the fingerprint resolution of the binary vector, particularly if one of the DNA fragments of binary vector to check is very short and therefore faint. (2) The use of minigel with ≤0.5 cm thickness and relatively a short period (1 h) may also contribute to the sharpness of DNA bands on the gel. (3) The alkaline lysis method may have contributed to the DNA yield. (4) We also tested growing the *A. tumefaciens* bacteria in YEP medium without the use of the binary vector selection antibiotics, and the prepared DNA showed a decreased density in the binary vector, and in some clones, the binary vector was totally lost. Therefore, the prepared DNA has enriched Ti plasmid DNA and is suitable to verify whether the Ti plasmid is still in correct fingerprints as in the control strain. The gel staining was usually with ethidium bromide, but recently, this toxic compound was replaced by GelRed (http://www.biotium.com) without losing any detection sensitivity of the DNA in the gel. This protocol is suitable for DNA prep from both the binary plasmid and the Ti plasmid.

14.5.3 A Qiagen Kit Method for Agrobacterial Plasmid DNA Preparation

- Grow *A. tumefaciens* with appropriate antibiotics, depending on the plasmids, in 5 ml YEP at 28 °C for approximately 24 h till $OD_{600} = 1–1.5$.
- Spin down the bacteria at 10,000 g for 5 min.
- Discard supernatant (remove as much supernatant as possible).
- Start the QIAGEN plasmid mini kit (Cat. 12125) by adding the alkaline buffer for bacterial lysis and then follow the kit's instruction for purification using Qiagen columns and isopropanol precipitation.
- Use the DNA from 0.4 to 0.5 ml culture for each digestion.
- Digest DNA in a 15-μl total volume with 8 μl restriction enzyme, 1× buffer, at 37 °C for 2 h.
- Run a minigel at 103 V for 50 min, stain with GelRed, and photograph to check the fingerprints (Fig. 14.6).
- Note that this protocol works well for the binary vector plasmid but has low yield for the Ti plasmid (Li, this study).

14.6 Isolation of Nuclei for DNA Extraction

14.6.1 Introduction to the Isolation of Plant Cell Nuclei

In molecular biology and genomic research, it is often necessary to isolate nuclear DNA. The first step of nuclear DNA prep is often the isolation of nuclei. Compared to chloroplasts and mitochondria, the nuclei are much larger and heavier and therefore can precipitate at a low speed in the centrifuge. After low-speed centrifugation to enrich nuclei, further purification of the nuclei can be done through

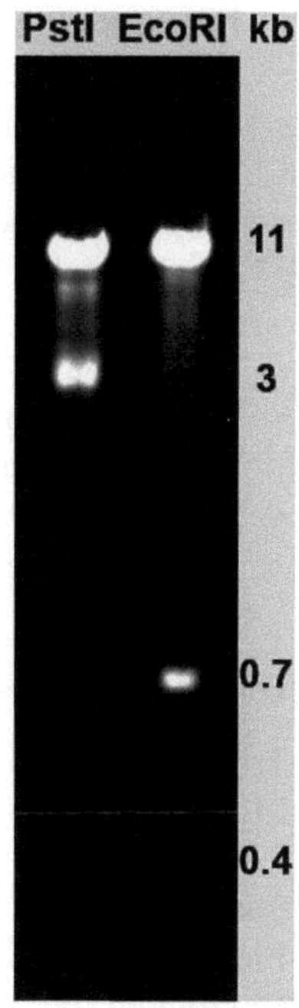

Fig. 14.6 Restriction of a plasmid prep from LBA4404 that carries a Ti helper plasmid and a binary vector. Digested with PstI. The DNA amount corresponds to 0.5 ml YEP culture. *YEP* yeast extract peptone (Li, unpublished)

centrifugation on a sucrose gradient (Ulrich and Ulrich 1991; Lutz et al. 2011), Percoll density gradient (Willmitzer and Wagner 1981; Kiss and Solymosy 1987; Sikorskaite et al. 2013), or directly by using a high concentration of sucrose buffer during the extraction process (Hernandez et al. 1996). To minimize the influence of cell walls, protoplasts were used for nuclei preparation (Ulrich and Ulrich 1991). Sucrose is less expensive than Percoll, and the sucrose gradient method can allow extraction of a larger amount of DNA.

14.6.2 A Sucrose Gradient Protocol for Isolating Nuclei

The author has used the following protocol modified from Huguet et al. (1975) and obtained high-quality DNA from leaves. This nuclear prep method used Triton X-100 to reduce contamination from chloroplasts and mitochondria and purified the enriched nuclei by laying a suspension of nuclei on top of a concentrated sucrose solution during centrifugation. The following protocol was slightly modified by reducing the sucrose concentration from 2.2 to 2 M and fine-tuning the description of the temperature, filtration, and solution volumes:

- Chill Honda buffer (Honda et al., 1966) on ice.
- Add 0.5 % Triton X-100.
- Grind leaves (e.g., 10 g) in the chilled buffer (10 ml/g tissue).

- Filter through three layers of Miracloth.
- Centrifuge at 350 g for 10 min.
- Wash the nuclei by suspending the pellet and recentrifuging at 350 g for 10 min.
- Repeat this washing step again.
- Suspend the pellet (enriched nuclei) in 4 ml of 15 mM Tris-HCl, pH 7.5.
- Lay the pellet solution on a 2-M sucrose (in 15 mM Tris-HCl, pH 7.5) uniform gradient.
- Centrifuge at 40,000 g for 45 min.
- Suspend the pellet in a buffer of 15 mM Tris-HCl, pH 8.0, 10 mM EDTA and extract DNA using a standard chloroform–phenol extraction method.
- If a Qiagen DNAeasy kit or a magnetic bead-binding method (Mehle et al. 2013) is used, the pellet suspension buffer should be the one provided by the commercial kit.

Honda Buffer (Honda et al. 1966)

Tris-HCl, pH 7.8,	25 mM
$MgCl_2$	2 mM
$CaCl_2$	2 mM
Sucrose	0.25 M
Ficoll	2.5%
Dextran	5%
Beta-mercaptoethanol	5 mM

14.7 Chloroplast DNA Extraction and Restriction Fragment Length Polymorphism Analysis

14.7.1 Introduction to Chloroplast DNA Extraction

Chloroplast DNA (often abbreviated as cpDNA, or ctDNA) preps are often contaminated with a considerable amount of nuclear DNA. This reduces the clarity of the restriction DNA bands on electrophoresis gels. Various cpDNA preparation protocols were published in the 1980s; many using cesium chloride (CsCl) gradient ultracentrifugation mainly for reducing nuclear DNA contamination (Shoemaker et al. 1984; Mourad and Polacco 1988; Martin et al. 1989; Mourad and Polacco 1989). Since it is relatively difficult to isolate highly purified cpDNA, sometimes people have to use Southern hybridization of total cellular DNA with cpDNA probes when comparing cpDNA between different species such as wild potato species (Nakagawa et al. 2000). A high salt method can be used to enrich cpDNA and reduce nuclear DNA sequence coverage during second-generation sequencing, but the purity is not enough to clearly detect cpDNA fragments on electrophoretic gels in enzymatic restriction analysis (Shi et al. 2012). Improved protocols are still required to generate clearly readable restrictive fragments of cpDNA.

14.7.2 Chloroplast DNA Preparation

The following protocol leads to very readable restriction fingerprints of cpDNA. The chloroplast isolation was modified from Kolodner and Tewari (1975) by Vedel and Mathieu (1983), while the DNA extraction step was modified in the author's laboratory at AAFC-Fredericton.

- Place rapidly growing plants in the dark for 1–2 days at 18–20 °C.
- Take 10 g of leaves with petioles removed.
- Wash the leaves with distilled water and put the leaves to a blender.
- Add 200 ml extraction buffer (Buffer A) to a prechilled blender.
- Blend by pulsing for three intervals: for 1 s, 1 s, and then 5 s.
- Pause briefly between pulses.
- Filter through two layers of Miracloth or a nylon tissue with 25 μm pores without squeezing the materials during filtering.
- Centrifuge at 200 g for 10 min (to remove nuclei) and transfer the supernatant into a new centrifuge tube.
- Centrifuge at 100 g for 10 min to pellet chloroplasts (keep the pellet).
- Wash the pellet twice with Buffer B and recentrifuge to collect chloroplasts (keep the pellet).
- Suspend the chloroplasts in Buffer C.
- Add 1 mg/ml pronase.
- Incubate for 3 h at 20 °C for enzymatic reaction.
- Treat with equal volume of a phenol-chloroform-isoamyl alcohol solution (25:24:1), mix by repeatedly inverting the tube, and centrifuge to get supernatant.
- Treat the supernatant one to two times with chloroform, each time centrifuge to recover the supernatant.
- Add one tenth volume of 3 M sodium acetate (pH 5.2, adjusted with acetic acid) or 5 M ammonium acetate (pH 5.5; this is because dissolving 57.81 g of ammonium acetate in water to a final volume of 100 ml, the pH will likely be 5.5).
- Precipitate at −20 °C with 1:1 volume of isopropanol or 2.5 volume of ethanol.
- Pellet the DNA, wash with chilled 70 % ethanol, dry the pellet, and dissolve in buffer TE for restriction analysis.

Using this protocol, restriction fragment length polymorphism was clearly detected on an electrophoresis gel between eggplant (*Solanum melongena*) and its wild relative *Solanum khasianum* (Fig. 14.7).

(*Note:* The http://www.sciencegateway.org/tools/rotor.htm website can be used to convert between revolution per minute and gram for centrifugation).

Buffer A

Tris 0.05 M
Mannitol 0.5 M
EDTA 0.003 M
Bovine serum albumin (BSA): 0.1 % (W/V)
β-mercaptoethanol 0.001 M
pH 8.

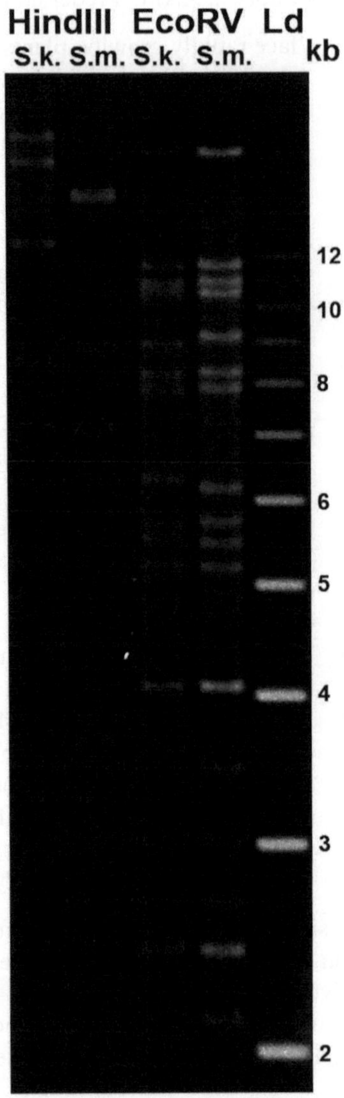

Fig. 14.7 Chloroplast DNA restriction analysis of eggplant (*Solanum melongena)* and its wild relative *Solanum khasianum* (Li, XQ, unpublished). The molecular marker is BRL 1-kb ladder. Note that the cpDNA fragments are clearly visible on the electrophoresis gel. *cpDNA* chloroplast DNA

Buffer B

Tris 0.05 M
Sucrose 0.3 M
EDTA: 0.02 M
pH 8

Buffer C

Tris 0.05 M
EDTA: 0.02 M
pH 8

Buffer TE

Tris-HCl 10 mM
EDTA 1 mM
pH 8.0

14.8 Mitochondrial DNA Extraction and Restriction Fragment Length Polymorphism Analysis

14.8.1 Introduction to Mitochondrial DNA Preparation

Plant mitochondrial DNA (mtDNA) preparation contains several steps, including differential centrifugation to enrich mitochondria and then extract the DNA. The DNA extraction used to be done first by phenol–chloroform treatment and then by CsCl gradient separation (Day 1997). However, the manual removal of each of the three close bands (nuclear, chloroplast, and mitochondrial DNAs) is challenging from a small CsCl gradient tube, under UV, and in a dark room. Likely because of nuclear DNA contamination, Southern blot hybridization is needed to check or compare the DNA samples (Day 1997). Some commercial kits can be used for DNA extraction of animal mitochondria (Valach et al. 2008). However, these kits usually do not address how to avoid nuclear DNA contamination. DNase treatment of isolated mitochondria prior to the lysing of the mitochondria can greatly reduce nuclear DNA contamination (Kolodner and Tewari 1972). At the stage of DNA extraction from mitochondria, it is possible to use a multiple precipitation method to remove the DNase and other proteins if mtDNA prep is from suspension cell cultures and some mitochondria-rich coleoptiles (Wilson and Chourey 1984). However, mtDNA can be seriously degraded from some leafy cotyledons such as tobacco and peanut cotyledons and therefore protein denaturation is done with the toxic chemical diethyldithiocarbamic acid (Wilson and Chourey 1984). This chapter describes alternative effective protocols for plant mitochondrial DNA preparation.

14.8.2 Mitochondrial DNA Preparation Protocol

The following protocol, modified from Kolodner and Tewari (1972), can obtain highly purified and undegraded mitochondrial DNA from both cultured cells and flowers, tubers, and leaves.

- Sample composed of 30 g of young leaves.
- Blend for about 8 s in 200 ml pre-chilled Buffer A (see the chloroplast DNA preparation section).
- Filter using two layers of Miracloth or a nylon tissue (about 25–100 μm pores).
- Centrifuge at 1500 g for 10 min to eliminate cell fragments, nuclei, and chloroplasts.
- Collect the supernatant and recentrifuge at 1500 g for 10 min.
- Collect the supernatant.
- Centrifuge at 10,000 g for 15 min to pellet mitochondria.
- Resuspend the pellet in 30 ml Buffer A and homogenize with a homogenizer potter.
- Add 0.4 ml of 1 M $MgCl_2$ and 3 mg DNaseI.
- Mix well and set on ice for 1 h.
- Stop the reaction by adding three volumes of Buffer B.
- Centrifuge at 10,000 g for 15 min, and discard the supernatant.
- Wash and pellet twice with Buffer B.
- Resuspend the pellet in 1 ml of Buffer C.
- Add 30 mg of sarkosyl (lauroyl sarcosinate) predissolved in 0.3 ml Buffer C.
- Shake for 1 h on a platform shaker at a speed of approximately 200 rpm.
- Add CsCl 1.65 g with 1/10,000 V/V of GelRed (http://www.biotium.com, 10,000X stock solution).
- Fill a plastic tube of "Quick-Seal (11 × 32 mm).
- Centrifuge using the centrifuge TL 100 with the TLV 100K rotor at 9000 rpm for 3 h at 18 °C.
- Collect the mitochondrial DNA band with a syringe under UV (if the DNase treatment stage worked well, there will be only one band, i.e., the mtDNA, in CsCl gradient).
- Dilute the DNA solution with two volumes of buffer TE.
- Centrifuge at 100,000 rpm; the present author used the centrifuge TLA 100 with the TLV 100K rotor for 1.5 h, decal 6, at 18 °C.
- Carefully pour off the supernatant.
- Dissolve the pellet in TE and mix well.
- The DNA is ready for restriction analysis.

This protocol can give very clean mtDNA and very clearly distinguishable fingerprints for mtDNA restriction analysis (Fig. 14.8). The fingerprints are identical between two preps from different plants, in terms of the number and the size of each DNA fragment. There is no detectable level of nuclear contamination. It is known that the mitochondrial genome consists of a large number of subgenomic molecules, and some of these molecules are at very low copy numbers in the population. If there is any obvious contamination from nuclear DNA, these low copy number

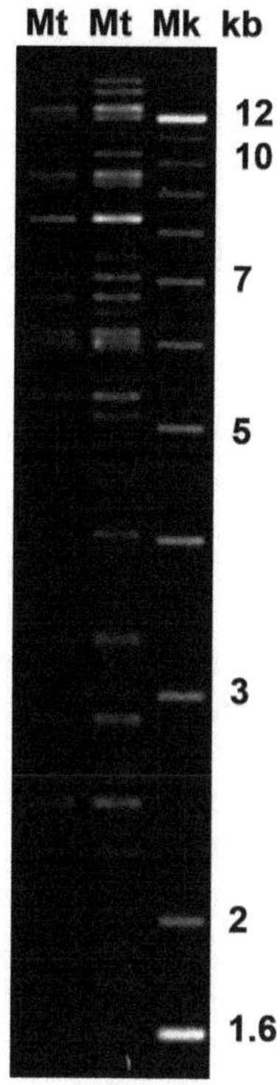

Fig. 14.8 Restriction analysis of *Nicotiana sylvestris* mitochondrial DNA (Li X-Q unpublished; Mk: BRL 1-kb ladder)

subgenomic molecules of mitochondrial genome data can easily be mixed up with the nuclear one and create difficulties for interpretation of the results. The high-purity mitochondrial DNA preparation method described in this chapter should be very useful in preparing mitochondrial DNA for mitochondrial genome sequencing because nuclear DNA contamination is minimal.

14.9 Total DNA/RNA Preparation for Electrophoretic and Northern Analysis

14.9.1 Introduction to Total RNA Prep

Total DNA/RNA preparations can be used for studying the ratio between nuclear DNA and nonnuclear organellar DNA as well as the overall transcriptome. Various total DNA preparation methods are available, including several commercial kits such as the QIAGEN DNeasy kits and RNeasy kits. However, most of these commercial kits do not describe the components. Therefore, there is no information about how effective these kits are in hydrolyzing mitochondrial and chloroplast membranes for organellar DNA/RNA release. The following protocol uses strong detergents which denature proteins and membranes to ensure DNA release from all organelles. The nucleic acid extraction from this protocol worked well in restriction analysis, RNA electrophoresis, single-nucleotide polymorphism (SNP) analysis, and complementary DNA (cDNA) library construction for sequencing.

14.9.2 A Protocol for Preparation of Total Nucleic Acid DNA/RNA

- Grind plant tissues into a powder using liquid nitrogen and a mortar and pestle.
- Add 3 ml ice-cold isolation buffer (0.1 M Tris-HCl, pH 8.0, 50 mM EDTA, 1.3 M NaCl, 1 % sarcosyl, 1 % SDS, 0.5 % Triton X-100, 1 % PVP, 10 mM DTT, and 60 mM β-mercaptoethanol).
- Extract the resulting suspension once with a 25:24:1 phenol-chloroform-isoamyl alcohol solution, and once with chloroform.
- Add sodium acetate to a concentration of 0.3 M.
- Precipitate the nucleic acids with 2.5 volumes of ethanol.
- Dissolve the samples in 100 μl of RNase-free water or TE buffer (10 mM Tris-HCl, pH 8.0, 1 M EDTA).
- If the double-stranded DNA (dsDNA) concentration is needed, one can use Quant-iT™ PicoGreen® dsDNA reagent from Invitrogen (http://www.lifetechnologies.com) to determine the dsDNA concentration (http://www.ncbi.nlm.nih.gov/pubmed/15560134). For northern hybridization analysis, the total nucleic acids can be used directly. If used in second generation sequencing, DNA can be removed from the total nucleic acids by treatment with RNase-free DNase.

The protocol can be used to prepare total nucleic acids (total DNA or total RNA). If the protocol is used to run RNA electrophoresis or Northern hybridization, diethylpyrocarbonate (DEPC) treatment of water, tips, and tubes should be used after the phenol–chloroform treatment step.

The method for DEPC treatment of water is as follows: Add DEPC to 0.1 %, let stand in fume hood overnight, and autoclave twice before use.

14.10 Enriched Mitochondrial RNA Preparation

14.10.1 Introduction to Enriched Mitochondrial RNA Preparation

RNA preparation from mitochondria is required for studies of mitochondrial gene expression such as in analysis of CMS of plants (Singh and Brown 1991; Singh et al. 1996; L'Homme et al. 1997; Li et al. 1998). The procedure of mitochondrial preparation is largely as modified from Kemble (1987), and the mtRNA high salt method was modified from Pérez et al. (1990) by Singh and Brown (1991).

14.10.2 High Salt Washing Method in Enriched Mitochondrial RNA Preparation

- Grind plant tissues (3–10 g) in five volumes of high salt buffer (prechilled on ice).
- Filter through two layers of Miracloth into a flask prechilled on ice.
- Centrifuge the filtrate at 2500 g for 5 min (Sorvall SS34 rotor at 4600 rpm).
- Discard pellet.
- Collect the supernatant and recentrifuge at 2500 g for 5 min.
- Collect supernatant and centrifuge at 17,200 g (12,000 rpm if using Sorvall SS 34 rotor) for 30 min to pellet mitochondria
- Suspend the pellet in Buffer 1 and recentrifuge at 17,200 g for 30 min.
- Suspend the pellet in 800 μl of buffer TE.
- Immediately add 200 μl of 10 % sarkosyl (in TE).
- Shake vigorously for 1 min or vortex for 30 s.
- Extract once in phenol, once in phenol–chloroform-isoamyl alcohol, and once in chloroform.
- Precipitate nucleic acids with one tenth v/v 3 M NaOAC (pH 5.2) and 2.5 volumes of ethanol overnight.
- All further handling/suspensions of the RNA must be in DEPC-treated water.

High Salt Buffer TENB

50 mM Tris pH 8.0
25 mM EDTA
1.3 M NaCl
0.1 % BSA
56 mM beta-mercaptoethanol
pH 8.0

One example of the use of this mtRNA is the Northern analysis of mitochondrial gene expression of plant CMS (see Fig. 14.9).

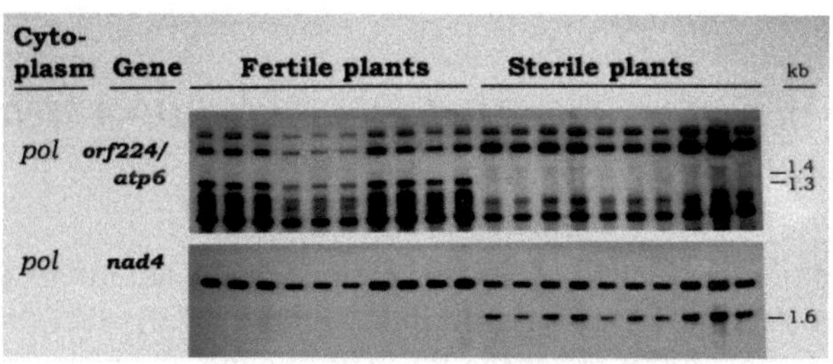

Fig. 14.9 Northern hybridization of mitochondrial RNA probed with the genes *orf224/atp6* and *nad4* for comparison of gene expression patterns between fertile and cytoplasmic male sterile brassica plants that have Polima CMS cytoplasm (Li et al. 1998). Note that the RNA bands are very sharp, suggesting that the mtRNA prepared by the high salt method was intact without significant degradation. *CMS*: cytoplasmic male sterility

14.11 High-Resolution DNA Melting Analysis for Studying Gene Expression

14.11.1 Introduction to DNA High Melting Analysis

For gene expression analysis, in addition to high-throughput genomic approaches, there are also several laboratory methods for routine analysis, including reverse transcriptase polymerase chain reaction (RT-PCR), real-time RT-PCR, blot hybridization, etc. Real-time quantitative RT-PCR (real-time qRT-PCR) analysis is one of the most widely used methods, but it is mainly designed to measure the density of the mRNA from the gene, i.e., based on transcript copy number. However, real-time qRT-PCR does not tell whether the two amplified DNA segments have any sequence differences. The melting temperature of a dsDNA fragment is defined as the temperature where 50 % of the DNA is single stranded (Santalucia 1998). Since a guanine–cytosine (GC) pair is stronger than an adenine–thymine (AT) pair, the DNA melting temperature is strongly influenced by the GC–AT ratio of the DNA region under analysis. The LightScanner system is currently used in genomic DNA genotyping (Cho et al. 2008; Liew et al. 2010; Lochlainn et al. 2011; Refsgaard et al. 2012). Since the RT-PCR products are in the form of DNA, their DNA melting temperature can also be detected using a LightScanner. The LightScanner produces better scanning specificity than current popular real-time qPCR instruments (Herrmann et al. 2007). The dsDNA binding dyes LCGreen® Plus offers better sensitivity and specificity in the DNA melting analysis than does SYBR Green (Wittwer et al. 2003; Herrmann et al. 2006; Herrmann et al. 2007).

The current author's research team has previously developed a method to detect the minor differences within the PCR-amplified segments using a LightScanner (Idaho Technology Inc., Salt Lake City, UT) to measure the DNA melting temperature (Yuan et al. 2009). The LightScanner-based high-resolution DNA melting (HRM) analysis was employed to study potato nuclear gene-derived transcript abundance and detected allelic dynamics of gene expression within various tissues, organs, genotypes, and treatments.

14.11.2 Description of the High-Resolution DNA Melting Analysis Method

The PCR conditions and HRM analysis described below are from a previous report (Yuan et al. 2009). The reverse transcription and RT-PCR stages are as standard RT-PCR (Yuan et al. 2009). After the RT-PCR cycling, the plate carrying the amplified DNA was analyzed by the LightScanner (following instructions provided by LightScanner) to measure the melting temperature. Melting curve analysis was carried out with LightScanner software (version 2.0). The melting temperature measurement process is very fast, taking about 15–20 min, and can sensitively detect

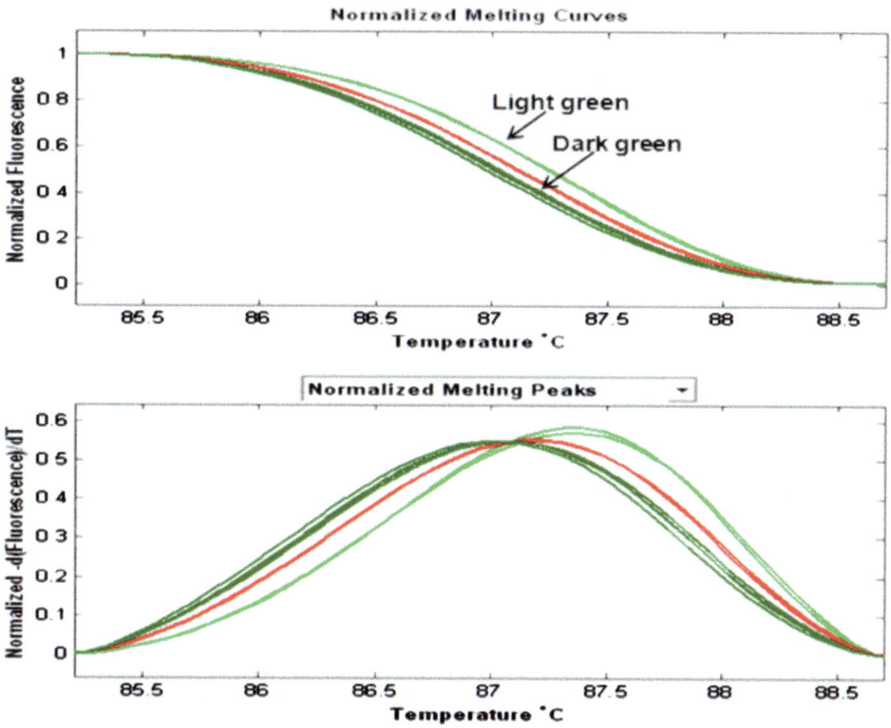

Fig. 14.10 High-resolution melting curves of a SNP region (CA548B) of an acid invertase locus. *Light green* curve (marked by the *arrow*): young roots; *red* curve: stolons; and *dark green* curves: mature tuber and vascular ring. The PCR reactions for a tissue were amplified in duplicate. Melting curve normalization was carried out with LightScanner software on the LightScanner (Idaho Inc., UT). Note that the HRM curves of the two repeats of each tissue are very similar and that young root curves are different from the rest. *SNP*: single-nucleotide polymorphism, *PCR*: polymerase chain reaction, *HRM*: high-resolution DNA melting (Adapted from Yuan et al. 2009)

the DNA melting temperature differences between the simplified DNA fragments (Fig. 14.10).

14.12 Transcriptome Electrophoretic Fingerprinting

14.12.1 Introduction to RNA Electrophoresis

There are various approaches in RNA analysis, including Northern hybridization, reverse-transcriptase real-time quantitative polymerase chain reaction (RT-qPCR), microarray, and second-generation DNA sequencing. Each of these approaches has its advantages and disadvantages. Northern hybridization and RT-qPCR are usually used to study one to a few genes only. Microarray and second-generation sequencing are transcriptome approaches but need very expensive equipment or are expensive to service out the work to genomics companies, particularly if the number of RNA samples is large. Here, we describe a routine laboratory method called "transcriptome electrophoretic fingerprinting (TEF)."

14.12.2 Transcriptome Electrophoretic Fingerprinting Protocol

The following protocol is modified from Finnegan and Brown (Finnegan and Brown 1986).

Set Up of the Electrophoresis Apparatus

- Check whether the required accessories of the V16 Vertical Gel Electrophoresis Apparatus (Life Technologies, Cat. No. 21070–032) are in place. These accessories include the V16 electrophoresis apparatus, power cords, 3-mm comb with 20 teeth, 3-mm spacers, glass plates, and binder clips.
- Clean the glass plate with 3 % H_2O_2 solution for 20 min, then rinse with sterile (autoclaved) water.
- Set up the glass plates on the apparatus with 3-mm spacers between the two glass plates.

Preparation of a Strong Agarose Gel Portion for the Bottom of the Gel

- Lean the apparatus against something to generate about a 45° angle. Pour approximately 300 ml of 1.5 % agarose in 1X TBE to the bottom of the electrophoresis buffer box until about 2 cm of gel forms in between the two glass plates.
- Let the bottom gel solidify for 1 h at room temperature.

Preparation of the Electrophoresis Gel

- Dissolve 22.5 g urea in some RNase-free water to make the final volume to 40 ml.
- Melt 0.75 g agarose in 7.5 ml 10X TBE and 27.5 ml RNA-free water.
- Immediately combine agarose and urea solutions.
- Let the 75 ml urea–agarose solution completely cool at room temperature. This solution will not solidify at room temperature.
- Pour the urea–agarose solution into the space between the two glass plates that has already solidified 1.5 % agarose on the bottom.
- Put a comb (3 mm × 20 teeth) into the space between the glass plates on top of the gel.
- Let the gel stand overnight at 4 °C. Reserve any surplus urea–agarose solution at room temperature. In case the gel shrinks upon standing over night, add some of the surplus urea–agarose solution (shrinkage usually occurs when the gel solution is poured while warm).
- (*Note:* The final concentration in the gel is 1 % agarose, 5 M urea, 1X TBE. This gel is good for resolving RNA molecules from 800 to 3000 bases).

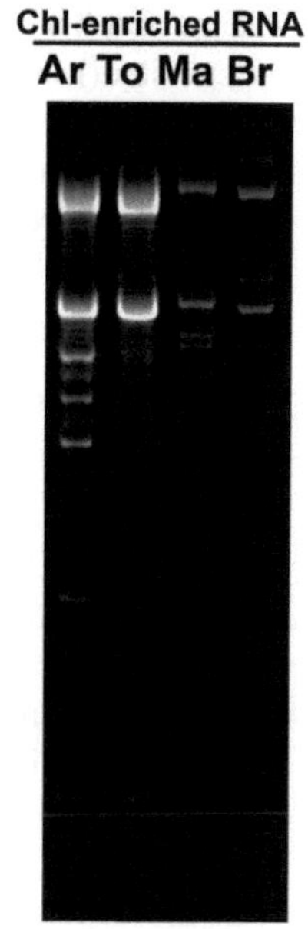

Fig. 14.11 Transcriptome electrophoresis fingerprints from total foliar RNA, chloroplast-enriched RNA, and mitochondria-enriched RNA. M: RNA size ladder, 0.24–9.5-kb RNA ladder (BRL Life Technologies, now: Invitrogen Life Technologies). Total RNA: Ar: *Arabidopsis thaliana*, To: tobacco, Ma: maize, and Br: *Brassica napus*. Note that the chloroplast RNA TEFs are different between species, suggesting genetic control of gene expression at the TEF level. *TEF* transcriptome electrophoretic fingerprinting (Li, unpublished)

Loading of the RNA Samples

- Prepare a loading dye: 10 M urea, 1.5 mM EDTA (pH 8.0), 0.05 % each of xylene, cyanol, and bromophenol blue (BioRad).
- Prepare the RNA sample and loading dye at a 2.5 v/v, or directly dissolve the RNA into the loading dye.
- Heat the RNA/dye solution at 85–90 °C for 1 min.
- Immediately set it on ice.
- Pour 1X TBE buffer into the electrophoresis apparatus.
- Take off the comb very slowly 5 min after adding the electrophoresis buffer.
- Wash the urea precipitations in the wells by pipetting the electrophoresis buffer near the wells.
- Load the samples.

Electrophoresis and Photographing

- Run gel at 120–130 V, for approximately 3 h at room temperature or 4 °C until xylene cyanol is 75 % of the way to the gel bottom. Wash the gel four times, 15 min each, with dH_2O.
- Stain using GelRed.
- Photograph with a gel documentation system. Blot the gel, if needed, for use in Northern hybridization.

An Example Result and Discussion of Transcriptome Electrophoretic Fingerprints After cpDNA prep using a method of high-salt, and not jointly with a commercial kit, the image of TEFs of chloroplast-enriched RNA is shown in Fig. 14.11.

A large number of polymorphisms were detected in the chloroplast TEFs of different species (Fig. 14.11). The *Arabidopsis* chloroplast TEF had very dense bands, whereas the maize chloroplast TEF was characterized by many relatively uniform bands. The two largest rRNA bands of *Arabidopsis* and tobacco were similar, but they differed greatly from those in maize and brassica.

14.13 Concluding Remarks

This chapter described 11 laboratory-level methods: three microscopic analysis methods, five DNA analysis methods (for nuclear, chloroplast, and mitochondrial DNA preparation), and three RNA analysis methods. Some of these methods can also be used to verify one another's results. For example, the in situ hybridization approach and the rhodamine staining method were used in the present author's finding that maize root caps are rich in mitochondria but have downregulated gene expression for the mitochondrial *atp*A gene. The organelle DNA prep methods can be used jointly with genome sequencing to study whether a sequence scaffold is from the nuclear genome or a cytoplasmic organelle genome. The transcriptome electrophoresis fingerprinting may be used in studying gene expression but also in genetically grouping the analyzed organisms. These methods can be useful in routine analysis of organelles, DNA, and RNA and can be useful in validating high-throughput sequencing or microarray results.

Acknowledgments The author thanks Mr. Muhammad Haroon of Agriculture and Agri-Food Canada for verification of the agrobacterial prep methods and Professor Danielle Donnelly of McGill University for the internal review of this chapter. Part of the work was conducted in the Laboratoire d'Amélioration des Plantes (Professeur Yves Demarly) de la Université de Paris-Sud-XI, the Laboratoire de Génétique Moléculaire des Plantes (Dr. Fernand Vedel) du Centre National de la Recherche Scientifique de France, the Departments of Botany and Plant Pathology (Professor Thomas K. Hodges) and Biological Sciences (Professor Stanton B. Gelvin) of Purdue University, and the Department of Biology of McGill University (Professor Gregory G. Brown). All the remaining parts were conducted in the author's laboratory (Cell Biotechnology) at the Department of Biology of Peking University and the authors' laboratory (Molecular Genetics) at the Potato Research Centre of Agriculture and Agri-Food Canada, Fredericton.

References

Bimboim HC, Doly J (1979) A rapid alkaline extraction procedure for screening recombinant plasmid DNA. Nucleic Acids Res 7:1513–1523

Brennicke A, Grohmann L, Hiesel R, Knoop V, Schuster W (1993) The mitochondrial genome on its way to the nucleus: different stages of gene transfer in higher plants. FEBS Lett 325:140–145

Chase CD, Pring DR (1985) Circular plasmid DNAs from mitochondria of *Sorghum bicolor*. Plant Mol Biol 5:303–311

Chazotte B (2011) Labeling mitochondria with rhodamine 123. Cold Spring Harb Protoc 6:892–894

Chilton MD, Currier TC, Farrand SK, Bendich AJ, Gordon MP, Nester EW (1974) *Agrobacterium tumefaciens* DNA and PS8 bacteriophage DNA not detected in crown gall tumors. Proc Natl Acad Sci U S A 71:3672–3676

Cho MH, Ciulla D, Klanderman BJ, Raby BA, Silverman EK (2008) High-resolution melting curve analysis of genomic and whole-genome amplified DNA. Clin Chem 54:2055–2058

Currier TC, Nester EW (1976) Isolation of covalently closed circular DNA of high molecular weight from bacteria. Anal Biochem 76:431–441

Day J (1997) Isolation of nuclear, chloroplast and mitochondrial DNA from plants. Biochem Edu 25:41–43

Dewey RE, Levings Iii CS, Timothy DH (1986) Novel recombinations in the maize mitochondrial genome produce a unique transcriptional unit in the Texas male-sterile cytoplasm. Cell 44:439–449

Ferrandiz C, Sessions A (2008) Nonradioactive in situ hybridization of RNA probes to sections of plant tissues. Cold Spring Harb Protoc 3. doi:10.1101/pdb.prot4943

Finnegan PM, Brown GG (1986) Autonomously replicating RNA in mitochondria of maize plants with S-type cytoplasm. Proc Natl Acad Sci U S A 83:5175–5179

Gambier RM, Mulcahy DL (1994) Confocal laser scanning microscopy of mitochondria within microspore tetrads of plants using rhodamine 123 as a fluorescent vital stain. Biotech Histochem 69:311–316

Gould J, Devey M, Hasegawa O, Ulian EC, Peterson G, Smith RH (1991) Transformation of zea mays L. using *Agrobacterium tumefaciens* and the shoot apex. Plant Physiol 95:426–434

Hall AM, Rhodes GJ, Sandoval RM, Corridon PR, Molitoris BA (2013) In vivo multiphoton imaging of mitochondrial structure and function during acute kidney injury. Kidney Int 83:72–83

Hernandez S, Sanchez M, Guzman P, Simpson J (1996) Isolation of high molecular weight DNA from plant nuclei. Methods Mol Cell Biol 5:349–352

Herrmann MG, Durtschi JD, Bromley LK, Wittwer CT, Voelkerding KV (2006) Amplicon DNA melting analysis for mutation scanning and genotyping: cross-platform comparison of instruments and dyes. Clin Chem 52:494–503

Herrmann MG, Durtschi JD, Wittwer CT, Voelkerding KV (2007) Expanded instrument comparison of amplicon DNA melting analysis for mutation scanning and genotyping. Clin Chem 53:1544–1548

Hiei Y, Komari T (2008) Agrobacterium-mediated transformation of rice using immature embryos or calli induced from mature seed. Nat Protoc 3:824–834

Honda SI, Hongladarom T, Laties GG (1966) A new isolation medium for plant organelles. J Exp Bot 17:460–472

Hood EE, Gelvin SB, Melchers LS, Hoekema A (1993) New *Agrobacterium* helper plasmids for gene transfer to plants. Transgenic Res 2:208–218

Hooykaas PJJ (1989) Transformation of plant cells via *Agrobacterium*. Plant Mol Biol 13:327–336

Huguet T, Jouanin L, Bazetoux S (1975) Occurrence of palindromic sequences in wheat DNA. Plant Science Letters 5:379–385

Karamian R, Ranjbar M (2008) Plant regeneration from *Onobrychis subnitens* Bornm. hypocotyl explants via somatic embryogenesis and organogenesis. Acta Biol Cracov Ser Bot 50:13–18

Kemble RJ (1987) A rapid, single leaf, nucleic acid assay for determining the cytoplasmic organelle complement of rapeseed and related *Brassica* species. Theor Appl Genet 73:364–370

Kiss T, Solymosy F (1987) Isolation of high molecular weight plant nuclear DNA suitable for use in recombinant DNA technology. Acta Biochim Biophysic Hung 22:1–5

Kolodner R, Tewari KK (1972) Physicochemical characterization of mitochondrial DNA from pea leaves. Proc Natl Acad Sci U S A 69:1830–1834

Kolodner R, Tewari KK (1975) The molecular size and conformation of the chloroplast DNA from higher plants. Biochim Biophys Acta 402:372–390

Langdale J (2001) In situ hybridization protocol. Oxford. p. http://www.plants.ox.ac.uk/langdale-lab/protocols/histology/in_site_hyb_general.pdf. Accessed 15 Feb 2015

L'Homme Y, Stahl RJ, Li X-Q, Hameed A, Brown GG (1997) *Brassica nap* cytoplasmic male sterility is associated with expression of a mtDNA region containing a chimeric gene similar to the pol CMS- associated *orf224* gene. Curr Genet 31:325–335

Li X-Q (2011) Functional genomics: transcriptomics. In: Bradeen JM, Kole C (eds) Genetics, genomics and breeding of potato. Enfield: Science Publishers. pp 207–228

Li X-Q (2015) Bioinformatic approaches for analysis of gene direction, chromosome base composition, mRNA polyadenylation, and protein network. In: Li X-Q, Donnelly D, Jensen TG (eds) Somatic genome manipulation: advances, methods and applications. Springer, New York (In Press)

Li X-Q, Demarly Y (1996) Somatic embryogenesis and plant regeneration in. *Medicago suffruticosa*. Plant Cell Tiss Org Cult 44:79–81

Li X-Q, Liu CN, Ritchie SW, Peng J, Gelvin SB, Hodges TK (1992) Factors influencing *Agrobacterium*-mediated transient expression of gusA in rice. Plant Mol Biol 20:1037–1048

Li X-Q, Stahl R, Brown GG (1995) Rapid micropreps and minipreps of Ti plasmids and binary vectors from *Agrobacterium tumefaciens*. Transgenic Res 4:349–351

Li X-Q, Zhang M, Brown GG (1996) Cell-specific expression of mitochondrial transcripts in maize seedlings. Plant Cell 8:1961–1975

Li X-Q, Jean M, Landry BS, Brown GG (1998) Restorer genes for different forms of *Brassica* cytoplasmic male sterility map to a single nuclear locus that modifies transcripts of several mitochondrial genes. Proc Natl Acad Sci U S A 95:10032–10037

Li X-Q, Zhang J, Luo S, Liu G, Murphy A, Leclerc Y, Xing T (2011) Effects of sampling methods on starch granule size measurement of potato tubers under a light microscope. Int J Plant Biol 2:e5

Liew M, Wittwer C, Voelkerding KV (2010) Nucleotide extension genotyping by high-resolution melting. J Mol Diagn 12:731–738

Liu CN, Li X-Q, Gelvin SB (1992) Multiple copies of virG enhance the transient transformation of celery, carrot and rice tissues by *Agrobacterium tumefaciens*. Plant Mol Biol 20:1071–1087

Liu X, Kim YJ, Müller R, Yumul RE, Liu C, Pan Y, Cao X, Goodrich J, Chen X (2011) AGAMOUS terminates floral stem cell maintenance in *Arabidopsis* by directly repressing WUSCHEL through recruitment of Polycomb Group proteins. Plant Cell 23:3654–3670

Lochlainn ST, Amoah S, Graham NS, Alamer K, Rios JJ, Kurup S, Stoute A, Hammond JP, Østergaard L, King GJ, White PJ, Broadley MR (2011) High Resolution Melt (HRM) analysis is an efficient tool to genotype EMS mutants in complex crop genomes. Plant Methods 7:43.

Lutz KA, Wang W, Zdepski A, Michael TP (2011) Isolation and analysis of high quality nuclear DNA with reduced organellar DNA for plant genome sequencing and resequencing. BMC Biotechnol 11:54.

Martin M, Tomás R, Cruz Izquierdo M, Prádanos C, Sabater B (1989) Preparation of chloroplast DNA from barley and lettuce and comparison of restriction fragments. Biol Plant 31:67–71

Mehle N, Nikolić P, Rupar M, Boben J, Ravnikar M, Dermastia M (2013) Automated DNA extraction for large numbers of plant samples. Methods Mol Biol 938:139–145

Meyerowitz EM (1987) In situ hybridization to RNA in plant tissue. Plant Mol Biol Rep 5:242–250

Mileshina D, K. NA, Weber-Lotfi F, Gualberto J, Dietrich A (2015) Mitochondrial genetic manipulation. In: X-Q Li X-Q, D Donnelly D, TG Jensen TG, (eds)editors. Somatic genome manipulation: advances, methods and applications. Springer, New York (In Press)

Mourad G, Polacco ML (1988) Mini-preparation of highly purified chloroplast DNA from maize. Plant Mol Biol Rep 6:193–199

Mourad G, Polacco ML (1989) Mini-preparation of highly purified chloroplast DNA from maize. Plant Mol Biol Rep 7:78–84

Nakagawa K, Uehara H, Hosaka K (2000) Chloroplast DNA variation in the wild potato species, *Solanum acaule* and *S. albicans*. Euphytica 116:197–202

Nassar AMK, Kubow S, Donnelly DJ (2015) Somatic embryogenesis for potato (*Solanum tuberosum L.*) improvement. In: Li X-Q, Donnelly D, Jensen TG (eds) Somatic genome manipulation: advances, methods and applications. Springer, New York (In Press)

Pelletier G, Budar F (2015) *Brassica* Ogu-INRA cytoplasmic male sterility: an example of successful plant somatic fusion for hybrid seed production. In: Li X-Q, Donnelly D, Jensen TG (eds) Somatic genome manipulation: advances, methods and applications. Springer, New York (In Press)

Pérez C, Bonavent JF, Bervillé A (1990) Preparation of mitochondrial DNAs from sunflowers (*Helianthus annuus L.*) and from beets (*Beta vulgaris*) using a medium with a high ionic strength. Plant Mol Biol Rep 8:104–113

Petit PX (1992) Flow cytometric analysis of rhodamine 123 fluorescence during modulation of the membrane potential in plant mitochondria. Plant Physiol 98:279–286

Qu H, Ma F, Li Q (2008) Comparative analysis of mitochondrial fragments transferred to the nucleus in vertebrate. J Genet Genom 35:485–490

Raineri DM, Bottino P, Gordon MP, Nester EW (1990) *Agrobacterium*-mediated transformation of rice (*Oryza sativa L.*). Bio/Technology 8:33–38

Refsgaard L, Olesen MS, Møller DV, Christiansen M, Haunsø S, Svendsen JH, Christensen AH (2012) Mutation analysis of the candidate genes SCN1B-4B, FHL1, and LMNA in patients with arrhythmogenic right ventricular cardiomyopathy. ApplTransl Genom 1:44–46

Ritchie SW, Lui CN, Sellmer JC, Kononowicz H, Hodges TK, Gelvin SB (1993) *Agrobacterium tumefaciens*-mediated expression of gusA in maize tissues. Transgenic Res 2:252–265

Roche (2008) DIG application manual for nonradioactive in situ hybridization, 4th edn. https://lifescience.roche.com/wcsstore/RASCatalogAssetStore/Articles/05353122001_08.08.pdf. Accessed 15 Feb 2015

Rokka V-M (2015) Protoplast technology in genome manipulation of potato through somatic cell fusion. In: Li X-Q, Donnelly D, Jensen TG (eds) Somatic genome manipulation: advances, methods and applications. Springer, New York (In Press)

Rousseau-Gueutin M, Ayliffe MA, Timmis JN (2011) Conservation of plastid sequences in the plant nuclear genome for millions of years facilitates endosymbiotic evolution. Plant Physiol 157:2181–2193

Santalucia JJ (1998) A unified view of polymer, dumbbell, and oligonucleotide DNA nearest-neighbor thermodynamics. Proc Natl Acad Sci U S A 95:1460–1465

Sciaky D, Montoya AL, Chilton MD (1978) Fingerprints of *Agrobacterium* Ti plasmids. Plasmid 1:238–253

Shi C, Hu N, Huang H, Gao J, Zhao YJ, Gao LZ (2012) An improved chloroplast DNA extraction procedure for whole plastid genome sequencing. PLoS ONE 7:e31468

Shoemaker RC, Palmer RG, Atherly AG (1984) Preparation of chloroplast DNA-enriched DNA samples from soybean. Plant Mol Biol Rep 2:15–20

Sikorskaite S, Rajamäki ML, Baniulis D, Stanys V, Valkonen JPT (2013) Protocol: optimised methodology for isolation of nuclei from leaves of species in the Solanaceae and Rosaceae families. Plant Methods 9:31.

Singh M, Brown GG (1991) Suppression of cytoplasmic male sterility by nuclear genes alters expression of a novel mitochondrial gene region. Plant Cell 3:1349–1362

Singh M, Hamel N, Menassa R, Li X-Q, Young B, Jean M, Landry BS, Brown GG (1996) Nuclear genes associated with a single *Brassica* CMS restorer locus influence transcripts of three different mitochondrial gene regions. Genetics 143:505–516

Slusarenko AJ (1990) A rapid miniprep for the isolation of total DNA from *Agrobacterium tumefaciens*. Plant Mol Biol Rep 8:249–252

Toki S (1997) Rapid and efficient *Agrobacterium*-mediated transformation in rice. Plant Mol Biol Rep 15:16–21

Traas J (2008) Whole-mount in situ hybridization of RNA probes to plant tissues. Cold Spring Harbor Protocols 3. doi:10.1101/pdb.prot4944.

Ulrich I, Ulrich W (1991) High-resolution flow cytometry of nuclear DNA in higher plants. Protoplasma 165:212–215

Valach M, Tomaska L, Nosek J (2008) Preparation of yeast mitochondrial DNA for direct sequence analysis. Curr Genet 54:105–109

Vedel F, Mathieu C (1983) Physical and gene mapping of chloroplast DNA from normal and cytoplasmic male sterile (radish cytoplasm) lines of. *Brassica napus*. Curr Genet 7:13–20

Wang CN, Chen YJ, Chang YC, Wu CH (2008) A step-by-step optimization guide for applying tissue specific RNA in-situ hybridization to non-model plant species. Taiwania 53:383–393

Willmitzer L, Wagner KG (1981) The isolation of nuclei from tissue-cultured plant cells. Exp Cell Res 135:69–77

Wilson AJ, Chourey PS (1984) A rapid inexpensive method for the isolation of restrictable mitochondrial DNA from various plant sources. Plant Cell Rep 3:237–239

Wise AA, Liu Z, Binns AN (2006) Three methods for the introduction of foreign DNA into *Agrobacterium*. Methods Mol Biol 343:43–53

Wittwer CT, Reed GH, Gundry CN, Vandersteen JG, Pryor RJ (2003) High-resolution genotyping by amplicon melting analysis using LCGreen. Clin Chem 49:853–860

Wu S, Song Y, Li Z, Wu Z, Han J, Han S (2012) Covalent labeling of mitochondria with a photostable fluorescent thiol-reactive rhodamine-based probe. Anal Methods 4:1699–1703

Xiang H, Li X-Q (2015) A pipeline for refining the unique mapping and repetitive sequence detection using two halves of each illumina sequence read. Plant Mol Biol Rep (In Press)

Yuan J, Haroon M, Lightfoot D, Pelletier Y, Liu Q, Li X-Q (2009) A high-resolution melting approach for analyzing allelic expression dynamics. Curr Issues Mol Biol 11 (Suppl. 1):i1–9

Zhang J, Murphy A, Liu G, Bizimungu B, Liu Q, Leclerc Y, Xing T, Li X-Q (2011a) Starch granule size variation and relationship with tuber dry matter content in heritage potato varieties. Sci Hort 130:503–509

Zhang JZ, Ai XY, Sun LM, Zhang DL, Guo WW, Deng XX, Hu CG (2011b) Molecular cloning and functional characterization of genes associated with flowering in citrus using an early-flowering trifoliate orange (*Poncirus trifoliata* L. Raf.) mutant. Plant Mol Biol 76:187–204

Chapter 15
Bioinformatic Approaches for Analysis of Gene Direction, Chromosome Base Composition, mRNA Polyadenylation, and Protein Network

Xiu-Qing Li

15.1 Introduction

Bioinformatics, in addition to laboratory methods, is a powerful tool to study somatic genome structure, evolution, and function. Bioinformatics uses huge data sets composed of sequences. Genomics tools for handling these sequence data sets are critical to this research. For example, these tools have helped in several recent studies of somatic genome variation, such as developmental variation in vertebrate (Smith et al. 2009; Smith et al. 2010; Smith et al. 2012), environmentally induced flax genome variation (Oh and Cullis 2003; Wang et al. 2012), and RNA-mediated somatic genome rearrangement in ciliates (Fang et al. 2012; Bracht et al. 2013; Fang and Landweber 2013). Before complete genome sequences are available, genome and gene expression studies are unusually on a few genes at a time. Extrapolating knowledge learned from individual case studies to whole genomes or kingdoms is largely based on speculation. Bioinformatic tools enable analysis of entire genomes, the transcriptomes, protein networks, and multiple specie comparisons. Analysis of the genome sequences of large numbers of species is a powerful tool to survey differences between species, subkingdoms, and kingdoms. This large-scale bioinformatic analysis can often correct faulty conclusions drawn from over-generalized individual studies. For example, it was widely believed that mRNA poly(A) site preference is for CA dinucleotides, based on studies of some genes and a few animals species. However, after analysis of many species, it was shown that the preference is for TA, CA, or GA dinucleotides, depending on the species (Li and Du 2013). In another example, individual studies lead to an impression that many genes

X.-Q. Li (✉)
Potato Research Centre, Agriculture and Agri-Food Canada, 850 Lincoln Road,
P.O. Box 20280, Fredericton, New Brunswick, E3B 4Z7, Canada
e-mail: Xiu-Qing.Li@agr.gc.ca; lixiuqing2008@gmail.com

X.-Q. Li et al. (eds.), *Somatic Genome Manipulation*,
DOI 10.1007/978-1-4939-2389-2_15

are in the same direction clusters on chromosomes. However, from bioinformatic analysis, gene direction was shown to have clear evolutionary variation between kingdoms; predominantly in the same direction within bacteria and archaeans but in opposite direction within fungi (Li and Du 2012). Bioinformatic tools are needed to conduct comparative analysis of many species.

In the author's laboratory, in addition to experimental protocols (Li 2015), several bioinformatics and computational approaches have been either tested or developed and will be elaborated in this chapter. These include analysis of gene direction evolution (Li and Du 2012), repetitive DNA sequences (Xiang and Li 2015), chromosome base composition (Li and Du 2014b), polyadenylation sites [poly(A) sites] (Li and Du 2013; Li and Du 2014a), protein domains (Li et al. 2011), and protein network (Du et al. 2012).

15.2 A Statistical Method for Comparative Study of Gene Direction Between Chromosomes

In most chromosomes, there are always some genes that have neighbors in the same direction and together they appear like a cluster. Clustered genes seem to have similar function or coordinated expression (Hurst et al. 2004), and were therefore studied comparatively between species using various technological approaches such as protein sequence BLAST searches, gene ontology assignments, and phylogenetic tree reconstruction (Tungsuchat-Huang et al. 2010). Similar clusters sometimes exist in closely related species. The clusters in different species can be found in similar or different regions. Such clusters vary greatly in length and frequencies among chromosomes. Testing gene direction for comparing between chromosomes or genomes was a statistical challenge.

We developed the following method to compare gene directions between chromosomes or genomes. The direction of any given intergenic region is totally determined by the directions of two genes at each of the two ends of the intergenic region. There are only four scenarios for an intergenic region: FF, BB, FB, and BF if F represents the forward direction of a gene, and B represents the backward direction of a gene. The ratio of these four possibilities (FF:BB:FB:BF) should be 1:1:1:1 if the gene direction is totally random. Therefore, these ratios can be used to statistically test whether the gene direction of a chromosome is significantly different from the random distribution by simply conducting a *chi-square test* against the default 1:1:1:1 model. This method can also be used to compare between genomes by comparing the real ratios of the four scenarios in each species.

Each gene's direction on the chromosome can be found in the gb or gbk file of chromosome sequences in the National Center for Biotechnology Information (NCBI, www.ncbi.nlm.nih.gov) nucleotide database or the genome database. For comparison between chromosomes, the complete sequences of chromosomes, or at least very large scaffolds, should be used.

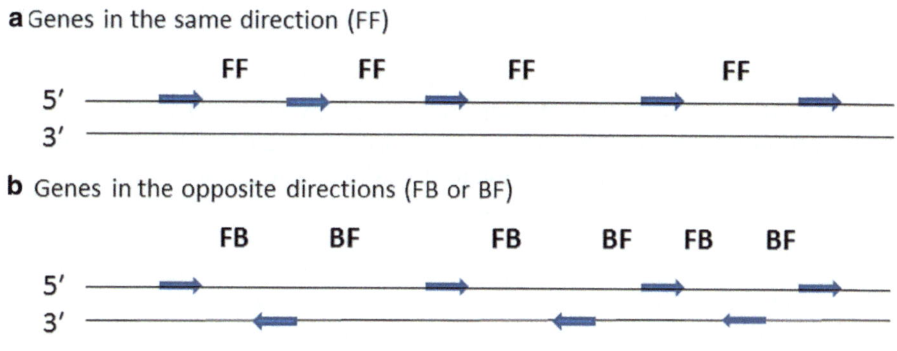

Fig. 15.1 Gene direction and the intergenic regions on chromosomes. *Arrow*: gene direction. F: *forward* direction; B: *backward* direction

Tandem arrangement of genes (Fig. 15.1a) does not necessarily mean the genes act as a functional unit. It simple means that in that region genes are not located on the opposite strand. If a chromosome is characterized by having opposite direction genes, it means that both strands have genes (Fig. 15.1b). Gene direction differences between chromosomes or genomes may indicate the differences in the extent of gene recombination between repetitive DNA sequences during the evolution. The author's collaborative research team has recently developed a bioinformatic pipeline to analyze repetitive DNA sequences in Illumina sequence reads (Xiang and Li 2015) and has confirmed repetitive DNA amount variation among chromosomes in plants (Li et al. unpublished). Gene direction analysis suggests that more primitive organisms such as bacteria and archaeans use mainly one strand to locate the genes but higher organisms such as fungi, plants, and animals use both DNA strands (Li and Du 2012).

15.3 Approaches for Genome and Chromosome Base Compositional Analysis

One of the most studied properties of genomic DNA is the GC content. This author prefers to call it the C + G content because C and G contents are not always equal and because GC can be potentially confused with the dinucleotide GC. There are various C + G content estimation methods, including annealing profiles (Nellåker et al. 2009), density bands on CsCl gradient formed during centrifugation (Wöstemeyer and Burmester 1986; Ahn and Winter 2005), flow cytometry (Šmarda et al. 2012), and/or sequence-based genome GC content comparison (Li and Du 2014b).

However, data analysis requires extra precaution because each of these methods has its weakness. The melting–annealing speed of DNA is also influenced by the degree of repetition of sequences such as satellite sequences (Appels et al. 1981; Cionini et al. 1985; Nellåker et al. 2009) in addition to its general positive correlation with C + G content. The resolution of the method of DNA density bands on the CsCl gradient is not very high because of the limit of the centrifuge tube length and because the DNA bands on the CsCl solution are often quite close to each other. The flow cytometry with 4′,6-diamidino-2-phenylindole (DAPI) staining can be significantly influenced by the staining procedure, including the concentration of DAPI, staining duration, and amount of washing. The genome sequencing approach has the highest resolution because it reaches the single base level. However, the complete genome sequence is only available for relatively few higher plants and animals at present. This may explain why the base composition estimations to date have mainly been done using the annealing temperature approach; with only two dicot and one monocot species compared (Šmarda and Bureš 2012). Minor difference can be created due to the difference in the degree of closing for the genome sequence gaps. Experimenters and data interpreters should keep in mind the advantages and disadvantages of these methods at both the experimental design and the interpretation stages.

Annealing temperature and flow cytometry approaches can generate whole genome level estimates but usually do not reveal the base composition of each chromosome individually. The chromosome sequence approach can reveal both genome information and interchromosomal differences. Some species do have significant differences in interchromosomal base composition (Li and Du 2014b).

Another extra precaution relates to correlation analysis and interpretation of the level of correlation. It is known that C + G content is positively correlated with chromosome (genome) size in bacteria (Nishida 2012) and some other prokaryotes (Musto et al. 2006) when pooling species together for a correlative analysis. However, the degree of correlation can be considerably increased if a small genome has a very low C + G content and a large genome has a very large C + G content. The general positive correlation in a kingdom does not mean there is such a positive correlation in every subkingdom or phyla. For example, in the archaeal Euryarchaeota phylum, some species have one larger chromosome (2.7–3.1 Mb) and one smaller chromosome (288–525 kb). In a data analysis conducted by the author, the large chromosomes were found to have higher C + G contents than the smaller chromosomes, which appears to support a positive correlation between chromosome size and C + G content. However, when we compared the correlation between chromosome length and C + G content within the large chromosome group, it was clear that the C + G content was considerably lower when the chromosome size increased (Fig. 15.2; Li and Du 2014b). The large chromosomes of these species likely shared the same evolutionary history before these species differentiated. Similarly, there was no tendency towards C + G content increase in accordance with chromosome size increase in the small chromosome group (Fig. 15.2). The large and small chromosomes likely do not share the same evolutionary origin. Interchromosomal comparisons within each chromosome group enabled the authors to conclude that the C + G content was not correlated with chromosome size between species within this phylum (Li and Du 2014b).

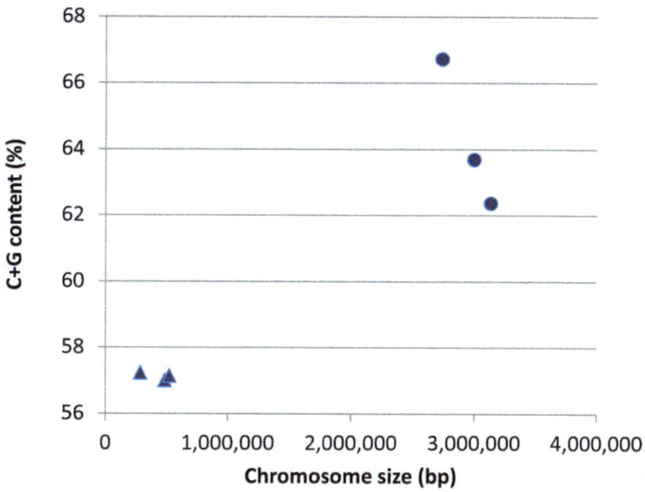

Fig. 15.2 Chromosome size and C + G content of the three species of the archaeal Euryarchaeota phylum. The three species are *Halorubrum lacusprofundi, Haloarcula hispanica,* and *Haloarcula marismortui*. Each species has one large chromosome (*round dots* in the figure) and one small chromosome (*triangle* marks in the figure). Note that even through the three large chromosomes have much larger C + G contents than the three smaller chromosomes, it is clear that there is either no positive correlation or there is a negative correlation between C + G content and chromosome size in the group of the three large chromosomes (Modified from Li and Du 2014b)

15.4 Technical Highlights in RNA Polyadenylation Site Analysis

Polyadenylation plays a critical role in gene expression. There are several genomic and bioinformatics approaches for studying poly(A) sites, including clustering the expressed sequence tags (ESTs), mapping mRNA and ESTs to genomes, and mapping second-generation sequence reads to the reference genomes (references). Each approach has its advantages and disadvantages. The mRNA sequence approach has its advantage in sequence quality because most sequences in the NCBI mRNA database were verified by sequencing from both the 5 prime and 3 prime ends. The disadvantage of this mRNA sequence approach is that the data yield is less than with the second-generation sequencing approach. The NCBI mRNA sequences approach is preferred than the Illumina TruSeq approach in poly(A) site mapping when the NCBI mRNA sequence number is sufficiently large, because TruSeq read-based poly(A) site mapping has a nonignorable level fault poly(A) tails due to the RNA/cDNA fragmentation-step during the sequencing library construction and the oligo-dT internal priming during sequencing (Li 2014; Li and Du 2014a). Compared to mRNA sequences in the NCBI nucleotide database, ESTs have relatively more data, and the second-generation sequence reads have massive numbers of sequences. ESTs have a better quality than TruSeq short reads for the poly(A) tail sequences due to the use of longer oligo dT sequences and a higher annealing temperature during cDNA library construction for sequencing (Li and Du 2014a). However, the sequences from ESTs and the second-generation sequences are single reads without verification from another direction. The setting of thresholds and suitable statistical methods is critical in eliminating errors when using EST or second-generation sequencing reads in poly(A) site mapping (Li 2014; Li and Du 2014a).

The EST-based approach can yield results about the polyadenylation sites that reflect the activity of genes and transcript copy numbers in the actual cell mRNA population. However, the EST approach is not ideal for comparing between species because some genes may be over- or under-represented because of gene expression activity differences. To solve the problem, we have used a unique mRNA and unique poly(A) sites approach (Li and Du 2013). By counting a group of same sequence (for the poly(A) site region) as one unique sequence, eliminating all the redundant sequences, and the mapping with zero tolerance to mismatch during cDNA-genome sequencing alignment, the bias caused by overexpression of certain genes in ESTs can be largely minimized (Li and Du 2014a).

The major DNA/RNA sequence databases including GenBank, RefSeq, EMBL, DDBJ, and PDB can be accessed from the National Center for Biotechnology Information site (http://www.ncbi.nlm.nih.gov/nuccore). As 3′ truncation is sometimes possible, the 3′ end of mRNA sequences from NCBI is not always the poly(A) site. Some of the poly(A) tail-like multiple A's at the 3′ end could be from internal priming during cDNA library construction (Beaudoing and Gautheret 2001; Nam et al. 2002) or during sequencing. This is particularly the case for ESTs, as internal priming can account for about 12% of EST poly(A) tails (Nam et al. 2002). Some rRNA can sometimes be polyadenylated (Slomovic et al. 2006) and not all the EST sets submitted to NCBI have had the rRNA ESTs pre-eliminated; therefore contamination of rRNA in the mRNA EST sequences can be an issue. In contrast, artificial poly(A) sites from internal priming can be largely eliminated because the mRNA sequences in the NCBI Nucleotide database have usually been verified by repeated sequencing from both the 5′ and 3′ ends of cDNA clones. To eliminate the internal priming issue and to avoid poor sequences, the author considered an mRNA as polyadenylated with relatively high confidence only when an mRNA sequence meets the following three criteria (Li and Du 2013): (1) the mRNA sequence upstream of the poly(A) tail must have at least 100 bases and have no N's; (2) the mRNA has a poly(A) tail at the 3′ end; and (3) the pure poly(A) tail must have at least 12 A's. The disadvantage of using mRNA sequences is the lack of species for which sufficiently large numbers of mRNA sequences are available, greatly limiting interspecies comparisons.

The mRNA nucleotide that is directly attached to the poly(A) tail can be called "the poly(A) tail attachment position of the poly(A) site" and the pre-mRNA nucleotide that corresponds to the first adenosine of the poly(A) tail can be called "the poly(A) tail starting position of the poly(A) site" (Li and Du 2013). Poly(A) sites can be classified into two groups in terms of the nucleotide type at the poly(A) tail starting position: A-type poly(A) sites, which have a pre-mRNA adenosine at the poly(A) tail starting position, and non-A-type poly(A) sites, which do not have an adenosine at the pre-mRNA poly(A) tail starting position.

The poly(A) tail attachment position and the starting position correspond exactly to the 5′ nucleotide and the 3′ nucleotide covering the cleavage site (bond), respectively, in the non-A-type poly(A) site; and therefore the poly(A) sites can be mapped precisely and accurately onto the genome (Li and Du 2013). However, when the poly(A) site is an adenosine, it can be challenging to map the poly(A) sites precisely because it is not known whether the adenosine is the nucleotide used to attach the poly(A) tail or the nucleotide to be replaced by the poly(A) tail. The issue is even more complex if the poly(A) site is located in a pre-RNA multiple adenosine stretch, because the oligo(dT) at the cDNA library construction stage can primer at different locations of the multiple A's. The mRNA–genome sequence alignment approach alone cannot determine which adenosine of the multiple A's is the real poly(A) site without additional tools to assist with identification. Further research is required to determine the poly(A) site more accurately in these multiple adenosine situations.

In mRNA–genome sequence mapping, the poly(A) tail starting position is usually found to be an adenosine (Li and Du 2013). Is this a sequence mapping artifact or is the adenosine really more frequent than other types of nucleotides at the poly(A) site position? The following calculation approach was used to answer this question by estimating the theoretical A-type polyadenylation [poly(A)] site frequency in the random model. If the A nucleotide percentage in mRNA is p, the A-type poly(A) site from the alignment will be $p+p(1-p)=p(2-p)$, where $(1-p)$ is the non-A nucleotide content. The multiple-A or multiple-non-A sequences do not alter the A site or non-A site probability in this random model, because both A and non-A have a random chance in this aspect within their nucleotide content ranges. Regardless of whether the adenosine is replaced or is extended in generating the poly(A) tail, the theoretical value of the A-type poly(A) tail in the alignment is still the same, but $p(1-p)$ is counted from the upstream A in the replacement model and counted from the downstream A directly adjacent to the poly(A) site in the extended model (Li and Du 2013).

15.5 Allele Comparison for Protein Domains

Different alleles are often expressed in different tissues or developmental stages. These differentially expressed alleles sometimes have variation on the functional domain region, particularly in a family of genes. The analysis of the selective loss of cysteine residues and disulphide bonds in a potato proteinase inhibitor II family (Li et al. 2011) is used in an example for multiple approaches to analyze such functional domain variants.

A mRNA sequence, such as the GenBank accession EF469204 for a proteinase inhibitor variant cDNA identified from a diploid potato clone (Li et al. 2011), can be translated into amino acid sequences using http://insilico.ehu.es/translate (Bikandi et al. 2004), and then can be used in PSI- and PHI-BLAST (Altschul et al. 1997; Altschul et al. 2005) to identify similar proteins and conserved known domains. The domain's known structure can be viewed with Cn3D 4.1 (an NCBI's helper application, http://www.ncbi.nlm.nih.gov/Structure/CN3D/cn3d.shtml). The variation between the mRNA-translated sequences and the amino acid sequence of the known structure can be directly compared. The amino acid changes at the activity site of the protein may have a greater chance of changing the protein's function.

The phylogeny trees of the amino acid sequences can be reconstructed using a phylogenetics software such as PAUP* (http://paup.csit.fsu.edu/), MEGA (www.megasoftware.net), or the Clustal and PROTDIS options in the software BiEdit Sequence Alignment Editor (Hall 1999; http://www.mbio.ncsu.edu/bioedit/bioedit.html). Their amino acid sequences can be aligned using Clustal (Thompson et al. 1997)-ProML option first, and then used to reconstruct the *protein maximum likelihood tree* (available in BioEdit). The trees can also be reconstructed using the "ClustalW Multiple Alignment" and 1000 "bootstraps NJ tree" option to do alignment first and then generate the phylogenetic tree with Protdist Neighbor Phylogenetic Tree Option (PROTDIST; an option in BioEdit). Further improvement of the tree after adjusting sequence alignment based on the NJ tree can be done using Bayesian phylogenetic analysis. The outtree files can be saved as .txt files, viewed using TreeView (http://taxonomy.zoology.gla.ac.uk/rod/rod.html), and then saved as graphics. The tree can be reformatted as unrooted tree, slanted cladogram, or rectangular cladogram. The out-group of the cladogram can be defined as well by an option in TreeView. The Ka/Ks calculator tool described by Liberles's group (Liberles 2001; Anisimova and Liberles 2007) can be used to calculate the rate of nonsynonymous substitutions (Ka) and the rate of synonymous substitutions (Ks).

15.6 Protein Network Analysis

Proteomics approaches can determine the length of amino acid sequences of proteins from somatic tissue. Some protein interaction databases are available from the literature, such as protein interaction directional scores (PIDS) for predicting the direction of individual protein interaction and signal flow (Liu et al. 2009). A positive PIDS for protein A to protein B in a signaling pathway means that the informational interaction direction is from A to B. As PIDS is for individual interaction, PIDS score does not give the summarized rating of the functional role of a protein, nor the ranked importance of a protein in the cellular protein network. The relative ranking of all the proteins in the cellular protein network requires the use of a suitable network analysis approach.

The Google PageRank algorithm (Brin and Page 1998) is powerful in providing high-quality search result rankings for websites (Brin and Page 1998), journals (Ball 2006) and in nondirectional ranking of proteins in protein networks

(Weston et al. 2004). After adjusting appropriate settings for forward and reverse rankings, the Google PageRank can also be used by the author's team in protein ranking in a network with signal/information flow directions (Du et al. 2012). Systematic difference has been identified between signal emission and signal receiving in the cellular protein network. The top-ranked proteins involved in signal transduction have been identified. These eight proteins (ACVR1, CDC42, RAC1, RAF1, RHOA, TGFBR1, TRAF2, TRAF6) are evolutionarily very conserved and are involved in many pathways, including signaling pathways, cell division, and cancer pathways (Du et al. 2012). Most of these top-ranked proteins are involved in cancer pathways and play key roles in inhibiting or stimulating cancerous cell growth (Du et al. 2012). Therefore, the PageRank can be an approach not only for assisting basic research in signal transduction in a protein network but also for generating information useful for applied research in drug development.

Du et al. (2012) conducted forward ranking and reverse ranking using the PageRank approach (Brin and Page 1998), which can be explained as follows: Let A be the adjacency matrix of the protein network, and let N be obtained from A by dividing each entry in A by its out-degree, then the basic PageRank (PR) satisfies $PR = N^t PR$, where PR is the eigenvector of matrix N^t (the transpose of A), corresponding to the eigenvalue 1. In the scaled version of PageRank, let α be a scale factor (a.k.a. damping factor) between 0 and 1. Use a new matrix $G = (g_{ij})_{n \times n}$ to replace $N = (n_{ij})_{n \times n}$ as follows: $g_{ij} = \alpha n_{ij} + \frac{1-\alpha}{n}$, where n is the dimension of A. Hence, the scaled PageRank (PR) satisfies $PR = G^t PR$. See Du et al. (2012) for more detail for the settings of scale factor alpha and the methodological difference in forward and reverse rankings.

Higher values are givens to proteins that receive signals from large network nodes in forward ranking, but given to proteins that send information to large network nodes in reversing ranking. A protein's ranking position is determined by the global evaluation of the protein's role, not necessarily the specific interaction in the network. The order of reverse ranking positions is not a simple reverted list of the forward ranking positions.

15.7 Concluding Remarks

This chapter describes methods or technical highlights for analyzing gene directions, genome size, chromosome base composition, genome size and C + G content correlation, RNA polyadenylation sites, allele comparison, protein domains, and protein networks. The methods can be used in analyzing both somatic and nonsomatic genomes, as long as the data preparation and settings are appropriately defined. Further research is required to understand the mechanisms of evolution of somatic genome base composition, to localize RNA poly(A) sites more accurately for the adenosine-type poly(A) sites, to reveal the regulation of poly(A) site selection during individual development, and to predict the function of variants of protein domains in different somatic cells. Many species have no sexual reproduction stage but still have clearly evolved in both the structure and function of their genomes. A combination of both experimental and bioinformatic approaches is required to study these somatic genomes.

Acknowledgments The author sincerely thanks Professor Donglei Du, a mathematician at University of New Brunswick, for all his assistance in mathematics and computer programing.

References

Ahn IY, Winter CE (2005) Determination of DNA base composition by small scale acrylamide-CsCl gradient centrifugation. J Biochem Biophys Methods 63:155–160

Altschul SF, Madden TL, Schäffer AA, Zhang J, Zhang Z, Miller W, Lipman DJ (1997) Gapped BLAST and PSI-BLAST: a new generation of protein database search programs. Nucleic Acids Res 25:3389–3402

Altschul SF, Wootton JC, Gertz EM, Agarwala R, Morgulis A, Schäffer AA, Yu YK (2005) Protein database searches using compositionally adjusted substitution matrices. FEBS J 272:5101–5109

Anisimova M, Liberles DA (2007) The quest for natural selection in the age of comparative genomics. Heredity 99:567–579

Appels R, Dennis ES, Smyth DR, Peacock WJ (1981) Two repeated DNA sequences from the heterochromatic regions of rye (*Secale cereale*) chromosomes. Chromosoma 84:265–277

Ball P (2006) Prestige is factored into journal ratings. *Nature* 439:770–771.

Beaudoing E, Gautheret D (2001) Identification of alternate polyadenylation sites and analysis of their tissue distribution using EST Data. Genome Res 11:1520–1526

Bikandi J, San Millan R, Rementeria A, Garaizar J (2004) In silico analysis of complete bacterial genomes: PCR, AFLP-PCR and endonuclease restriction. Bioinformatics 20:798–799

Bracht JR, Fang W, Goldman AD, Dolzhenko E, Stein EM, Landweber LF (2013) Genomes on the edge: programmed genome instability in ciliates. Cell 152:406–416

Brin S, Page L (1998) The anatomy of a large-scale hypertextual web search engine. Comput Net ISDN Sys 30:107–117

Cionini PG, Bassi P, Cremonini R, Cavallini A (1985) Cytological localization of fast renaturing and satellite DNA sequences in *Vicia faba*. Protoplasma 124:106–111

Du D, Lee CF, Li X-Q (2012) Systematic differences in signal emitting and receiving revealed by PageRank analysis of a human protein interactome. PLoS ONE 7(9):e44872

Fang W, Landweber LF (2013) RNA-mediated genome rearrangement: hypotheses and evidence. Bioessays 35:84–87

Fang W, Wang X, Bracht JR, Nowacki M, Landweber LF (2012) Piwi-interacting RNAs protect DNA against loss during oxytricha genome rearrangement. Cell 151:1243–1255

Hall TA (1999) BioEdit: a user-friendly biological sequence alignment editor and analysis program for Windows 95/98/NT. Nucl Acids Symp Ser 41:95–98

Hurst LD, Pál C, Lercher MJ (2004) The evolutionary dynamics of eukaryotic gene order. Nat Rev Genet 5:299–310

Li X-Q (2014) Comparative analysis of the base compositions of the pre-mRNA 3′ cleaved-off region and the mRNA 3′ untranslated region relative to the genomic base composition in animals and plants. PLoS ONE 9: e99928.

Li X-Q (2015) Laboratory methods for investigating nuclear and cytoplasmic genomes and transcriptome. In: Li X-Q, Donnelly D Jensen TG (eds) Somatic genome manipulation: advances, methods and applications. Springer, New York, Chapter 14 (In press)

Li X-Q, Du D (2012) Gene direction in living organisms. Sci Rep 2:982

Li X-Q, Du D (2013) RNA polyadenylation sites on the genomes of microorganisms, animals, and plants. PLoS ONE 8:e79511

Li X-Q, Du D (2014a) Motif types, motif locations and base composition patterns around the RNA polyadenylation site in microorganisms, plants and animals. BMC Evol Biol 14:162

Li X-Q, Du D (2014b) Variation, evolution, and correlation analysis of C + G content and genome or chromosome size in different kingdoms and phyla. PLoS ONE 9:e88339

Li X-Q, Zhang T, Donnelly D (2011) Selective loss of cysteine residues and disulphide bonds in a potato proteinase inhibitor II family. PLoS ONE 6 6(4):e18615.

Liberles DA (2001) Evaluation of methods for determination of a reconstructed history of gene sequence evolution. Mol Biol Evo 18:2040–2047

Liu W, Li D, Wang J, Xie H, Zhu Y, He F (2009) Proteome-wide prediction of signal flow direction in protein interaction networks based on interacting domains. Mol Cell Proteomics 8:2063–2070

Musto H, Naya H, Zavala A, Romero H, Alvarez-Valín F, Bernardi G (2006) Genomic GC level, optimal growth temperature, and genome size in prokaryotes. Biochem Biophys Res Commun 347:1–3

Nam DK, Lee S, Zhou G, Cao X, Wang C, Clark T, Chen J, Rowley JD, Wang SM (2002) Oligo(dT) primer generates a high frequency of truncated cDNAs through internal poly(A) priming during reverse transcription. Proc Natl Acad Sci U S A 99:6152–6156

Nellåker C, Li F, Uhrzander F, Tyrcha J, Karlsson H (2009) Expression profiling of repetitive elements by melting temperature analysis: variation in HERV-W gag expression across human individuals and tissues. BMC Genomics 10:532.

Nishida H (2012) Evolution of genome base composition and genome size in bacteria. Front Microbiol 3:420

Oh TJ, Cullis CA (2003) Labile DNA sequences in flax identified by combined sample representational difference analysis (csRDA). Plant Mol Biol 52:527–536

Slomovic S, Laufer D, Geiger D, Schuster G (2006) Polyadenylation of ribosomal RNA in human cells. Nucleic Acids Res 34:2966–2975

Šmarda P, Bureš P (2012) The variation of base composition in plant genomes. In: Wendel JF, Greilhuber J, Doležel J, Leitch IJ (eds) Plant genome diversity. Springer, Wien, pp 209–235

Šmarda P, Bureš P, Šmerda J, Horová L (2012) Measurements of genomic GC content in plant genomes with flow cytometry: a test for reliability. New Phytol 193:513–521

Smith JJ, Antonacci F, Eichler EE, Amemiy CT (2009) Programmed loss of millions of base pairs from a vertebrate genome. Proc Natl Acad Sci U S A 106:11212–11217

Smith JJ, Stuart AB, Sauka-Spengler T, Clifton SW, Amemiya CT (2010) Development and analysis of a germline BAC resource for the sea lamprey, a vertebrate that undergoes substantial chromatin diminution. Chromosoma 119:381–389

Smith JJ, Baker C, Eichler EE, Amemiya CT (2012) Genetic consequences of programmed genome rearrangement. Curr Biol 22:1524–1529

Thompson JD, Gibson TJ, Plewniak F, Jeanmougin F, Higgins DG (1997) The CLUSTAL_X windows interface: flexible strategies for multiple sequence alignment aided by quality analysis tools. Nucleic Acids Res 25:4876–4882

Tungsuchat-Huang T, Sinagawa-García SR, Paredes-López O, Maliga P (2010) Study of plastid genome stability in tobacco reveals that the loss of marker genes is more likely by gene conversion than by recombination between 34-bp loxP repeats. Plant Physiol 153:252–259

Wang Z, Hobson N, Galindo L, Zhu S, Shi D, McDill J, Yang L, Hawkins S, Neutelings G, Datla R, Lambert G, Galbraith DW, Grassa CJ, Geraldes A, Cronk QC, Cullis C, Dash PK, Kumar PA, Cloutier S, Sharpe AG, Wong GKS, Wang J, Deyholos MK (2012) The genome of flax (*Linum usitatissimum*) assembled de novo from short shotgun sequence reads. Plant J 72:461–473

Weston J, Elisseeff A, Zhou D, Leslie CS, Noble WS (2004) Protein ranking: from local to global structure in the protein similarity network. Proc Natl Acad Sci U S A 101:6559–6563

Wöstemeyer J, Burmester A (1986) Structural organization of the genome of the zygomycete *Absidia glauca*: evidence for high repetitive DNA content. Curr Genet 10:903–907

Xiang H, Li X-Q (2015) A pipeline for refining the unique mapping and repetitive sequence detection using two halves of each Illumina sequence read. Plant Mol Biol Rep (In Press)

Index

X.-Q. Li et al. (eds.), *Somatic Genome Manipulation*,
DOI 10.1007/978-1-4939-2389-2

MIX
Papier aus verantwortungsvollen Quellen
Paper from responsible sources
FSC® C105338
FSC
www.fsc.org

If you have any concerns about our products,
you can contact us on
ProductSafety@springernature.com

In case Publisher is established outside the EU,
the EU authorized representative is:
Springer Nature Customer Service Center GmbH
Europaplatz 3, 69115 Heidelberg, Germany

Printed by Libri Plureos GmbH
in Hamburg, Germany